STRUCTURAL BIOLOGY OF BACTERIAL PATHOGENESIS

STRUCTURAL BIOLOGY OF BACTERIAL PATHOGENESIS

Edited by

Gabriel Waksman

Institute of Structural Molecular Biology,
Birkbeck College and University College London,
London, United Kingdom

Michael Caparon

Department of Molecular Microbiology,
Washington University School of Medicine,
St. Louis, Missouri

Scott Hultgren

Department of Molecular Microbiology,
Washington University School of Medicine,
St. Louis, Missouri

ASM PRESS

Washington, D.C.

Address editorial correspondence to ASM Press, 1752 N St. NW, Washington, DC 20036-2904, USA

Send orders to ASM Press, P.O. Box 605, Herndon, VA 20172, USA
Phone: (800) 546-2416 or (703) 661-1593
Fax: (703) 661-1501
E-mail: books@asmusa.org
Online: www.asmpress.org

Library of Congress Cataloging-in-Publication Data

Structural biology of bacterial pathogenesis / edited by Gabriel Waksman, Michael Caparon, Scott Hultgren.
    p. ; cm.
  Includes bibliographical references and index.
  ISBN 1-55581-301-1 (hardcover)
 1. Virulence (Microbiology) 2. Bacteria—Ultrastructure. 3. Molecular microbiology.
  [DNLM: 1. Bacteria—pathogenicity. 2. Bacterial Infections—microbiology. 3. Bacterial Physiology. 4. Virulence Factors. QW 730 S927 2005] I. Waksman, Gabriel. II. Caparon, Michael. III. Hultgren, Scott.

  QR175.S76 2005
  571.9'93—dc22

                      2004030413

10 9 8 7 6 5 4 3 2 1

Cover illustration: Structure of the PapD-PapK chaperone-subunit complex. PapD is shown in white ribbon representation; PapK is shown in surface representation, color coded according to electrostatic potential, with red used for the most negatively charged regions and blue used for the most positively charged regions.

CONTENTS

CONTRIBUTORS

Elizabeth A. Campbell • The Rockefeller University, 1230 York Ave., Box 224, New York, NY 10021

Michael Caparon • Department of Molecular Microbiology, Washington University School of Medicine, St. Louis, MO 63110-1093

Cecilia P. C. Chiu • Department of Biochemistry and Molecular Biology, University of British Columbia, Vancouver, British Columbia, Canada V6T 1Z3

Robert T. Clubb • Department of Chemistry and Biochemistry, Molecular Biology Institute, and UCLA-DOE Institute for Genomics and Proteomics, University of California, Los Angeles, Los Angeles, CA 90095-1570

Kevin M. Connolly • Department of Chemistry and Biochemistry, Molecular Biology Institute, and UCLA-DOE Institute for Genomics and Proteomics, University of California, Los Angeles, Los Angeles, CA 90095-1570

Shane E. Cotter • Edward Mallinckrodt Department of Pediatrics, Washington University School of Medicine, 660 South Euclid Ave., St. Louis, MO 63110

Seth A. Darst • The Rockefeller University, 1230 York Ave., Box 224, New York, NY 10021

Karen Dodson • Department of Molecular Microbiology, Washington University, 4940 Parkview Pl., St. Louis, MO 63110

Katrina T. Forest • Department of Bacteriology, University of Wisconsin—Madison, Madison, WI 53706

Scott J. Hultgren • Department of Molecular Microbiology, Washington University School of Medicine, 4940 Parkview Pl., St. Louis, MO 63110

Erich Lanka • Max-Planck-Institut für Molekulare Genetik, Ihnestraße 73, Dahlem, D-14195 Berlin, Germany

Jodi B. Lubetsky • Center for Advanced Biotechnology and Medicine, Howard Hughes Medical Institute, Department of Biochemistry, University of Medicine and Dentistry of New Jersey—Robert Wood Johnson Medical School, Piscataway, NJ 08854

Frederic G. Sauer • Section of Microbial Pathogenesis, Boyer Center for Molecular Medicine, Yale University School of Medicine, 295 Congress Ave., New Haven, CT 06536

Savvas N. Savvides • Laboratory for Protein Biochemistry, Ghent University, K. L. Ledeganckstraat 35, B-9000 Ghent, Belgium

Gunnar Schröder • Max-Planck-Institut für Molekulare Genetik, Ihnestraße 73, Dahlem, D-14195 Berlin, Germany, and Division of Molecular Microbiology, Biozentrum, University of Basel, Klingelbergstraße 50-70, CH-4056 Basel, Switzerland

Craig L. Smith • Department of Molecular Microbiology, Washington University, 4940 Parkview Pl., St. Louis, MO 63110

C. Erec Stebbins • Laboratory of Structural Microbiology, The Rockefeller University, New York, NY 10021

Joseph W. St. Geme III • Edward Mallinckrodt Department of Pediatrics and Department of Molecular Microbiology, Washington University School of Medicine, 660 South Euclid Ave., St. Louis, MO 63110

Ann M. Stock • Center for Advanced Biotechnology and Medicine, Howard Hughes Medical Institute, Department of Biochemistry, University of Medicine and Dentistry of New Jersey—Robert Wood Johnson Medical School, Piscataway, NJ 08854

Natalie C. J. Strynadka • Department of Biochemistry and Molecular Biology, University of British Columbia, Vancouver, British Columbia, Canada V6T 1Z3

Neeraj K. Surana • Edward Mallinckrodt Department of Pediatrics, Washington University School of Medicine, 660 South Euclid Ave., St. Louis, MO 63110

Liang Tong • Department of Biological Sciences, Columbia University, New York, NY 10027

Rodney K. Tweten • Microbiology and Immunology, BMSB-1053, 940 Stanton L. Young Blvd., The University of Oklahoma Health Sciences Center, Oklahoma City, OK 73104

Gabriel Waksman • Institute of Structural Molecular Biology, Birkbeck and University College London, Malet St., London WC1E 7HX, United Kingdom

Hye-Jeong Yeo • Edward Mallinckrodt Department of Pediatrics, Washington University School of Medicine, 660 South Euclid Ave., St. Louis, MO 63110

Calvin K. Yip • Department of Biochemistry and Molecular Biology, University of British Columbia, Vancouver, British Columbia, Canada V6T 1Z3

INTRODUCTION

Recent years have seen a rapid increase in structural information for proteins implicated in bacterial pathogenesis. From structures involved in adhesion and host recognition to those describing elements of bacterial secretion systems, this explosion in the field of structural microbiology has led to spectacular advances in our understanding of bacterial pathogenesis. To our knowledge, this book is the first attempt at compiling the structural biology work that has taken place in the field of bacterial pathogenesis in recent years. It is only an attempt because the repertoire of structural successes in this field is now so large that recruiting all the people responsible for it to write chapters turned out to be impossible. Moreover, it would have made for a rather unwieldy book. We have thus excluded the structural biology of toxins, as this area has already been covered in other books published by ASM Press. Instead, we have focused on a few areas that represent elements of the basic paradigm of bacterial pathogenesis. Bacteria enter the host and use sophisticated sensory pathways to upregulate expression of virulence factors. They recognize and bind to host receptors by using a diverse array of adhesins whose display on the bacterial cell surface often requires a dedicated assembly pathway. Once bound to the host cell, a bacterium begins to manipulate host cell behavior, using specialized secretion systems to deliver effector proteins to the host-pathogen interface. The actions of these effectors then provoke the host's response.

Bacterial pathogens actively probe the environment in which they live. Successful virulence relies on bacterial sensing of environmental cues. This is achieved by molecular mechanisms, the most important of which are described in the first two chapters. In chapter 1, Campbell and Darst provide an overview of the molecular basis of regulation of bacterial transcription by anti-σ factors. For bacteria, gene regulation is controlled primarily at the level of transcription initiation, by controlling the ability of σ factors to recognize promoter sequences. Anti-σ factors are an example of proteins involved in linking various cellular processes and signal transduction pathways to the control of σ factor function leading to gene regulation. Some anti-σ factors, notably those that interact with the class of σ factors known as the extracellular function σ factors, have been implicated in bacterial pathogenesis. In chapter 2, Lubetsky and Stock describe another system used by bacteria to sense and respond to their environment, the ubiquitous two-component system. Coordination of expression of virulence factors and changes in housekeeping functions that occur when bacteria migrate from a free-living state to association with a host is essential, and two-component signaling systems are commonly involved in many of these processes. Two-component systems are also attractive targets for design of antibiotics, and this issue is discussed in great and fascinating detail in chapter 2.

Early events in infection must include bacterial attachment and host recognition. Chapter 3, by Smith et al., provides an overview of what is known about the structural biology of adhesion molecules, while chapter 4, by Yip et al., focuses on one particular interaction which determines adhesion of enteropathogenic *Escherichia coli*, the interaction of intimin with the bacterially encoded Tir receptor. This is a remarkable system in which the bacterium uses a type III secretion system to deliver its own receptor to the targeted eukaryotic cells.

Recognition of receptors requires that adhesion molecules be exposed to the host cell membrane, often at a tip of a pilus structure. Thus, assembly of adhesins often requires specialized secretion machineries. Chapter 5, by Sauer et al., provides the structural details of our understanding of the chaperone-usher pathway. This system, which is geared for the production of adhesive pili, is probably the structurally best-documented system involved in bacterial pathogenesis. Moreover, because pilus biogenesis by chaperone-usher pathways is a polymerization process involving protein monomer subunits that are structurally truncated, its molecular basis could have been unraveled only by using structural biology approaches. Other important adhesive fibers are the type IV pili, which are assembled by a specialized transport machinery that is related to type II secretion systems. In chapter 6, Forest provides an overview of type IV pilus structure and assembly. Finally, the mechanisms leading to display of adhesion molecules at the surface of gram-positive bacteria are addressed by Connolly and Clubb in chapter 7, which focuses on the role and structure of sortases in that process.

The next four chapters are devoted to the structural biology of general secretion systems. There are six secretion systems in bacteria that have been well characterized to date: types I to V, which operate mostly in gram-negative bacteria, and the recently discovered injectosome in gram-positive bacteria. Recent structural progress has been made in type III, IV, and V secretion as well as the injectosome, and thus we are covering those systems most extensively. We start this section of the book with a chapter that could also have found a place in the previous group of chapters. Chapter 8, by Surana et al., describes the use of the type V secretion system (otherwise known as autotransporters) to display bacterial adhesions at the cell surface of *Haemophilus influenzae*. Autotransporters are fascinating molecules that possess within their primary structure the necessary sequence encoding their own machinery for export through the outer membrane of gram-negative bacteria. Recent structural advances have provided clues as to how this happens. In chapter 9, Stebbins provides an overview of the rapid progress made in defining the molecular basis of type III secretion assembly and effector function. The structural biology of type III effectors has provided insights into the various means that bacteria have developed to hijack host functions and subvert them to their advantage. A general theme has emerged: type III secretion effectors are molecular mimics of existing host proteins and thus utilize their ability to bind to host proteins to inhibit or trigger cellular functions that increase the pathogen's chances of surviving the onslaught of host defenses to eventually proceed with successful infection.

Chapter 10, by Schröder et al., describes another secretion system of gram-negative bacteria, the type IV secretion system. The type IV secretion system distinguishes itself by being used for export of both proteins and DNAs. It is ancestrally related to bacterial conjugation systems and also exports virulence factors such the protein CagA of *Helicobacter pylori*, the causative agent of gastric ulcers, and the pertussis toxin of *Bordetella pertussis*, which is responsible for whooping cough. This group of chapters concludes with a chapter (chapter 11, by Tweten and Caparon) on the newly discovered injectosome in gram-negative bacteria. The injectosome appears to be similar to type III secretion machinery in that both systems appear to be geared for injection of virulence factors directly into the host cells.

Successful bacterial infections mobilize an arsenal of bacterial effectors which serves to overcome host defenses. What do those defenses consist of, and what is their molecular basis? These questions are the focus of the last chapter of the book (chapter 12, by Tong), which describes recent advances in the structural biology of Toll-like receptors. These receptors directly sense the presence of bacteria and thus trigger the first-line defenses against bacterial pathogens.

We hope that this book will provide a flavor of the enormously productive contribution that structural biology has made to our understanding of bacterial diseases. Many more chapters could have been written on the subject. Type I secretion, for example, has seen extraordinary advances in recent years. We would have loved to include a chapter on recent advances in type II secretion. Toxin structures could provide the subject of an entire book. The pace of discovery in these areas has been impressive, and we expect it to accelerate even further. Two factors contribute to this trend: a larger number of structural biology research groups interested in bacterial pathogenesis and the advent of fast methods for structure determination. However, as hinted in chapter 9, one discipline should make a resounding entry into the field in the next few years: high-resolution electron microscopy (cryo-electron microscopy). Most of the systems under study will be unraveled only when visualization of the protein complexes that they form is achieved. Thus, a combined effort in the areas of biochemistry, to purify such complexes, and crystallography and cryo-electron microscopy, to visualize them, holds the key to the future.

Gabriel Waksman
Michael Caparon
Scott Hultgren

Structural Biology of Bacterial Pathogenesis
Edited by G. Waksman et al.
© 2005 ASM Press, Washington, D.C.

Chapter 1

Regulation of Bacterial Transcription by Anti-σ Factors

Elizabeth A. Campbell and Seth A. Darst

σ FACTORS: CONDUCTORS OF GENE EXPRESSION

In bacteria, gene expression is regulated primarily at the step of transcription initiation. The DNA-dependent RNA polymerase (RNAP), the central enzyme of transcription, comprises an evolutionarily conserved, 400-kDa catalytic core of five subunits ($\alpha_2\beta\beta'\omega$). Promoter-specific initiation requires an additional polypeptide, the sigma (σ) factor, which interacts with the core RNAP to form the initiation-competent holoenzyme (Burgess et al., 1969; Travers and Burgess, 1969; Gross et al., 1998; Murakami and Darst, 2003).

The transcription cycle begins when the σ factor associates with core RNAP to form the holoenzyme, which then locates promoters through sequence-specific interactions between elements of σ and the promoter DNA (Murakami et al., 2002). For group 1 or primary σ factors, promoter sequences are characterized by two hexamers of consensus DNA sequence: the Pribnow box, or the -10 element, centered at about -10 with respect to the transcription start site at $+1$, and the -35 element (reviewed by Gross et al. [1998]). The -10 and -35 elements are usually separated by 17 bp of relatively nonconserved sequence (Hawley and McClure, 1983; Harley and Reynolds, 1987). After promoter recognition, σ factors play a major role in melting the DNA to form the open complex (reviewed by Darst et al. [1997] and Gross et al. [1998]). The σ factors may also play roles in other steps of initiation.

Most bacterial σ factors belong to a homologous family closely related to *Escherichia coli* σ^{70}, with distinct regions of highly conserved sequence (Color Plate 1A [see color insert]) (Lonetto et al., 1992; Gruber and Bryant, 1997). Structural analysis reveals that group 1 σ factors comprise four flexibly linked domains, $\sigma_{1.1}$, σ_2, σ_3 and σ_4, containing conserved regions 1.1, 1.2 to 2.4, 3.0 to 3.1, and 4.1 to 4.2, respectively (Color Plate 1) (Campbell et al., 2002b; A. Shekhtman, O. Muzzin, C. A. Olson, S. A. Darst, D. Cowburn, and T. Muir, unpublished data). A region inserted between regions 1.2 and 2.1 is not

Elizabeth A. Campbell and Seth A. Darst • The Rockefeller University, 1230 York Ave., Box 224, New York, NY 10021.

conserved in size, sequence, or structure (σ_{nc}) (Color Plate 1A). Alternative σ factors lack $\sigma_{1.1}$ and σ_{nc} and sometimes also lack σ_3.

Functions have been ascribed to some of the σ regions on the basis of genetic, biochemical, and structural studies. All four σ structural domains interact with core RNAP, and the interactions are spread throughout most of the conserved regions (Joo et al., 1997; Sharp et al., 1999; Mekler et al., 2002; Murakami et al., 2002a; Vassylyev et al., 2002). Region 2.4 recognizes the -10 element sequence (Kenney et al., 1989; Siegele et al., 1989; Zuber et al., 1989; Daniels et al., 1990; Waldburger et al., 1990; Tatti et al., 1991), while region 2.3 contains residues that melt and stabilize the transcription bubble (Helmann and Chamberlin, 1988; Jones and Moran, 1992; Juang and Helmann, 1994, 1995; deHaseth and Helmann, 1995; Murakami et al., 2002b). Region 3.0 recognizes the "extended -10" motif found immediately upstream of the -10 element at some promoters (Barne et al., 1997; Murakami et al., 2002b). Region 4.2 directly interacts with the -35 element through its helix-turn-helix motif (Gardella et al., 1989; Kenney et al., 1989; Siegele et al., 1989; Campbell et al., 2002b).

The X-ray structure of *Thermus aquaticus* core RNAP revealed a molecule shaped like a crab claw, with a 27-Å-wide internal channel. The molecule is about 150 Å long (from the back to the tip of the claws), 115 Å tall, and 110 Å wide. The enzyme active site is located on the back wall of the channel, where an essential Mg^{2+} ion is chelated. In holoenzyme, the σ structural domains, σ_2, σ_3, and σ_4 ($\sigma_{1.1}$ is missing from the available structures) lay spread out across the upstream face of the RNAP crab claw (Color Plate 1B) (Murakami et al., 2002b; Vassylyev et al., 2002). The promoter-binding determinants of σ, σ_2 (-10 element), and σ_4 (-35 element) are solvent exposed and spaced appropriately to interact with their cognate DNA elements. The σ_3 and σ_4 domains are separated by 45 Å in the holoenzyme. This distance is spanned by an extended 33-residue linker, comprising primarily σ region 3.2 (the $\sigma_{3.2}$-loop), which loops into the RNAP active-site channel and then winds its way out through the RNA exit channel. The high-resolution structures of σ and the holoenzyme indicate that the σ factor is a highly flexible polypeptide, composed of defined structural and functional domains that are capable of undergoing large rearrangements relative to each other.

The studies described above were performed with group 1, or primary, σ factors, which transcribe genes necessary for exponential growth under favorable conditions. However, bacteria must be able to adapt quickly to environmental challenges. Almost all bacteria contain alternative σ factors that direct the transcription of specialized regulons which help the organism adjust to environmental challenges such as nutrient deprivation, high cell density, sporulation, flagellar synthesis, and temperature and chemical stresses (reviewed by Helmann and Chamberlin [1988], Gross et al. [1992], Gross et al. [1996], and Wosten [1998]). While the details of the initiation steps may vary among different types of σ factors, the broad outline of the process is likely to be conserved since all σ factors must perform a common and central set of functions: recruiting the RNAP to the transcription start site and guiding the formation of the transcription bubble to form the transcription-competent open complex. Many alternative σ factors are held in an inactive state by specific regulatory proteins known as anti-σ factors (reviewed by Brown and Hughes [1995] and Hughes and Mathee [1998]), which allow cytoplasmic pools of the alternative σ factors to be maintained, facilitating short response times through inactivation of the anti-σ factor. The rest of this chapter focuses on cognate anti-σ/σ pairs for which structural studies have provided insights into function and regulation.

REGULATION OF *BACILLUS* SPORULATION BY σ^F
AND THE ANTI-σ FACTOR SpoIIAB

Many members of the *Bacillus* and *Clostridium* genera (such as *Bacillus anthracis*) are dangerous pathogens because of their ability to form spores, a form of the organism that is extremely resistant to a wide variety of challenges such as desiccation, high pressure, radiation, heat, freezing, harsh chemicals, and starvation (Setlow and Johnson, 1997; Wilson et al., 2002), making them easy to spread and difficult to eliminate. These pathogens are in the spore form when they infect their host via contaminated food, soil, or air. On entry into the host, the spores encounter favorable conditions for growth and proceed to germinate into vegetative cells that can multiply and establish an infection.

The complex process of spore development has been extensively studied at the molecular and cellular level using *B. subtilis* as a model. Spore formation begins with an asymmetric cell division, yielding a small forespore and a larger mother cell, separated by a septum (Piggot and Coote, 1976; Losick et al., 1986; Errington, 1993). The forespore initially lies beside the mother spore but is then enveloped and nurtured by the mother cell. The mother cell eventually lyses and releases the mature spore. Each cell type contains one copy of the same chromosomal DNA, yet they follow separate developmental fates due to differential gene expression within each cell. The morphological destinies of the two cells are regulated by a series of sporulation-specific σ factors (Losick and Pero, 1981; Driks and Losick, 1991; Margolis et al., 1991). Once a septum is established, σ^F (encoded by the *spoIIAC* gene) is the first σ factor to initiate the cascade of forespore-specific gene expression by inducing the expression of other sporulation-specific σ factors in the forespore, which in turn direct the expression of downstream genes necessary for the spore to complete its development (Errington, 1993; Duncan et al., 1994; Levin and Losick, 1994).

Although present in both the forespore and the mother cell (Min et al., 1993; Patridge and Errington, 1993), σ^F is active only in the forespore (Margolis et al., 1991; Bylund et al., 1994; Levin and Losick, 1994). The compartment specific activity of σ^F is regulated by members of the *spoIIA* operon to which it also belongs. Coexpressed with the gene encoding σ^F (*spoIIAC*) are the genes of its negative regulator, the anti-σ factor SpoIIAB, and the anti-anti-σ factor SpoIIAA (Schmidt et al., 1990; Duncan and Losick, 1993; Min et al., 1993; Alper et al., 1994). SpoIIAB functions as an anti-σ factor holding σ^F in an inactive complex in an ATP-stimulated manner (Duncan and Losick, 1993; Min et al., 1993; Alper et al., 1994). SpoIIAB is also a serine kinase that phosphorylates SpoIIAA (Min et al., 1993; Alper et al., 1994; Diederich et al., 1994), which functions as an anti-anti-σ in the forespore by attacking and dissociating the SpoIIAB-σ^F complex (Alper et al., 1994). Under normal conditions, SpoIIAA attack of the SpoIIAB-σ^F complex leads to the transfer of the γ-phosphate of ATP in the SpoIIAB-binding pocket to the target serine of SpoIIAA. However, neither ATP hydrolysis nor the kinase reaction is necessary for SpoIIAA to induce the release of σ^F from SpoIIAB (Diederich et al., 1994; Campbell et al., 2002a; Ho et al., 2003). Instead, a mechanism of gradual displacement of σ^F from SpoIIAB by SpoIIAA (initially termed the zipper mechanism and later renamed the docking model) due to steric clashes was proposed (Campbell et al., 2002a; Ho et al., 2003). In addition, unphosphorylated SpoIIAA binds to SpoIIAB in an ADP-dependent manner, antagonizing the role of SpoIIAB as an anti-σ factor by preventing SpoIIAB from reassociating with σ^F (Alper et al., 1994; Diederich et al., 1994). SpoIIAA oscillates between the phosphorylated

(P-AA) and unphosphorylated forms by the actions of the kinase SpoIIAB, the phosphatase SpoIIE, and other factors (Duncan et al., 1995; Arigoni et al., 1996; Feucht et al., 1996; Dworkin and Losick, 2001). It is the equilibrium between these two forms of SpoIIAA that determines the amount of free σ^F able to bind RNAP and initiate the cascade leading to the mature spore.

SpoIIAB was the first anti-σ factor to be crystallized, and this was accomplished with SpoIIAB in complex with σ^F (Campbell et al., 2002a) by using proteins from the thermophile *B. stearothermophilus*. The crystal structure (Color Plate 2 [see color insert]) revealed a SpoIIAB dimer (magenta) bound to one σ^F molecule (blue), confirming the stoichiometry of 2SpoIIAB to σ^F determined from solution measurements (Campbell and Darst, 2000). Although the complex was purified as the ATP-bound form, the crystals were formed only after ATP had hydrolyzed to ADP (shown in stick form in Color Plate 2A), representing the low-affinity form of the complex and possibly explaining why most of σ^F was disordered and not visible in the structure. Although sequence analysis indicates that σ^F contains three of the σ^{70} family structural domains (σ_2, σ_3, and σ_4), only σ_3 was ordered in the structure with SpoIIAB. Analysis of the crystals confirmed that σ_2 and σ_4 were present in the crystals but were disordered. This is consistent with the flexibility of the linker regions but inconsistent with genetic (Decatur and Losick, 1996) and proteolytic protection (Campbell and Darst, 2000) data indicating that all three σ domains interact with SpoIIAB. However, these studies probed the high-affinity complex of SpoIIAB(ATP) with σ^F, while the crystal structure revealed the low-affinity SpoIIAB(ADP) complex, which may explain the lack of strong binding of σ_2 and σ_4 to SpoIIAB. The ordered portion of σ^F (residues 102 to 158) is bound to one face of the SpoIIAB dimer, directly at the dimer interface, making asymmetric contacts with each SpoIIAB monomer (Color Plate 2).

The anti-σ SpoIIAB is a serine kinase that bears no similarity in primary sequence or structure to the Hanks-type serine kinases found in bacteria and eukaryotes (Kennelly and Potts, 1996; Zhang, 1996). Rather, the ATP-binding domain of SpoIIAB belongs to a family of histidine kinases of the bacterial His-Asp two-component signal transduction systems (Duncan and Losick, 1993; Min et al., 1993) and ATPases of the GHKL superfamily (Dutta and Inouye, 2000). SpoIIAB lacks the His box, which includes the conserved His residue that is autophosphorylated in the histidine kinases (reviewed by Stock et al. [2000]), but contains the conserved N, G1, G2, and G3 motifs (Min et al., 1993). Indeed, the structural core of SpoIIAB has the ATP-binding Bergerat fold (Bergerat, 1997) found in the histidine kinases EnvZ (Tanaka et al., 1998) and CheA (Bilwes et al., 1999; Bilwes et al., 2001) and in the ATP-binding domains of the more functionally diverse GHKL superfamily.

The visible portion of σ^F (comprising only 20% of the entire molecule) consists of a compact, three-helix domain. The structure of this domain is nearly identical to the fold of the homologous region from *T. aquaticus* σ^A (Campbell et al., 2002a; Campbell et al., 2002b). Based on structural studies of σ factors more evolutionarily distant from group 1 σ factors than σ^F is, we can infer with great certainty that the structures of σ_2^F and σ_4^F are identical to the corresponding domains of σ^A as well and that the interactions with core RNAP in the σ^F holoenzyme are conserved. This allows us to map the core-binding surface onto σ_3^F (Color Plate 2A; shaded gray on σ_3^F in Color Plate 3B). The SpoIIAB dimer binds σ_3^F across the core-binding interface. Thus, binding of SpoIIAB to σ^F occludes a core-binding interface of the σ factor, explaining the anti-σ activity of SpoIIAB.

The structure, in combination with genetic and biochemical studies, provided insight into how the anti-anti-σ factor SpoIIAA displaces σ^F from SpoIIAB. Genetic experiments identified a residue on SpoIIAB (AB-Arg20) that was critical for interaction with both SpoIIAA and σ^F, as well as a residue that was important for interacting with only SpoIIAA (AB-Glu104) (Garsin et al., 1998). How can a residue critical for induced release by SpoIIAA interact with σ^F at the same time? This puzzle was resolved by the structure and later supported by biochemical and genetic studies (Campbell et al., 2002a; Ho et al., 2003). A docking model of induced release was proposed based on the following arguments. (i) SpoIIAA has at least three sites of interaction on SpoIIAB: Glu104, the ATP-binding pocket (represented by Thr49), and Arg20 (these residues are yellow in Color Plate 2B). (ii) The structural asymmetry of the 2SpoIIAB/σ^F interface means that one molecule of the AB dimer is more exposed for AA to dock. The docking model proposes that the σ_3^F domain bound to the SpoIIAB dimer interacts with, and occludes, Arg20 of one AB monomer (labeled AB1 in Color Plate 2B) but that Arg20 of AB2 is relatively free for interaction with SpoIIAA. Furthermore, the asymmetric position of σ_3^F on the SpoIIAB dimer results in partial occlusion of the nucleotide-binding pocket of AB1, but the pocket of AB2 is more accessible (Color Plate 2B). Consistent with this model is the observation that a heterodimer of SpoIIAB with Arg20 mutated in only one monomer is competent to bind σ^F but is impervious to SpoIIAA attack (Ho et al., 2003). This fits with the view that Arg20 of AB1 interacts with σ^F and that SpoIIAA initially interacts with the available Arg20 of the AB2. More evidence for this model comes from observations of structures of SpoIIAA in complex with SpoIIAB, confirming that SpoIIAA interacts with Arg20 of SpoIIAB and introduces a steric clash with σ_3^F to induce displacement (Masuda et al., 2004). Thus, the unusual stoichiometry of the complex (2SpoIIAB-σ^F) and the resulting asymmetry of the 2SpoIIAB-σ^F interaction are critical to its proper regulation in the sporulation pathway.

REGULATION OF THE PERIPLASMIC STRESS RESPONSE BY THE ECF σ FACTOR σ^E AND THE ANTI-σ RseA

The periplasmic space, between the inner (cytoplasmic) and outer membranes of gram-negative bacteria, is a more dense and oxidizing environment than the cytoplasm, is believed to lack ATP, and is more exposed to the external environment. Thus, it is not surprising that the periplasm contains its own set of protein folding catalysts and proteases distinct from the cytoplasm (reviewed by Ravio and Silhavy [2001]). By far the largest group of σ^{70}-type σ factors, the group 4 or extracytoplasmic function (ECF) σ factors, primarily regulate responses to periplasmic stress (Ravio and Silhavy, 2001; Helmann, 2002). ECF σ factors are found in a wide variety of bacterial genomes, including many pathogens, and the different ECF σ factors within a bacterial species often outnumber all other σ factors combined (Helmann, 2002). Many ECF σ factors are regulated by anti-σ factors encoded within the same operon as the σ factor (reviewed by Ravio and Silhavy [2001]).

A signal transduction pathway involving ECF σ factors related to *Escherichia coli* σ^E plays an important role in pathogenesis in some organisms. In *Pseudomonas aeruginosa,* an opportunistic pathogen in patients with cystic fibrosis, the σ^E operon regulates mucoidy, a virulence determinant (Martin et al., 1993a; Martin et al., 1993b; Martin et al., 1993c). In

Salmonella enterica serovar Typhimurium, the causative agent of typhoid fever, σ^E is necessary for virulence and immunogenicity (Humphreys et al., 1999). In *Mycobacterium tuberculosis,* the causative agent of tuberculosis, σ^E is upregulated during intracellular growth within macrophages (Jensen-Cain and Quinn, 2001), and deletions in the gene lead to impairment of growth and survival in macrophages (Manganelli et al., 2001). A deletion of σ^E in *Vibrio cholerae,* the causative agent of cholera, leads to highly attenuated virulence (Kovacikova and Skorupski, 2002).

In *E. coli,* σ^E responds to periplasmic stress by activating the expression of proteins that mitigate the deleterious effects of misfolded proteins in the periplasm (De Las Penas et al., 1997a, 1997b; Missiakas et al., 1997; Dartigalongue et al., 2001). Initially, σ^E was identified by its requirement for the cell to survive extreme temperatures (Erickson and Gross, 1989; Raina et al., 1995; Rouviere et al., 1995), but it was subsequently found to be essential for normal cell growth (De Las Penas et al., 1997a). Like many ECF σ factors, the gene encoding σ^E, *rpoE,* is found in an operon with its regulatory genes, *rseA, rseB,* and *rseC* (De Las Penas et al., 1997b; Missiakas et al., 1997). RseA is an inner membrane protein with an N-terminal cytoplasmic domain that binds σ^E and serves as the anti-σ domain, and a C-terminal periplasmic domain that senses stress in the periplasm (De Las Penas et al., 1997b; Missiakas et al., 1997). The signal from the periplasm is communicated to the transcriptional apparatus through a signal transduction pathway that relieves inhibition of σ^E by its anti-σ factor RseA (Connolly et al., 1997; Danese and Silhavy, 1997). Under inducing conditions, RseA is rapidly degraded and σ^E is released into the cytoplasm, where it is free to bind core RNAP and induce its regulons (Ades et al., 1999).

The crystal structure of the amino-terminal 90 amino acids of the cytoplasmic domain of the anti-σ RseA (which is sufficient for binding and inhibiting σ^E and is referred to below as RseA-N) in complex with *E. coli* σ^E was determined to 2-Å resolution (Campbell et al., 2003). Although the ECF σ factors are the most divergent in sequence compared to other σ^{70} family members (less than 20% sequence identity with large gaps and deletions), σ^E comprises two independently folded, globular α-helical domains that correspond closely in structure to σ_2 and σ_4 (Color Plate 3 [see color insert]; σ^E_2 is shown in green, and σ^E_4 is shown in yellow). This is an important point that strongly suggests that all σ^{70} family members have identical folds within the structural domains. ECF σ factors lack σ_3, and the two domains (σ^E_2 and σ^E_4) are connected by an α-helix and flexible linker (red, partly disordered, and drawn as a dotted line in Color Plate 3).

The structure of the RseA-N/σ^E complex confirmed the 1:1 stoichiometry determined by biochemical methods (Campbell et al., 2003). The ordered segment of RseA-N (drawn in magenta ribbons in Color Plate 3) folds into a unique, four-helix domain that is sandwiched between σ^E_2 and σ^E_4. The interface with each σ^E domain is extensive; the σ^E_2/RseA-N interface buries 2,120 Å^2 of protein surface, while the σ^E_4/RseA-N interface buries 1,685 Å^2, yielding a total buried surface area of 3,805 Å^2. Almost 50% of the RseA-N surface is buried in the interaction with σ^E. Similar to the σ^F/SpoIIAB complex, the interfaces between σ^E and RseA are dominated by hydrophobic interactions, with polar and ionic interactions dispersed around the perimeter of each interface.

A σ^E holoenzyme model was generated, based on the σ^A holoenzyme structure and the close structural similarity of corresponding σ^E and σ^A domains (Murakami et al., 2002a, 2002b; Vassylyev et al., 2002). This allowed the identification of regions on the σ^E surface that would be involved in core RNAP contacts (shaded gray on σ^E in Color Plate 3B).

Despite its small size, RseA-N efficiently occludes each of the core-binding regions. At the same time, the anti-σ factors holds σ^E_2 and σ^E_4 together in a conformation that is not amenable to forming holoenzyme (compare to σ^A in Color Plate 1B).

σ^E is activated by relief of RseA inhibition through an ordered pathway of proteolytic degradation (Ades et al., 1999). The PDZ domain of the protease DegS, exposed in the periplasm, binds to unfolded outermembrane proteins, initiating the signal transduction cascade by activating DegS, which cleaves RseA within its periplasmic domain (Alba et al., 2001; Walsh et al., 2003). Subsequently, YaeL, a member of the SpoIVFB-S2P family of membrane-embedded proteases (Rudner et al., 1999; Brown et al., 2000), cleaves the truncated RseA within its transmembrane helix (Alba et al., 2002; Kanehara et al., 2002), releasing the cytoplasmic domain (approximately amino acids 1 to 100) of RseA, still in the complex with σ^E, into the cytosol. RseA-cyto [RseA(1–100)] is sufficient for anti-σ activity (De Las Penas et al., 1997b; Missiakas et al., 1997), and the crystal structure indicates that just RseA-N [RseA(1–66)] is sufficient for σ^E binding and inhibition since the C-terminal 24 residues are disordered in the crystal structure. Thus, further degradation of RseA-N would be required to release σ^E activity. In the structure, RseA-N is almost completely protected by σ^E, suggesting that complete degradation of RseA may require the processive unfolding/proteolytic activity found in the ClpAP/ClpXP proteases, where the exposed C-terminal segment of RseA-cyto could tag the peptide for degradation (Alba et al., 2002; Flynn et al., 2003). Once RseA was completely degraded, σ^E would be free to bind RNAP and initiate transcription of the stress response regulons.

REGULATION OF FLAGELLAR SYNTHESIS BY THE σ FACTOR FliA AND THE ANTI-σ FlgM

The bacterial flagellum provides motility necessary for chemotaxis. Proper functioning and regulation of flagellar assembly is important for virulence in pathogens that colonize the gastrointestinal tract, such as *Salmonella* (Schmitt et al., 1994; Schmitt et al., 1996a, 1996b), *Yersinia* (Kapatral et al., 1996; Young et al., 1999, 2000), and *Vibrio* (Klose and Mekalanos, 1997; Prouty et al., 2001). Flagellar synthesis, which has been well studied in *S. enterica* serovar Typhimurium and *E. coli,* is tightly controlled by a complex regulatory system organized into a transcriptional hierarchy of three classes (reviewed by Hughes and Mathee [1998]). Environmental stimuli induce class 1 genes (Silverman and Simon, 1974; Komeda et al., 1975), whose products activate the transcription of class 2 genes (Liu and Matsumura, 1994). Class 2 gene expression produces the proteins required for assembly of the hook–basal-body (HBB) structure, as well as two transcriptional regulators: FliA and FlgM (Kutsukake et al., 1990), which regulate the transcription of class 3 genes encoding proteins necessary for the completion of flagellar assembly. These include the motor force generators and the flagellin subunit that polymerizes to form the flagellar filament (reviewed by Macnab [1995]).

FliA (also known as σ^{28}, σ^F in *E. coli,* or σ^D in *B. subtilis;* hereafter called σ^{FliA} [Ohnishi et al., 1990; Liu and Matsumura, 1995]) is a group 3 σ factor required for flagellar class 3 gene expression (Kutsukake and Iino, 1994; Liu and Matsumura, 1996; Kutsukake, 1997). The activity of σ^{FliA}, and thus of class 3 gene expression, is strongly repressed until completion of HBB assembly through negative regulation by the anti-σ factor FlgM (Gillen and Hughes, 1991a, 1991b; Ohnishi et al., 1992). Once the HBB is completely assembled, FlgM is secreted through the lumen of the HBB (Hughes et al., 1993; Kutsukake, 1994; Karlinsey et al., 2000), depleting cytoplasmic FlgM and releasing σ^{FliA} to bind RNAP and direct class 3 gene expression.

A genetic screen identified mutants of σ^{FliA} that were insensitive to FlgM inhibition. Substitutions were located in σ^{FliA} domains σ^{FliA}_2, σ^{FliA}_3, and σ^{FliA}_4, suggesting that all three domains interact with FlgM (Kutsukake, 1994; Chadsey and Hughes, 2001). Biochemical studies indicated that FlgM functions by preventing the formation of the σ^{FliA} holoenzyme. In addition, kinetic studies showed that the presence of FlgM could increase the dissociation rate of σ^{FliA} from the holoenzyme, leading to the proposal that FlgM could destabilize the σ^{FliA} holoenzyme through its interactions with σ^{FliA}_4 (Chadsey et al., 1998; Chadsey and Hughes 2001).

The crystal structure of the σ^{FliA}-FlgM complex from the thermophilic bacteria *Aquifex aeolicus* (Sorenson et al., 2004) confirmed that FlgM binds to σ^{FliA} in a 1:1 stoichiometry and interacts with domains σ^{FliA}_2 and σ^{FliA}_4 (Color Plate 4 [see color insert]). In the structure of the complex, the three σ^{FliA} domains pack into a tight, compact unit, with the extended FlgM molecule wrapped around the outside. Interestingly, the loop that connects σ^{FliA}_2 and σ^{FliA}_3 was not devoid of secondary structure (as it is in the σ^A holoenzyme structure [Color Plate 1B]) but, rather comprises α-helices that extend from the second α-helix of region 3.1 (shown in red in Color Plate 4). Modeling of σ^{FliA}_2 and σ^{FliA}_4 onto the corresponding domains of σ^A in the context of the holoenzyme suggests that this connector region must undergo a conformational change from helix to loop in order to avoid clashes in the RNAP main channel.

Although there is no structural similarity between RseA and FlgM, like RseA on σ^E, FlgM occludes the primary core-binding regions on σ^{FliA}_2 and σ^{FliA}_4, supporting the previous finding that FlgM prevents σ^{FliA} from binding core RNAP (Kutsukake and Iino, 1994). It is also possible that σ^{FliA}_4 may transiently dissociate from core RNAP (enzyme "breathing") and allow FlgM to bind, increasing its local concentration. FlgM binding would prevent σ^{FliA}_4 from reassociating with RNAP, thus destabilizing the overall σ^{FliA} interaction with core RNAP. This model (which we call the docking model since it is similar to SpoIIAA displacement of σ^F from SpoIIAB) would support the observation that FlgM destabilizes the σ^{FliA} holoenzyme (Chadsey and Hughes, 2001).

GENERAL THEMES OF ANTI-σ REGULATION

The first insights into σ-factor structure came from limited proteolysis to define structural domains (Lowe et al., 1979; Gribskov and Burgess, 1983; Chang and Doi, 1990; Cannon et al., 1995; Chen and Helmann, 1995; Severinova et al., 1996) and subsequent crystal structures of σ_2, σ_3, and σ_4 (Malhotra et al., 1996; Campbell et al., 2002b). Intact σ factors have only been crystallized in complex with binding partners, either core RNAP (Murakami et al., 2002a, 2002b; Vassylyev et al., 2002) or anti-σ factors (Campbell et al., 2002a; Campbell et al., 2003; Sorensen et al., 2004). This fact and the structures themselves reveal the remarkable flexibility common to all σ^{70} family members. The discrete structural domains of σ undergo major rearrangements relative to each other, depending on the binding partner.

The current structures allow us to postulate that anti-σ factors generally function by directly occluding important core RNAP-binding determinants of σ, as well as by holding the flexibly linked σ domains in a position that is not compatible with holoenzyme formation. This is achieved by multipartite interactions, where the anti-σ factor interacts with two, and often three, of the σ structural domains simultaneously. Thus, FlgM and RseA not only obscure the core-binding determinants of their cognate σ factors but also tie up the σ factors in a conformation that could not form the holoenzyme. Although this is not visualized in the structure, genetic studies indicate that SpoIIAB simultaneously binds three structural domains of σ^F (Decatur and Losick, 1996), suggesting that SpoIIAB operates though this mechanism as well.

The multipartite interactions between σ factors and anti-σ factors may appear redundant. However, because all the domains of σ interact with core RNAP, occlusion of only one interaction by an anti-σ factor could allow the other domains of σ to bind to RNAP. This could lead to the formation of ternary complexes of core RNAP, σ, and anti-σ factor, latched onto one σ domain. Although unlikely to allow holoenzyme function, such complexes would be undesirable since they would titrate functional core RNAP away from the transcription cycle.

Other factors such as AsiA (Stevens, 1977; Colland et al., 1998; Severinova et al., 1998) and Rsd (Jishage and Ishihama, 1998, 1999; Jishage et al., 2001) have been called anti-σ factors because they also inhibit normal σ-factor function. The available data indicate that these factors interact with their cognate σ factor by binding only one domain, σ_4 (Colland et al., 1998; Severinova et al., 1998; Dove and Hochschild, 2001; Jishage et al., 2001). However, the true function of AsiA is not simply as an anti-σ factor but also as a coactivator of T4 middle transcription (Ouhammouch et al., 1995; Hinton et al., 1996). The inhibition of σ^{70} by Rsd is modest, implying that it also has an alternative biological function. It has been proposed that AsiA and Rsd be categorized in a new functional class of proteins called σ appropriators, since they modify the function of σ and holoenzymes rather than completely block it (Westblade et al., 2004).

Most members of the σ^{70} family are highly conserved in sequence within the structural domains, except for the ECF σ factors. The structure of an anti-σ factor in complex with its cognate ECF σ revealed that RseA functions as an anti-σ factor by preventing σ^E from binding to core RNAP (Campbell et al., 2003). In addition to providing the details of the σ/anti-σ factor interaction, this structure provided insight into the structure of ECF σ, whose sequence divergence left one wondering about structural conservation between this family and the more highly conserved members of the σ^{70} family. The structure of an ECF σ factor revealed that despite the sequence variability, all σ^{70} family members contain at least two main domains (σ_2 and σ_4) of highly conserved structure (Campbell et al., 2003). These domains represent the critical elements for the basic function common to all σ factors, which is that of binding core RNAP, and for mediating promoter recognition and open-complex formation. The structural conservation of these domains among all members of the σ^{70} family indicates that functional constraints allow little room for structural evolution of σ factors.

Unlike the σ factors, the anti-σ factors are highly diverse in structure, with no similarity in sequence or structure. Nevertheless, they target the same highly conserved σ domains. The structural conservation of σ factors reflects the common and central functions that all σ factors must perform. Despite their structural conservation, σ^{70} family members are regulated by a wide range of structurally and evolutionarily unrelated anti-σ factors. For example, SpoIIAB is a dimeric serine kinase (Dutta and Inouye, 2000; Campbell et al.,

2002a), FlgM is mostly unfolded in the absence of σ^{FliA} (Daughdrill et al., 1997), while RseA is an inner membrane protein (De Las Penas et al., 1997b; Missiakas et al., 1997). In addition, each anti-σ is regulated differently: SpoIIAB is held in an inactive complex by the anti-anti-σ factor SpoIIAA (Alper et al., 1994; Diederich et al., 1994); FliA is exported through the HBB (Hughes et al., 1993; Kutsukake, 1994; Karlinsey et al., 2000), and RseA is degraded by a series of proteases (Ades et al., 1999; Alba et al., 2001, Alba et al., 2002; Kanehara et al., 2002). The diversity of anti-σ-factor structure and regulation reflects the wide variety of stimuli to which anti-σ factors respond and the wide variety of regulatory mechanisms that control their function. Alternative σ factors must redirect the activity of RNAP in response to changes in the environment so that the bacteria can adapt and survive. Functional constraints on the σ factors themselves do not allow for much structural variability. The evolution of structurally and functionally diverse anti-σ factors provide much more flexibility for the regulation of transcription initiation. Thus, we propose that the functional and structural diversity of anti-σ factors reflects the need for bacteria to relay a wide variety of environmental cues to the core transcriptional apparatus via regulation of the structurally conserved σ factors.

REFERENCES

Ades, S. E., L. E. Connolly, B. M. Alba, and C. A. Gross. 1999. The *Escherichia coli* σ^E-dependent extracytoplasmic stress response is controlled by the regulated proteolysis of an anti-sigma factor. *Genes Dev.* **13:**2449–2461.

Alba, B. M., J. A. Leeds, C. Onufryk, C. H. Lu, and C. A. Gross. 2002. DegS and YaeL participate sequentially in the cleavage of RseA to activate the σ^E-dependent extracytoplasmic stress response. *Genes Dev.* **16:**2156–2168.

Alba, B. M., H. J. Zhong, J. C. Pelayo, and C. A. Gross. 2001. *degS* (*hhoB*) is an essential *Escherichia coli* gene whose indispensable function is to provide σ^E activity. *Mol. Microbiol.* **40:**1323–1333.

Alper, S., L. Duncan, and R. Losick. 1994. An adenosine nucleotide switch controlling the activity of a cell type-specific transcription factor in *B. subtilis. Cell* **77:**195–205.

Arigoni, F., L. Duncan, S. Alper, R. Losick, and P. Stragier. 1996. SpoIIE governs the phosphorylation state of a protein regulating transcription factor σ^F during sporulation in *Bacillus subtilis. Proc. Natl. Acad. Sci. USA* **93:**3238–3242.

Barne, K. A., J. A. Bown, S. J. W. Busby, and S. D. Minchin. 1997. Region 2.5 of the *Escherichia coli* RNA polymerase σ^{70} subunit is responsible for the recognition of the 'extended-10' motif at promoters. *EMBO J.* **16:**4034–4040.

Bergerat, A. 1997. An atypical topoisomerase II from Archaea with implications for meiotic recombination. *Nature* **386:**414–417.

Bilwes, A. M., L. R. Alex, B. R. Crane, and M. I. Simon. 1999. Structure of CheA, a signal-transducing histidine kinase. *Cell* **96:**131–141.

Bilwes, A. M., C. M. Quezada, L. R. Croal, B. R. Crane, and M. I. Simon. 2001. Nucleotide binding of the histidine kinase CheA. *Nat. Struct. Biol.* **8:**353–360.

Brown, K. L., and K. T. Hughes. 1995. The role of anti-sigma factors in gene regulation. *Mol. Microbiol.* **16:**397–404.

Brown, M. S., J. Ye, R. B. Rawson, and J. L. Goldstein. 2000. Regulated intramembrane proteolysis: a control mechanism conserved from bacteria to humans. *Cell* **100:**391–398.

Burgess, R. R., A. A. Travers, J. J. Dunn, and E. K. F. Bautz. 1969. Factor stimulating transcription by RNA polymerase. *Nature* **221:**43–44.

Bylund, J., L. Zhang, M. Haines, M. Higgins, and P. Piggot. 1994. Analysis by fluorescence microscopy of the development of compartment-specific gene expression during sporulation of *Bacillus subtilis. J. Bacteriol.* **176:**2898–2905.

Campbell, E., J. Tupy, T. Gruber, S. Wang, M. Sharp, C. Gross, and S. Darst. 2003. Crystal structure of *Escherichia coli* σ^E with the cytoplasic domain of its anti-σ RseA. *Mol. Cell* **11:**1067–1078.

Campbell, E. A., and S. A. Darst. 2000. The anti-σ factor SpoIIAB forms a 2:1 complex with σ^F, contacting multiple conserved regions of the σ factor. *J. Mol. Biol.* **300:**17–28.

Campbell, E. A., S. Masuda, J. L. Sun, O. Muzzin, C. A. Olson, S. Wang, and S. A. Darst. 2002a. Crystal structure of the *Bacillus stearothermophilus* anti-σ factor SpoIIAB with the sporulation σ factor σ^F. *Cell* **108:**795–807.

Campbell, E. A., O. Muzzin, M. Chlenov, J. L. Sun, C. A. Olson, O. Weinman, M. L. Trester-Zedlitz, and S. A. Darst. 2002b. Structure of the bacterial RNA polymerase promoter specificity σ subunit. *Mol. Cell* **9:**527–539.

Cannon, W., S. Missailidis, C. Smith, A. Cottier, S. Austin, M. Moore, and M. Buck. 1995. Core RNA polymerase and promoter DNA interactions of purified domains of σ^N: bipartite functions. *J. Mol. Biol.* **248:**781–803.

Chadsey, M. S., and K. T. Hughes. 2001. A multipartite interaction between *Salmonella* transcription factor σ^{28} and its anti-sigma factor FlgM: implications for σ^{28} holoenzyme destablilization through stepwise binding. *J. Mol. Biol.* **306:**915–929.

Chadsey, M. S., J. E. Karlinsey, and K. T. Hughes. 1998. The flagellar anti-σ factor FlgM actively dissociates *Salmonella typhimurium* σ^{28} RNA polymerase holoenzyme. *Genes Dev.* **12:**3123–3136.

Chang, B.-Y., and R. H. Doi. 1990. Overproduction, purification, and characterization of *Bacillus subtilis* RNA polymerase σ^A factor. *J. Bacteriol.* **172:**3257–3263.

Chen, Y. F., and J. D. Helmann. 1995. The *Bacillus subtilis* flagellar regulatory protein σ^D: overproduction, domain analysis and DNA-binding properties. *J. Mol. Biol.* **249:**743–753.

Colland, F., G. Orsini, E. Brody, H. Buc, and A. Kolb. 1998. The bacteriophage T4 AsiA protein: a molecular switch for sigma 70-dependent promoters. *Mol. Microbiol.* **27:**819–829.

Connolly, L., A. De Las Penas, B. M. Alba, and C. A. Gross. 1997. The response to extracytoplasmic stress in *Escherichia coli* is controlled by partially overlapping pathways. *Genes Dev.* **11:**2012–2021.

Danese, P. N., and T. J. Silhavy. 1997. The sigma (E) and the Cpx signal transduction systems control the synthesis of periplasmic protein-folding enzymes in *Escherichia coli*. *Genes Dev.* **11:**1183–1193.

Daniels, D., P. Zuber, and R. Losick. 1990. Two amino acids in an RNA polymerase σ factor involved in the recognition of adjacent base pairs in the −10 region of a cognate promoter. *Proc. Natl. Acad. Sci. USA* **87:**8075–8079.

Darst, S. A., J. W. Roberts, A. Malhotra, M. Marr, K. Severinov, and E. Severinova. 1997. Pribnow box recognition and melting by *Escherichia coli* RNA polymerase, p. 27–40. *In* F. Ekstein and D. M. J. Lilley (ed.), *Nucleic Acids and Molecular Biology.* Springer-Verlag, London, United Kingdom.

Dartigalongue, C., D. Missiakas, and S. Raina. 2001. Characterization of the *Escherichia coli* σ^E regulon. *J. Biol. Chem.* **276:**20866–20875.

Daughdrill, G. W., M. S. Chadsey, J. E. Karlinsey, K. T. Hughes, and F. W. Dahlquist. 1997. The C-terminal half of the anti-sigma factor, FlgM, becomes structured when bound to its target σ^{28}. *Nat. Struct. Biol.* **4:**285–291.

Decatur, A. L., and R. Losick. 1996. Three sites of contact between the *Bacillus subtilis* transcription factor σ^F and its antisigma factor SpoIIAB. *Genes Dev.* **10:**2348–2358.

deHaseth, P. L., and J. D. Helmann. 1995. Open complex formation by *Escherichia coli* RNA polymerase: the mechanism of polymerase-induced strand separation of double helical DNA. *Mol. Microbiol.* **16:**817–824.

De Las Penas, A., L. Connolly, and C. A. Gross. 1997a. σ^E is an essential sigma factor in *Escherichia coli*. *J. Bacteriol.* **179:**6862–6864.

De Las Penas, A., L. Connolly, and C. A. Gross. 1997b. The σ^E-mediated response to extracytoplasmic stress in *Escherichia coli* is transduced by RseA and RseB, two negative regulators of σ^E. *Mol. Microbiol.* **24:**373–385.

Diederich, B., J. F. Wilkinson, T. Magnin, S. M. A. Najafi, J. Errington, and M. D. Yudkin. 1994. Role of interactions between SpoIIAA and SpoIIAB in regulating cell-specific transcription factor σ^F of *Bacillus subtilis*. *Genes Dev.* **8:**2653–2663.

Dove, S. L., and A. Hochschild. 2001. Bacterial two-hybrid analysis of interactions between region 4 of the σ^{70} subunit of RNA polymerase and the transcriptional regulators Rsd from *Escherichia coli* and AlgQ from *Pseudomonas aeruginosa*. *J. Bacteriol.* **183:**6413–6421.

Driks, A., and R. Losick. 1991. Compartmentalized expression of a gene under the control of sporulation transcription factor σ^E of *Bacillus subtilis*. *Proc. Natl. Acad. Sci. USA* **88:**9934–9938.

Duncan, L., S. Alper, and R. Losick. 1994. Establishment of cell type specific gene expression during sporulation in *Bacillus subtilis*. *Curr. Opin. Genet. Dev.* **4:**630–636.

Duncan, L., S. Alper, and R. Losick. 1995. Activation of cell-specific transcription by a serine phosphatase at the site of asymmetric division. *Science* **270:**641–644.

Duncan, L., S. Alper, and R. Losick. 1996. SpoIIAA governs the release of the cell-type-specific transcription factor σ^F from its anti-sigma factor SpoIIAB. *J. Mol. Biol* **260:**147–164.

Duncan, L., and R. Losick. 1993. SpoIIAB is an anti-σ facor that binds to and inhibits transcription by regulatory protein σ^F from *Bacillus subtilis. Proc. Natl. Acad. Sci. USA* **90:**2325–2329.

Dutta, R., and M. Inouye. 2000. GHKL, an emergent ATPase/kinase superfamily. *Trends Biochem. Sci.* **25:**24–28.

Dworkin, J., and R. Losick. 2001. Differential gene expression governed by chromosomal spatial asymmetry. *Cell* **107:**339–346.

Erickson, J. W., and C. A. Gross. 1989. Identification of the σ^E subunit of *Escherichia coli* RNA polymerase: a second alternate σ factor involved in high-temperature gene expression. *Genes Dev.* **3:**1462–1471.

Errington, J. 1993. Sporulation in *Bacillus subtilis*: regulation of gene expression and control of morphogenesis. *Microbiol. Rev.* **57:**1–33.

Feucht, A., M. D. Duncan, and J. Errington. 1996. Bifunctional protein required for asymmetric cell division and cell-specific transcription in *Bacillus subtilis. Genes Dev.* **10:**794–803.

Flynn, J. M., S. B. Neher, Y. I. Kim, R. T. Sauer, and T. A. Baker. 2003. Proteomic discovery of cellular substrates of the ClpXP protease reveals five classes of signals. *Mol. Cell* **11:**671–683.

Gardella, T., T. Moyle, and M. M. Susskind. 1989. A mutant *Escherichia coli* sigma 70 subunit of RNA polymerase with altered promoter specificity. *J. Mol. Biol.* **206:**579–590.

Garsin, D. A., D. M. Paskowitz, L. Duncan, and R. Losick. 1998. Evidence for common sites of contact between the antisigma factor SpoIIAB and its partner SpoIIAA and the developmental transcription factor σ^F in *Bacillus subtilis. J. Mol. Biol.* **284:**557–568.

Gillen, K. L., and K. T. Hughes. 1991a. Molecular characterization of *flgM*, a gene encoding a negative regulator of flagellin synthesis in *Salmonella typhimurium. J. Bacteriol.* **173:**6435–6459.

Gillen, K. L., and K. T. Hughes. 1991b. Negative regulatory loci coupling flagellin synthesis to flagellar assembly in *Salmonella typhimurium. J. Bacteriol.* **173:**2301–2310.

Gribskov, M., and R. R. Burgess. 1983. Overexpression and purification of the sigma subunit of *Escherichia coli* RNA polymerase. *Gene* **26:**109–118.

Gross, C. A., C. Chan, A. Dombroski, T. Gruber, M. Sharp, J. Tupy, and B. Young. 1998. The functional and regulatory roles of sigma factors in transcription. *Cold Spring Harbor Symp. Quant. Biol.* **63:**141–155.

Gross, C. A., C. L. Chan, and M. A. Lonetto. 1996. A structure/function analysis of *Escherichia coli* RNA polymerase. *Philos. Trans. R. Soc. Lond. Ser. B* **351:**475–482.

Gross, C. A., M. Lonetto, and R. Losick. 1992. Bacterial sigma factors, p. 129–176. *In* K. Yamamoto and S. McKnight (ed.), *Transcriptional Regulation.* Cold Spring Harbor Laboratory, Cold Spring Harbor, N.Y.

Gruber, T. M., and D. A. Bryant. 1997. Molecular systematic studies of eubacteria, using sigma70-type sigma factors of group 1 and group 2. *J. Bacteriol.* **179:**1734–1747.

Harley, C. B., and R. P. Reynolds. 1987. Analysis of *E. coli* promoter sequences. *Nucleic Acids Res.* **15:**2343–2361.

Hawley, D. K., and W. R. McClure. 1983. Compilation and analysis of *Escherichia coli* promoter DNA sequences. *Nucleic Acids Res.* **11:**2237–2255.

Helmann, J. D. 2002. The extracytoplasmic function (ECF) sigma factors. *Adv. Microb. Physiol.* **46:**47–110.

Helmann, J. D., and M. J. Chamberlin. 1988. Structure and function of bacterial sigma factors. *Annu. Rev. Biochem.* **57:**839–872.

Hinton, D. M., R. March-Amegadzie, J. S. Gerber, and M. Sharma. 1996. Characterization of pretranscription complexes made at a bacteriophage T4 middle promoter: involvement of the T4 MotA activator and the T4 AsiA protein, a sigma 70 binding protein, in the formation of the open complex. *J. Mol. Biol.* **256:**235–248.

Ho, M., K. Carniol, and R. Losick. 2003. Evidence in support of a docking model for the release of the transcription factor σ^F from the antisigma factor SpoIIAB in *Bacillus subtilis. J. Biol. Chem.* **278:**20898–20905.

Hughes, K. T., K. L. Gillen, M. J. Semon, and J. E. Karlinsey. 1993. Sensing structural intermediates in bacterial flagellar assembly by export of a negative regulator. *Science* **262:**1277–1280.

Hughes, K. T., and K. Mathee. 1998. The Anti-sigma factors. *Annu. Rev. Microbiol.* **52:**231–286.

Humphreys, S., A. Stevenson, A. Bacon, A. Weinhardt, and M. Roberts. 1999. The alternative sigma factor, σ^E is critically important for virulence of *Salmonella typhimurium. Infect. Immun.* **67:**1560–1568.

Jensen-Cain, D., and F. Quinn. 2001. Differential expression of *sigE* by *Mycobacterium tuberculosis* during intracellular growth. *Microb. Pathog.* **30:**271–278.

Jishage, M., D. Dasgupta, and A. Ishihama. 2001. Mapping of the Rsd contact site on the sigma 70 subunit of *Escherichia coli* RNA polymerase. *J. Bacteriol.* **183:**2952–2956.

Jishage, M., and A. Ishihama. 1998. A stationary phase protein in *Escherichia coli* with binding activity to the major sigma subunit of RNA polymerase. *Proc. Natl. Acad. Sci. USA* **95:**4953–4958.

Jishage, M., and A. Ishihama. 1999. Transcriptional organization and in vivo role of the *Escherichia coli rsd* gene, encoding the regulator of RNA polymerase sigma D. *J. Bacteriol.* **181:**3768–3776.

Jones, C. H., and C. P. J. Moran. 1992. Mutant σ factor blocks transition between promoter binding and initiation of transcription. *Proc. Natl. Acad. Sci. USA* **89:**1958–1962.

Joo, D. M., N. Ng, and R. Calender. 1997. A sigma32 mutant with a single amino acid change in the highly conserved region 2.2 exhibits reduced core RNA polymerase affinity. *Proc. Natl. Acad. Sci. USA* **94:**4907–4912.

Juang, Y. L., and J. D. Helmann. 1994. A promoter melting region in the primary sigma factor of *Bacillus subtilis:* identification of functionally important aromatic amino acids. *J. Mol. Biol.* **235:**1470–1488.

Juang, Y.-L., and J. D. Helmann. 1995. Pathway of promoter melting by *Bacillus subtilis* RNA polymerase at a stable RNA promoter: effects of temperature, δ protein, and σ factor mutations. *Biochemistry* **34:**8465–8473.

Kanehara, K., K. Ito, and Y. Akiyama. 2002. YaeL (EcfE) activates the σ^E pathway of stress response through a site-2 cleavage of anti-σ^E, RseA. *Genes Dev.* **16:**2156–2168.

Kapatral, V., J. W. Olson, J. C. Pepe, V. L. Miller, and S. A. Minnich. 1996. Temperature-dependent regulation of *Yersinia enterocolitica* class III flagellar genes. *Mol. Microbiol.* **19:**1061–1071.

Karlinsey, J. E., J. Lonner, K. L. Brown, and K. T. Hughes. 2000. Translation/secretion coupling by type III secretion systems. *Cell* **102:**487–497.

Kennelly, P. J., and M. Potts. 1996. Fancy meeting you here! A fresh look at "prokaryotic" protein phosphorylation. *J. Bacteriol.* **178:**4759–4764.

Kenney, T. J., K. York, P. Youngman, and C. P. J. Moran. 1989. Genetic evidence that RNA polymerase associated with σA factor uses a sporulation-specific promoter in *Bacillus subtilis. Proc. Natl. Acad. Sci. USA* **86:**9109–9113.

Klose, K. E., and J. J. Mekalanos. 1997. Differential regulation of multiple flagellins in *Vibrio cholerae. J. Bacteriol.* **180:**303–316.

Komeda, U., H. Suzuki, J. I. Ishidsu, and T. Iino. 1975. The role of cAMP in flagellation of *Salmonella typhimurium. Mol. Gen. Genet.* **142:**289–298.

Kovacikova, G., and K. Skorupski. 2002. The alternative sigma factor σ^E plays an important role in intestinal survival and virulence in *Vibrio cholerae. Infect. Immun.* **70:**5355–5362.

Kutsukake, K. 1994. Excretion of the anti-sigma factor through a flagellar substructure couples flagellar gene expression with flagellar assembly in *Salmonella typhimurium. Mol. Gen. Genet.* **243:**805–812.

Kutsukake, K. 1997. Autogenous and global control of the flagellar master operon, *flhD,* in *Salmonella typhimurium. Mol. Gen. Genet.* **254:**440–448.

Kutsukake, K., and T. Iino. 1994. Role of the FliA-FlgM regulatory system on the transcriptional control of the flagellar regulon and flagellar formation in *Salmonella typhimurium. J. Bacteriol.* **176:**3598–3605.

Kutsukake, K., Y. Ohya, and T. Iino. 1990. Transcriptional analysis of the flagellar regulon of *Salmonella typhimurium. J. Bacteriol.* **172:**741–747.

Levin, P., and R. Losick. 1994. Characterization of a cell division gene from *Bacillus subtilis* that is required for vegetative and sporulation septum formation. *J. Bacteriol.* **176:**1451–1459.

Liu, X., and P. Matsumura. 1994. The FlhD/FlhC complex, a transcriptional activator of the *Escherichia coli* flagellar class II operons. *J. Bacteriol.* **176:**7345–7351.

Liu, X., and P. Matsumura. 1995. An alternative sigma factor controls transcription of flagellar class-III operons in *Escherichia coli*: gene sequence, overproduction, purification, and characterization. *Gene* **164:**81–84.

Liu, X., and P. Matsumura. 1996. Differential regulation of multiple overlapping promoters in flagellar class II operons in *Escherichia coli. Mol. Microbiol.* **21:**613–615.

Lonetto, M., M. Gribskov, and C. A. Gross. 1992. The σ70 family: sequence conservation and evolutionary relationships. *J. Bacteriol.* **174:**3843–3849.

Losick, R., and J. Pero. 1981. Cascades of sigma factors. *Cell* **25:**582–584.

Losick, R., P. Youngman, and P. Piggot. 1986. Genetics of endospore formation in *Bacillus subtilis. Annu. Rev. Genet.* **20:**625–669.

Lowe, P. A., D. A. Hager, and R. R. Burgess. 1979. Purification and properties of the σ subunit of *Escherichia coli* DNA-dependent RNA polymerase. *Biochemistry* **18:**1344–1352.

Macnab, R. M. 1995. Flagella and motility, p. 123–145. *In* F. C. Neidhardt, R. Curtiss III, J. L. Ingraham, E. C. C. Lin, K. B. Low, B. Magasanik, W. S. Reznikoff, M. Riley, M. Schaechter, and H. E. Umbarger (ed.), Escherichia coli *and* Salmonella: *Cellular and Molecular Biology,* 2nd ed. American Society for Microbiology, Washington, D.C.

Malhotra, A., E. Severinova, and S. A. Darst. 1996. Crystal structure of a σ^{70} subunit fragment from *Escherichia coli* RNA polymerase. *Cell* **87:**127–136.

Manganelli, R., M. Voskuil, G. K. Schoolnik, and I. Smith. 2001. The *Mycobacterium turberculosis* ECF sigma factor σ^E: role in global gene expression and survival in macrophages. *Mol. Microbiol.* **41:**423–437.

Margolis, P. S., A. Driks, and R. Losick. 1991. Establishment of cell type by compartmentalized activation of a transcription factor. *Science* **254:**562–565.

Martin, D., B. Holloway, and V. Deretic. 1993a. Characterization of a locus determining the mucoid status of *Pseudomonas aeruginosa. J. Bacteriol.* **175:**1153–1164.

Martin, D., M. Schurr, M. Mudd, and V. Deretic. 1993b. Differentiation of *Pseudomonas aeruginosa* into the alginate-producing form: inactivation of *mucB* causes conversion to mucoidy. *Mol. Microbiol.* **9:**497–506.

Martin, D., M. Schurr, M. Mudd, J. Govan, and B. Holloway. 1993c. Mechanism of conversion to mucoidy in *Pseudomonas aeruginosa* infecting cystic fibrosis patients. *Proc. Natl. Acad. Sci. USA* **90:**8377–8381.

Masuda, S., K. S. Murakami, S. Wang, O. C. Anders, J. Donigian, F. Leon, S. A. Darst, and E. A. Campbell. 2004. Crystal structures of the ADP and ATP bound forms of the *Bacillus* anti-sigma factor SpoIIAB in complex with the anti-anti-sigma SpoIIAA. *J. Mol. Biol.* **340:**941–956.

Mekler, V., E. Kortkhonjia, J. Mukhopadhyay, J. Knight, A. Revyakin, A. N. Kapanidis, W. Niu, Y. W. Ebright, R. Levy, and R. H. Ebright. 2002. Structural organization of bacterial RNA polymerase holoenzyme and the RNA polymerase-promoter open complex. *Cell* **108:**599–614.

Min, K. T., C. M. Hilditch, B. Diederich, J. Errington, and M. D. Yudkin. 1993. σ^F, the first compartment-specific transcription factor of *B. subtilis,* is regulated by an anti-σ factor that is also a protein kinase. *Cell* **74:**735–742.

Missiakas, D., M. P. Mayer, M. Lemaire, C. Georgopoulos, and S. Raina. 1997. Modulation of the *Escherichia coli* σ^E (RpoE) heat-shock transcription-factor activity by the RseA, RseB and RseC proteins. *Mol. Microbiol.* **24:**355–371.

Murakami, K., and S. A. Darst. 2003. Bacterial RNA polymerases: the wholo story. *Curr. Opin. Struct. Biol.* **13:**31–39.

Murakami, K., S. Masuda, and S. A. Darst. 2002a. Structural basis of transcription initiation: RNA polymerase holoenzyme at 4 Å resolution. *Science* **269:**1280–1284.

Murakami, K., S. Masuda, E. A. Campbell, O. Muzzin, and S. A. Darst. 2002b. Structural basis of transcription initiation: an RNA polymerase holoenzyme-DNA complex. *Science* **296:**1285–1290.

Ohnishi, K., K. Kutsukake, H. Suzuki, and T. Iino. 1990. Gene *fliA* encodes an alternative σ factor specific for flagellar operons in *Salmonella typhimurium. Mol. Gen. Genet.* **221:**139–147.

Ohnishi, K., K. Kutsukake, H. Suzuki, and T. Lino. 1992. A novel transcriptional regulation mechanism in the flagellar regulon of *Salmonella typhimurium:* an antisigma factor inhibits the activity of the flagellum-specific sigma factor, sigma F. *Mol. Microbiol.* **6:**3149–3157.

Ouhammouch, M., K. Adelman, S. R. Harvey, G. Orsini, and E. N. Brody. 1995. Bacteriophage T4 MotA and AsiA proteins suffice to direct *Escherichia coli* RNA polymerase to initiate transcription at T4 middle promoters. *Proc. Natl. Acad. Sci. USA* **92:**1451–1455.

Patridge, S., and J. Errington. 1993. Importance of morphological events and intercellular interactions in the regulation of prespore-specific gene expression during sporulation in *Bacillus subtilis. Mol. Microbiol.* **8:** 945–955.

Piggot, P., and J. Coote. 1976. Genetic aspects of bacterial endospore formation. *Bacteriol. Rev.* **40:**908–962.

Prouty, M. G., N. E. Correa, and K. E. Klose. 2001. The novel sigma54- and sigma28-dependent flagellar gene transcription hierarchy of *Vibrio cholerae. Mol. Microbiol.* **39:**1595–1609.

Raina, S., D. Missiakas, and C. Georgopoulos. 1995. The *rpoE* gene encoding the σ^E (σ^{24}) heat shock sigma factor of *Escherichia coli. EMBO J.* **14:**1043–1055.

Ravio, T. L., and T. J. Silhavy. 2001. Periplasmic stress and ECF sigma factors. *Annu. Rev. Microbiol.* **55:**591–624.

Rouviere, P. E., A. De Las Penas, J. Mescas, Z. L. Chin, K. E. Rudd, and C. A. Gross. 1995. *rpoE,* the gene encoding the second heat-shock sigma factor, σ^E, in *Escherichia coli. EMBO J.* **14:**1032–1042.

Rudner, D. Z., P. Fawcett, and R. Losick. 1999. A family of membrane-embedded metalloproteases involved in regulated proteolysis of membrane-associated transcription factors. *Proc. Natl. Acad. Sci. USA* **96:**14765–14770.

Schmidt, R., P. Margolis, L. Duncan, R. Coppolecchia, C. J. Moran, and R. Losick. 1990. Control of developmental transcriptional factor σ^F by sporulation regulatory proteins SpoIIAA and SpoIIAB in *Bacillus subtilis. Proc. Natl. Acad. Sci. USA* **87:**9221–9225.

Schmitt, C. K., S. C. Darnell, and A. D. O'Brien. 1996a. The attenuated phenotype of a *Salmonella typhimurium flgM* mutant is related to expression of FliC flagellin. *J. Bacteriol.* **178:**2911–2915.

Schmitt, C. K., S. C. Darnell, and A. D. O'Brien. 1996b. The *Salmonella typhimurium flgM* gene, which encodes a negative regulator of flagella synthesis and is involved in virulence, is present and functional in other *Salmonella* species. *FEMS Microbiol. Lett.* **135:**281–285.

Schmitt, C. K., S. C. Darnell, V. L. Tesh, B. A. Stocker, and A. D. O'Brien. 1994. Mutation of *flgM* attenuates virulence of *Salmonella typhimurium,* and mutation of *fliA* represses the attenuated phenotype. *J. Bacteriol.* **176:**368–377.

Setlow, P., and E. A. Johnson. 1997. Spores and their significance, p. 30–65. *In* M. P. Doyle, L. R. Beuchat, and T. J. Montville (ed.), *Food Microbiology: Fundamentals and Frontiers.* American Society for Microbiology, Washington, D.C.

Severinova, E., K. Severinov, and S. A. Darst. 1998. Inhibition of *Escherichia coli* RNA polymerase by bacteriophage T4 AsiA. *J. Mol. Biol.* **279:**9–18.

Severinova, E., K. Severinov, D. Fenyö, M. Marr, E. N. Brody, J. W. Roberts, B. T. Chait, and S. A. Darst. 1996. Domain organization of the *Escherichia coli* RNA polymerase σ^{70} subunit. *J. Mol. Biol.* **263:**637–647.

Sharp, M. M., C. L. Chan, C. Z. Lu, M. T. Marr, S. Nechaev, E. W. Merritt, K. Severinov, J. W. Roberts, and C. A. Gross. 1999. The interface of sigma with core RNA polymerase is extensive, conserved, and functionally specialized. *Genes Dev.* **13:**3015–3026.

Siegele, D. A., J. C. Hu, W. A. Walter, and C. A. Gross. 1989. Altered promoter recognition by mutant forms of the sigma 70 subunit of *Escherichia coli* RNA polymerase. *J. Mol. Biol.* **206:**591–603.

Silverman, M., and M. Simon. 1974. Characterizaton of *Escherichia coli* flagellar mutants that are insenstitive to catabolite repression. *J. Bacteriol.* **120:**1196–1203.

Sorenson, M. K., S. S. Ray, and S. A. Darst. 2004. Crystal structure of the flagellar sigma/anti-sigma complex sigma(28)/FlgM reveals an intact sigma factor in an inactive conformation *Mol. Cell* **14:**127–138.

Stevens, A. 1977. Inhibition of DNA-enzyme binding by an RNA polymerase inhibitor from T4 phage-infected *Escherichia coli. Biochim. Biophys. Acta* **475:**193–196.

Stock, J. B., V. L. Robinson, and P. N. Goudreau. 2000. Two-component signal transduction. Annu. Rev. Biochem. **69:**183–215.

Tanaka, T., S. Saha, C. Tomomori, R. Ishima, D. Liu, D. I. Tong, H. Park, R. Dutta, L. Qin, M. B. Swindells, et al. 1998. NMR structure of the histidine kinase domain of the *E. coli* osmosensor EnvZ. *Nature* **396:**88–92.

Tatti, K. M., C. H. Jones, and C. P. J. Moran. 1991. Genetic evidence for interaction of sigma E with the spoIIID promoter in *Bacillus subtilis. J. Bacteriol.* **173:**7828–7833.

Travers, A. A., and R. R. Burgess. 1969. Cyclic re-use of the RNA polymerase sigma factor. *Nature* **222:**537–540.

Vassylyev, D. G., S. Sekine, O. Laptenko, J. Lee, M. N. Vassylyeva, S. Boruhkhov, and S. Yokoyama. 2002. Crystal structure of a bacterial RNA polymerase holoenzyme at 2.6 Å resolution. *Nature* **417:**712–719.

Waldburger, C., T. Gardella, R. Wong, and M. M. Susskind. 1990. Changes in conserved region 2 of *Escherichia coli* sigma 70 affecting promoter recognition. *J. Mol. Biol.* **215:**267–276.

Walsh, N., B. Alba, B. Baundana, C. Gross, and R. Sauer. 2003. OMP peptide signals initiate the envelope-stress response by activating DegS protease via relief of inhibition mediated by its PDZ domain. *Cell* **113:**61–71.

Westblade, L., L. Ilag, A. Powel, A. Kolb, C. Robinson, and S. Busby. 2004. Studies of the *Escherichia coli* Rsd-sigma70 complex. *J. Mol. Biol.* **335:**685–692.

Wilson, M., R. McNab, and B. Henderson. 2002. *Bacterial Disease Mechanisms: an Introduction to Cellular Microbiology.* Cambridge University Press, Cambridge, United Kingdom.

Wosten, M. M. 1998. Eubacterial sigma-factors. *FEMS Microbiol. Rev.* **22:**127–150.

Young, G., J. L. Badger, and V. L. Miller. 2000. Motility is required to initiate host cell invasion by *Yersinia enterocolitica. Infect. Immun.* **68:**4323–4326.

Young, G. M., D. H. Schmiel, and V. L. Miller. 1999. A new pathway for the secretion of virulence factors by bacteria: the flagellar export apparatus functions as a protein-secretion system. *Proc. Natl. Acad. Sci. USA* **96:**6456–6461.

Zhang, C. C. 1996. Bacterial signalling involving eukaryotic-type protein kinases. *Mol. Microbiol.* **20:**9–15.

Zuber, P., J. Healy, H. L. Carter III, S. Cutting, C. P. Moran, Jr., and R. Losick. 1989. Mutation changing the specificity of an RNA polymerase sigma factor. *J. Mol. Biol.* **206:**605–614.

Structural Biology of Bacterial Pathogenesis
Edited by G. Waksman et al.
© 2005 ASM Press, Washington, D.C.

Chapter 2

Two-Component Signal Transduction and Chemotaxis

Jodi B. Lubetsky and Ann M. Stock

TWO-COMPONENT SIGNALING

As unicellular organisms, bacteria are commonly faced with major perturbations in their surroundings. In order to survive and thrive, bacteria must be able to adapt quickly to continuously changing environments. Hence, they contain numerous signaling systems designed to detect environmental signals and elicit appropriate responses involving alterations in gene expression or activation of certain gene products. Two-component signaling systems, the principal signaling strategy in bacteria, couple environmental signals to cellular responses and play a key role in bacterial adaptability. These well-conserved systems are involved in a large variety of responses including osmoregulation, chemotaxis, metabolic regulation, and virulence factor expression.

There are several reasons why two-component systems appear to be attractive targets for the development of new antibiotics. Two-component proteins are present in almost all gram-positive and gram-negative bacteria but are absent from animals. Two-component signaling systems are abundant, with 20 to 40 different systems typically present in a bacterial genome. Importantly, two-component-mediated bacterial adaptability is often essential for the establishment of successful infections. Specifically, in addition to changes in housekeeping functions that occur when bacteria migrate from a free-living state to association with a host, bacteria coordinate the expression of virulence factors that enable the pathogen to grow within the host organism. Two-component signaling systems are commonly involved in many of these processes.

This chapter provides a review of the architecture and structures of histidine kinases (HKs) and response regulators (RRs) and a description of a well-characterized two-component system, the bacterial chemotaxis signaling pathway. In addition, the potential of two-component systems as drug targets and the progress that has been made with inhibitor design and development are discussed.

Jodi B. Lubetsky and Ann M. Stock • Center for Advanced Biotechnology and Medicine, Howard Hughes Medical Institute, Department of Biochemistry, University of Medicine and Dentistry of New Jersey—Robert Wood Johnson Medical School, Piscataway, NJ 08854.

STRUCTURAL CHARACTERIZATION OF HISTIDINE
KINASES AND RESPONSE REGULATORS

Two-component signal transduction systems are structured around two conserved proteins: an HK and an RR (for reviews, see Hoch and Silhavy [1995], Inouye and Dutta [2003], Robinson et al. [2000], and Stock et al. [2000]). HKs typically function as sensors of the cellular environment. The majority of HKs are transmembrane proteins with variable extracellular domains involved in sensing different chemical or physical stimuli specific to each pathway. The conserved kinase core of the HK catalyzes ATP-dependent autophosphorylation at an invariant His residue, providing a high-energy phosphoryl group poised for transfer. RRs are typically multidomain proteins with highly conserved regulatory domains and variable effector domains. The conserved regulatory domain of the RR catalyzes the transfer of the phosphoryl group from the phospho-His of the HK to its own conserved Asp side chain (Color Plate 5 [see color insert]). The resulting high-energy acyl phosphate activates the RR, allowing the effector domain to elicit a specific output response. In most systems, the response is regulation of expression of a specific set of genes. However, some RRs control responses other than transcriptional regulation, which contribute in different ways to the adaptation of the organism to environmental changes.

The simplest of the phosphotransfer pathways involves the single-step transfer of phosphoryl groups between a His of an HK and an Asp of a downstream RR (Color Plate 5A). The phosphotransfer pathway can also be more complex, involving components that contain multiple phosphodonor and acceptor sites (Color Plate 5B). Whereas simple phosphotransfer pathways predominate in prokaryotes, the more complex phosphorelays predominate in eukaryotes. Like most signaling domains, the conserved domains of two-component systems are modular. There is considerable variability in the arrangements of domains within proteins and the assembly of these proteins into pathways.

Regulation of two-component pathways can occur at numerous points and is often complex, involving multiple targets of regulation. In many systems, auxiliary proteins, in addition to HKs and RRs, act to influence the level of RR phosphorylation. Many fundamentally different schemes of regulation have been identified in different systems. The basic two-component pathway provides a versatile mechanism that can be embellished and optimized in an almost infinite number of ways to meet the specific needs of individual signaling systems.

Histidine Kinases

HKs play a role in linking the extracellular environment to the intracellular adaptive response. The diversity in HKs is created through the combination of modular sensing, catalytic, and auxiliary domains. This architecture allows an individual HK to be optimized for its role in a specific signaling system. The majority of HKs are dimeric transmembrane proteins, consisting of an N-terminal periplasmic sensing domain and a cytoplasmic C-terminal kinase core. The periplasmic domain senses the input stimuli that reflect the many different environmental signals. However, not all HKs are membrane bound. Soluble HKs are regulated by intracellular stimuli, including interactions with cytoplasmic domains of other receptor proteins.

Fluctuations in stimuli directly or indirectly modulate signaling by regulating the activities of the HK core. These fluctuations can lead to changes in the autophosphorylation activity of the HK. In addition, many HKs possess a highly regulated phosphatase activity that returns RRs to their unphosphorylated state. Although the details of HK phosphatase activity are not understood, evidence suggests that for many HKs, it is independent of the reverse phosphotransfer reaction.

The kinase core consists of a catalytic ATP-binding domain and a dimerization domain. The structure of the HK catalytic domain is well conserved, but its fold is distinct from that of the Ser/Thr/Tyr kinase family. Rather, the HK fold is similar to the folds found in the ATPase domains of the chaperone Hsp90, the DNA topoisomerase gyrase B, and the DNA mismatch repair protein MutL (Bilwes et al., 1999; Tanaka et al., 1998). This domain has an α/β fold consisting of three α-helices layered on a five-strand β-sheet (Color Plates 5D and 6 [see color insert]). The catalytic domain contains several highly conserved regions that are located in both rigid and flexible regions of the structure. The N box is positioned on the central helix of $\alpha 1$, with the G1 box on the adjacent strand. The F and G2 boxes are part of a polypeptide segment that extends away from the rest of the molecule and consists of a short α helix, $\alpha 3$, followed by a long loop (Color Plate 6A).

Several structures of the catalytic domain in complex with either ADP or nucleotide analogues provide insight into the structural features involved in its activity (Bilwes et al., 2001; Tanaka et al., 1998). In each of the structures, nucleotide binding is associated with conformational changes in the flexible loop between $\alpha 2$ and $\alpha 3$, the ATP lid (Color Plate 6B). On nucleotide binding, the ATP lid changes from a poorly defined region to a more ordered closed conformation. Specificity is provided through the diversity in the length and conformation of the lid. The ATP-binding domain of *Thermotoga maritima* HK CheA binds nucleotides in a similar orientation as several ATPases. However, residues involved in specific interactions differ, making the hydrophobic nucleotide-binding pocket of CheA a potential target for specifically designed inhibitors (Bilwes et al., 2001).

The dimerization domain is the second of the two conserved domains in the kinase core. In *Escherichia coli* HK EnvZ, the dimerization domains form a four-helix bundle composed of two identical helix-turn-helix subunits that create a symmetric homodimer with two active-site His residues per structural unit (Color Plate 5C) (Tomomori et al., 1999). However, the conserved His residues are not always located in the dimerization domain but, instead, can be located in separate His-containing phosphotransfer (HPt) domains.

In eukaryotes, HPt domains generally function as separate phosphotransfer proteins independent of the HK core. In prokaryotes, HPt domains are often contained within hybrid kinases. Hybrid kinases are more complex members of the HK family that contain multiple phosphodonor and phosphoacceptor sites. For example, the *E. coli* hybrid HK ArcB contains two N-terminal transmembrane regions, a kinase core, an RR-like regulatory domain, and an HPt domain (Ishige et al., 1994). HPt domains also have a four-helix bundle topology, but they can be composed of either one or two chains. The structures of the monomeric HPt domains of CheA (Zhou et al., 1995), ArcB (Ikegami et al., 2001), and Ypd1 (Song et al., 1999; Xu and West, 1999) are four-helix bundles that provide one active-site His residue per structural unit (Color Plate 7A [see color insert]). The dimeric HPt domain of Spo0B resembles the His containing dimerization domain of EnvZ, with two active-site His residues per structural unit (Varughese et al., 1998) (Color Plate 7B).

There is limited information about the orientation of the dimerization and ATP-binding domains in typical HKs. The atypical HK CheA provides the only insight into an intact kinase core structure (see Color Plate 10D) (Bilwes et al., 1999). The structure, consisting of the C-terminal region of CheA, includes a dimerization domain (P3), an ATP-binding domain (P4), and an auxiliary protein-protein interaction domain (P5). However, the dimerization domain lacks the conserved His residues for phosphorylation, which, instead, are located remotely in an N-terminal HPt domain (P1) attached by a flexible linker in the intact protein. The remaining domain of CheA, P2, serves as an RR-binding domain. In a typical HK, the His-containing dimerization and ATP-binding domains are adjacent to each other. If the arrangement of the dimerization and ATP-binding domains observed in CheA were maintained in a typical kinase, the His residue would be positioned too far away from the kinase active site for His phosphorylation to occur without a substantial conformational change.

In the absence of high-resolution structural information, cross-linking studies have been used to probe the orientation between the catalytic and dimerization domains in the typical HK *E. coli* EnvZ (Cai et al., 2003). The proposed model of domain orientations in EnvZ differs significantly from the domain orientations observed in the CheA crystal structure (Bilwes et al., 1999). In the model of EnvZ, the ATP-binding domain is positioned to interact with the intersubunit interface of the dimerization domains, whereas in CheA, each ATP-binding domain makes intramolecular contacts with only a single subunit of the dimerization domain. The domain arrangement proposed for EnvZ provides a structural explanation for the intermolecular phosphorylation that occurs between monomers within an HK dimer (Ninfa et al., 1993; Swanson et al., 1993; Wolfe and Stewart, 1993; Yang and Inouye, 1991).

Response Regulators

The majority of RRs are transcription factors composed of an N-terminal regulatory domain and a C-terminal DNA-binding domain. The conserved regulatory domain functions as a phosphorylation-activated switch with distinct conformations in the unphosphorylated and phosphorylated states. The regulatory domain has enzymatic activity, catalyzing Mg^{2+}-dependent phosphoryl transfer from the HK and autodephosphorylation of its own phospho-Asp, with half-lives ranging from seconds to hours, the latter approximating the chemical half-life of an acyl phosphate. RRs can be classified into three major subfamilies based on homology of their DNA-binding effector domains: the OmpR/PhoB winged-helix domains, the NarL/FixJ four-helix domains, and the NtrC/DctD ATPase-coupled helix-turn-helix domains (Stock et al., 1989b). The remaining RRs are either transcription factors with DNA-binding domains of miscellaneous folds or proteins that regulate cellular activities through enzymatic activity or protein-protein interactions. Over the last several years, an increasing number of structures of regulatory domains, effector domains, and full-length RRs have been determined, giving us a clearer picture of the mechanism of phosphorylation-induced activation.

The three-dimensional structures of numerous regulatory domains have been determined, and all share the same fold (Color Plate 5E). The regulatory domain, consisting of approximately 120 amino acid residues, forms a five-strand parallel sheet flanked by two α-helices on one face and three on the other. This domain is characterized by a set of conserved residues, four of which cluster in an active site at the C-terminal edge of the β-sheet: the site of Asp phosphorylation, a pair of acidic residues involved in metal binding, and a Lys that forms a salt bridge with the phosphate. Two additional conserved residues, a

Ser/Thr in β4 and a Phe/Tyr in β5, play direct roles in the propagation of conformational changes that accompany phosphorylation.

An important area of research focuses on understanding how phosphorylation of the conserved regulatory domain affects the activities of the structurally and functionally diverse effector domains. The short lifetime of the phosphorylated state has made structural characterization of the activated states of regulatory domains challenging. Using a variety of techniques, structural studies of several activated regulatory domains, including those of FixJ, NtrC, CheY, and Spo0A, have provided insight into the mechanism of activation. A crystal structure of the regulatory domain of phospho-FixJ was determined by eliminating autodephosphorylation activity by the removal of the active-site Mg^{2+} with EDTA (Birck et al., 1999). NtrC was maintained in a steady state of phosporylation by the presence of a low-molecular-weight phosphodonor during nuclear magnetic resonance spectroscopy analysis (Kern et al., 1999). Several structures of activated CheY were determined by using different chemical analogues: a stable covalent phosphonomethyl thioether (Halkides et al., 2000) and a noncovalent beryllofluoride adduct that provides a mimic of the phosphorylated state (Cho et al., 2000). Crystals of phosphorylated Spo0A from the hyperthermophile *Bacillus stearothermophilus* were obtained fortuitously without any attempt to phosphorylate the protein prior to crystallization (Lewis et al., 1999).

In each of these structural studies, the activated and unactivated regulatory domains exhibit minor differences in the positioning of secondary-structure elements that accompany the movements of several specific side chains. In the phosphorylated regulatory domain structures, the side chain of the conserved Ser/Thr in α4 is reoriented to form a hydrogen bond with an oxygen of the phosphate. The cavity vacated by the rotated Ser/Thr is filled by the side chain of the conserved Phe/Tyr, which is reoriented inward, toward the site of phosphorylation (Color Plate 8 [see color insert]). In all cases, the major structural changes are localized to the C-terminal half of the regulatory domain. In specific RRs, additional changes occur in other regions of the domain. Thus, while fundamentally similar alterations are induced by phosphorylation, there is some diversity with respect to the location and magnitude of the structural changes that occur in different RRs.

A representative structure of an effector domain from each of the DNA-binding RR subfamilies has been determined. All transcription factor effector domains contain a helix-turn-helix DNA-binding motif; however, the folds in each subfamily are distinct. The OmpR effector domain, a representative member of the largest subfamily of RRs, has a modified winged-helix fold. The winged-helix core consists of three α-helices and a C-terminal β-hairpin. An additional N-terminal four-strand antiparallel β-sheet that forms an integral part of the domain distinguishes the OmpR/PhoB subfamily from all other winged-helix domains (Kondo et al., 1997; Martinez-Hackert and Stock, 1997). The DNA-binding domain of NarL, a representative RR of the NarL/FixJ family, is a four-helix bundle, whose middle helices form the classic helix-turn-helix motif (Baikalov et al., 1996; Baikalov et al., 1998; Maris et al., 2002). The NtrC/DctD subfamily of RRs is more complex than is either of the OmpR/PhoB or NarL/FixJ subfamilies, since NtrC homologs contain three domains: an N-terminal regulatory domain, a central ATPase domain, and a C-terminal DNA-binding domain. The DNA-binding domain consists of four helical segments, two of which dimerize to form a four-helix bundle. The ATPase domain, found in activators of the σ^{54} RNA polymerase holoenzyme, is composed of α/β- and α-helical subdomains (Pelton et al., 1999).

A key mechanistic question being addressed by structural experiments with RRs is how phosphorylation-induced structural alterations in the regulatory domain effect changes in the activities of the structurally and functionally diverse effector domains. Although there are numerous structures of isolated regulatory and effector domains, only a few structures of full-length RRs (e.g., CheB [Djordjevic et al., 1998a], NarL [Baikalov et al., 1996; Baikalov et al., 1998], DrrB [Buckler et al., 2002], and DrrD [Robinson et al., 2003]) have been determined (Color Plate 9 [see color insert]). Despite the overall similarity in the structures of their regulatory domains, all proteins use a different subset of the α4-β5-α5 face of the regulatory domain for interdomain contacts, implying different mechanisms of intramolecular communication between the N- and C-terminal domains. For example, the structures of NarL and CheB both exhibit an extensive interdomain interface in which the regulatory domains play an inhibitory role, obstructing access to the catalytic and binding sites in the effector domains. Although the structure of DrrB reveals a large interface as well, there is no obvious structural evidence of inhibition. The regulatory domain does not obstruct access to the recognition helix or present steric conflicts when DrrB is docked onto DNA (Robinson et al., 2003). DrrD has a much smaller region of interdomain contact that is most probably the consequence of crystal packing. Again, there is no evidence of structural inhibition, and it appears that the domains of DrrD lack a fixed orientation in the unphosphorylated state. Nonetheless, biochemical studies suggest that for some members of the OmpR/PhoB subfamily, to which DrrB and DrrD belong, the unphosphorylated regulatory domain does inhibit DNA binding by the effector domain (Ellison and McCleary, 2000).

In addition to intramolecular interactions such as the inhibitory interactions discussed above, RRs use a variety of other mechanisms to regulate effector domain activity. One such mechanism is oligomerization. Unactivated RRs exist in a variety of oligomerization states that can change on phosphorylation. For example, on phosphorylation, RRs such as FixJ (Da Re et al., 1999), PhoB (Fiedler and Weiss, 1995; McCleary, 1996), and DrrB (Robinson et al., 2003) dimerize; in contrast, NtrC (Wyman et al., 1997) and ArcA (Jeon et al., 2001) dimers form higher-order oligomers, estimated to be octamers. It is becoming apparent that there is great diversity in the mechanisms of RR activation, with fundamentally different strategies of intramolecular and intermolecular regulatory mechanisms being employed by different RRs, even those within a single structurally related subfamily.

CHEMOTAXIS, AN EXTENSIVELY CHARACTERIZED TWO-COMPONENT SYSTEM

Although chemotaxis represents a nontraditional two-component system that has been directly linked to pathogenesis in only a few bacterial species, it is one of the most extensively studied and well-characterized two-component signaling systems and thus warrants mention in this chapter. Chemotaxis proteins were among the first HK and RRs for which biochemical activities were defined and for which three-dimensional structures were determined. In addition, the three-dimensional structures of almost all of the remaining chemotaxis proteins have been determined, including several protein complexes, giving a fairly complete structural description of a two-component signaling pathway.

Bacteria alter their swimming behavior in response to changes in stimuli such as chemicals, pH, and temperature. In the enteric bacterium *E. coli*, the chemotaxis signaling pathway controls the direction of flagellar rotation; counterclockwise flagellar rotation produces smooth-swimming behavior, and clockwise rotation produces tumbling, allowing reorientation. By controlling the frequency of tumbling, cells can achieve a biased random walk that allows efficient migration along chemical gradients toward favorable conditions. Although there are differences in the details of the pathway between species, bacteria that exhibit taxis share a number of the same key components in their signal transduction pathway (for reviews, see Bourret and Stock [2002] and Bren and Eisenbach [2000]).

The bacterial chemotaxis pathway is built around a bifurcated two-component phosphotransfer system involving a single HK, CheA, and two RRs, CheY and CheB (Fig. 1). The central branch of the pathway, involving phosphotransfer from CheA to CheY, controls flagellar rotation, while the other branch, involving phosphotransfer from CheA to CheB, provides a feedback loop that attenuates the motor response and contributes to adaptation. The autophosphorylation activity of the HK CheA is controlled by transmembrane chemotaxis receptors known as methylated chemotaxis proteins (MCPs). The signaling activity of the chemoreceptors is influenced by both the ligand occupancy of the periplasmic sensing domain and the covalent modification state of the cytoplasmic signaling domain. Importantly, this allows signaling to reflect changes in ligand occupancy rather than absolute bound ligand concentrations.

The number of different MCPs and the specific ligands they detect vary in each species. However, all have in common a highly conserved cytoplasmic domain that undergoes reversible methylation at specific glutamate residues. MCPs that have been characterized are homodimers irrespective of the presence or absence of ligand. The cytoplasmic domain monomers consist of an antiparallel coiled coil that dimerizes to form a four-helix

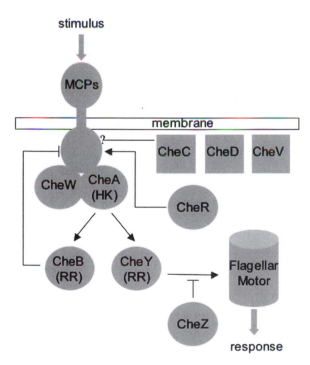

Figure 1. Schematic diagram of bacterial chemotaxis signal transduction. External stimuli trigger changes in protein modifications and protein-protein interactions that lead to behavioral responses generated by the flagellar motors, as described in the text. The proteins present in the *E. coli* chemotaxis system are shown as circles. CheC, CheD, and CheV (squares) are absent in *E. coli* but are present in a large number of other bacteria.

structure (Kim et al., 1999). The crystal structure of *E. coli* Tsr reveals a trimeric association of dimers. The physiological significance of this higher-order structure has not been determined; however, receptors are known to cluster at the poles of the cell (Maddock and Shapiro, 1993). The sensing domains of chemoreceptors are more variable than the cytoplasmic domains (for a review, see Falke and Hazelbauer [2001]). Sensing domains are typically periplasmic, but some transmembrane receptors have cytoplasmic sensing domains. Other chemoreceptors are completely cytoplasmic, lacking transmembrane anchors.

The CheA kinase is coupled to the chemotaxis receptors through the protein CheW. The solution structure of *T. maritima* CheW reveals two five-strand β-barrels that surround a hydrophobic core (Color Plate 10A [see color insert]) (Griswold et al., 2002). On the basis of nuclear magnetic resonance spectroscopy chemical shift data and mutational studies, the face of one of these domains was determined to be involved in the interaction with CheA (Boukhvalova et al., 2002; Griswold et al., 2002), whereas an adjacent surface may serve as an interface for receptor binding (Boukhvalova et al., 2002).

As described in the HK section (see above), the CheA monomer is made up of five domains: the HPt-like domain (P1), the RR-binding and recognition domain (P2), the dimerization domain (P3), the ATP-binding kinase domain (P4), and the CheW interaction coupling domain (P5). CheA autophosphorylates at a conserved histidine residue in the P1 domain, a monomeric five-helix bundle (Color Plate 10B) (Zhou et al., 1995). The flexibility in the linkers between the P1 domain and the kinase domain allows for ATP-dependent *trans*-phosphorylation between monomers of the dimer (Wolfe and Stewart, 1993). The P2 domain serves as an interaction site for RRs CheY and CheB. The structure of P2, a four-strand antiparallel β-sheet with two crossing helices, has been determined both alone and in complex with CheY (Color Plate 10C) (McEvoy et al., 1996; McEvoy et al., 1998; Welch et al., 1998). The helices serve as the site of interaction with the α4-β5-α5 site of CheY. The structure of the P3-P5 domains reveals an HK dimerization domain consisting of a pair of antiparallel helices that form a four-helix bundle within the dimer, an ATP-binding catalytic domain with an α/β fold common to all HKs, and a regulatory/coupling domain consisting of two twisted five-strand barrels similar to the fold of CheW (Color Plate 10D) (Bilwes et al., 1999).

CheY is the diffusible signal that shuttles between the receptor complex and the flagellar motor complex. Unlike most RRs, which contain both an N-terminal regulatory domain and a C-terminal effector domain, CheY lacks an effector domain. It catalyzes the transfer of the phosphoryl group from CheA to its own conserved Asp residue. It has a three-dimensional structure common to all N-terminal regulatory domains of multidomain RRs: a five-strand β-sheet doubly wound with five α-helices (Color Plates 5E, 8, and 10E) (Stock et al., 1989a; Volz and Matsumura, 1991). Phosphorylation of CheY reduces its affinity for the P2 domain of CheA and increases its affinity for interaction with FliM, the switch component of the flagellar rotor, thus promoting tumbly behavior (Welch et al., 1993, 1994).

Decay of the CheY signal occurs by dephosphorylation, a reaction that is accelerated by the phosphatase CheZ. Although CheZ is present in *E. coli,* it is not present in most bacteria. It is a dimer composed of two helices from each monomer, which form an extended four-helix bundle structure (Color Plate 10E) (Zhao et al., 2002). The C terminus of CheZ is a long, flexible tail that terminates in a 13-residue helix that functions to tether phosphorylated CheY, allowing low-affinity contacts between the active sites of CheY and

CheZ. The active site for dephosphorylation is composed of residues from both CheY and CheZ.

Receptor methylation levels affect CheA autophosphorylation rates and contribute to adaptation in the pathway. Receptor methylation levels are controlled by methylesterase CheB and methyltransferase CheR. CheR constitutively methylates chemotaxis receptors at multiple glutamate residues within the cytoplasmic signaling domains. CheB, which demethylates the chemotaxis receptors, is the primary locus of regulation. Phosphorylation of CheB by the phosphodonor CheA enhances CheB methylesterase activity. Demethylation of the receptors decreases CheA autophosphorylation, resulting in decreased phosphorylation of CheY and less interaction with FliM. Thus, the bacterium adapts to its prestimulus swimming behavior. Although the RRs CheY and CheB are phosphorylated on the same timescale, the adaptation pathway occurs in a delayed fashion relative to the phosphorylation-mediated motor response as a result of the lower rates of methylation and demethylation (Segall et al., 1986).

CheR is a two-domain protein with a helical N-terminal domain that interacts with the methylation region of the receptor and an α/β C-terminal domain with a fold common to class I *S*-adenosylmethionine-dependent methyltransferases (Color Plate 10F) (Djordjevic and Stock, 1997). A small β-sheet subdomain is inserted into the C-terminal domain of CheR. The subdomain functions as a site of interaction with a pentapeptide recognition sequence that caps a long, flexible tail at the C termini of the receptors (Djordjevic and Stock, 1998). The pentapeptide tethers CheR and facilitates a weak interaction between the CheR active site and the methylation region of the receptors (Wu et al., 1996).

CheB, also a two-domain protein, contains an N-terminal regulatory domain that has a fold similar to that of CheY and a C-terminal effector domain that serves to demethylate the chemotaxis receptors (Color Plate 9A). The crystal structure of CheB reveals that the N-terminal domain packs against the active site of the C-terminal domain, restricting active-site access (Djordjevic et al., 1998). Phosphorylation of the N-terminal domain of CheB relieves the inhibition of the C-terminal domain. The N-terminal domain provides an activating role as well, since phosphorylated CheB has higher methylesterase activity than does the C-terminal domain of CheB alone (Anand et al., 1998; Anand and Stock, 2002). The binding of CheB and CheY to CheA is competitive and presumably involves a similar or sterically overlapping surface on the P2 domain of CheA (Li et al., 1995). However, CheY and CheB presumably use different faces for interaction with P2, since the α4-β5-α5 surface utilized by CheY is buried in the interdomain interface of CheB.

Three protein components, CheC, CheD, and CheV, are absent from *E. coli* but are present in a large number of other bacteria. The functions of these three proteins are poorly understood, but all appear to play some role in receptor adaptation (Karatan et al., 2001; Rosario et al., 1995; Rosario and Ordal, 1996). CheC and CheD are involved in regulating MCP methylation. *Bacillus subtilis* CheC interacts with several chemotaxis system components including CheA and has a sequence similar to several flagellar proteins including the N terminus of *B. subtilis* FliY (Kirby et al., 2001), a phosphatase that enhances hydrolysis of phospho-CheY (Szurmant et al., 2003). The structure of *T. maritima* CheC has recently been determined, and CheC has been shown to have phospho-CheY phosphatase activity (Park et al., 2004). *B. subtilis* CheD functions as a deamidase, catalyzing the amide hydrolysis of glutaminyl side chains of the chemoreceptor McpA (Kristich and Ordal, 2002).

TWO-COMPONENT SYSTEMS AS TARGETS FOR
ANTIMICROBIAL DRUG THERAPY

The emergence of multiple drug resistance is an increasing problem. Bacteria have evolved resistance mechanisms that include pumping the drug out of the cell, inactivating the drug, modifying the drug target so that it is no longer susceptible, or using an alternative pathway to bypass the drug-targeted pathway through the acquisition of mutations in the bacterial chromosome or plasmids containing resistance genes. This resistance necessitates the development of new classes of antimicrobial drugs (for a review, see McDevitt et al. [2002]).

Several features make the proteins of two-component signaling systems attractive antimicrobial therapeutic targets. Two-component signaling is highly prevalent in bacteria and common to both gram-positive and gram-negative organisms. Bacterial genomes typically encode 20 to 40 different two-component regulatory systems. For example, the *E. coli* genome encodes a total of 62 two-component system HKs and RRs (Mizuno, 1997). In contrast, a relatively small number of two-component pathways have been identified in eukaryotes (for a review, see Chang and Stewart [1998]), and no two-component proteins have yet been identified in animals. Furthermore, as discussed in this section, two-component systems are commonly involved in the regulation of expression of virulence factors, and some two-component systems are involved in antibiotic resistance mechanisms.

On the other hand, there are some significant challenges in targeting two-component proteins for antibiotic development. Most two-component systems are not essential (Fabret and Hoch, 1998; Throup et al., 2000); loss of an individual pathway may compromise growth but typically is not bactericidal. The use of broad-spectrum drugs that simultaneously target several two-component pathways, thereby severely compromising growth, might circumvent this problem. However, it is questionable whether such drugs can be designed, given the low level of amino acid sequence identity (typically 20 to 30%) between different HKs and RRs.

Two-component systems contain two conserved proteins that can be targeted for inhibition; each poses different obstacles. From a molecular perspective the HK is the preferred target, but from a physiological perspective the RR is superior. Protein active sites are traditional inhibitor targets. The HK contains a well-defined active site that binds ATP, and the pharmaceutical industry is experienced in designing protein kinase inhibitors. In contrast, the active site of the RR is a shallow dimple with minimal substrate recognition features. In addition to the phospho-His of HKs, RRs can utilize many different small molecules as phosphodonors, including phosphoimidazole, carbamoyl phosphate, acetyl phosphate, and phosphoramidate, suggesting that substrate recognition is limited to only the phosphoryl group (Lukat et al., 1992). Alternatively, recognition surfaces outside the active sites could potentially be targeted for either HKs or RRs, but inhibition of protein-protein interactions has rarely been a successful drug design strategy.

While the HK is the more attractive target for inhibitor design, loss of HK activity rarely produces a null phenotype. There are many potential explanations for the activity of a specific RR in the absence of its cognate kinase. Some RRs have low but appreciable activity in their unphosphorylated states. Alternatively, intracellular small-molecule phosphodonors such as acetyl phosphate (Feng et al., 1992; McCleary and Stock, 1994; McCleary et al., 1993) or cross talk from noncognate HKs (Fisher et al., 1995) can also contribute to RR activity. Thus, inhibiting a specific HK may not be sufficient to significantly compromise growth of the bacteria.

Despite the challenges noted above, two-component proteins have been actively pursued as targets for new classes of antimicrobial drugs. The relatively low level of sequence identity between different two-component proteins bodes poorly for identification of a generic inhibitor, and it seems more likely that successful drugs will be targeted against specific proteins. The following sections discuss several classes of two-component systems that are attractive drug targets.

Essential Genes

Essential genes are appealing targets for antimicrobial therapy. However, there are only a few occurrences of essential two-component system genes. The two-component system YycF-YycG is essential in several nonpathogenic and pathogenic organisms including *B. subtilis, Staphylococcus aureus,* and *Streptococcus pneumoniae.* The YycF-YycG system is involved in modulating the expression of the *ftsAZ* operon implicated in cell division (Fukuchi et al., 2000). In *B. subtilis,* YycF-YycG is the only essential system of the 35 two-component systems (Fabret and Hoch, 1998). Neither the RR YycF nor the HK YycG can be inactivated without loss of viability. In *S. aureus,* a gene encoding the RR YycF was identified as the locus of a point mutation resulting in a temperature-sensitive lethal phenotype. Insertional inactivation of either *yycF* or *yycG* is lethal (Martin et al., 1999). In *S. pneumoniae,* the RR YycF is the only essential component of the 14 putative RRs and 13 putative HKs. Unlike the situation in *B. subtilis* and *S. aureus,* inactivation of the HK YycG has little effect on the growth of *S. pneumoniae* (Throup et al., 2000).

Several essential two-component proteins have been found in other bacteria. The HK CckA and the RR CtrA, which are involved in bacterial cell cycle regulation, are both essential in nonpathogenic *Caulobacter crescentus* (Jacobs et al., 1999; Quon et al., 1996). The RR MtrA from *Mycobacterium tuberculosis,* the bacterium responsible for tuberculosis, is an essential component that plays a role in transcriptional activation during infection of phagocytic cells (Zahrt and Deretic, 2000). Several RRs in the gastric pathogen *Helicobacter pylori* are also known to be essential (Beier and Frank, 2000).

Pathogenesis

Numerous two-component pathways are directly involved in virulence. These include systems that mediate bacterial adherence, cell motility, cell-to-cell communication, toxin expression, and adaptation to the host. These tightly regulated systems are necessary for the initiation and maintenance of infection and the survival of bacteria within the host organism. The following are a small sampling of virulence-associated systems.

For many pathogenic bacteria, motility plays important roles during infection of the host (for a review, see Josenhans and Suerbaum [2002]). In some pathogens, motility is required only in the initial phases of infection, since the motility apparatus is involved in adherence to the host. In other pathogens, motility is needed both to establish and to maintain an infection. Some pathogens require not only motility but also directed movement (chemotaxis) throughout the infection process. In many other pathogens, the role of motility remains unclear. Two-component regulatory systems are involved in many aspects of motility, including the regulation of expression of genes encoding the motility apparatus and chemotaxis. Examples of the former include the PilS-PilR pilus formation system in *Pseudomonas aeruginosa* and the PilA-PilB pilus production system in *Neisseria gonorrhoeae*. The well-conserved CheA-CheY-CheB components that mediate chemotaxis are discussed in the previous section.

For bacteria to successfully initiate and maintain an infection, bacterial cells must reach a critical cell density and coordinate the expression of virulence factors at the infection site. Known as quorum sensing, this cell-to-cell communication is dependent on diffusible secreted signals that activate signaling pathways that regulate gene expression (for a review, see Miller and Bassler [2001]). Quorum-sensing systems can be based on either of two different molecular strategies. The quorum-sensing systems of gram-positive bacteria typically involve peptide autoinducers that function as signals for two-component signaling pathways. One well-characterized two-component quorum-sensing system is the Agr system of *S. aureus,* a global regulator of bacterial virulence genes. The *agr* locus encodes a set of four proteins, AgrA to AgrD, that are necessary for transcriptional activation of two promoters. Density-dependent accumulation of an extracellular cyclic thiolactone peptide, autoinducing peptide (AIP), is derived by processing of AgrD by AgrB. Once a critical binding concentration is reached, AIP triggers the activation of the HK AgrC, which in turn activates the AgrA RR transcription factor that controls virulence factor expression (Ji et al., 1995, 1997; Lyon et al., 2002; Mayville et al., 1999). The specificity of the AIP/AgrC interaction has enabled the development of an inhibitor of virulence factor expression by using a truncated version of a naturally occurring thiolactone peptide (Lyon et al., 2000).

Drug Resistance

Two-component systems have been identified that are either directly involved in drug resistance or involved in the transfer of resistance genes between organisms. In *S. pneumoniae,* expression of the penicillin-binding protein that mediates penicillin resistance is under the control of a two-component system, CiaH-CiaR. Mutations in the sensor HK CiaH lead to increased resistance to β-lactams (Hakenbeck et al., 1999). Another well-known example of a two-component system involved in antibiotic resistance is the VanS-VanR system, which regulates vancomycin resistance in enterococci (for a review, see Walsh et al. [1996]). Vancomycin is a glycopeptide antibiotic that noncovalently binds to the C-terminal D-Ala–D-Ala motif in the peptidoglycan strands in the bacterial cell wall, blocking transglycosylation and transpeptidation. The HK VanS and the RR VanR regulate the transcription of genes encoding the proteins responsible for conversion of the D-Ala–D-Ala motif to D-Ala–D-Lac, an alteration that results in resistance to the activity of vancomycin.

Many of the antibiotics used today are derived from natural sources, and resistance mechanisms have evolved in the antibiotic-producing organisms. Resistance genes can be transferred through conjugative plasmid transfer from donor to recipient bacteria (for a review, see Grohmann et al. [2003]). Some two-component systems are involved in regulating plasmid transfer. For example, in *Bacteroides,* the self-transfer of tetracycline resistance elements is regulated by tetracycline. Disruption of the RteA-RteB two-component system eliminates plasmid transfer, suggesting that this system is involved in tetracycline regulation of tetracycline resistance elements (Stevens et al., 1993).

STRATEGIES OF INHIBITION

Chemical library screening is the most common approach to the identification of novel antimicrobial inhibitors. Large chemical libraries, assembled by pharmaceutical companies, are screened against a model biochemical assay or a specific target system in order to identify lead compounds that exhibit inhibitory activity. Lead compounds are then further optimized by using structure-activity relationship studies to develop either broad-based inhibitors that target conserved system components or specific inhibitors that target a particular system. An alternative method used to develop drugs is the rational design of inhibitors based on known biochemical and structural data for a specific biological system.

Most two-component system inhibitors described to date have been identified through chemical library screening (for reviews, see Barrett and Hoch [1998], Macielag and Goldschmidt [2000], Matsushita and Janda [2002], and Stephenson and Hoch [2002]). However, this has led to compounds that appear to operate by multiple mechanisms and have poor selectivity. In the following sections we describe the results of chemical library screening and initial attempts at rational design of inhibitors.

Chemical Library Screening

Much of the published work on two-component system inhibitors has explored the use of chemical library screening with the intention of developing broad-based inhibitors that target the fairly well-conserved autophosphorylation and phosphotransfer domains of HKs. For example, the R. W. Johnson Pharmaceutical Research Institute screened its chemical libraries for inhibitors of two-component systems by using the *B. subtilis* KinA-Spo0F and/or the *E. coli* NRII-NRI proteins in a model biochemical assay. Several groups of compounds were identified that inhibited autophosphorylation and phosphotransfer activities, including RWJ-49815 (a member of the hydrophobic tyramine family) (Barrett et al., 1998), diaryltriazoles (Sui et al., 1998), salicylanilides (Hlasta et al., 1998; Macielag et al., 1998), 6-oxa isosteres of anacardic acid (Kanojia et al., 1999), benzimidazoles (Weidner-Wells et al., 2001), benzoxazines, and bis-phenols (Hilliard et al., 1999). Other research groups, including those at Parke-Davis (Domagala et al., 1998) and TerraGen Discovery (Trew et al., 2000), as well as Hoch's group at the Scripps Research Institute (Strauch et al., 1992), have also performed library screening with both synthesized and natural compounds to identify two-component system inhibitors.

Two of the most potent R. W. Johnson inhibitors, the trityl RWJ-49815 and the salicylanide Closantel, have been characterized further (Stephenson et al., 2000). Both compounds cause a structural alteration in the kinase that leads to aggregation of the isolated C-terminal domain of the HK KinA. The authors suggest that the compounds intercalate into the hydrophobic core of the dimerization domain, disrupting the four-helix bundle and thereby exposing hydrophobic surfaces that promote aggregation.

Representative inhibitors from the above classes have also been examined in mechanistic studies, specifically in tests for antimicrobial activity against *S. aureus* and for their effects on bacterial and red blood cell membrane integrity, bacterial viability, and macromolecular synthesis (Hilliard et al., 1999). Only the benzimidazoles and trityl compounds exhibited strong correlations between enzymatic inhibition and antibacterial activity. Furthermore, 23 of the 24 inhibitors studied affected membrane integrity in either bacteria or erythrocytes. Additionally, 11 of 12 compounds tested caused greatly reduced cellular incorporation of RNA, DNA, and protein precursors. Hilliard and colleagues hypothesize that bacterial killing may be due to multiple mechanisms, some of which are independent of HK inhibition. These studies illustrate some of the obstacles that occur in trying to identify general inhibitors and suggest that greater success might be achieved by using a more selective inhibitor identification method targeted against specific two-component systems.

Library screening targeted to a specific two-component system has been applied to the alginate system of *P. aeruginosa,* an opportunistic respiratory pathogen that is the primary cause of mortality in cystic fibrosis patients. In pulmonary infections, *P. aeruginosa* converts to a nontractable mucoid phenotype due to overproduction of the exopolysaccharide alginate. Alginate production is regulated by AlgR1, an RR that controls the expression of genes required for alginate synthesis, and AlgR2, an HK. Chakrabarty and colleagues screened a library of 25,000 synthetic compounds for inhibition of transcriptional activation of the *algD* promoter (Roychoudhury et al., 1993). Fifteen inhibitors were identified, and two classes were investigated further. The halophenyl isothiazolone inhibitors inhibit the kinase activity of AlgR2, whereas the imidazolium salt compounds inhibit the DNA-binding activity of AlgR1 (Roychoudhury et al., 1993). The compounds were also tested for inhibition of other two-component kinase activities. The AlgR2 kinase inhibitors were found to be relatively specific for AlgR2. Surprisingly, the inhibitors of the RR AlgR1 also inhibit the autophosphorylation of CheA, NRII, and KinA. While this suggests that these inhibitors might function through a nonspecific mechanism, the authors argue against this on the basis of the low concentrations of compounds required for inhibition.

Rational Inhibitor Design

Although a number of inhibitors have been identified through chemical library screening, this strategy has mostly produced compounds that are poorly selective and have undefined mechanisms of action. The availability of structural information for two-component proteins enables a rational design approach for the development of specific inhibitors.

Theoretically, inhibitors can be directed toward a number of different targets within two-component signaling pathways. Structural studies have facilitated the identification of targets that appear to be most amenable for design of inhibitors. Targeting a catalytic active site or a ligand-binding site is usually more straightforward than blocking an interaction between two proteins. In two-component systems, the HK active site is an attractive target since it has a well-defined cleft for nucleotide binding. In contrast, the shallow active site of the RR that recognizes a phosphoryl group for transfer is significantly less attractive as a target.

The most definitive study of the ATP-binding pocket of HKs is the analysis of CheA. Structures of the ATP-binding domain of *T. maritima* CheA in complex with various nucleotides have been determined (Bilwes et al., 2001). The structures reveal a nucleotide-binding pocket that is highly amenable for inhibitor development. There is potential for great selectivity in inhibitor binding, since residues involved in nucleotide interactions in CheA differ from residues involved in nucleotide interactions in other ATP-binding domains. However, there are no published reports of success in obtaining nucleotide analog inhibitors. The difficulty in crystallizing HK domains has undoubtedly limited the exploitation of this region for inhibitor development.

An alternative strategy is to extrapolate information from proteins with similar folds. For example, the antifungal antibiotic radicocol, which is known to inhibit the ATPase activity of Hsp90, was tested for inhibitory activity of the structurally similar HK and HK-like proteins (Besant et al., 2002). Radicocol was found to inhibit yeast HK Sln1 and branched-chain α-keto acid dehydrogenase kinase, a mammalian Ser/Thr kinase having a fold homologous to the HKs.

Only a few other examples of rationally designed inhibitors of two-component systems have been reported. One interesting example involves the AgrC-AgrA quorum-sensing system, in which the ligand-binding site of the HK sensing domain was targeted for inhibition. In the Agr system, an extracellular cyclic thiolactone peptide, AIP, triggers the activation of the HK, ArgC. A truncated version of AIP was developed as an inhibitor of virulence factor expression on the basis of detailed studies of the mechanism of AIP-receptor interaction (Lyon et al., 2000). Lack of knowledge of the specific ligands that interact with most HK-sensing domains precludes this from being a general strategy for the development of HK inhibitors.

FUTURE DIRECTIONS

To date, an extremely limited number of inhibitor design studies have been reported. Most of the two-component system inhibitors described in the literature have been obtained by high-throughput screening methods that target HK autophosphorylation or phosphotransfer. This strategy has yielded inhibitors that are nonspecific and/or mechanistically not well understood. Structural and biochemical studies of both HKs and RRs have revealed that there are multiple mechanisms of two-component system activation, implying the need for inhibitor development on a system-specific basis. The continued generation of additional structures of two-component proteins should greatly facilitate this process. A more rational structure-based drug design approach may lead to the successful development of inhibitors of two-component signaling systems that are fundamental to bacterial regulation.

REFERENCES

Anand, G. S., P. N. Goudreau, and A. M. Stock. 1998. Activation of methylesterase CheB: evidence of a dual role for the regulatory domain. *Biochemistry* **37:**14038–14047.

Anand, G. S., and A. M. Stock. 2002. Kinetic basis for the stimulatory effect of phosphorylation on the methylesterase activity of CheB. *Biochemistry* **41:**6752–6760.

Baikalov, I., I. Schröder, M. Kaczor-Grzeskowiak, D. Cascio, R. P. Gunsalus, and R. E. Dickerson. 1998. NarL dimerization? Suggestive evidence from a new crystal form. *Biochemistry* **37:**3665–3676.

Baikalov, I., I. Schröder, M. Kaczor-Grzeskowiak, K. Grzeskowiak, R. P. Gunsalus, and R. E. Dickerson. 1996. Structure of the *Escherichia coli* response regulator NarL. *Biochemistry* **35:**11053–11061.

Barrett, J. F., R. M. Goldschmidt, L. E. Lawrence, B. Foleno, R. Chen, J. P. Demers, S. Johnson, R. Kanojia, J. Fernandez, J. Bernstein, L. Licata, A. Donetz, S. Huang, D. J. Hlasta, M. J. Macielag, K. Ohemeng, R. Frechette, M. B. Frosco, D. H. Klaubert, J. M. Whiteley, L. Wang, and J. A. Hoch. 1998. Antibacterial agents that inhibit two-component signal transduction systems. *Proc. Natl. Acad. Sci. USA* **95:**5317–5322.

Barrett, J. F., and J. A. Hoch. 1998. Two-component signal transduction as a target for microbial anti-infective therapy. *Antimicrob. Agents Chemother.* **42:**1529–1536.

Beier, D., and R. Frank. 2000. Molecular characterization of two-component systems of *Helicobacter pylori.* *J. Bacteriol.* **182:**2068–2076.

Besant, P. G., M. V. Lasker, C. D. Bui, and C. W. Turck. 2002. Inhibition of branched-chain alpha-keto acid dehydrogenase kinase and Sln1 yeast histidine kinase by the antifungal antibiotic radicicol. *Mol. Pharmacol.* **62:**289–296.

Bilwes, A. M., L. A. Alex, B. R. Crane, and M. I. Simon. 1999. Structure of CheA, a signal-transducing histidine kinase. *Cell* **96:**131–141.

Bilwes, A. M., C. M. Quezada, L. R. Croal, B. R. Crane, and M. I. Simon. 2001. Nucleotide binding by the histidine kinase CheA. *Nat. Struct. Biol.* **8:**353–360.

Birck, C., L. Mourey, P. Gouet, B. Fabry, J. Schumacher, P. Rousseau, D. Kahn, and J.-P. Samama. 1999. Conformational changes induced by phosphorylation of the FixJ receiver domain. *Struct. Fold. Des.* **7:**1505–1515.

Boukhvalova, M., R. VanBruggen, and R. C. Stewart. 2002. CheA kinase and chemoreceptor interaction surfaces on CheW. *J. Biol. Chem.* **277:**23596–23603.

Bourret, R. B., and A. M. Stock. 2002. Molecular information processing: lessons from bacterial chemotaxis. *J. Biol. Chem.* **277:**9625–9628.

Bren, A., and M. Eisenbach. 2000. How signals are heard during bacterial chemotaxis: protein-protein interactions in sensory signal propagation. *J. Bacteriol.* **182:**6865–6873.

Buckler, D. R., Y. Zhou, and A. M. Stock. 2002. Evidence of intradomain and interdomain flexibility in an OmpR/PhoB homolog from *Thermotoga maritima. Structure* **10:**153–164.

Cai, S. J., A. Khorchid, M. Ikura, and M. Inouye. 2003. Probing catalytically essential domain orientation in histidine kinase EnvZ by targeted disulfide crosslinking. *J. Mol. Biol.* **328:**409–418.

Chang, C., and R. C. Stewart. 1998. The two-component system. Regulation of diverse signaling pathways in prokaryotes and eukaryotes. *Plant Physiol.* **117:**723–731.

Cho, H. S., S. Y. Lee, D. Yan, X. Pan, J. S. Parkinson, S. Kustu, D. E. Wemmer, and J. G. Pelton. 2000. NMR structure of activated CheY. *J. Mol. Biol.* **297:**543–551.

Da Re, S., J. Schumacher, P. Rousseau, J. Fourment, C. Ebel, and D. Kahn. 1999. Phosphorylation-induced dimerization of the FixJ receiver domain. *Mol. Microbiol.* **34:**504–511.

Djordjevic, S., P. N. Goudreau, Q. Xu, A. M. Stock, and A. H. West. 1998. Structural basis for methylesterase CheB regulation by a phosphorylation-activated domain. *Proc. Natl. Acad. Sci. USA* **95:**1381–1386.

Djordjevic, S., and A. M. Stock. 1998. Chemotaxis receptor recognition by methyltransferase CheR. *Nat. Struct. Biol.* **5:**446–450.

Djordjevic, S., and A. M. Stock. 1997. Crystal structure of the chemotaxis receptor methyltransferase CheR suggests a conserved structural motif for binding *S*-adenosylmethionine. *Structure* **5:**545–558.

Domagala, J. M., D. Alessi, M. Cummings, S. Gracheck, L. Huang, M. Huband, G. Johnson, E. Olson, M. Shapiro, R. Singh, Y. Song, R. Van Bogelen, D. Vo, and S. Wold. 1998. Bacterial two-component signalling as a therapeutic target in drug design. Inhibition of NRII by the diphenolic methanes (bisphenols). *Adv. Exp. Med. Biol.* **456:**269–286.

Ellison, D. W., and W. R. McCleary. 2000. The unphosphorylated receiver domain of PhoB silences the activity of its output domain. *J. Bacteriol.* **182:**6592–6597.

Fabret, C., and J. A. Hoch. 1998. A two-component signal transduction system essential for growth of *Bacillus subtilis:* implications for anti-infective therapy. *J. Bacteriol.* **180:**6375–6383.

Falke, J. J., and G. L. Hazelbauer. 2001. Transmembrane signaling in bacterial chemoreceptors. *Trends Biochem. Sci.* **26:**257–265.

Feng, J., M. R. Atkinson, W. McCleary, J. B. Stock, B. L. Wanner, and A. J. Ninfa. 1992. Role of phosphorylated metabolic intermediates in the regulation of glutamine synthetase synthesis in *Escherichia coli. J. Bacteriol.* **174:**6061–6070.

Fiedler, U., and V. Weiss. 1995. A common switch in activation of the response regulators NtrC and PhoB: phosphorylation induces dimerization of the receiver modules. *EMBO J.* **14:**3696–3705.

Fisher, S. L., W. Jiang, B. L. Wanner, and C. T. Walsh. 1995. Cross-talk between the histidine protein kinase VanS and the response regulator PhoB. *J. Biol. Chem.* **270:**23143–23149.

Fukuchi, K., Y. Kasahara, K. Asai, K. Kobayashi, S. Moriya, and N. Ogasawara. 2000. The essential two-component regulatory system encoded by *yycF* and *yycG* modulates expression of the *ftsAZ* operon in *Bacillus subtilis. Microbiology* **146:**1573–1583.

Griswold, I. J., H. Zhou, R. V. Swanson, L. P. McIntosh, M. I. Simon, and F. W. Dahlquist. 2002. The solution structure and interactions of CheW from *Thermotoga maritima. Nat. Struct. Biol.* **9:**121–125.

Grohmann, E., G. Muth, and M. Espinosa. 2003. Conjugative plasmid transfer in gram-positive bacteria. *Microbiol. Mol. Biol. Rev.* **67:**277–301.

Hakenbeck, R., T. Grebe, D. Zahner, and J. B. Stock. 1999. β-Lactam resistance in *Streptococcus pneumoniae:* penicillin-binding proteins and non-penicillin-binding proteins. *Mol. Microbiol.* **33:**673–678.

Halkides, C. J., M. M. McEvoy, E. Casper, P. Matsumura, K. Volz, and F. W. Dahlquist. 2000. The 1.9 Å resolution crystal structure of phosphono-CheY, an analogue of the active form of the response regulator, CheY. *Biochemistry* **39:**5280–5286.

Hilliard, J. J., R. M. Goldschmidt, L. Licata, E. Z. Baum, and K. Bush. 1999. Multiple mechanisms of action for inhibitors of histidine protein kinases from bacterial two-component systems. *Antimicrob. Agents Chemother.* **43:**1693–1699.

Hlasta, D. J., J. P. Demers, B. D. Foleno, S. A. Fraga-Spano, J. Guan, J. J. Hillard, M. J. Macielag, K. A. Ohemeng, C. M. Sheppard, Z. Sui, G. C. Webb, M. A. Weidner-Wells, H. Werblood, and J. F. Barret. 1998. Novel inhibitors of bacterial two-component systems with gram positive antibacterial activity: pharmacophore identification based on the screening hit closantel. *Bioorg. Med. Chem. Lett.* **8:**1923–1928.

Hoch, J. A., and T. J. Silhavy (ed.). 1995. *Two-Component Signal Transduction.* ASM Press, Washington, D.C.

Ikegami, T., T. Okada, I. Ohki, J. Hirayama, T. Mizuno, and M. Shirakawa. 2001. Solution structure and dynamic character of the histidine-containing phosphotransfer domain of anaerobic sensor kinase ArcB from *Escherichia coli. Biochemistry* **40:**375–386.

Inouye, M., and R. Dutta (ed.). 2003. *Histidine Kinases in Signal Transduction.* Academic Press, Inc., San Diego, Calif.

Ishige, K., S. Nagasawa, S. Tokishita, and T. Mizuno. 1994. A novel device of bacterial signal transducers. *EMBO J.* **13:**5195–5202.

Jacobs, C., I. J. Domian, J. R. Maddock, and L. Shapiro. 1999. Cell cycle-dependent polar localization of an essential bacterial histidine kinase that controls DNA replication and cell division. *Cell* **97:**111–120.

Jeon, Y., Y. S. Lee, J. S. Han, J. B. Kim, and D. S. Hwang. 2001. Multimerization of phosphorylated and non-phosphorylated ArcA is necessary for the response regulator function of the Arc two-component signal transduction system. *J. Biol. Chem.* **276:**40873–40879.

Ji, G., R. Beavis, and R. P. Novick. 1997. Bacterial interference caused by autoinducing peptide variants. *Science* **276:**2027–2030.

Ji, G., R. C. Beavis, and R. P. Novick. 1995. Cell density control of staphylococcal virulence mediated by an octapeptide pheromone. *Proc. Natl. Acad. Sci. USA* **92:**12055–12059.

Josenhans, C., and S. Suerbaum. 2002. The role of motility as a virulence factor in bacteria. *Int. J. Med. Microbiol.* **291:**605–614.

Kanojia, R. K., W. Murray, J. Bernstein, J. Fernandez, B. D. Foleno, H. Krause, L. Lawrence, G. Webb, and J. F. Battett. 1999. 6-Oxa isosteres of anacardic acids as potent inhibitors of bacterial histidine protein kinase (HPK)-mediated two-component regulatory systems. *Bioorg. Med. Chem. Lett.* **9:**2947–2952.

Karatan, E., M. M. Saulmon, M. W. Bunn, and G. W. Ordal. 2001. Phosphorylation of the response regulator CheV is required for adaptation to attractants during *Bacillus subtilis* chemotaxis. *J. Biol. Chem.* **276:**43618–43626.

Kern, D., B. F. Volkman, P. Luginbuhl, M. J. Nohaile, S. Kustu, and D. E. Wemmer. 1999. Structure of a transiently phosphorylated switch in bacterial signal transduction. *Nature* **40:**894–898.

Kim, K. K., H. Yokota, and S.-H. Kim. 1999. Four-helical-bundle structure of the cytoplasmic domain of a serine chemotaxis receptor. *Nature* **400:**787–792.

Kirby, J. R., C. J. Kristich, M. M. Saulmon, M. A. Zimmer, L. F. Garrity, I. B. Zhulin, and G. W. Ordal. 2001. CheC is related to the family of flagellar switch proteins and acts independently from CheD to control chemotaxis in *Bacillus subtilis. Mol. Microbiol.* **42:**573–585.

Kondo, H., A. Nakagawa, J. Nishihira, Y. Nishimura, T. Mizuno, and I. Tanaka. 1997. *Escherichia coli* positive regulator OmpR has a large loop structure at the putative RNA polymerase interaction site. *Nat. Struct. Biol.* **4:**28–31.

Kristich, C. J., and G. W. Ordal. 2002. *Bacillus subtilis* CheD is a chemoreceptor modification enzyme required for chemotaxis. *J. Biol. Chem.* **277:**25356–25362.

Lewis, R. J., J. A. Brannigan, K. Muchová, I. Barák, and A. J. Wilkinson. 1999. Phosphorylated aspartate in the structure of a response regulator protein. *J. Mol. Biol.* **294:**9–15.

Li, J., R. V. Swanson, M. I. Simon, and R. M. Weis. 1995. The response regulators CheB and CheY exhibit competitive binding to the kinase CheA. *Biochemistry* **34:**14626–14636.

Lukat, G. S., W. R. McCleary, A. M. Stock, and J. B. Stock. 1992. Phosphorylation of bacterial response regulator proteins by low molecular weight phospho-donors. *Proc. Natl. Acad. Sci. USA* **89:**718–722.

Lyon, G. J., P. Mayville, T. W. Muir, and R. P. Novick. 2000. Rational design of a global inhibitor of the virulence response in *Staphylococcus aureus,* based in part on localization of the site of inhibition to the receptor-histidine kinase, AgrC. *Proc. Natl. Acad. Sci. USA* **97:**13330–13335.

Lyon, G. J., J. S. Wright, A. Christopoulos, R. P. Novick, and T. W. Muir. 2002. Reversible and specific extracellular antagonism of receptor-histidine kinase signaling. *J. Biol. Chem.* **277:**6247–6253.

Macielag, M. J., J. P. Demers, S. A. Fraga-Spano, D. J. Hlasta, S. G. Johnson, R. M. Kanojia, R. K. Russell, Z. Sui, M. A. Weidner-Wells, H. Werblood, B. D. Foleno, R. M. Goldschmidt, M. J. Loeloff, G. C. Webb, and J. F. Barrett. 1998. Substituted salicylanilides as inhibitors of two-component regulatory systems in bacteria. *J. Med. Chem.* **41:**2939–2945.

Macielag, M. J., and R. Goldschmidt. 2000. Inhibitors of bacterial two-component signalling systems. *Expert Opin. Investig. Drugs* **9:**2351–2369.

Maddock, J. R., and L. Shapiro. 1993. Polar location of the chemoreceptor complex in the *Escherichia coli* cell. *Science* **259:**1717–1723.

Maris, A. E., M. R. Sawaya, M. Kaczor-Grzeskowiak, M. R. Jarvis, S. M. Bearson, M. L. Kopka, I. Schroder, R. P. Gunsalus, and R. E. Dickerson. 2002. Dimerization allows DNA target site recognition by the NarL response regulator. *Nat. Struct. Biol.* **9:**771–778.

Martin, P. K., T. Li, D. Sun, D. P. Biek, and M. B. Schmid. 1999. Role in cell permeability of an essential two-component system in *Staphylococcus aureus. J. Bacteriol.* **181:**3666–3673.

Martinez-Hackert, E., and A. M. Stock. 1997. The DNA-binding domain of OmpR: crystal structure of a winged-helix transcription factor. *Structure* **5:**109–124.

Matsushita, M., and K. D. Janda. 2002. Histidine kinases as targets for new antimicrobial agents. *Bioorg. Med. Chem.* **10:**866–867.

Mayville, P., G. Ji, R. Beavis, H. Yang, M. Goger, R. P. Novick, and T. W. Muir. 1999. Structure-activity analysis of synthetic autoinducing thiolactone peptides from *Staphylococcus aureus* responsible for virulence. *Proc. Natl. Acad. Sci. USA* **96:**1218–1223.

McCleary, W. R. 1996. The activation of PhoB by acetylphosphate. *Mol. Microbiol.* **20:**1155–1163.

McCleary, W. R., and J. B. Stock. 1994. Acetyl phosphate and the activation of two-component response regulators. *J. Biol. Chem.* **269:**31567–31572.

McCleary, W. R., J. B. Stock, and A. J. Ninfa. 1993. Is acetyl phosphate a global signal in *Escherichia coli*? *J. Bacteriol.* **175:**2793–2798.

McDevitt, D., D. J. Payne, D. J. Holmes, and M. Rosenberg. 2002. Novel targets for the future development of antibacterial agents. *J. Appl. Microbiol.* **92:**28S–34S.

McEvoy, M. M., A. C. Hausrath, G. B. Randolph, S. J. Remington, and F. W. Dahlquist. 1998. Two binding modes reveal flexibility in kinase/response regulator interactions in the bacterial chemotaxis pathway. *Proc. Natl. Acad. Sci. USA* **95:**7333–7338.

McEvoy, M. M., D. R. Muhandiram, L. E. Kay, and F. W. Dahlquist. 1996. Structure and dynamics of a CheY-binding domain of the chemotaxis kinase CheA determined by nuclear magnetic resonance spectroscopy. *Biochemistry* **35:**5633–5640.

Miller, M. B., and B. L. Bassler. 2001. Quorum sensing in bacteria. *Annu. Rev. Microbiol.* **55:**165–199.

Mizuno, T. 1997. Compilation of all genes encoding two-component phosphotransfer signal transducers in the genome of *Escherichia coli. DNA Res.* **4:**161–168.

Ninfa, E. G., M. R. Atkinson, E. S. Kamberov, and A. J. Ninfa. 1993. Mechanism of autophosphorylation of *Escherichia coli* nitrogen regulator II (NRII or NtrB): *trans*-phosphorylation between subunits. *J. Bacteriol.* **175:**7024–7032.

Park, S. Y., X. Chao, G. Gonzalez-Bonet, B. D. Beel, A. M. Bilwes, and B. R. Crane. 2004. Structure and function of an unusual family of protein phosphatases: the bacterial chemotaxis proteins CheC and CheX. *Mol. Cell* **16:**563–574.

Pelton, J. G., S. Kustu, and D. E. Wemmer. 1999. Solution structure of the DNA-binding domain of NtrC with three alanine substitutions. *J. Mol. Biol.* **292:**1095–1110.

Quon, K. C., G. T. Marczynski, and L. Shapiro. 1996. Cell cycle control by an essential bacterial two-component signal transduction protein. *Cell* **84:**83–93.

Robinson, V. L., D. R. Buckler, and A. M. Stock. 2000. A tale of two components: a novel kinase and a regulatory switch. *Nat. Struct. Biol.* **7:**628–633.

Robinson, V. L., T. Wu, and A. M. Stock. 2003. Structural analysis of the domain interface in DrrB, a response regulator of the OmpR/PhoB subfamily. *J. Bacteriol.* **185:**4186–4194.

Rosario, M. M., J. R. Kirby, D. A. Bochar, and G. W. Ordal. 1995. Chemotactic methylation and behavior in *Bacillus subtilis:* role of two unique proteins, CheC and CheD. *Biochemistry* **34:**3823–3831.

Rosario, M. M. L., and G. W. Ordal. 1996. CheC and CheD interact to regulate methylation of *Bacillus subtilis* methyl-accepting chemotaxis proteins. *Mol. Microbiol.* **21:**511–518.

Roychoudhury, S., N. A. Zielinski, A. J. Ninfa, N. E. Allen, L. N. Jungheim, T. I. Nicas, and A. M. Chakrabarty. 1993. Inhibitors of two-component signal transduction systems: inhibition of alginate gene activation in *Pseudomonas aeruginosa. Proc. Natl. Acad. Sci. USA* **90:**965–969.

Segall, J. E., S. M. Block, and H. C. Berg. 1986. Temporal comparisons in bacterial chemotaxis. *Proc. Natl. Acad. Sci. USA* **83:**8987–8991.

Song, H. K., J. Y. Lee, M. G. Lee, J. Moon, K. Min, J. K. Yang, and S. W. Suh. 1999. Insights into eukaryotic multistep phosphorelay signal transduction revealed by the crystal structure of Ypd1p from *Saccharomyces cerevisiae. J. Mol. Biol.* **293:**753–761.

Stephenson, K., and J. A. Hoch. 2002. Virulence- and antibiotic resistance-associated two-component signal transduction systems of Gram-positive pathogenic bacteria as targets for antimicrobial therapy. *Pharmacol. Ther.* **93:**293–305.

Stephenson, K., Y. Yamaguchi, and J. A. Hoch. 2000. The mechanism of action of inhibitors of bacterial two-component signal transduction systems. *J. Biol. Chem.* **275:**38900–38904.

Stevens, A. M., N. B. Shoemaker, L. Y. Li, and A. A. Salyers. 1993. Tetracycline regulation of genes on *Bacteroides* conjugative transposons. *J. Bacteriol.* **175:**6134–6141.

Stock, A. M., J. M. Mottonen, J. B. Stock, and C. E. Schutt. 1989a. Three-dimensional structure of CheY, the response regulator of bacterial chemotaxis. *Nature* **337:**745–749.

Stock, A. M., V. L. Robinson, and P. N. Goudreau. 2000. Two-component signal transduction. *Annu. Rev. Biochem.* **69:**183–215.

Stock, J. B., A. J. Ninfa, and A. M. Stock. 1989b. Protein phosphorylation and regulation of adaptive responses in bacteria. *Microbiol. Rev.* **53:**450–490.

Strauch, M. A., D. deMendoza, and J. A. Hoch. 1992. *cis*-Unsaturated fatty acids specifically inhibit a signal-transducing protein kinase required for initiation of sporulation in *Bacillus subtilis. Mol. Microbiol.* **6:**2909–2917.

Sui, Z., J. Guan, D. J. Hlasta, M. J. Macielag, B. D. Foleno, R. M. Goldschmidt, M. J. Loeloff, G. C. Webb, and J. F. Barret. 1998. SAR studies of diaryltriazoles against bacterial two-component regulatory systems and their antibacterial activities. *Bioorg. Med. Chem. Lett.* **8:**1929–1934.

Swanson, R. V., R. B. Bourret, and M. I. Simon. 1993. Intermolecular complementation of the kinase activity of CheA. *Mol. Microbiol.* **8:**435–441.

Szurmant, H., M. W. Bunn, V. J. Cannistraro, and G. W. Ordal. 2003. *Bacillus subtilis* hydrolyzes CheY-P at the location of its action, the flagellar switch. *J. Biol. Chem.* **278:**48611–48616.

Tanaka, T., S. K. Saha, C. Tomomori, R. Ishima, D. Liu, K. I. Tong, H. Park, R. Dutta, L. Qin, M. B. Swindells, T. Yamazaki, A. M. Ono, M. Kainosho, M. Inouye, and M. Ikura. 1998. NMR structure of the histidine kinase domain of the *E. coli* osmosensor EnvZ. *Nature* **396:**88–92.

Throup, J. P., K. K. Koretke, A. P. Bryant, K. A. Ingraham, A. F. Chalker, Y. Ge, A. Marra, N. G. Wallis, J. R. Brown, D. J. Holmes, M. Rosenberg, and M. K. Burnham. 2000. A genomic analysis of two-component signal transduction in *Streptococcus pneumoniae. Mol. Microbiol.* **35:**566–576.

Tomomori, C., T. Tanaka, R. Dutta, H. Park, S. K. Saha, Y. Zhu, R. Ishima, D. Liu, K. I. Tong, H. Kurokawa, H. Qian, M. Inouye, and M. Ikura. 1999. Solution structure of the homodimeric core domain of *Escherichia coli* histidine kinase EnvZ. *Nat. Struct. Biol.* **6:**729–734.

Trew, S. J., S. K. Wrigley, L. Pairet, J. Sohal, P. Shanu-Wilson, M. A. Hayes, S. M. Martin, R. N. Manohar, M. I. Chicarelli-Robinson, D. A. Kau, C. V. Byrne, E. M. Wellington, J. M. Moloney, J. Howard, D. Hupe, and E. R. Olson. 2000. Novel streptopyrroles from *Streptomyces rimosus* with bacterial protein histidine kinase inhibitory and antimicrobial activities. *J. Antibiot.* **53:**1–11.

Varughese, K. I., Madhusudan, X. Z. Zhou, J. M. Whiteley, and J. A. Hoch. 1998. Formation of a novel four-helix bundle and molecular recognition sites by dimerization of a response regulator phosphotransferase. *Mol. Cell* **2:**485–493.

Volz, K., and P. Matsumura. 1991. Crystal structure of *Escherichia coli* CheY refined at 1.7 Å resolution. *J. Biol. Chem.* **266:**15511–15519.

Walsh, C. T., S. L. Fisher, I. S. Park, M. Prahalad, and Z. Wu. 1996. Bacterial resistance to vancomycin: five genes and one missing hydrogen bond tell the story. *Chem. Biol.* **3:**21–28.

Weidner-Wells, M. A., K. A. Ohemeng, V. N. Nguyen, S. Fraga-Spano, M. J. Macielag, H. M. Werblood, B. D. Foleno, G. C. Webb, J. F. Barret, and D. J. Hlasta. 2001. Amidino benzimidazole inhibitors of bacterial two-component systems. *Bioorg. Med. Chem. Lett.* **11:**1545–1548.

Welch, M., N. Chinardet, L. Mourey, C. Birck, and J.-P. Samama. 1998. Structure of the CheY-binding domain of histidine kinase CheA in complex with CheY. *Nat. Struct. Biol.* **5:**25–29.

Welch, M., K. Oosawa, S.-I. Aizawa, and M. Eisenbach. 1994. Effects of phosphorylation, Mg^{2+}, and conformation of the chemotaxis protein CheY on its binding to the flagellar switch protein FliM. *Biochemistry* **33:**10470–10476.

Welch, M., K. Oosawa, S.-I. Aizawa, and M. Eisenbach. 1993. Phosphorylation-dependent binding of a signal molecule to the flagellar switch of bacteria. *Proc. Natl. Acad. Sci. USA* **90:**8787–8791.

Wolfe, A. J., and R. C. Stewart. 1993. The short form of the CheA protein restores kinase activity and chemotactic ability to kinase-deficient mutants. *Proc. Natl. Acad. Sci. USA* **90:**1518–1522.

Wu, J., J. Li, G. Li, D. G. Long, and R. M. Weis. 1996. The receptor binding site for the methyltransferase of bacterial chemotaxis is distinct from the sites of methylation. *Biochemistry* **35:**4984–4993.

Wyman, C., I. Rombel, A. K. North, C. Bustamante, and S. Kustu. 1997. Unusual oligomerization required for activity of NtrC, a bacterial enhancer-binding protein. *Science* **275:**1658–1661.

Xu, Q., and A. H. West. 1999. Conservation of structure and function among histidine-containing phosphotransfer (HPt) domains as revealed by the crystal structure of YPD1. *J. Mol. Biol.* **292:**1039–1050.

Yang, Y., and M. Inouye. 1991. Intermolecular complementation between two defective mutant signal transducing receptors. *Proc. Natl. Acad. Sci. USA* **88:**11057–11061.

Zahrt, T. C., and V. Deretic. 2000. An essential two-component signal transduction system in *Mycobacterium tuberculosis. J. Bacteriol.* **182:**3832–3838.

Zhao, R., E. J. Collins, R. B. Bourret, and R. E. Silversmith. 2002. Structure and catalytic mechanism of the *E. coli* chemotaxis phosphatase CheZ. *Nat. Struct. Biol.* **9:**570–575.

Zhou, H., D. F. Lowry, R. V. Swanson, M. I. Simon, and F. W. Dahlquist. 1995. NMR studies of the phosphotransfer domain of the histidine kinase CheA from *Escherichia coli:* assignments, secondary structure, general fold, and backbone dynamics. *Biochemistry* **34:**13858–13870.

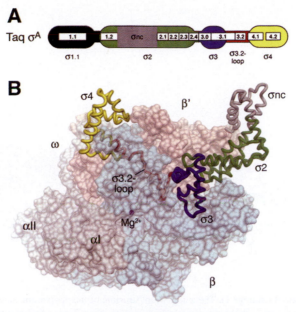

Color Plate 1 (chapter 1). All stretched out: structural domains of σ in the context of the RNAP holoenzyme. (A) Modular architecture of group 1 σ. Shown is a schematic representation of the primary sequence of *T. aquaticus* σ^A. Thick bars indicate structural domains, while thin bars represent linkers connecting the domains (Campbell et al., 2002b). Distinct structural elements discussed in the text are color coded and labeled beneath (σ$_{1.1}$, black; σ$_2$, green; σ$_3$, blue; σ$_{3.2}$ loop, red; σ$_4$, yellow). Evolutionarily conserved regions of sequence (Lonetto et al., 1992; Gruber and Bryant, 1997) are denoted by white boxes and labeled in white. The nonconserved region found only in group 1 σ factors is labeled "σ$_{nc}$" and shaded gray. (B) Structure of the RNAP holoenzyme (Murakami et al., 2002a; Vassylyev et al., 2003). The core enzyme is drawn as a transparent molecular surface, with the subunits labeled and colored as follows: β′, pink; β, cyan; αI, αII, and ω, gray. σ^A is drawn as an α-carbon backbone worm as found in the holoenzyme structure. The structural domains of σ are labeled and colored as in panel A. The σ$_{3.2}$ loop (connecting σ$_3$ and σ$_4$) threads behind a domain of β known as the β-flap.

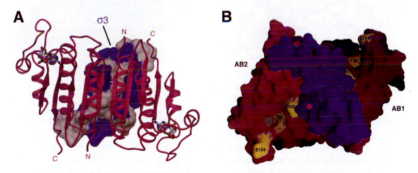

Color Plate 2 (chapter 1). Beauty in asymmetry: the sporulation transcription factor σ^F bound to the anti-σ factor SpoIIAB. (A) The domain of *B. stearothermophilus* σ^F (σ$_3$^{F}) visible in the crystal structure is shown as a molecular surface and colored blue. The surface of σ$_3$^{F} that interacts with core RNAP is colored gray. The SpoIIAB dimer, drawn in ribbons and colored magenta, extensively interacts with the core-binding surface of σ$_3$^{F}. The nucleotide, ADP in this structure, is drawn in stick form and sits in the serine kinase domain of the anti-σ factor. (B) This view is flipped 180° from that in panel A to show the asymmetric interaction of σ$_3$^{F} and the SpoIIAB dimer. SpoIIAB and σ$_3$^{F} are both drawn as molecular surfaces and colored as in panel A. Residues of SpoIIAB that are known to be critical for displacement and interaction with the anti-anti-σ factor SpoIIAA are colored yellow on SpoIIAB. The asymmetry of the interaction led to a docking model, which predicted that SpoIIAA displaces σ^F by docking onto the SpoIIAB monomer labeled AB2, where Arg20, a residue critical for displacement, was exposed (Campbell et al., 2002a; Ho et al., 2003). The docking of SpoIIAA would occur via interactions with Glu104, the ATP-binding pocket (shown in grey and represented by Thr 49), and Arg20.

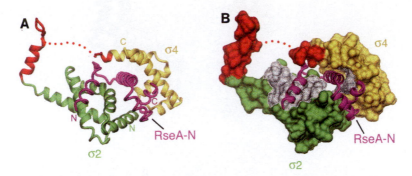

Color Plate 3 (chapter 1). The anti-σ wrap: structure of the periplasmic stress response ECF σ factor, σ^E, with the cytoplasmic domain of its anti-σ RseA. (A) Ribbon diagram of *E. coli* σ^E with the anti-σ domain of RseA. Domains of σ^E are colored as follows: σ^E_2, green; σ^E_4, yellow; σ^E_2-σ^E_4 linker, red; RseA (N-terminal 66 residues), magenta (shown as a coil). The anti-σ domain is sandwiched between σ^E_2 and σ^E_4. (B) Surface representation of σ^E, color coded as in panel A but with the core-binding surfaces shaded grey on σ^E_2 and σ^E_4. RseA, which extensively occludes the core-interacting surfaces of σ^E, is shown as a magenta ribbon.

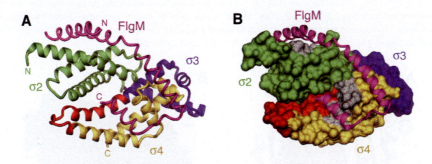

Color Plate 4 (chapter 1). All tied up: structure of the flagellar σ^{FliA} with its anti-σ FlgM. (A) σ^{FliA} is depicted as a ribbon structure with the domains colored as follows: σ^{FliA}_2, green; σ^{FliA}_3, blue; σ^{FliA}_3-σ^{FliA}_4 linker, red; σ^{FliA}_4, yellow. FlgM, shown as a magenta coil, threads around the compacted σ factor, interacting extensively with σ^{FliA}_2 and σ^{FliA}_4. (B) Same view as shown in panel A, with σ^{FliA} shown as molecular surface and FlgM shown as a ribbon. Colors are the same as in panel A, except that the surfaces of σ^{FliA} that interact with core RNAP are shaded grey. FlgM effectively occludes the core-binding interfaces on both σ^{FliA}_2 and σ^{FliA}_4.

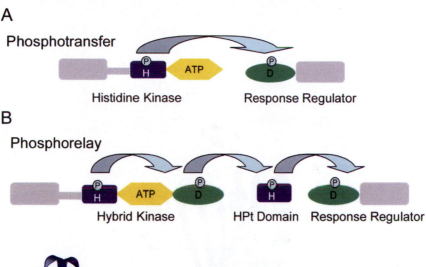

A

Phosphotransfer

Histidine Kinase   Response Regulator

B

Phosphorelay

Hybrid Kinase   HPt Domain   Response Regulator

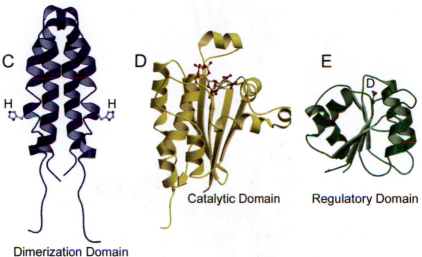

C

Dimerization Domain

D

Catalytic Domain

E

Regulatory Domain

Color Plate 5 (chapter 2). Conserved domains of two-component pathways. (A) The simplest of the two-component pathways, a phosphotransfer pathway, consists of a single-step transfer of a phosphoryl group between a His of an HK and an Asp of a downstream RR. A typical HK contains a variable periplasmic sensing domain (gray) and a conserved kinase core consisting of a dimerization domain (blue) and a catalytic ATP-binding domain (yellow). The RR contains a conserved N-terminal regulatory domain (green) and a variable C-terminal effector domain (gray). (B) Two-component pathways can also be more complex, involving components that contain multiple phosphodonor and acceptor sites. These phosphorelay systems typically involve hybrid kinases that contain a histidine kinase core together with an RR regulatory domain, a separate His-containing phosphotransfer domain, and a separate RR. However, the conserved domains are modular and are found in a variety of different arrangements in various phosphorelay pathways. (C to E) Ribbon images are shown for representative conserved domains: the His-containing dimerization domain (blue) with the His residue that is phosphorylated depicted in ball-and-stick representation (PDB code 1JOY) (C), the catalytic ATP-binding domain (yellow) with the bound nucleotide shown in ball-and-stick representation (PDB code 1I5D) (D), and the Asp-containing regulatory domain (green) with the Asp residue that is phosphorylated depicted in ball-and-stick representation (PDB code 2CHE) (E).

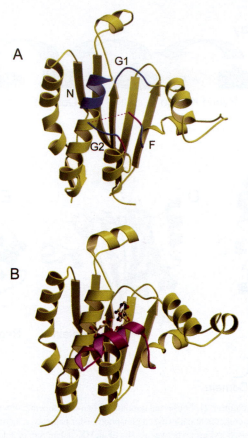

Color Plate 6 (chapter 2). ATP-binding domain of chemotaxis HK CheA. (A) The crystal structure of the nucleotide-binding domain of CheA (residues 354-542; PDB code 1B3Q) is shown as a ribbon representation, with labels designating the highly conserved motifs (cyan) that surround the nucleotide-binding site: the N box, the G1 box, the F box, and the G2 box. The F-box and G-box motifs lie within a disordered region known as the ATP lid (magenta). (B) In the crystal structure of CheA with ADPCP-Mg^{2+} bound (PDB code 1I58), the ATP lid (magenta) adopts a more ordered, closed conformation relative to the poorly defined corresponding region present in the structure of CheA without bound nucleotide. The ATP analog is shown in ball-and-stick representation.

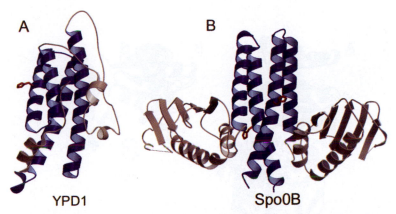

Color Plate 7 (chapter 2). His-containing phosphotransfer (HPt) domains. HPt domains are represented by *Saccharomyces cerevisiae* YPD1 (PDB code 1C02) (A) and *B. subtilis* Spo0B (PDB code 1IXM) (B). The HPt proteins contain a four-helix bundle (blue) that can be composed of either one chain with one active His residue (red) per structural unit, as in YPD1, or two chains with two active His residues per structural unit, as in the Spo0B dimer. Additional structural elements (brown) are variable among HPt proteins.

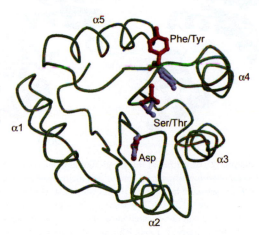

Color Plate 8 (chapter 2). Conserved activation mechanism of the RR regulatory domain. Side chains of the conserved Asp, Ser/Thr, and Phe/Tyr are displayed in conformations observed in unphosphorylated (red) and phosphorylated (blue) RRs on a representative backbone (green, with α-helices labeled). In RRs phosphorylated at the conserved Asp, the side chain of the Ser/Thr is reoriented to form a hydrogen bond with a phosphate oxygen. The side chain of Phe/Tyr is positioned inward, toward the site of phosphorylation, filling the cavity vacated by the rotated Ser/Thr. The repositioning of these residues is involved in the propagation of a subtle conformational change to the α4-β5-α5 face of the regulatory domain.

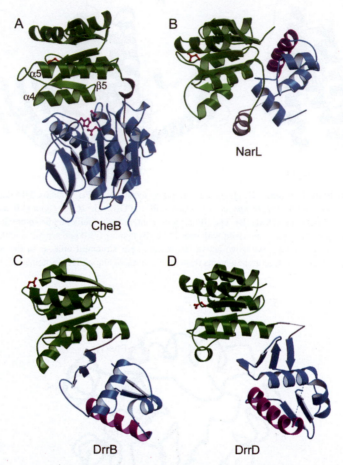

Color Plate 9 (chapter 2). Structures of full-length multidomain RRs. Ribbon representations of *Salmonella enterica* serovar Typhimurium methylesterase CheB (PDB code 1A2O) (A), *E. coli* transcription factor NarL (PDB code 1A04) (B), *T. maritima* transcription factor DrrB (PDB code 1P2F) (C) *T. maritima* transcription factor DrrD (PDB code 1KGS) (D) are aligned with similar orientations of the regulatory domain. The regulatory domain (green) is highly conserved among RRs, and all contain a conserved Asp (red) as the site of phosphorylation. The effector domain (cyan) of the methylesterase CheB contains a catalytic triad (magenta) at its active site. The transcription factor effector domains (cyan) of NarL, DrrB, and DrrD contain a recognition helix (magenta) that binds DNA. The regulatory and effector domains are joined by linkers (gray), which are often disordered in the crystal structures (dashed lines). The interdomain interface varies in size and involves different subsets of the α4-β5-α5 face of the regulatory domain, implying the presence of different mechanisms of intramolecular communication in different RRs.

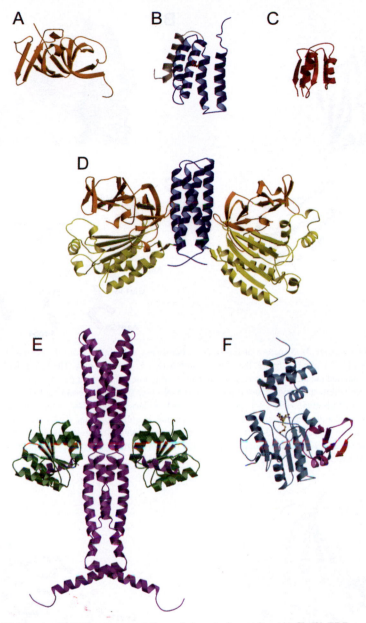

Color Plate 10 (chapter 2). Ribbon representations of chemotaxis proteins. (A) CheW (PDB code 1K0S) couples CheA kinase to the chemotaxis receptors. (B) The P1 domain of CheA (coordinates provided by F. W. Dahlquist) is an HPt-like domain containing a four-helix bundle (blue) and an additional helix (brown) that contains the site of His phosphorylation (red). (C) The P2 domain of CheA (PDB code 1FWP) serves as an interaction site for the RRs CheY and CheB (structures shown in Color Plates 5E and 9A, respectively). The helices serve as an interaction site for the α4-β5-α5 face of CheY. (D) The dimerization domain (blue), catalytic kinase domain (yellow), and CheW interaction/coupling domain (orange) comprise the C-terminal half of CheA (PDB code 1B3Q). To date, this is the only structure of an intact kinase core. (E) The phosphatase CheZ (purple) accelerates the dephosphorylation of the RR CheY (green) in some species of bacteria. In the CheY-CheZ complex (PDB code 1KMI), two monomers of CheZ dimerize to form a four-helix bundle. An additional C-terminal helix of CheZ that caps a long, disordered tail tethers CheY through interaction with the α4-β5-α5 face and facilitates an otherwise weak interaction between the active sites of CheY and CheZ. (F) The methyltransferase CheR (PDB code 1BC5) bound to the product of the methylation reaction, *S*-adenosylhomocysteine (shown in ball-and-stick representation) is shown bound to a peptide (red) that corresponds to a recognition sequence (NWETF) at the C termini of chemotaxis receptors. The pentapeptide-binding β-subdomain (magenta) is a specialized insertion into the methyltransferase fold, which serves to tether CheR to the chemotaxis receptors.

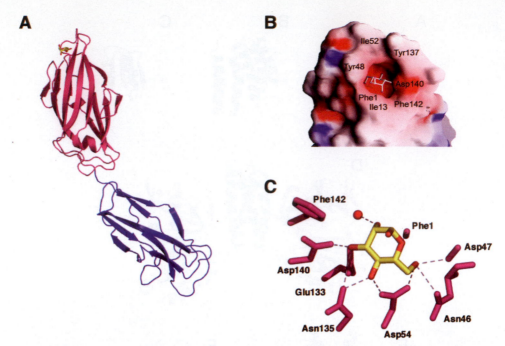

Color Plate 11 (chapter 3). Structure of FimH. (A) Coordinates were taken from the crystal structure of the FimCH complex (1KLF). In this figure, the ribbon representation is of FimH only. The lectin-binding domain is shown in magenta, and the pilin domain is shown in blue. D-Mannose is located at the top of the molecule, where carbon atoms are yellow and oxygen atoms are red. (B) Molecular surface representation with electrostatic potential surface (Nicholls et al., 1991), with positively charged residues shown in blue, negatively charged residues shown in red, neutral and hydrophobic residues shown in white. Residues defining the hydrophobic ridge around the mannose-binding pocket are labeled. (C) Mannose-binding site, with FimH residues shown in magenta. Mannose residues are shown with carbon atoms in yellow and oxygen atoms in red. Panel B adapted from Hung et al. (2002) with permission.

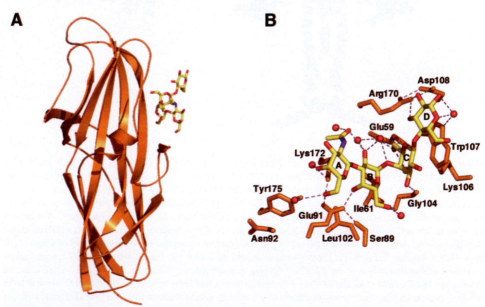

Color Plate 12 (chapter 3). Structure of PapGII. (A) The ribbon representation of the PapGII lectin-binding domain is shown in orange. The globoside receptor is shown with carbon atoms in yellow, oxygen atoms in red, and nitrogen atoms in blue. (B) The PapGII-binding site is shown with PapGII residues in orange. The globoside polysaccharide is shown with carbon atoms in yellow, oxygen atoms in red, and nitrogen atoms in blue.

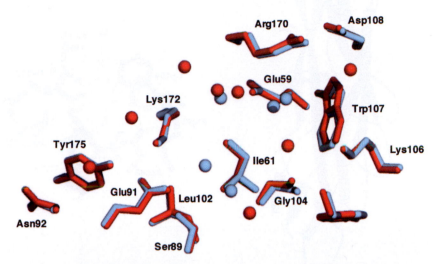

Color Plate 13 (chapter 3). Superposition of receptor-bound and apo PapGII-binding sites. The receptor-bound structure is shown in red, while the apo structure is show in cyan. Solvent molecules are shown as red or cyan spheres.

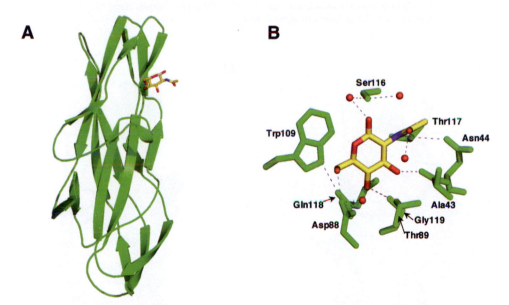

Color Plate 14 (chapter 3). Structure of GafD. (A) The ribbon representation of the GafD lectin-binding domain is shown in green. The GlcNAc receptor is shown with carbon atoms in yellow, oxygen atoms in red, and nitrogen atoms in blue. (B) The GafD-binding site is shown with GafD residues shown in green. The GlcNAc is shown with carbon atoms in yellow, oxygen atoms in red, and nitrogen atoms in blue.

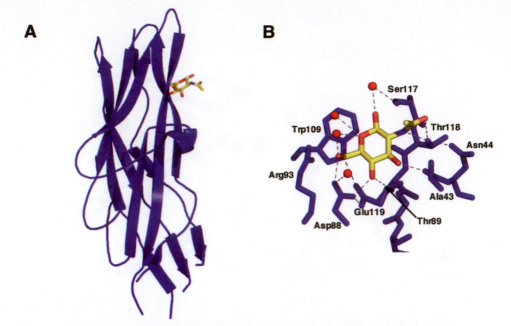

Color Plate 15 (chapter 3). Structure of F17a-G. (A) The ribbon representation of the lectin-binding domain of F17a-G is depicted in blue. GlcNAc is shown with carbon atoms in yellow, oxygen atoms in red, and nitrogen atoms in blue. (B) The F17a-G-binding site with F17a-G residues shown in green. The GlcNAc is shown with carbon atoms in yellow, oxygen atoms in red, and nitrogen atoms in blue.

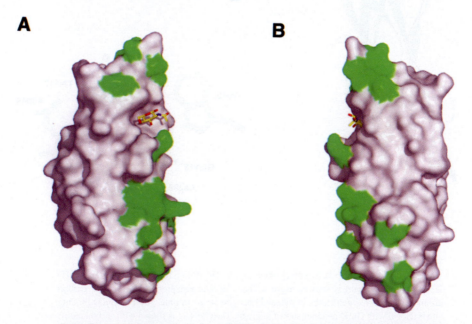

Color Plate 16 (chapter 3). Variants of F17a-G. (A) The front side of the molecule containing the natural variants of the F17a-G lectin-binding domain is shown in green. (B) The back side of the molecule is depicted. GlcNAc is shown in the binding site, with carbon atoms in yellow, oxygen atoms in red, and nitrogen atoms in blue.

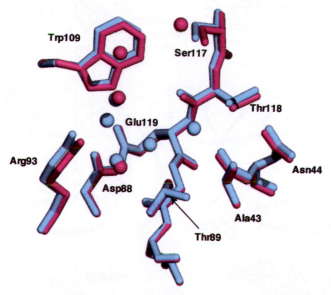

Color Plate 17 (chapter 3). Superposition of receptor-bound and apo-F17a-G binding sites. The receptor-bound structure is shown in magenta, while the apo structure is show in cyan. Solvent molecules are shown as magenta or cyan spheres.

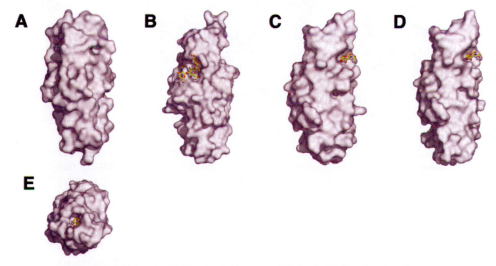

Color Plate 18 (chapter 3). Structural alignment of the lectin-binding domains. Structural alignment was performed by using a three-dimensional protein structure comparison and alignment program (Shindyalov and Bourne, 1998). The lectin-binding domain is shown as a molecular surface, while the receptor is shown as sticks with carbon atoms in yellow, oxygen atoms in red, and nitrogen atoms in blue. (A) FimH; (B) PapGII; (C) GafD; (D) F17a-G; (E) Fim rotated 90° to show the mannose-binding pocket.

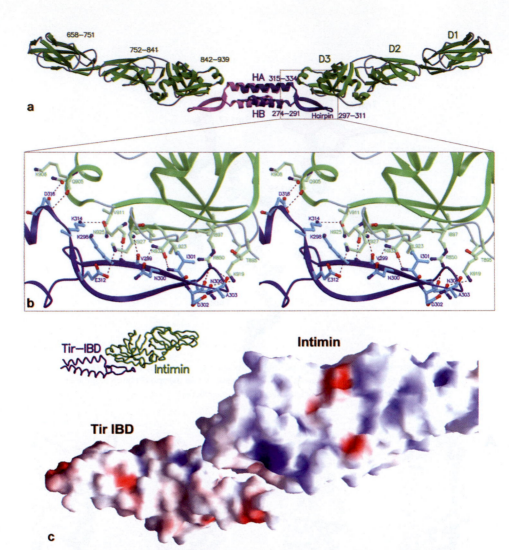

Color Plate 19 (chapter 4). Structure of the EPEC Tir IBD in complex with intimin. (a) The crystal structure of the complex reveals an antiparallel Tir IBD-intimin dimer. The binding of Tir IBD is mediated by the C-terminal-most domain of intimin, D3. Tir IBD binding does not dramatically influence the structural integrity of intimin (the Cα positions of the isolated intimin structure can be superimposed onto that of the complex structure with a root mean squared deviation of 0.67 Å). (b) Close-up view of the Tir-binding pocket on intimin. Intimin binds specifically to the β-hairpin and the N terminus of helix HA of the Tir IBD dimer. The binding interface is primarily hydrophobic, but electrostatic interactions, such as the salt bridge between Lys927 of intimin and Glu312 of the Tir IBD, and backbone interactions also contribute to the specificity and affinity of the interaction. (c) Complementary electrostatics between Tir IBD and intimin. An electrostatic surface potential representation of the Tir IBD-intimin complex as computed by GRASP (Nicholls et al., 1991) is shown here, with red representing negatively charged regions and blue representing positively charged regions. The Tir IBD has an overall negative charge, but the dimerization interface is predominantly hydrophobic. Intimin has a complementary net positive charge with a positively charged tip pointing towards the α-helices of the Tir IBD dimer.

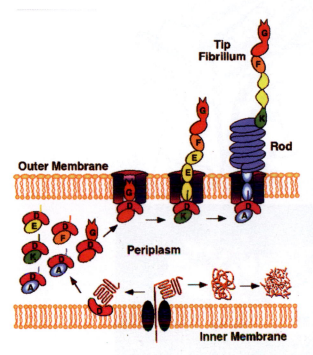

Color Plate 20 (chapter 5). Fiber assembly by the chaperone-usher pathway. A schematic of P pilus assembly is shown. Single letters indicate the corresponding Pap protein. Fiber subunits enter the bacterial periplasm via the Sec system (YEG). In absence of the chaperone, subunits misfold, aggregate, and are proteolytically degraded. The chaperone (PapD) forms a soluble complex with each subunit (including the PapG adhesin, PapF, PapE, PapK, and PapA). The chaperone facilitates the folding of the subunit, stabilizes it, and caps its interactive surfaces while priming it for polymerization. Chaperone-subunit complexes are targeted to the outer membrane usher (PapC), where chaperone dissociation permits subunit incorporation into the growing fiber. The pilus adopts its quaternary structure outside the cell.

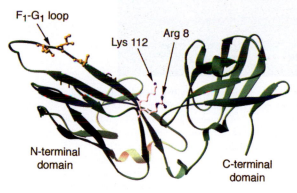

Color Plate 21 (chapter 5). The chaperone. A ribbon diagram (Carson, 1997) of the crystal structure of PapD is shown. The N- and C-terminal domains, the F_1-G_1 loop, and the F_1 and G_1 strands of the chaperone (green) are labeled. The conserved basic residues in the cleft (Arg8 and Lys112) are shown in white (carbon) and blue (nitrogen) ball-and-stick representation. The alternating hydrophobic residues of the F_1-G_1 loop are in yellow.

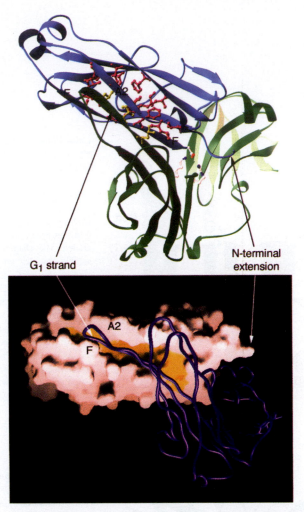

G$_1$ strand

N-terminal extension

A2

F

Color Plate 22 (chapter 5). Donor strand complementation. Two representations of the PapD-PapK chaperone-subunit complex are shown. In the upper panel, PapD is in green and PapK is in cyan. The chaperone G$_1$ strand lies between the subunit A2 and F strands and completes its Ig fold. The alternating hydrophobic residues (yellow) of the G$_1$ strand contribute to the hydrophobic core (magenta) of the subunit. The clamping interaction in the chaperone cleft (red, white, and blue ball-and-stick structure) can be appreciated. In the crystal structure, residues 1 to 8 of PapK (the majority of the N-terminal extension) are disordered and not visible. The N-terminal extension label indicates residue 9 of PapK. In the lower panel, PapD is depicted as a worm and PapK is shown in surface representation (Nicholls et al., 1991). The view is looking into the subunit groove. The subunit hydrophobic core is in yellow. The exposed position of the subunit N-terminal extension (residue 9 is indicated) in the complex and the wedging action of the chaperone, which holds the A and F strands away from each other at one end of the groove (the right side of the complex as shown) can be appreciated.

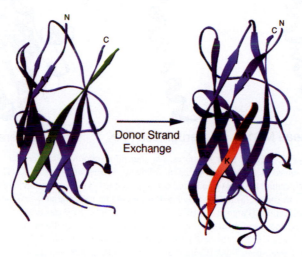

Color Plate 23 (chapter 5). Donor strand exchange. The PapD-PapE chaperone-subunit complex (left; only the G_1 strand [green] of the chaperone is shown) and a complex of the PapE subunit (blue) bound to a peptide corresponding to the N-terminal extension (residues 1 to 11) of PapK (red) are shown. Strands are labeled. Note the reversal in orientation and shift in register of the complementing strands on donor strand exchange. Note also the positions of the N and C termini (corresponding to the beginning of the A strand and the end of the F strand, respectively, on either side of the groove) of the subunit in each complex. The displacement of the chaperone allows the A and F strands to move together at this end of the groove, closing it and sealing the N-terminal extension in place. In the chaperone-subunit complex, the loops at the end of the subunit away from its N and C termini are (bottom of view as shown) disordered. The adoption of a more compact, ordered state by the subunit after donor strand exchange can be appreciated.

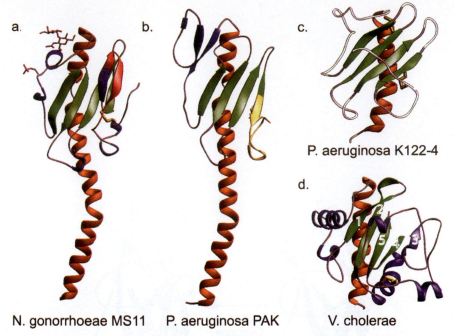

Color Plate 24 (chapter 6). Type IV pilin structures. The four pilin subunit structures that have been solved to high resolution are presented in the same orientation, following superposition of α1-C. The structurally conserved α1-helix (orange), central β-sheet (green), and disulfide-bonded cysteines (yellow) are flanked by species-specific structural elements (blue). (a) *N. gonorrhoeae* PilE, including hypervariable residues (red) and saccharide and phosphate modifications (Parge et al., 1995; Forest et al.,1999). (b) *P. aeruginosa* PAK PilA, including the structurally conserved and main-chain-dominated receptor-binding loop (yellow) (Hazes et al., 2000; Craig et al., 2003). (c) *P. aeruginosa* K122-4 pilin with poorly defined regions represented by a white coil (Keizer et al., 2001). (d) *V. cholerae* classical strain TcpA (Craig et al., 2003); the β-strands are numbered to highlight the type IVb pilin connectivity.

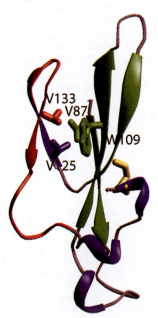

Color Plate 25 (chapter 6). Stabilization of the *N. gonorrhoeae* D-region by hydrophobic packing of conserved β2 Val87, β3 Trp109, and β5 Val125 with β6 Val133 in the hypervariable sequence. Side chain colors, including the yellow disulfide, are the same as in Fig. 2a. For clarity, α1, the αβ loop, β1, and the C-terminal tail (residues 152 to 158) are not included in the view, which is rotated 90° counterclockwise about the vertical axis from its orientation in Fig. 2a.

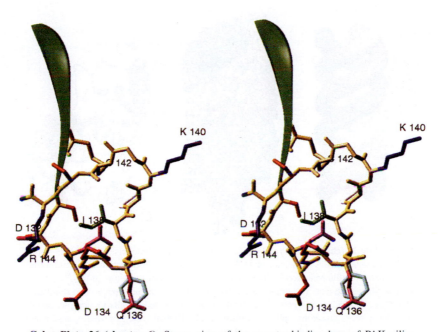

Color Plate 26 (chapter 6). Stereo view of the receptor-binding loop of PAK pilin. Residues 131 to 144 (ball-and-stick representation with main chain atoms in yellow and side chain atoms colored by residue) form a binding surface composed largely of main chain atoms. The disulfide bond (yellow) connects this D-region loop to β4 (green ribbon). Every other side chain is labeled.

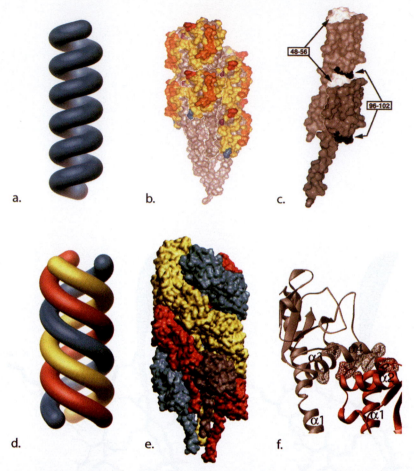

Color Plate 27 (chapter 6). Type IV pilus filament assembly models. (a) One-start helix with a diameter of 60 Å and a pitch of 41 Å, based on PAK fiber diffraction and used for the *N. gonorrhoeae* filament model. (b) Surface representation of two turns of the *N. gonorrhoeae* filament model. The surface is decorated by hypervariable (orange) and semivariable (yellow) residues and posttranslational modifications (red, saccharide; purple, phosphoserine; blue, glycerophosphate [Stimson et al., 1996]), which prevent exposure of invariant (grey) amino acids except in the bottom five subunits. A single subunit is outlined in white. Reprinted from Forest et al. (1999) with permission. (c) Longitudinal filament-packing interactions of residues 96 to 102 on residues 48 to 56 are supported by end-binding antipeptide antibodies. These two monomers are in the same orientation as the rightmost two in panel b, with the four intervening subunits in the helix removed for clarity. Reprinted from Forest and Tainer (1997b) with permission. (d) Schematic representation of TCP, a 3-start helix with a diameter of 80 Å and a pitch of 45 Å. (Reprinted from Craig et al. (2003) with permission. (e) All-atom surface for 18 subunits of the TCP filament model, representing one complete turn of each strand. Colors match the schematic in panel a, except for a single grey surface to show the boundaries of one monomer. (f) Packing interactions of α3 T122 and α4 Leu176, Ile179, Val182, and Leu185 side chains from the magenta subunit in panel b, with α2 side chains Pro58, Gly72, and Leu76 from the red subunit below it.

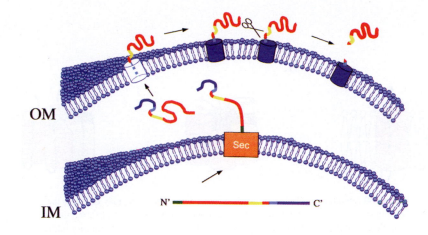

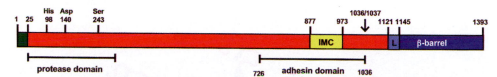

Color Plate 28 (chapter 8). Structure and secretion of Hap. The N-terminal signal sequence is shown in green, the Hap$_S$ passenger domain is shown in red, the Hap$_\beta$ translocator domain is shown in dark blue, the putative intramolecular chaperone region at the C terminus of Hap$_S$ is shown in yellow, and the α-helix (linker) N-terminal to the β-barrel
domain is shown in pale blue. The signal sequence targets the preprotein to the inner membrane (IM) and is then cleaved. The C terminus then targets the protein to the outer membrane (OM) and forms a β-barrel with a channel, allowing translocation of the passenger
domain to the bacterial surface. Ultimately, the protein undergoes cleavage via an intermolecular event, mediated by the catalytic triad consisting of His98, Asp140, and Ser243,
resulting in extracellular release of Hap$_S$.

Color Plate 29 (chapter 8). Structure of the translocator domain of the *N.
meningitidis* NalP autotransporter. The β-strands of the NalP β-barrel are
shown in shades of blue, and the α-helix that crosses the pore is shown in red
(PDB code 1uyn).

Color Plate 30 (chapter 8). Models of architecture of Hap$_\beta$. (A) Hap$_\beta$ may reside in the outer membrane as a monomer with a single central pore. (B) Hap$_\beta$ may form a multimer with a single common pore. (C) Hap$_\beta$ may form a multimer, with each subunit forming a separate pore.

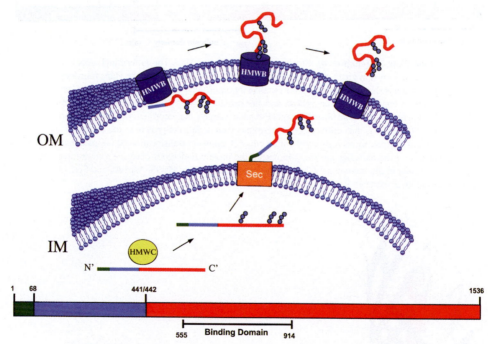

Color Plate 31 (chapter 8). Structure and secretion of HMW1. The atypical N-terminal signal sequence of HMW1 is shown in green, the secretion domain is shown in pale blue, and the mature protein is shown in red. The HMW1 preproprotein interacts with HMW1C (shown in yellow) in the cytoplasm, resulting in glycosylation (shown as blue circles). Subsequently, the signal sequence targets the preproprotein to the inner membrane (IM) and is then cleaved. The secretion domain then targets the proprotein to the outer membrane (OM), interacting with the HMW1B outer membrane translocator and undergoing cleavage. Ultimately, the mature protein is translocated across the outer membrane and localized on the bacterial surface. Most surface protein is tethered to the bacterial surface, but a small amount is released extracellularly.

Color Plate 32 (chapter 8). Structure of the secretion domain of *B. pertussis* FHA. The secretion domain of FHA is a right-handed parallel β-helix (PDB code 1rwr). The regions shown in green and gold are relatively poorly conserved with respect to other TpsA proteins. The region in green serves to cap the N terminus of FHA, and the region in gold extends laterally from the core β-helix. The NPNL and NPNGI motifs are shown in red and form type I β-turns.

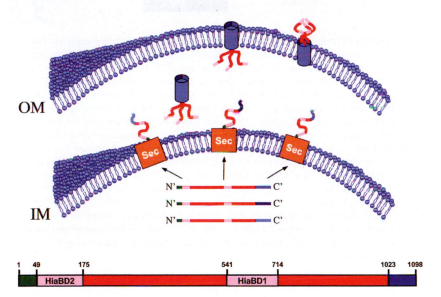

Color Plate 33 (chapter 8). Structure and secretion of Hia. The atypical N-terminal signal sequence is shown in green; the passenger domain is shown in red, with binding domains in pink; and the translocator domain is shown in dark blue. The signal sequence targets the preprotein to the inner membrane (IM) and is then cleaved. Subsequently, the C terminus forms a trimeric structure in the outer membrane (OM) and facilitates translocation of the passenger domain to the bacterial surface. The functional passenger domain is trimeric.

A.
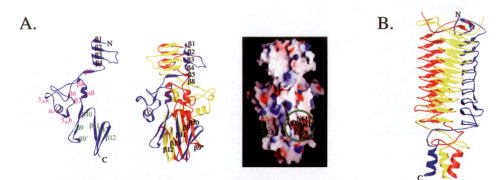
B.

Color Plate 34 (chapter 8). Molecular features of Hia primary binding domain and YadA collagen binding domain. (A) Structure of HiaBD1 (primary binding domain) and location of the binding pockets on the HiaBD1 trimer surface. (Left) Ribbon diagram of the HiaBD1 monomer structure. Secondary structural elements are labeled from β1 to β13 for β-strands, αA and αB for α-helices, and 3_{10}a and 3_{10}b for 3_{10}-helices, with labels in black, magenta, and green for domains 1, 2, and 3, respectively. (Middle) Ribbon diagram of the HiaBD1 trimer structure. The three subunits are shown in blue, yellow, and red. Only selected secondary-structure elements are labeled. (Right) Electrostatic-potential surface of HiaBD1 trimer. The view represents the same orientation as that in the ribbon diagram of the trimer (middle). The green circles delineate the receptor-binding regions of Hia, containing residues N617, D618, A620, V656, E668, and E678. Note that the entire binding pocket is formed by a single subunit, and thus three identical binding sites are related by a 120° rotation, as illustrated by the solid circle for the red subunit and the dashed circle for the blue subunit. (B) Ribbon diagram of the trimeric YadA collagen-binding domain (PDB code 1p9h). The three subunits are shown in blue, yellow, and red. The trimeric structure is composed of head and neck regions, and the collagen-binding head region adopts a novel nine-coil left-handed parallel β-roll.

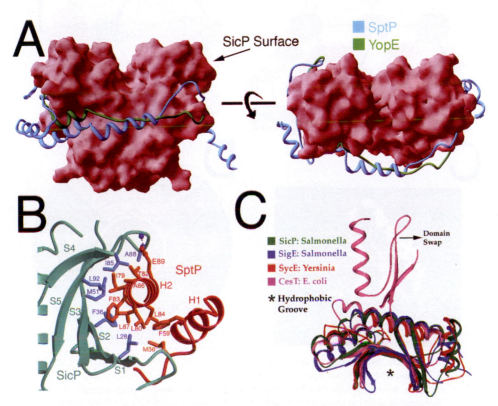

Color Plate 35 (chapter 9). Type III secretion chaperones. (A) Comparison of the SptP-SicP structure of *Salmonella* with the YopE-SycE structure of *Yersinia*. To create the figure, the chaperones SicP and SycE were aligned from the cocrystal structures with their cognate virulence factors. The molecular surface of SicP is shown, and the SptP and YopE polypeptides are drawn as ribbons in cyan and green, respectively. Two different orientations are shown for clarity. (B) An α-helix of SptP (red with orange side chains) inserts into the hydrophobic groove formed from the highly twisted β-sheet of an outer SicP molecule (aqua with cyan side chains). The hydrophobic groove surrounds the helix, contacting almost all SptP side chains along the length of the helix. The SptP contacts are primarily hydrophobic and are centered on a phenylalanine stacking between residues Phe83 of SptP and Phe36 of SicP. (C) Alignment of four secretion chaperones whose structures have been determined. The hydrophobic groove is marked by an asterisk.

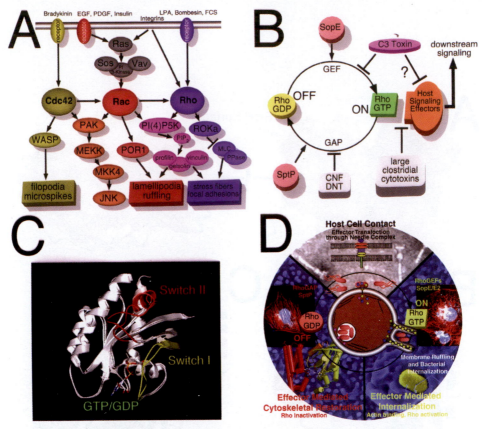

Color Plate 36 (chapter 9). Rho GTPases and bacterial cytoskeletal modulation. (A) Schematic of the complex signaling cascades affected by the Rho GTPases Cdc42, Rac, and Rho. Arrows indicate directions of information flow, and colors roughly segregate the influence of different signaling pathways. This depiction is extremely simplified and is missing many elements, especially of cross talk, in this network. EGF, epidermal growth factor; PDGF, platelet-derived growth factor; LPA, lysophosphatidic acid. (B) The regulation of the Rho GTPases at the biochemical level is depicted schematically. The influence of GEFs and GAPs on the nucleotide state is shown, and the function of bacterial toxins (e.g., C3 and CNF) and effectors (e.g., SopE and SptP) is illustrated. (C) The molecular switch of the small GTPases is illustrated by the GTP- and GDP-bound forms of the Ras oncogene. Switch I (yellow) and switch II (red) are shown in their different conformations in each of the structures with different nucleotide. (D) Overview of the full-circle "yin and yang" of host cell manipulation by *Salmonella*. The type III secretion system is shown to inject virulence proteins (small colored circles) into the host cell (depicted by the thick grey membrane). Actin filaments (black lines) are recruited to the site of bacterial attachment (right side), and actin is polymerized directly by actin-binding proteins (yellow) or indirectly by RhoGTPase activation (by SopE, for example), leading to membrane ruffles (actin-stained cell inset [red] and electron micrograph [blue] with attached bacterium surrounded by ruffles [yellow]). Subsequent shutdown of RhoGTPase signaling (due to SptP) returns the cells to a normal cytoskeletal structure (cell inset with regained actin stress fiber network). At this later stage, *Salmonella* organisms reside within a complicated vacuole and replicate inside the cell.

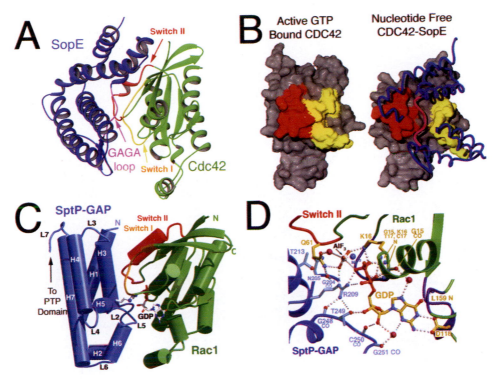

Color Plate 37 (chapter 9). Structural basis of bacterial GEF and GAP activities. (A) Structure of the complex of SopE and Cdc42. SopE is shown as a blue polypeptide, and Cdc42 is shown in green. The conserved GAGA loop of SopE (magenta) inserts between switch I (yellow) and switch II (red) of the GTPase, displacing them and impairing their ability to coordinate the nucleotide, which is not bound in the complex. (B) Surface comparison of active Cdc42 and SopE-disrupted Cdc42. The portions of the molecular surface corresponding to switches I and II are in yellow and red, respectively. SopE is shown as a blue ribbon, with the GAGA loop in magenta (residues in the center of SopE are omitted so as not to block the view of the GAGA loop). The switch regions are found in the complex to be in different conformations relative to that in the active form of the GTPase (on the left). (C) The AlF_3 transition state complex reveals that the SptP NH_2-terminal four-helix bundle GAP domain (blue) inserts a catalytic arginine (shown in white with nitrogens colored blue) into the active site of the GTPase Rac1 (shown in green with the bound GDP nucleotide). The key regulatory and enzymatic elements are also shown, with switch I (orange) and switch II (red), but the tyrosine phosphatase domain of SptP (see Color Plate 38) is omitted for clarity. (D) SptP positions Gln61 of Rac1 through molecular contacts and inserts Arg209 into the active site to stabilize the transition state. Hydrogen bonds are indicated by white dotted lines or as smaller gray dotted lines for weak bonds. AlF_3 is shown with the fluorides colored brown and the aluminum colored grey. The magnesium ion and water molecules are shown as large blue and magenta spheres, respectively. GDP carbon bonds are shown in yellow. A large "W" indicates the nucleophilic water molecule positioned by Gln61 of Rac1. The phosphate-binding loops (P-loop) and guanine-binding loops (G-loops) of Rac1 are shown in purple. The bonds proposed to form during the phosphoryl transfer are shown as solid white lines.

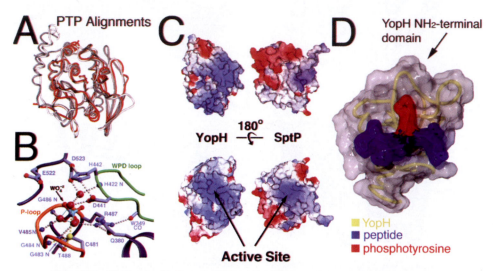

Color Plate 38 (chapter 9). Bacterial PTPs. (A) Alignment of the bacterial PTP enzymes (SptP and YopH) together with the eukaryotic PTP1B. The structures align very well, especially in the active site (the key cysteine residue is shown), with SptP differing from the other two by lacking an NH_2-terminal helix (on the far left in the image). (B) Representative depiction of the PTP active site. Shown is the SptP active site with coordinated water molecules and an atom of tungstate (phosphate mimic and PTP inhibitor) bound by the key active-site residues, Asp441, Cys481, and Arg487. (C) Comparison of the aligned phosphatase domains of SptP and YopH illustrated with a molecular surface colored by electrostatic potential such that red is negative (acidic) and blue is positive (basic). A "front" and "back" view are given, which are related by the 180° rotation about the axis indicated. (D) Molecular surface representation of the complex of the NH_2-terminal domain of YopH (grey) with a phosphotyrosine (red)-containing peptide (blue). The YopH and peptide surfaces are shown partially transparent to reveal the peptide (white) and the main chain of the folded NH_2-terminal domain (yellow).

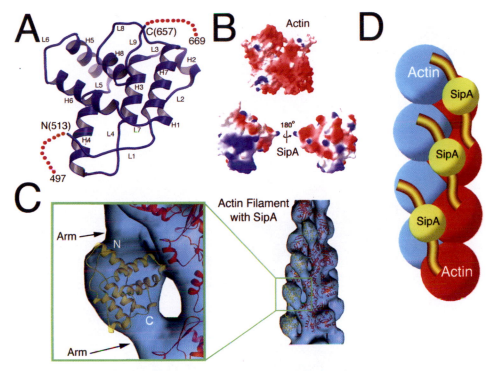

Color Plate 39 (chapter 9). Structure and function of an actin stapler. (A) The structure of SipA$^{497-669}$ is shown as a ribbon diagram and reveals that the domain possesses a globular, and not an extended, conformation. The termini visible and modeled in the structure are denoted N and C, respectively, with the corresponding final amino acid number noted in parentheses. Red dots represent peptide at the termini present in the crystals (indicated by the final residue number) but which were not modeled due to disorder. (B) The surface electrostatics of SipA reveal a polarized molecular surface (shown such that blue indicates a basic or positively charged region and red indicates an acidic or negatively charged region). One side of the molecule possesses a large basic patch, whereas the other side is primarily acidic. An actin monomer is shown above SipA, colored by electrostatic potential. (C) Image analysis of electron micrographs of filaments formed from G-actin and SipA$^{497-669}$ reveals an additional (nonactin) globular mass with extended, nonglobular density. Placed over the actin density is a model of the F-actin filament (red) generated from actin monomers (for clarity, certain portions of the actin structures were omitted). The SipA$^{497-669}$ crystal structure (yellow) is modeled into the globular core of the nonactin density. In the enlargement of the SipA density in the maps, the N and C termini of the modeled crystal structure are located near the arm density and labeled. (D) The actin filament is shown as two entwined polymers (one in red, the other in cyan) of actin monomers (shown as spheres and labeled) . SipA is drawn as a smaller yellow sphere with two extended arms projecting from opposite sides of the molecule such that they contact actin protomers in opposing strands, as observed in the EM densities.

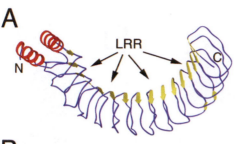

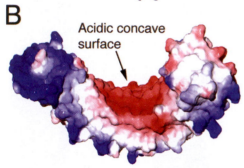

Color Plate 40 (chapter 9). The acidic cradle of the LRR-containing YopM. (A) Ribbon diagram showing the structure of the LRR of YopM. The two helices capping the repeats are shown in red, and the inner, concave face points up. (B) The surface electrostatics of YopM (oriented identically to the image in panel A) show that the relative charge distribution makes the inner surface more acidic than the outer surface, particularly the ends. The concave inner surface of LRR-containing proteins is thought to represent the binding site for other proteins.

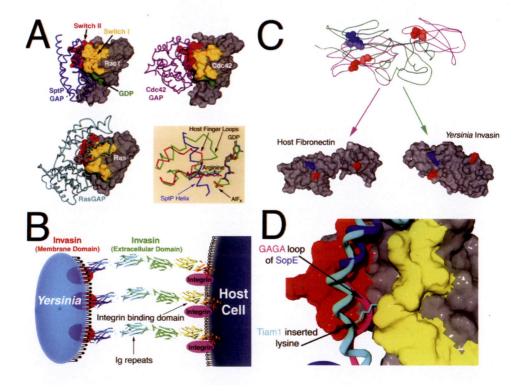

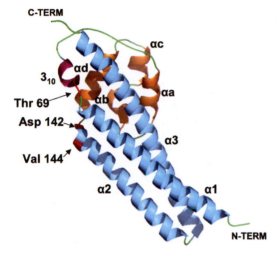

C-TERM

αc

3₁₀ αd

Thr 69

Asp 142 αb αa

Val 144 α3

α2 α1

N-TERM

Color Plate 42 (chapter 10). Structure of TraC. Ribbon diagram showing the TraC helical bundle structure. The small helical appendage at the side of the helical core is shown in orange. Marked in red are the positions of three amino acids that were shown to play an important role for function and that may be part of an extended patch on the surface of TraC which may participate in interactions with other proteins.

Color Plate 41 (chapter 9). Structural mimicry in bacterial virulence. (A) A comparison of the SptP GAP domain and host GAP enzymes reveals that they engage the GTPase in a similar manner but that SptP is differentiated by its smaller size and unique fold. The GTPases are shown as molecular surfaces colored gray, except for the switch I (orange) and switch II (red) regulatory domains, and the nucleotide GDP (green). In the lower right-hand corner of this panel, an alignment of the GTPase-bacterial GAP structures is shown, with an emphasis on the catalytic arginine and the protein scaffolding that presents this residue to the GTPase active site. The nucleotide and AlF_3 moieties are also shown. (B) Invasin, with an outer membrane-anchoring domain (schematically shown in red) and an integrin-binding extracellular domain (structure shown as a ribbon diagram colored from blue at the N terminus through the spectrum to red at the C terminus), binds the β_1-integrin receptor, links the bacterium to the host, and induces integrin mediated signaling that leads to bacterial internalization. Ig, immunoglobulin. (C) The integrin-binding region of invasin mimics the host integrin ligand fibronectin. At the top are shown the molecular surfaces for these two domains. Shown in red and blue are the locations of integrin contacting aspartic acid and arginine residues, respectively. Below is shown the results of aligning the proteins (invasin in green and fibronectin in purple) based on minimizing the distance between these key integrin-contacting residues. (D) Shown is a superposition of the SopE-CDC42 structure alignment with the Tiam1-Rac1 structure. SopE (blue) inserts the GAGA loop (magenta) in a location where the host factor Tiam1 (cyan) inserts a lysine residue (cyan stick model representation) achieving a similar displacement of the switch I and II regions of the GTPases.

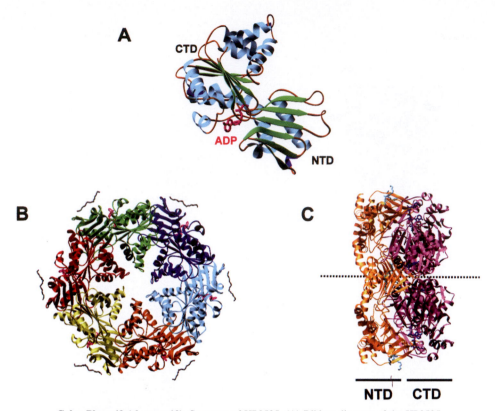

Color Plate 43 (chapter 10). Structure of HP0525. (A) Ribbon diagram of the HP0525 monomer. (B) Ribbon diagram of the HP0525 hexamer viewed down the large hole formed by the N-terminal domains. Shown in the periphery of the molecule are polyethylene glycol (PEG) molecules that were components of the crystallization condition and were found to interact specifically with the protein in the crystal. By virtue of this interaction, it has been proposed that HP0525 interacts with the inner membrane by using the molecular surface that interacts with PEG molecules in the crystal (Yeo et al., 2000). (C) Ribbon diagram of HP0525 viewed from the side, illustrating the double-ring structure formed by the stacking of the NTDs and the CTDs. (D) Alternating mode of nucleotide binding in the ATPγS-HP0525 complex. Shown are simulated-annealing electron density maps illustrating the relative binding of ATPγS in molecules A and B of the ATPγS-HP0525 complex. The β-phosphate position of ATPγS in site B is co-occupied by a sulfate ion. (E) Structural superposition of unliganded HP0525 (violet) and the ADP-HP0525 complex (gray). The notation A to F is arbitrary, with molecule A defined as the first subunit listed in the PDB coordinate files of the unbound and ATPγS-bound structures. The two hexameric assemblies were overlaid with respect to molecule A of ADP-HP0525. Molecule F exhibits the largest conformational change and is colored in green (apo-HP0525) and red (ADP-HP0525), respectively. (F) Structural variability of the NTDs in unliganded HP0525. The overlay was carried out with respect to the CTD of subunit A of ADP-HP0525 (red). Subunit F of apo-HP0525 (green) exhibits the largest rotation (~15° more open) about the hinge region between the NTD and CTD. The coloring scheme for all other subunits is as follows: apo-HP0525_A, dark gray; apo-HP0525_B, pink; apo-HP0525_C, orange; apo-HP0525_D, yellow; apo-HP0525_E, cyan. (G) Model for the mode of action of VirB11 ATPases. The N-terminal and C-terminal domains are represented in pink and light blue, respectively. NTDs locked in a rigid conformation by the binding of ATP and ADP are shown in cyan and yellow, respectively.

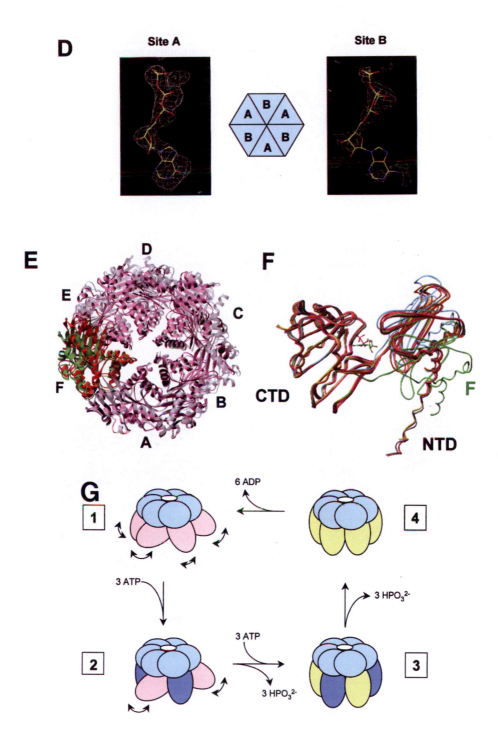

D

Site A   Site B

E

D

E   C

F   B

A

F

CTD   F

NTD

G

1   6 ADP   4

3 ATP

2   3 ATP   3 HPO₃²⁻   3

3 HPO₃²⁻

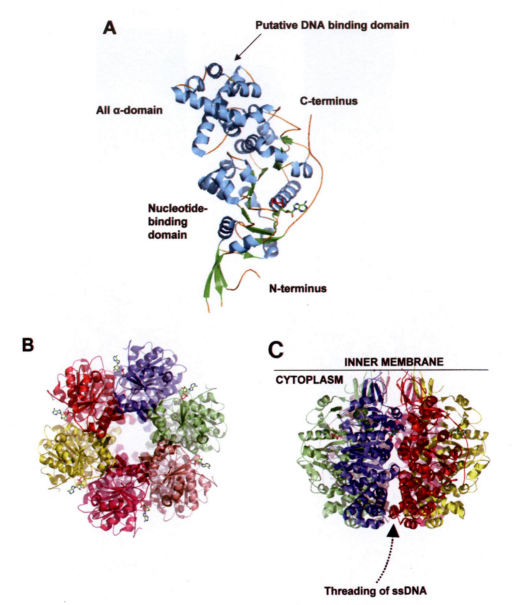

Color Plate 44 (chapter 10). Structure of TrwB. (A) Ribbon diagram of the TrwB monomer and domain organization. (B) Ribbon diagram of the TrwB hexamer viewed through the 22-Å mouth of the structure that is expected to face the bacterial membrane. (C) Side view of the sphere-shaped TrwB hexamer and relative positioning to the bacterial membrane.

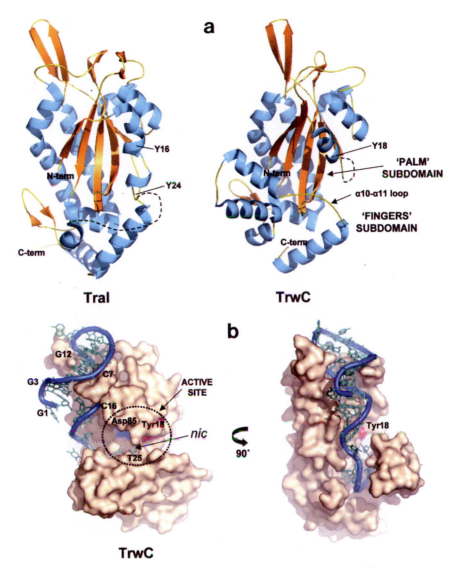

Color Plate 45 (chapter 10). Structure of the relaxases TraI (F) and TrwC (R388). (a) Ribbon diagrams of F-factor TraI and R388 TrwC relaxases (PDB entries 1P4D and 1OMH, respectively). The structures are presented in the same orientation for comparison purposes. Positions of catalytic residues Tyr18 (TrwC) and Tyr16-Tyr23 (TraI) are indicated. Disordered regions of the crystals that could not be traced are depicted as dashed lines. (b and c) Molecular surface diagrams of the crystallographically determined DNA binding cleft in R388 TrwC (b) and localization of residues important for DNA recognition in F factor TraI (c). Also shown are the catalytic residues Tyr18 (TrwC) and Tyr16 and Tyr23 (TraI) in magenta and the putative general bases Asp85 (TrwC) and Asp81 (TraI) in blue. The 25-mer DNA in complex with TrwC is marked to indicate the beginning and end of the sequence at G1 and T25, respectively. DNA segments G3 to C7 and G12 to C16 in the double-stranded hairpin are involved in Watson-Crick base-pairing. Also indicated are the region of the active site of TrwC (dotted circle) and the *nic* site immediately downstream of T25.

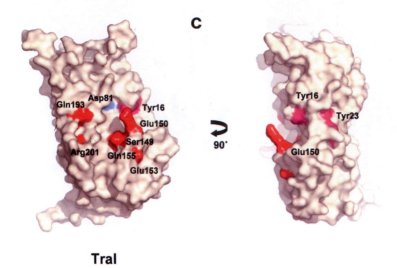

TraI

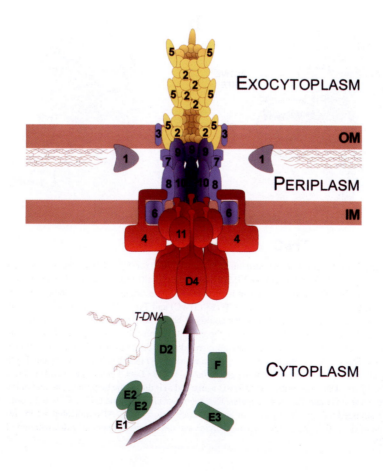

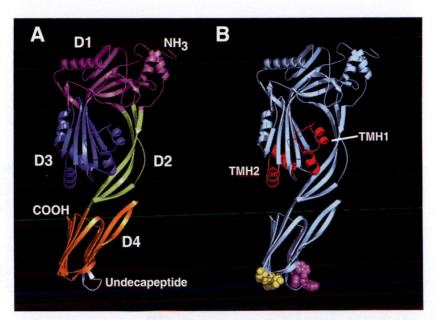

Color Plate 47 (chapter 11). Crystal structure of PFO. Shown are ribbon representations of the crystal structure of soluble PFO. (A) D1 to D4 are shown in magenta, green, blue, and orange, respectively. The location of the undecapeptide is in white. (B) Shown in red are the α-helical regions of D3 that form the two transmembrane β-hairpins (TMH1 and TMH2). Shown as space-filled atoms are the residues of D4 that anchor PFO to the membrane. The space-filled atoms colored in magenta are the tryptophans of the undecapeptide, and those in yellow are residues of the other hydrophobic loops in D4 shown to penetrate the membrane surface (Ramachandran et al., 2002). The representations in panels A and B were generated using PyMOL (DeLano, 2002).

Color Plate 46 (chapter 10). Architecture of the type IV secretion machinery. A model view of the type IV secretion apparatus composed of energizers (red), core complex components (blue), peptidoglycanases (grey), and surface-exposed components (yellow) is shown. The subunits are named according to the VirB-VirD4 secretion system of *A. tumefaciens*. Substrates secreted into host cells are exemplified by the *A. tumefaciens* T4SS substrates (green), one of which interacts with a nonsecreted chaperone (VirE1 [white]). IM and OM, inner membrane and outer membrane; the peptidoglycan is schematically represented at the periplasmic/outer membrane surface. Further explanations are given in the text.

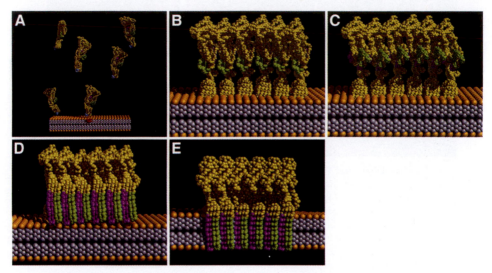

Color Plate 48 (chapter 11). Proposed model for the cytolytic mechanism of the CDCs. Shown is a proposed model of the binding and membrane insertion of a CDC based on the PFO crystal structure. In panels B to E, only D1, D3, and D4 are visible; D2 is located behind D3. The two transmembrane hairpins are shown in magenta (TMH1) and green (TMH2). (A) Membrane recognition and binding via D4; (B) assembly of the prepore oligomer (only six monomers are shown; a complete oligomer would be composed of 35 to 50 monomers in a circular complex); (C) movement of D3 away from the D2 backbone (directly behind D3) and rearrangement of the secondary structures of the D3 α-helices (in magenta and green) that ultimately form the two transmembrane β-hairpins; (D) proposed formation of a preinsertion transmembrane β-sheet; (E) insertion of the β-sheet to form the transmembrane β-barrel. The twist in the core β-sheet of D3 (Color Plate 47) probably relaxes, which would allow the two TMHs to move to opposite sides of their positions in the monomer and extend into the two transmembrane β-hairpins.

Gram-positive Injectosome

Gram-negative Type III Secretion

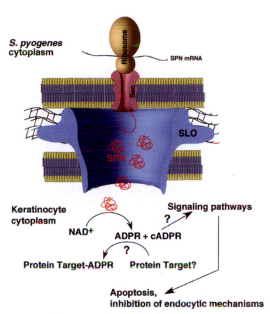

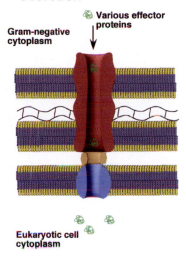

Color Plate 49 (chapter 11). Injectosome model. (Left) Proposed structure of the injectosome and the known and hypothetical intracellular effects of SPN. Although apoptosis and interruption of phagocytosis can be induced by the intact injectosome, the intracellular pathways by which this effect is communicated are not known, although both effects appear to be induced by the presence of SPN in the keratinocyte cytoplasm (Bricker et al., 2002). (Right) Schematic of the type III secretion system (for details, see chapter 9). Although the injectosome is functionally similar to the type III secretion system, the latter probably evolved from the flagellar assembly apparatus and is not related evolutionarily to the injectosome.

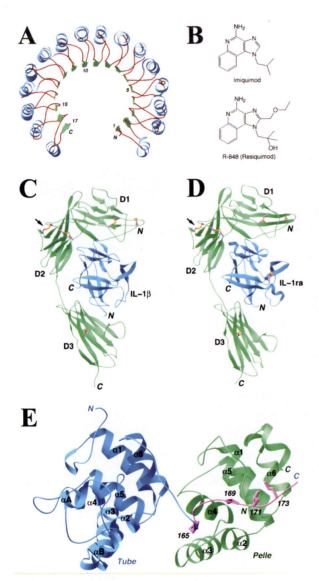

Color Plate 50 (chapter 12). Structures of TLRs and IL-1Rs. (A) Crystal structure of an LRR protein, the RNase inhibitor (Protein Data Bank entry 2BNH). The β-strands are shown as arrowed ribbons, and the α-helices are shown as coiled ribbons. The structure contains 17 β/α LRR repeats. (B) Chemical structures of the antiviral compounds imiquimod (top) and R-848 (resiquimod) (bottom), which activate TLR7 and TLR8. (C) Crystal structure of the complex between the extracellular domain of IL-1RI (green) and IL-1β (cyan) (Protein Data Bank entry 1ITB). The three Ig domains of the receptor are labeled D1, D2, and D3. The disulfide bridges are shown in purple for carbon atoms, and the bridge linking D1 and D2 is indicated by the arrow. (D) Crystal structure of the complex between the extracellular domain of IL-1RI (green) and IL-1ra (cyan) (Protein Data Bank entry 1IRA), shown in the same orientation for D1 and D2 as that used in panel C. Note the different positions of the D3 domains of the two complexes. (E) Crystal structure of the complex between the DD of *Drosophila* Pelle (green) and Tube (cyan, except for residues 165 to 173, which are shown in purple) (Protein Data Bank entry 1D2Z). Side chains from the C-terminal tail of Tube make crucial contacts with Pelle, and the structural observations are supported by functional studies.

```
TLR1     ARNIPLEELQRNLQFHAFISYSGH...........DSFWVKNELLPNLEKE...G
TLR2     AKRKPRKAPSRNICYDAFVSYSER...........DAYWVENLMVQELENFN.PP
TLR3     LGFKEIDRQTEQFEYAAYIIHAYK...........DKDWV.WEHFSSMEKED.QS
TLR4     LLAGCIKYGRGENIYDAFVIYSSQ...........DEDWVRNELVKNLEEGV.PP
Toll     LWFVTEEDLDKDKKFDAFISYSHK...........DQSFIEDYLVPQLEHGP.QK
IL-1RI   YDFLPIKAS.DGKTYDAYILYPKTVGEG....STSDCDIFVFKVLPEVLEKQC..G
IL-1RAcP AHFGTDETILDGKEYDIYVSYAR.........NAEEEEFVLLTLRGVLENEF..G
IL-18R   HLTRRDETLTDGKTYDAFVSYLKECRP.....ENGEEHTFAVEILPRVLEKHF..G
AcPL     TYQSKDQTLGDKKDFDAFVSYAKWSSFPSEATSSLSEEHLALSLFPDVLENKY..G
IL-1RAPL NHFGAEELDGDNKDYDAYLSYTKVDPDQW.NQETGEEERFALEILPDMLEKHY..G
MyD88    ITTLDDPLGHMPERFDAFICYCPS...........DIQFVQEMIRQLEQTN.YR
dMyD88   EDQRCVQMGQPLPRYNACVLYAEA...........DIDHATEIMNNLESER.YN
MAL      THASDSGSSRWSKDYDVCVCHSEE...........DLVAAQDLVSYLEGST.AS
TRIF     LFPSSLESSSEQKFYNFVILHARA...........DEHIALRVREKLEALG.VP
TIRP     VEEMFEEEAEEEVFLKFVILHAED...........DTDEALRVQNLLQDDFGIK
                            ●●●

S.S.     |βB|  BB  |----αB---|  |--βC-| |-αC-|  |αC'|
TLR1     MQICLHERNFVPGKSIVENIITCIEK..SYKSIFVLSPNFVQS..EWCHYEL.
TLR2     FKLCLHKRDIPGKWIIDNIIDSIEK..SHKTVFVLSENFVKS..EWCKYEL.
TLR3     LKFCLEERDFEAGVFELEAIVNSIKR..SRKIIFVITHHLLKD..PLCKRFKV
TLR4     FQLCLHYRDFIPGVAIAANIIHEGPH..SRKVIVVVSQHFIQS..RWCIFEY.
Toll     FQLCVHERDWLVGGHIPENIMRSVAD..SRRTIIVLSQNFIKS..EWARLEF.
IL-1RI   YKLFIYGRDDYVGEDIVEVINENVKK..SRRLIIILVRETSGFSWLGGSSEE.
IL-1RAcP YKLCIFDRDSLPGGIVTDETLSFIQK..SRRLLVVLSPNYVLQ.GTQALLEL.
IL-18R   YKLCIFERDVVPGGAVVDEIHSLIEK..SRRLIIVLSKSYMS...NEVRYEL.
AcPL     YSLCLLERDVAPGGVYAEDIVSIIKR..SRRGIFILSPNYVN...GPSIFEL.
IL-1RAPL YKLFIPDRDLIPTGTYIEDVARCVDQ..SKRLIIVMTPNYVVR.RGWSIFEL.
MyD88    LKLCVSDRDVLPGTCVWSIASELIEK.RCRRMVVVVSDDYLQS..KECDFQT.
dMyD88   LRLFLRHRDMLMGVFFEHVQLSHFMATRCNHLIVVLTEEFLRS..PENTYLV.
MAL      LRCFLQLRDATPGGAIVSELCQALSSSHCR..VLLITPGFLQD..PWCKYQM.
TRIF     DGATFCBDFQVPGRGELSCLQDAIDH..SAFIILLLTSNFDCR..LSLHQVN.
TIRP     PGIIF.ARMPCGRQHLQNLDDAVNG..SAWTILLLTENFLRD..TWCNFQF.
                 ●●       ●●

S.S.     |--αC"--|  |--βD-|  |----αD----|  |--βE-|
TLR1     YFAHHNLFHEGSNSLILILLEPIPQYSIPSSYHKLKSLMARRTYLEWPKEKSK
TLR2     DFSHFRLFDENNDAAILILLEPIEKKAIPQRFCKLRKIMNTKTYLEWPMDEAQ
TLR3     HHAVQQAIEQNLDSIILVFLEEIPDYKLNHALCLRRGMFKSHCILNWPVQKER
TLR4     EIAQTWQFLSSRAGIIFIVLQKVEKTLLRQQV.ELYRLLSRNTYLEWEDSVLG
Toll     RAAHRSALNEGRSRIIVIIYSDIGD..VEKLDEELKAYLKMNTYLKWGDP...
IL-1RI   QIAMYNALVQDGIKVVLLELEKIQDYEKMP..ESIKFIKQKHGAIRWSGDFTQ
IL-1RAcP KAGLENMASRGNINVILVQYKAVKE....TKVKELKRAKTVLTVIKWKGEKSK
IL-18R   ESGLHEALVERKIKIILIEFTPVTDFTFLPQSLKLLK...SHRVLKWKADKSL
AcPL     QAAVNLALDDQTLKLILIKFCYFQEPESLPHLVKKALR..VLPTVTWRGLKSV
IL-1RAPL ETRLRNMLVTGEIKVILIECSELRGIMNYQEVEALKHTIKLLTVIKWHGPKCN
MyD88    KFA..LSLSPGAHQKRLI..PIKYKAMKKEFPSILRFITVCDYTNPCTKS..
dMyD88   NFT..QKIQIENHTRKII...PILYKT.DMHIPQTLGIYTHIKYAGDSKLF..
MAL      LQA..LTEAPGA.EGCTI..PLLSGLSRAAYPPELRFMYYVDGRGPDG....
TRIF     QAM.MSNLTRQGSPDCVI...PFLPLESSPAQLSSDTASLLSGLVRLDEHS..
TIRP     YTSLMNSVNRQHKYNSVI...PMRPL.NNPLP.RERTPFALQTINALEEES..

S.S.     |---αE--|
TLR1     .....RGLFWANLRAAINIKLTEQAKK
TLR2     .....REGFWVNLRAAIKS
TLR3     .....IGAFRHKLQVALGSKNSVH
TLR4     .....RHIFWRRLRKALLDGKSWNPEGTVGTGCNWQEATSI
Toll     .......WFWDKLRFALPHRRPVG (97)
IL-1RI   GPQSAKTRFWKNVRYHMPVQRRSPSSKHQLLSPATKEKLQREAHVPLG
IL-1RAcP YPQ...GRFWKQLQVAMPVKKSPRRSSSDEQGLSYSSLKNV
IL-18R   SYN...SRFWKNLLYLMPAKTVKPGRDEPEVLPVLSES
AcPL     PPN...SRFWAKMRYHMPVKNSQG (33)
IL-1RAPL KLN...SKFWKRLQYEMPFKRIE (131)
MyD88    .......WFWTRLAKALSLP
dMyD88   .......NFWDKLARSLHDLDAF (156)
MAL      .....GFRQVKEAVMRCKLLQEGEGERDSATVSDLL
TRIF     ......QIFARKVANTFKPHRLQ (178)
TIRP     ......RGFPTQVERIFQESVYKTQQTIWKETRNMVQRQFIA
                 ●●
```

Color Plate 51 (chapter 12). Amino acid sequence alignment of representative TIR domains. Three different subfamilies of TIR domains are shown: the TLR subfamily (pink block), the IL-1R subfamily (green block), and the adaptor subfamily (blue block). The row labeled S.S. denotes the assignment of secondary-structure elements based on the crystal structure information. Residues in β-strands and α-helices are boxed in cyan and yellow, respectively. Residues in the three conserved motifs among the TIR domains are indicated by blue dots, except for the position of the Lps^d mutation, which is in red. Sequences from plant disease resistance (R) proteins are not shown.

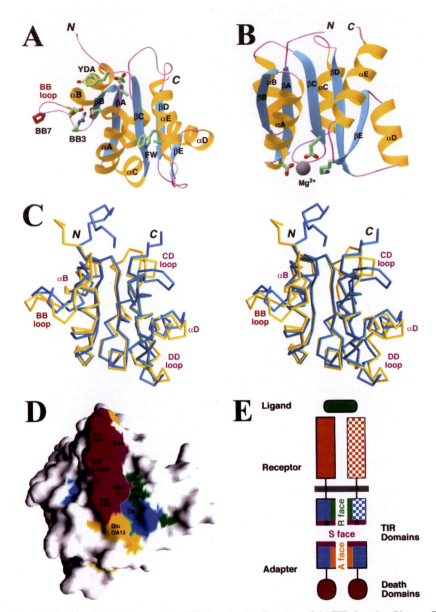

Color Plate 52 (chapter 12). Structures of the TIR domain. (A) Structure of the TIR domain of human TLR2. The side chains of conserved motifs 1 and 3 are shown and labeled YDA and FW. The BB loop corresponds to conserved motif 2. (B) Structure of CheY, viewed in the same orientation. (C) Structural overlap of the TIR domains of human TLR1 (cyan) and TLR2 (yellow). (D) Molecular surface of the TIR domain of human TLR1. The conserved surface patch for the BB loop is colored in magenta, except that the BB7 residue (the *Lps*[d] mutation site) is shown in red. (E) Schematic model of the interactions between the IL-1Rs and TLRs and the TIR-containing adaptor molecules. Three types of interfaces in the TIR domain signaling complex are indicated: R face, A face, and S face. The asterisk indicates the site of the *Lps*[d] mutation. The receptor is shown as a heterodimer, but it could also be a homodimer or oligomer.

Structural Biology of Bacterial Pathogenesis
Edited by G. Waksman et al.
© 2005 ASM Press, Washington, D.C.

Chapter 3

Sugar Recognition and Bacterial Attachment

Craig L. Smith, Karen Dodson, Gabriel Waksman, and Scott J. Hultgren

Bacteria utilize a number of ways to recognize and colonize host tissues. One strategy is to bind and make intimate contact with host cells. Binding allows bacteria to persist in advantageous locations where there may be high nutrient concentrations or to provide protection from hostile environments. Attachment can also trigger a cascade of signaling events in the bacteria, allowing them to subvert host functions during pathogenesis. However, this is not a one-way street. Signaling cascades activated in the host may allow the host to mount an effective immune response. Thus, a battle is waged between the invading bacteria and the host. A recurrent theme in bacterial pathogenesis is the interaction of a protein on the bacterial surface, called an adhesin, with surface elements or receptors, frequently carbohydrates, located on the surface of the target cells. Many bacteria harbor adhesins at the tip of adhesive organelles called pili or fimbriae. These are assembled through the chaperone-usher pathway, a pathway that is responsible for the assembly of over 100 pilus and nonpilus surface structures, which are involved in the pathogenesis of diseases such as cystitis, pyelonephritis, pneumonia, and bubonic plague. The specific stereochemical fit between the bacterial adhesin molecule and the receptor determines tissue and cell tropisms. Cryoelectron microscopy studies of pili, along with other biophysical methods such as X-ray crystallography and nuclear magnetic resonance spectroscopy on components of the chaperone-usher pathway, are providing invaluable information on pilus assembly and on host-pathogen interactions. In this chapter, we discuss the contribution of structural biology to our understanding of the molecular basis of bacterial attachment to glycoprotein and glycolipid host receptors.

ASSEMBLY OF PILI AND FIMBRIAE

The chaperone-usher pathway is involved in the biogenesis of more than 100 different pili structures in pathogenic gram-negative bacteria (Thanassi et al., 1998a). Information necessary for assembly, including pilus subunits (or pilins), regulatory elements, and

Craig L. Smith, Karen Dodson, and Scott J. Hultgren • Department of Molecular Microbiology, Washington University, 4940 Parkview Pl., St. Louis, MO 63110. *Gabriel Waksman* • Institute of Structural Molecular Biology, Birkbeck and University College London, Malet St., London WC1E 7HX, United Kingdom.

accessory factors, is arranged onto operons located on the bacterial chromosome. Pilus assembly is facilitated by a periplasmic chaperone and an outer membrane protein known as an usher. After transcription and translation, subunits are translocated by the general secretory pathway to the periplasm, where they interact with the periplasmic chaperone. The chaperone assists in the folding of subunits into their chaperone-bound conformations and prevents subunits from forming nonproductive interactions by capping interactive surfaces (Jones and Hultgren, 1996; Sauer et al., 2000b). Subunits travel through the outer membrane usher, where they are assembled into the final structure (Thanassi et al., 1998a). Details of the chaperone-usher pathway can be found in chapter 5 of this volume. Examples of pili and fimbriae that are assembled through the chaperone-usher pathway are type 1 pili, P pili, and G and F17 fimbriae.

Type 1 Pili

Escherichia coli expresses adhesive fibers called type 1 pili on its surface. Bacterium-host interactions mediated through type 1 pili are essential for the ability of uropathogenic *E. coli* (UPEC) to colonize the bladder and cause disease (Connell et al., 1996; Langermann et al., 1997; Mulvey et al., 1998; Thankavel et al., 1997). The type 1 pilus operon encodes the major structural subunit (FimA), a periplasmic chaperone (FimC), an outer membrane usher (FimD), two pilus adaptor proteins (FimG and FimF), and an adhesin (FimH). These pili are composed of composite fibers, which can be subdivided into two major parts. First, the pilus forms a helical rod that extends up to several micrometers from the outermembrane (Sauer et al., 2000a). Second, the thinner tip fibrillum extends from the rod at the distal end of the pilus rod, where it is capped by the adhesin. Kinetic studies of subunit-usher interactions have shown that the FimH adhesin, when bound to the chaperone FimC, has high affinity for the FimD usher while the other subunits (FimG, FimF, and FimA) associate with the usher with lower affinities (Saulino et al., 1998). It is this kinetic partitioning which ensures that the FimH adhesin is the first subunit of each pilus rod and allows type 1 pili to be built in a top-down fashion. The adaptor proteins FimG and FimF are added after FimH. Lastly, hundreds of FimA monomers are recruited to form the bulk of the rod.

Three-dimensional reconstruction data show that type 1 pili are right-handed helical rods with an external diameter of ~7 nm and a longitudinal cavity that is 2.0 to 2.5 nm in diameter (Hahn et al., 2002). This is in good agreement with the model proposed by Choudhury et al. in which consecutive major subunits were proposed to associate via head-to-tail interactions giving rise to the type 1 fiber (Choudhury et al., 1999; Hahn et al., 2002). When subunits are translocated across the inner membrane into the periplasm, they immediately bind to the periplasmic chaperone, FimC. FimC is composed of two immunoglobulin (Ig)-like domains (Choudhury et al., 1999; Pellecchia et al., 1999; Sauer et al., 1999). The Ig-like domains are arranged in the shape of a boomerang, with a short linker connecting the two domains. The pilin subunits are also composed of an Ig-like fold, except for two major modifications. Typically, Ig folds have seven to nine β-strands that pack against each other, forming a barrel-like structure (Bork et al., 1994). Pilin domains are missing their last (seventh) β-strand. This missing strand exposes the hydrophobic core to the solvent, making the domain unstable. It also produces a deep goove at the surface of the subunit. However, FimC can temporarily stabilize this domain by inserting its

G1-β-strand into the groove in a process termed donor strand complementation (Choudhury et al., 1999; Sauer et al., 1999). These chaperone-subunit complexes are now primed for assembly at the usher (Saulino et al., 2000; Saulino et al., 1998; Thanassi et al., 1998b; Thanassi et al., 2002). Also, the pilin domains of FimA, FimG, and FimF each have an amino-terminal extension composed of 15 to 20 residues that does not contribute to its Ig fold and is disordered in crystal structures (Choudhury et al., 1999; Sauer et al., 1999). Structural and biochemical data suggest that pilus assembly occurs through competitive displacement of the chaperone G1 strand by the N-terminal extension of the incoming subunit in a process termed donor stand exchange (Choudhury et al., 1999; Sauer et al., 1999). Chapter 5 provides a detailed account of the progress made in understanding the molecular basis of pilus biogenesis by the chaperone-usher pathway.

P Pili

Although P pili are assembled in a similar manner to type 1 pili (Choudhury et al., 1999; Hung and Hultgren, 1998; Jones and Hultgren, 1996; Sauer et al., 1999; Sauer et al., 2000b; Sauer et al., 2002; Saulino et al., 2000; Thanassi et al., 1998b), there are differences in the type of receptors they bind and the tissues they infect. P pili are encoded by the *pap* gene cluster and are made up of the major structural subunit of the pilus rod (PapA), a rod terminator involved in anchoring the pilus to the membrane (PapH), an outer membrane usher (PapC), a periplasmic chaperone (PapD), and adaptor subunits (PapK, PapE, and PapF). The tip fibrillum of P pili, containing the PapG adhesin and three adaptor proteins (PapK, PapE, and PapF), is longer than the fibrillum of type 1 pili. PapG, like FimH of type 1 pili, is located at the tip of P pili (Thanassi and Hultgren, 2000). Similar to type 1 pili, electron microscopy reconstructions of P pili determined that they are helical rods with thin fibrillar tips. The rods consist of 3.28 subunits per turn, with a pitch of 24.9 and a rise per subunit of 7.58 Å (Bullitt and Makowski, 1995). This is in good agreement with X-ray data (Gong and Makowski, 1992). P pili, like type 1 pili, are necessary for adherence and colonization in the urinary tract but are associated with kidney infections known as pyelonephritis (Jones and Hultgren, 1996). P pili are essential for pyelonephritic UPEC isolates to adhere to renal tissue and cause infection in monkeys (Roberts et al., 1994).

G and F17 Fimbriae

Gastrointestinal diseases are an important cause of mortality and morbidity in humans and domesticated animals (Merckel et al., 2003). Enterotoxigenic *E. coli* strains colonize the small intestine of their host and induce diarrhea by secreting enterotoxins (Nataro and Kaper, 1998).

G fimbriae and the closely related F17 fimbriae are expressed in enterotoxigenic *E. coli* strains and are associated with septicemia in farm animals (Saarela et al., 1995; Saarela et al., 1996). Both fimbriae bind to tissue found in the intestinal brush borders and mammalian basement membranes (Saarela et al., 1995). They are very similar in structure to type 1 and P pili. Like type 1 and P pili, G and F17 fimbriae are composed of a helical rod extending from the bacterial cell membrane. Attached to the rod is the flexible tip fibrillum with an open structure. At the end of the tip fibrillum is the GafD adhesin in G pili and the F17 adhesin in F17 pili.

LECTINS AS ADHESINS

To date, the adhesive interactions between host and pathogenic bacteria of over 300 different species have been investigated (Ofek et al., 2003). Overall, they can be divided into three major categories. The first involves the recognition of a protein on the surface of bacterium and a cognate protein located on the mucosal surface of a mammalian cell (Ofek et al., 2003). An example of this would be the interaction between fibronectin, which is a major component of the extracellular matrix, and fibronectin-binding proteins located on the suface of the bacterial cell (Hasty et al., 1994). The second type, and least well characterized, involves interactions between the hydrophobic moieties of proteins and lipids (Ofek et al., 2003). Lastly, the best-characterized interaction is that of a protein called a lectin with carbohydrates. Typically, the lectin is expressed on the bacterial surface while the carbohydrate receptor is located on the host surface. Mammalian tissues contain cells that are decorated with various carbohydrate linkages to protein and lipids and are involved in a variety of functions which include cell adhesion and cell-to-cell recognition and communication. Bacteria have learned to exploit the fact that mammalian tissues have glycoproteins and glycolipds in their membranes by expressing lectins at the tips of their pili. Recently, the crystal structures of FimH, PapGII, GafD, and F17a-G have been determined and are providing a wealth of information on lectin-host receptor interactions.

FimH

The crystal structure of FimCH was solved to 2.5-Å resolution. The structure revealed that FimH was folded into two domains (Color Plate 11 [see color insert]). The N-terminal lectin-binding domain has an overall ellipsoid shape with approximate dimensions of 53.2, 24.3, and 22.9 Å, with a jellyroll-like topology (Choudhury et al., 1999). The C-terminal pilin domain has an Ig-like topology, except that it is missing its C-terminal strand, making FimH unstable on its own. However, donor strand complementation with FimC creates a highly stable complex (Choudhury et al., 1999). This structure, along with the structure of PapDK, shows how periplasmic chaperones stabilize pilin subunits and has led to a proposal about how subunits can assemble into a cylindrical pilus rod or a linear fibrillum. At the distal end of the lectin-binding domain is a deep 459-Å^3 pocket (Laskowski, 1995; Merckel et al., 2003). Asp54 at the base of the pocket gives it a net negative charge (Color Plate 11B). This region was initially designated the mannose-binding pocket because a molecule of C-HEGA, a substituent of the crystallization buffer, was found in this site. Also, C-HEGA distantly resembles the physiologically relevant mannose (Choudhury et al., 1999; Knight et al., 1997). To understand the structural basis of the interaction of FimH with its receptor, Hung et al. (2002) determined the crystal structure of FimCH with mannose to 2.8-Å resolution. Mannose was found in the same pocket as the C-HEGA molecule, confirming the initial assignment of the mannose-binding site. The position, configuration, and orientation of the D-mannose were unambiguously defined by the electron density. Also, FimH selected out the alpha configuration around the reducing anomeric oxygen O1 of D-mannose (Hung et al., 2002). The adhesin-receptor interactions are extensive. All of the hydroxyl groups of mannose except O-1, which would normally be linked to other polysaccharides, interact with the receptor-binding pocket of FimH. Side chain interactions include Asp140, Asn135, Asp54, and Asn 46 (Color Plate 11C). Asp54 makes cooperative hydrogen-bonding interactions with O-4 and O-6 of mannose. Asn135 makes bifurcated hydrogen bonds with O-3 and O-4. Main chain interactions include the amide nitrogens of the amino-terminal residue Phe1 and Asp47. The amino-terminal group of Phe1 makes interactions with O-2 and O-5, while the amide nitrogen of Asp47 makes hydrogen-bonding interactions with O-6. Phe1 helps create the side of the binding pocket, and its aromatic ring makes side chain interactions with Gln133 and Phe144. Finally, the mannose pocket is surrounded by a hydrophobic ridge composed of residues Ile13, Phe1, Tyr48, Ile52, Tyr137, Asp140, and Phe142 (Color Plate 11B).

Functional analyses show that residues that interact with mannose are important in adherence to and invasion of bladder epithelial cells (Hung et al., 2002). Mutations of Asp54, Gln133, Asn135, and Asp140 abolish FimH function as determined by the hemagglutinin (HA) titer and bladder adherence assays. However, the Asn46Asp mutant could adhere to and invade 5637 bladder cells. Also, there was no significant change in the HA titer compared to that in the wild type. A plausible reason for this is that the carboxylic acid moiety of aspartic acid can make interactions similar to those of the asparagine without causing large structural perturbations. When a more severe mutation is introduced, such as Asn46Ala, FimH function is abolished. These mutants were also examined for their ability to bind mannose and trimannose (Hung et al., 2002). As in the HA titer and invasion assays, all FimH mutants were defective in mannose binding except for Asn46Asp. However, the Asn46Ala mutation is defective for mannose binding. Interestingly, although mutations of these binding site residues affected mannose binding, FimH retained the ability to bind trimannose. The increase in affinity for trimannose is probably due to the stacking of the hydrophobic face of the saccharides onto the hydrophobic ridge around the pocket formed by Ile13, Tyr48, Ile52, and Phe142 (Hung et al., 2002). It was suggested that a function of the hydrophobic ridge might be to direct the sugar into the pocket in a manner that facilitates polar interactions. The FimH-binding pocket is conserved among UPEC isolates. The *fimH* gene was sequenced from 200 UPEC strains and an enterohemorrhagic *E. coli* strain. The results show that there is little heterogeneity in the FimH sequences among the different UPEC isolates. In fact, all of the residues involved in mannose binding were invariant, suggesting that the pocket is highly conserved in nature and that even slight changes in the binding site could produce serious deleterious effects. For example, Hung et al. (2002) solved the crystal structure of the FimH mutant Gln133Asn with methyl-α-mannopyranoside to 3.0-Å resolution. The Q133N mutation not only leads to a loss of multiple FimH-mannose interactions but also results in different hydrogen-bonding networks and geometry of the mannose-binding pocket (Hung et al., 2002).

PapG

PapG exists as three variants: PapGI, PapGII, and PapGIII. This is based on their ability to bind the globoseries of the glycolipids GbO3, GbO4 (globoside), and GbO5 (Frosman antigen), respectively (Stromberg et al., 1990, 1991). The crystal structure of the lectin-binding domain of PapGII in complex with its receptor GalNAcβ1-3Galα1-4Galβ1-4Glc (globoside or GbO4) was determined to 1.8-Å resolution (Color Plate 12A [see color insert]) (Dodson et al., 2001). Also, the solution structure of PapGII was determined (Sung et al., 2001a; Sung et al., 2001b). PapGII has axial dimensions of approximately 62.7, 25.3, and 22.5 Å and is an elongated β-sandwich that topologically resembles a jellyroll barrel (Sung et al., 2001b). The receptor-binding pocket in PapGII is shallow and elongated, with a volume of approximately 284 $Å^3$ (Laskowski, 1995; Merckel et al., 2003). However, the solvent-accessible area of PapGII buried when the receptor binds is 716 $Å^2$ (Dodson et al., 2001). Residues Ser89, Lys106, Trp107, Asp108, Arg170, Lys172, and Tyr175 define the rim of the binding pocket, while the floor is made up of Glu59, Ile61, Glu91, Leu102, and Gly104 (Color Plate 12B). Each residue of globoside is involved in extensive direct and water-mediated hydrogen-bonding interactions. Functional analyses showed that Trp107 and Arg170 are particularly critical residues in globoside interactions. When these residues are changed to alanine, receptor binding is abolished (Dodson et al., 2001). Trp107 makes hydrophobic and aromatic contacts with the Galβ1-4Glc moiety of globoside. Arg170 interacts with globoside as well as the protein. It is within hydrogen-bonding distance of the O-2 and O-3 hydroxyls of residue D of globoside as well as the main chain carbonyls of residues Gly168 and Ala169 of PapGII (Color Plate 12B). Also, Arg170 makes a salt bridge with Asp108. Other mutations (Glu59, Glu91, Lys103, and Lys172) to alanine have lesser effects on hemagglutination. It has been shown that PapGI and PapGII can recognize Gal1α1-4Gal (galabiose) disaccharide, which makes up the core structure of both GbO3 and GbO4 (Striker et al., 1995; Stromberg et al., 1990, 1991). A major difference is that PapGII can recognize glucose when it is added to the reducing end of galabiose (Hultgren et al., 1989). With the solving of the crystal structure, we now understand the structural basis of this interaction. Using the PapGII coordinates, the sequence of PapGI was threaded onto the structure and the differences were analyzed. One major difference between PapGI and PapGII is the area surrounding the Glc residue (Dodson et al., 2001). In PapGII, the critical Arg170 residue interacts with the Glc residue. In the PapGI variant, the corresponding residue is a histidine and thus is probably unable to make this interaction. This observation supports previous biochemical observations (Striker et al., 1995). Interestingly, the residues surrounding the galabiose moiety of GbO4 are conserved, which explains why PapGI and PapGII bind GbO3 and GbO4 equally well in vitro. So, how do PapGI and PapGII discriminate between them in vivo? The answer may lie in the orientation and accessibility of the sugar residues when they are attached to the lipid and inserted in the cell membrane (Stromberg et al., 1991). Modeling studies predict that the Glc residue on GbO3 may be partially buried in the membrane. However, the same residue on GbO4 is more exposed to the extracellular environment and thus more accessible for binding by PapGII. Therefore, since PapGII makes interactions with the Glc residue, a partially buried Glc residue predicted in GbO3 would negatively affect PapGII binding to the receptor (Dodson et al., 1993; Striker et al., 1995)

The crystal structure of apo-PapGII was solved to determine if conformational changes occur when PapGII engages its receptor (Color Plate 13 [see color insert]). The structure of the unbound form is very similar to that of the bound form, with a root mean square deviation of 0.16 Å in backbone atom positions (Dodson et al., 2001). Residues involved in receptor binding have no significant changes in conformation. However, there are large changes in the water structure around the binding site. There are eight water-mediated interactions in the receptor-bound structure. The apo structure has five water-mediated interactions, and only one of those (W2) was found in the same place. Four other molecules were found in positions equivalent to those where atoms in the GbO4 receptor were located when bound to PapG. Interestingly, three water molecules were closely aligned with three hydroxyls known to be important for binding the Galα1-4Gal core. Before the receptor can bind, these water molecules must be displaced.

GafD and F17a-G

Recently, the crystal structures of the GafD adhesin domain and the F17a-G adhesin were determined with their receptor, GlcNAc. Since GafD and F17a-G adhesin have almost 90% sequence identity, it is not surprising that the structures are virtually identical. Each has the dimensions of approximately 64.0 by 32.1 by 26.7 Å (Color Plates 14A and 15A [see color insert]). The overall fold is similar to that in PapG and FimH, in that they are β-barrels with a jellyroll-like topology. Athough the binding pocket is located on its side as in PapG, it is in a different position. The pocket is shallow but larger, with an approximate volume of 329 Å^3 compared to 284 Å^3 for the PapG globoside-binding pocket (Laskowski, 1995; Merckel et al., 2003). Each —OH group of GlcNac makes hydrogen-bonding interactions with GafD. There are four side chain (Asp88, Thr89, Thr117 and Gln118) and two main chain (Gly119 and Ala43) interactions (Color Plate 14B). The binding site in F17a-G is very similar (Color Plate 15B). Surprisingly, GafD and F17a-G have significantly different solvent networks around the binding pocket. GafD has just one water-mediated hydrogen bond interaction through Ser116 (Color Plate 14B). However, F17a-G has three additional solvent interactions (Color Plate 15B). Lastly, Trp109 of both GafD and F17a-G makes hydrophobic contact with the sugar moiety such that the indole ring is parallel with the plane of the sugar (Buts et al., 2003; Merckel et al., 2003). Hydrogen-bonding interactions of Thr117 and Asn44 with the N-acetyl group (—NAc) of GlcNAc creates specificity of GlcNAc over glucose. Hemagglutination inhibition was seen with GlcNAc but not glucose, mannose, or glucosamine. Only GlcNAc possess an —NAc group (Merckel et al., 2003). Also, GafD and presumably F17a-G bind acetylated sugar with the following strengths: GlcNAc > ManNAc > GalNAc (Merckel et al., 2003). A possible reason for this is the position of the —OH groups of ManNAc and GalNAc, which would make less favorable interactions with the protein. Taken together, the binding pockets of GafD and F17a-G are designed to bind the acetylated sugar GlcNAc. There are five variants of F17a-G, and the differences are localized in areas away from the receptor-binding site (Color Plate 16 [see color insert]). Since all amino acids involved in binding are conserved, they have identical affinities for GlcNAc. It is striking that these residues are localized in two areas. This suggests that they may serve an undetermined function. A disulfide bridge exists between Cys53 and Cys110. Merckel et al. (2003) suggested that the disulfide bond adds rigidity to the binding site, which helps in the discrimination of GlcNAc versus other N-glycosylated sugars in GafD. However, F17b-G can still bind this sugar even though the Cys110 is replaced by a serine (Buts et al., 2003). In fact, F17b-G elution from an affinity column requires higher concentrations of GlcNAc than does F17a-G and F17d-G elution (Buts et al., 2003). This suggests that the disulfide bond does not enhance GlcNAc in F17b-G, and this is probably also the case for GafD. Like the PapG adhesin, the F17 apo crystal structure is almost identical to the ligand-bound structure. The root mean square deviation for Cα between the F17a-G bound and unbound structures is 0.36 Å. Thus, in F17 and probably in GafD no significant conformational changes occur when the receptor interacts with the binding pocket (Color Plate 17 [see color insert]). Also, as in the PapG apo structure, the solvent structure around the binding site was different. Water molecules mimic the O-3 and O-4 hydroxyl groups; several other water molecules form a network that stacks against the indole ring of Trp109. These water molecules must be expelled before the GlcNAc receptor can bind.

COMPARISONS AMONG THE BACTERIAL ADHESINS

The crystal structures of FimH, PapGII, GafD, and F17a-G provide an opportunity for understanding the molecular details of host-pathogen interactions as well as the evolutionary relationships between the different adhesins. Except for GafD and F17, there is little sequence homology between the molecules. However, each structure has a similar overall fold (Color Plate 18 [see color insert]). All four adhesin domains are shaped like elongated prolate ellipsoid molecules, with dimensions ranging from 50 to 70 Å by 25 to 30 Å by 25 to 30 Å. When all structures are aligned with respect to FimH, they have their N termini at the top of the molecules and their C termini at the bottom. Merckel et al. (2003) noticed a structural motif which they termed the switchback, which is composed of a short strand-turn-helix/strand-turn-strand and is present in all four lectin domains. The structural similarity, combined with the low sequence identity, indicates that these adhesin domains arose from a common ancestor and are related by divergent evolution.

Structural alignment also shows that the lectin-binding sites, although on the top half of each molecule, are located in different places on the protein (Color Plate 18). In FimH, the binding site is located at the tip of the molecule, while for PapG, GafD, and F17, the binding site is on the side. These adhesin structures seem to have found a balance between the requirements necessary to engage their cognate receptor and the structural restriction required for pilus assembly. A reason why these adhesins have similar dimensions may be that the proteins are secreted through the usher. Since the proteins are secreted in a folded state, they must be small enough to travel through the pore.

Urothelial plaques are composed of proteins called uroplakins, which line the umbrella cells of the bladder. The plaques cover the luminal surface of the bladder and are thought to function as part of a permeability barrier and help strengthen bladder epithelial cells. Uroplakins, which are made up of four proteins, Ia, Ib, II, and III, are assembled into 16-mm-diameter hexagonal complexes (Sun et al., 1996, 1999). It was recently shown that type 1 pili can make direct interactions with uroplakin particles (Min et al., 2002; Zhou et al., 2001); in particular, this interaction is mediated through FimH binding to uroplakin Ia. Uroplakin Ia is glycosylated with polysaccharides that have a terminal mannose. This suggests that FimH, with its binding site located at the top of the protein, sticks its lectin-binding domain into the uroplakin hexagonal structure, where it binds to the terminal mannose residue found on uroplakin Ia (Hung et al., 2002; Mulvey et al., 1998; Zhou et al., 2001).

The glycolipid globoside, which is located on the kidney epithelium, is a receptor for PapG. The binding site of PapGII adhesin is located on the side of the molecule. The pilin and receptor-binding domains of PapGII are arranged in a head-to-tail fashion with a short linker between them (Dodson et al., 1993). These would put the receptor-binding surfaces on the side of the pilus tip. Structural studies suggest that globoside would be in an approximately parallel orientation in relation to its ceramide group (Pascher et al., 1992; Stromberg et al., 1991). For the PapGII-binding site to interact with its receptor, the receptor-binding domain of PapGII must be orientated in a parallel fashion with the cell membrane so that it can bind to the receptor. P pili have a longer flexible tip fibrillum than FimH does. This longer fibrillium may allow for extra flexibility needed for the adhesin to make this side-on interaction with the receptor.

G and F17a-G fimbriae bind to intestinal brush borders and mammalian basement membranes which have glycoproteins that contain GlcNAc. The intestinal epithelium contains GlcNAc. Based on P pili and type 1 pili, it is likely that the G and F17a-G fimbriae have also evolved to maximize their ability to engage their receptors. Taken together, the pili, along with the adhesin at their tip, have also coevolved with their cognate mammalian receptors. The ability to interact with different tissue is facilitated by structural differences in the pilus rod and in the binding pocket of the adhesins.

RATIONAL VACCINE AND DRUG DESIGN

The detailed structural analysis of bacterial adhesins has demonstrated their central role in the host-pathogen interface during the infection process and has made them attractive targets for the development of new antimicrobial therapies. One strategy is to design receptor analogs that bind to the adhesin with high affinity for the local structure around the binding site. For example, FimH binds trimannose with a higher affinity than it binds mannose, possibly as a result of the hydrophobic ridge that surrounds the binding pocket. Receptor analogs could be designed to take advantage of these interactions. Also, GafD has a pocket adjacent to the binding pocket, which can accommodate an additional monosaccharide. Thus, a potential inhibitor might be a disaccharide composed of a GlcNAc-like moiety connected to another monosaccharide or monosaccharide-like moiety with different chemical substituents that can interact with the neighboring residues. The crystal structures of PapG and GbO4 give chemists the ability to rationally design inhibitors to block interactions. PapG exists in three molecular variants classified by their slightly different agglutination patterns. The structural requirement for class I and class II PapG adhesins, which bind to globoseries glycolipids, have been investigated with a large number of synthetic saccharides. The results have indicated that the galabiose disaccharide was required for binding both class I and class II adhesins. This allowed chemists to use galabiose as a core structure and to introduce substituents that interact with the lectin in a favorable manner and inhibit hemagglutination (Ohlsson et al., 2002).

The identification of FimH as the adhesive subunit of type 1 pili and the elucidation of the chaperone-usher pathway used to assemble these adhesive organelles have made it possible to purify large amounts of pure FimH and test its efficacy as a vaccine. The use of purified FimH as a vaccine, rather than whole type 1 pili, in which FimH is a minor component, significantly enhances the host immune response to the FimH adhesin. Antisera from mice inoculated with a FimCH chaperone-adhesin complex or with a naturally occurring truncated form of FimH (FimH$_t$) adhesin were able to block binding in vitro to bladder epithelial cells of 49 of 52 primary clinical *E. coli* isolates from patients with urinary tract infections. In contrast, antisera from mice inoculated with whole type 1 pili were poor inhibitors of binding, with less than 50% of the clinical isolates being blocked even at a 1:50 dilution of antisera. Nine weeks after primary immunization with a FimH-based vaccine, mice exhibited a 100- to 1,000-fold reduction in the number of bacteria recovered from their bladders compared with mice vaccinated with FimC alone as a control. The adhesin vaccine induced long-term serum IgG antibodies to FimH as well as functional inhibitory titers in vaginal secretions. Vaccinated animals showed substantial resistance to subsequent infection by type 1-piliated UPEC. However, it may be essential to include other critical antigens in a vaccine cocktail in order to increase the potency and efficacy of the vaccine to completely prevent acute urinary tract infections and/or eradicate bacterial bladder reservoirs.

CONCLUSIONS

Electron microscopy of the pilus structure along with crystal and nuclear magnetic resonance spectroscopy structures of the bacterial adhesins and other structural components of the chaperone-usher pathways have provided a detailed picture of and insights into pilus biogenesis and host-pathogen interactions. Structural analyses of the lectin-binding domains of FimH, PapGII, GafD, and F17a-G suggest that they have evolved from a common ancestral protein. Differences in pilus structure, along with differences in the lectin-binding domains, allow the bacterium to orient their adhesin in order to effectively and efficiently bind to their receptors. Also, these structures are being used as a platform to design novel inhibitors, which can bind to these domains and inhibit host-pathogen interactions. Highly purified adhesins can be used to create potent vaccines. Although considerable progress has been made in the last 10 years, there are still issues to address. Future work includes high-resolution analysis of the pilus structures by cryoelectron microscopy. These pictures, along with crystal and solution structures of pilus subunits, will allow us to build a high-resolution model of a pilus. Another important breakthrough for the field would be the structure of the integral membrane usher. A ternary structure of an usher with its chaperone and adhesin would give us a snapshot of the interactions necessary to initiate pilus biogenesis.

REFERENCES

Bork, P., L. Holm, and C. Sander. 1994. The immunoglobulin fold. Structural classification, sequence patterns and common core. *J. Mol. Biol.* **242:**309–320.

Bullitt, E., and L. Makowski. 1995. Structural polymorphism of bacterial adhesion pili. *Nature* **373:**164–167.

Buts, L., J. Bouckaert, E. De Genst, R. Loris, S. Oscarson, M. Lahmann, J. Messens, E. Brosens, L. Wyns, and H. De Greve. 2003. The fimbrial adhesin F17-G of enterotoxigenic *Escherichia coli* has an immunoglobulin-like lectin domain that binds N-acetylglucosamine. *Mol. Microbiol.* **49:**705–715.

Choudhury, D., A. Thompson, V. Stojanoff, S. Langermann, J. Pinkner, S. J. Hultgren, and S. D. Knight. 1999. X-ray structure of the FimC-FimH chaperone-adhesin complex from uropathogenic *Escherichia coli*. *Science* **285:**1061–1066.

Connell, I., W. Agace, P. Klemm, M. Schembri, S. Marild, and C. Svanborg. 1996. Type 1 fimbrial expression enhances *Escherichia coli* virulence for the urinary tract. *Proc. Natl. Acad. Sci. USA* **93:**9827–9832.

Dodson, K. W., F. Jacob-Dubuisson, R. T. Striker, and S. J. Hultgren. 1993. Outer-membrane PapC molecular usher discriminately recognizes periplasmic chaperone-pilus subunit complexes. *Proc. Natl. Acad. Sci. USA* **90:**3670–3674.

Dodson, K. W., J. S. Pinkner, T. Rose, G. Magnusson, S. J. Hultgren, and G. Waksman. 2001. Structural basis of the interaction of the pyelonephritic *E. coli* adhesin to its human kidney receptor. *Cell* **105:**733–743.

Gong, M., and L. Makowsk. 1992. Helical structure of P pili from *Escherichia coli*. Evidence from X-ray fiber diffraction and scanning transmission electron microscopy. *J. Mol. Biol.* **228:**735–742.

Hahn, E., P. Wild, U. Hermanns, P. Sebbel, R. Glockshuber, M. Haner, N. Taschner, P. Burkhard, U. Aebi, and S. A. Muller. 2002. Exploring the 3D molecular architecture of *Escherichia coli* type 1 pili. *J. Mol. Biol.* **323:**845–857.

Hasty, D. L., H. S. Courtney, E. V. Sokurenko, and I. Ofek. 1994. Bacteria-extracellular matrix interactions, p. 197–211. *In* P. Klemm (ed.), *Fimbriae: Adhesion, Genetics, Biogenesis, and Vaccines.* CRC Press, Inc., Boca Raton, Fla.

Hultgren, S. J., F. Lindberg, G. Magnusson, J. Kihlberg, J. M. Tennent, and S. Normark. 1989. The PapG adhesin of uropathogenic *Escherichia coli* contains separate regions for receptor binding and for the incorporation into the pilus. *Proc. Natl. Acad. Sci. USA* **86:**4357–4361.

Hung, C. S., J. Bouckaert, D. Hung, J. Pinkner, C. Widberg, A. DeFusco, C. G. Auguste, R. Strouse, S. Langermann, G. Waksman, and S. J. Hultgren. 2002. Structural basis of tropism of *Escherichia coli* to the bladder during urinary tract infection. *Mol. Microbiol.* **44:**903–915.

Hung, D. L., and S. J. Hultgren. 1998. Pilus biogenesis via the chaperone/usher pathway: an integration of structure and function. *J. Struct. Biol.* **124:**201–220.

Jones, C. H., and S. J. Hultgren. 1996. Structure, function, and assembly of adhesive P pili, p. 175–219. *In* H. L. T. Mobley and J. W. Warren (ed.), *Urinary Tract Infections: Molecular Pathogenesis and Clinical Management.* ASM Press, Washington, D.C.

Knight, S., M. A. Mulvey, and J. Pinkner. 1997. Crystallization and preliminary X-ray diffraction studies of the FimC-FimH chaperone-adhesin complex from Escherichia coli. *Acta Crystallogr. Ser. D* **D53:**207–210.

Langermann, S., S. Palaszynski, M. Barnhart, G. Auguste, J. S. Pinkner, J. Burlein, P. Barren, S. Koenig, S. Leath, C. H. Jones, and S. J. Hultgren. 1997. Prevention of mucosal *Escherichia coli* infection by FimH-adhesin-based systemic vaccination. *Science* **276:**607–611.

Laskowski, R. A. 1995. SURFNET: a program for visualizing molecular surfaces, cavities, and intermolecular interactions. *J. Mol. Graph.* **13:**323–330, 307–308.

Merckel, M. C., J. Tanskanen, S. Edelman, B. Westerlund-Wikstrom, T. K. Korhonen, and A. Goldman. 2003. The structural basis of receptor-binding by *Escherichia coli* associated with diarrhea and septicemia. *J. Mol. Biol.* **331:**897–905.

Min, G., M. Stolz, G. Zhou, F. Liang, P. Sebbel, D. Stoffler, R. Glockshuber, T. T. Sun, U. Aebi, and X. P. Kong. 2002. Localization of uroplakin Ia, the urothelial receptor for bacterial adhesin FimH, on the six inner domains of the 16 nm urothelial plaque particle. *J. Mol. Biol.* **317:**697–706.

Mulvey, M. A., Y. S. Lopez-Boado, C. L. Wilson, R. Roth, W. C. Parks, J. Heuser, and S. J. Hultgren. 1998. Induction and evasion of host defenses by type 1-piliated uropathogenic *Escherichia coli. Science* **282:**1494–1497.

Nataro, J. P., and J. B. Kaper. 1998. Diarrheagenic *Escherichia coli. Clin. Microbiol. Rev.* **11:**142–201.

Nicholls, A., K. A. Sharp, and B. Honig. 1991. Protein folding and association: insights from the interfacial and thermodynamic properties of hydrocarbons. *Proteins* **11:**281–296.

Ofek, I., D. L. Hasty, and R. J. Doyle. 2003. *Bacterial Adhesion to Animal Cell and Tissues.* ASM Press, Washington, D.C.

Ohlsson, J., J. Jass, B. E. Uhlin, J. Kihlberg, and U. J. Nilsson. 2002. Discovery of potent inhibitors of PapG adhesins from uropathogenic *Escherichia coli* through synthesis and evaluation of galabiose derivatives. *Chembiochem* **3:**772–779.

Pascher, I., M. Lundmark, P. G. Nyholm, and S. Sundell. 1992. Crystal structures of membrane lipids. *Biochim. Biophys. Acta.* **1113:**339–373.

Pellecchia, M., P. Sebbel, U. Hermanns, K. Wuthrich, and R. Glockshuber. 1999. Pilus chaperone FimC-adhesin FimH interactions mapped by TROSY-NMR. *Nat. Struct. Biol.* **6:**336–339.

Roberts, J. A., B. I. Marklund, D. Ilver, D. Haslam, M. B. Kaack, G. Baskin, M. Louis, R. Mollby, J. Winberg, and S. Normark. 1994. The Gal(alpha 1-4)Gal-specific tip adhesin of *Escherichia coli* P-fimbriae is needed for pyelonephritis to occur in the normal urinary tract. *Proc. Natl. Acad. Sci. USA* **91:**11889–11893.

Saarela, S., S. Taira, E. L. Nurmiaho-Lassila, A. Makkonen, and M. Rhen. 1995. The *Escherichia coli* G-fimbrial lectin protein participates both in fimbrial biogenesis and in recognition of the receptor N-acetyl-D-glucosamine. *J. Bacteriol.* **177:**1477–1484.

Saarela, S., B. Westerlund-Wikstrom, M. Rhen, and T. K. Korhonen. 1996. The GafD protein of the G (F17) fimbrial complex confers adhesiveness of *Escherichia coli* to laminin. *Infect. Immun.* **64:**2857–2860.

Sauer, F. G., M. Barnhart, D. Choudhury, S. D. Knight, G. Waksman, and S. J. Hultgren. 2000a. Chaperone-assisted pilus assembly and bacterial attachment. *Curr. Opin. Struct. Biol.* **10:**548–556.

Sauer, F. G., K. Futterer, J. S. Pinkner, K. W. Dodson, S. J. Hultgren, and G. Waksman. 1999. Structural basis of chaperone function and pilus biogenesis. *Science* **285:**1058–1061.

Sauer, F. G., M. A. Mulvey, J. D. Schilling, J. J. Martinez, and S. J. Hultgren. 2000b. Bacterial pili: molecular mechanisms of pathogenesis. *Curr. Opin. Microbiol.* **3:**65–72.

Sauer, F. G., J. S. Pinkner, G. Waksman, and S. J. Hultgren. 2002. Chaperone priming of pilus subunits facilitates a topological transition that drives fiber formation. *Cell* **111:**543–551.

Saulino, E. T., E. Bullitt, and S. J. Hultgren. 2000. Snapshots of usher-mediated protein secretion and ordered pilus assembly. *Proc. Natl. Acad. Sci. USA* **97:**9240–9245.

Saulino, E. T., D. G. Thanassi, J. S. Pinkner, and S. J. Hultgren. 1998. Ramifications of kinetic partitioning on usher-mediated pilus biogenesis. *EMBO J.* **17:**2177–2185.

Shindyalov, I. N., and P. E. Bourne. 1998. Protein structure alignment by incremental combinatorial extension (CE) of the optimal path. *Protein Eng.* **11:**739–747.

Striker, R., U. Nilsson, A. Stonecipher, G. Magnusson, and S. J. Hultgren. 1995. Structural requirements for the glycolipid receptor of human uropathogenic *Escherichia coli. Mol. Microbiol.* **16:**1021–1029.

Stromberg, N., B. I. Marklund, B. Lund, D. Ilver, A. Hamers, W. Gaastra, K. A. Karlsson, and S. Normark. 1990. Host-specificity of uropathogenic *Escherichia coli* depends on differences in binding specificity to Gal alpha 1-4Gal-containing isoreceptors. *EMBO J.* **9:**2001–2010.

Stromberg, N., P. G. Nyholm, I. Pascher, and S. Normark. 1991. Saccharide orientation at the cell surface affects glycolipid receptor function. *Proc. Natl. Acad. Sci. USA* **88:**9340–9344.

Sun, T. T., F. X. Liang, and X. R. Wu. 1999. Uroplakins as marker of urothelial differentiation. *Adv. Exp. Med. Biol.* **462:**7–18.

Sun, T. T., H. Zhao, J. Provet, U. Aebi, and X. R. Wu. 1996. Formation of asymmetric unit membrane during urothelial differentiation. *Mol. Biol. Rep.* **23:**3–11.

Sung, M. A., H. A. Chen, and S. Matthews. 2001a. Sequential assignment and secondary structure of the triple-labelled carbohydrate-binding domain of papG from uropathogenic *E. coli. J. Biomol. NMR* **19:**197–198.

Sung, M. A., K. Fleming, H. A. Chen, and S. Matthews. 2001b. The solution structure of PapGII from uropathogenic *Escherichia coli* and its recognition of glycolipid receptors. *EMBO Rep.* **2:**621–627.

Thanassi, D. G., and S. J. Hultgren. 2000. Assembly of complex organelles: pilus biogenesis in gram-negative bacteria as a model system. *Methods* **20:**111–126.

Thanassi, D. G., E. T. Saulino, and S. J. Hultgren. 1998a. The chaperone/usher pathway: a major terminal branch of the general secretory pathway. *Curr. Opin. Microbiol.* **1:**223–231.

Thanassi, D. G., E. T. Saulino, M. J. Lombardo, R. Roth, J. Heuser, and S. J. Hultgren. 1998b. The PapC usher forms an oligomeric channel: implications for pilus biogenesis across the outer membrane. *Proc. Natl. Acad. Sci. USA* **95:**3146–3151.

Thanassi, D. G., C. Stathopoulos, K. Dodson, D. Geiger, and S. J. Hultgren. 2002. Bacterial outer membrane ushers contain distinct targeting and assembly domains for pilus biogenesis. *J. Bacteriol.* **184:**6260–6269.

Thankavel, K., B. Madison, T. Ikeda, R. Malaviya, A. H. Shah, P. M. Arumugam, and S. N. Abraham. 1997. Localization of a domain in the FimH adhesin of *Escherichia coli* type 1 fimbriae capable of receptor recognition and use of a domain-specific antibody to confer protection against experimental urinary tract infection. *J. Clin. Investig.* **100:**1123–1136.

Zhou, G., W. J. Mo, P. Sebbel, G. Min, T. A. Neubert, R. Glockshuber, X. R. Wu, T. T. Sun, and X. P. Kong. 2001. Uroplakin Ia is the urothelial receptor for uropathogenic *Escherichia coli:* evidence from in vitro FimH binding. *J. Cell. Sci.* **114:**4095–4103.

Structural Biology of Bacterial Pathogenesis
Edited by G. Waksman et al.
© 2005 ASM Press, Washington, D.C.

Chapter 4

Host Receptors of Bacterial Origin

Calvin K. Yip, Cecilia P. C. Chiu, and Natalie C. J. Strynadka

An important step in the successful colonization and subsequent induction of disease by microbial pathogens is adherence to host surfaces (Finlay and Falkow, 1997). To bind to their host, pathogenic bacteria must overcome physical and mechanical barriers such as flushing and electrostatic repulsion. To achieve this, bacterial pathogens express ligands known as adhesins, which directly mediate their interaction with the host cell and/or the extracellular matrix. These adhesins, usually elongated proteins with elaborated structures on the bacterial surface, bind to a variety of host molecules including cell surface proteins, glycoproteins, glycolipids, and carbohydrates. Although a significant number of adhesins have been identified and characterized, the identification of their respective host receptors has been more difficult. The available examples, though, reveal significant structural diversity among bacterial adhesins and variation in the host molecules that they bind and suggest that the molecular mechanisms of host-pathogen interactions are quite complex. While the expression of adhesins allows a pathogen to bind to its host with higher affinity, such an interaction still relies on the presence and availability of their respective host receptors. The surprising discovery that tight adherence of certain bacterial pathogens to their hosts does not require host-derived receptors has revised our views of bacterial adherence. The ability of such pathogens to insert their own customized receptor into their host eliminates the need for a host receptor and theoretically could expand the range of hosts these pathogens can infect. In this chapter, we illustrate how specific pathogenic bacteria participate in such a sophisticated cross talk with their host by highlighting recent advances in our understanding of the structure and function of the first identified host receptor of bacterial origin—the translocated intimin receptor (Tir) from enteropathogenic *Escherichia coli* (EPEC) and its cognate adhesin intimin.

EPEC CAUSES THE ATTACHING AND EFFACING LESION

EPEC, an extracellular bacterial pathogen which infects humans by colonizing the intestinal mucosa, is a major cause of infantile diarrhea in developing countries and continues to be a major health threat worldwide (Clarke et al., 2003). The histopathological

Calvin K. Yip, Cecilia P. C. Chiu, and Natalie C. J. Strynadka • Department of Biochemistry and Molecular Biology, University of British Columbia, Vancouver, British Columbia, Canada V6T 1Z3.

hallmark of an EPEC infection is the formation of attaching and effacing (A/E) lesions, observed in infections of cultured epithelial cells as well as in small bowel biopsy specimens from infected patients. Characterized by localized "effacement" or destruction of brush border microvilli, intimate bacterial attachment, gross cytoskeletal reorganization of the intestinal epithelium, and formation of actin-rich "pedestals," the A/E lesion is associated with EPEC virulence (Fig. 1) (Donnenberg et al., 1997; Moon et al., 1983). A number of related bacterial pathogens known as the A/E pathogens, including enterohemorrhagic *E. coli* (EHEC), which causes Hamburger's disease, rabbit-enteropathogenic *E. coli,* and *Citrobacter rodentium,* which infects mice, are also capable of inducing A/E lesions. In spite of the diversity in host organisms and specific tissue types infected, the pathogenic mechanisms and adherence strategies used by these pathogens are highly similar.

All of the genetic factors required for EPEC to elicit the A/E lesion can be mapped to a 35-kbp pathogenicity island known as the locus of enterocyte effacement (LEE). The entire EPEC LEE, when cloned on a plasmid and subsequently introduced into the *E. coli* K-12 strain, can confer on this nonpathogenic bacterial strain the ability to produce A/E lesions (McDaniel and Kaper, 1997). Orthologous LEE pathogenicity islands have since been identified in other A/E pathogens. Although subtle differences exist among these LEE's, their sequence and organization are largely conserved, implying a high degree of functional similarity. Pathogenicity islands such as the LEE are thought to be acquired via horizontal gene transfer, and the G+C content of the LEE (38.4%), which is significantly different from the 50% observed for the *E. coli* chromosome, strongly implies different origins.

The EPEC LEE has been completely sequenced, and it contains 41 open reading frames arranged in at least five polycistronic operons (Fig. 2) (Elliott et al., 1998). Three of these operons (known as *LEE1, LEE2,* and *LEE3*) contain the *esc* and *sep* genes, which encode structural components of a functional type III secretion system. The *LEE4* operon encodes a number of Esp proteins, which are secreted by the type III secretion system. The *LEE5/Tir* operon contains three important genes directly involved in the intimate attachment of bacteria to host: *eae,* which encodes the outer membrane adhesion molecule known as intimin; *tir,* which encodes the translocated intimin receptor (Tir); and *cesT,* which encodes the type III chaperone of Tir. Expression of the various operons as well as of discrete LEE genes is transcriptionally regulated by at least one protein: the LEE-encoded global regulator Ler, a protein which shows similarity to the H-NS family of DNA-binding proteins and has been demonstrated to activate the transcription of LEE operons and genes (Elliott et al., 2000). Recently, it has been shown that another LEE-located gene, Orf11, is also required for LEE gene expression (Deng et al., 2004).

THE EPEC TYPE III SECRETION SYSTEM AND DELIVERY OF Tir INTO THE HOST

A significant portion of the LEE pathogenicity island encodes TTSS components and type III secreted proteins (Deng et al., 2004). It is therefore not surprising that this TTSS plays a crucial role in EPEC pathogenesis. Found in many gram-negative animal and plant pathogens, including EPEC and *Salmonella, Pseudomonas, Yersinia,* and *Chlamydia* spp., the TTSS is a membrane-spanning macromolecular complex, composed of approximately 20 distinct proteins, which allows these pathogenic bacteria to translocate virulence

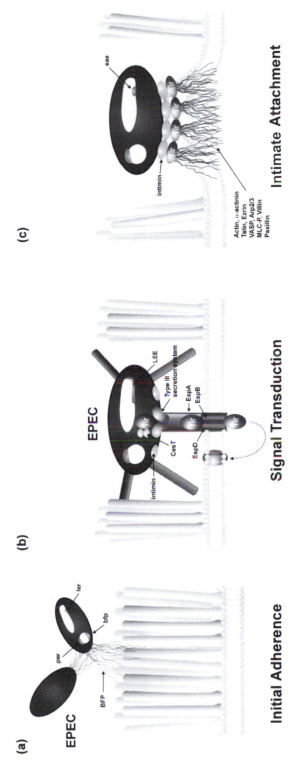

(a)

EPEC

Initial Adherence

(b)

EPEC

Signal Transduction

(c)

Intimate Attachment

Figure 1. Proposed steps in A/E lesion formation by EPEC. (a) The first step of EPEC infection involves localized adherence to the host intestinal epithelium, which does not require the LEE pathogenicity island and is mediated by the plasmid-encoded bundle forming pilus (BFP). (b) The second step involves assembly of the LEE-encoded TTSS and its translocon. Type III effectors including Tir are delivered into the host. Translocated Tir is phosphorylated and inserts into the host cell membrane by an unknown process. (c) The last step involves intimin binding to Tir. The clustering of Tir by intimin results in very intimate attachment. This binding also transmits a signal via the transmembrane domains and cytoplasmic tails of Tir, which results in recruitment of cytoskeletal proteins, massive reorganization of the host cytoskeleton, and, ultimately, formation of the actin-rich pedestal.

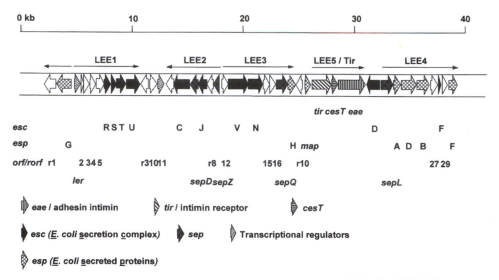

Figure 2. Genetic organization of the LEE pathogenicity island of EPEC. The EPEC LEE contains 41 open reading frames (ORFs) organized into at least five operons, designated *LEE1, LEE2, LEE3, LEE4,* and *LEE5/Tir.* The *esc* genes encode structural components of the TTSS, whereas the *esp* genes encode type III secreted/translocated proteins. The LEE-encoded regulator Ler is located in the *LEE1* operon.

proteins, known as effectors, directly into the cytoplasm of eukaryotic cells. Under specific conditions in the absence of host cells, these effectors can be secreted by the TTSS to the extracellular medium (Hueck, 1998). Many of the type III effectors structurally and functionally mimic eukaryotic factors and are capable of subverting key cellular processes to the benefit of the pathogens (Stebbins and Galan, 2001). Although the types of translocated effectors are quite divergent among different bacterial pathogens, genetic analyses and electron microscopy studies have shown that TTSSs are structurally and functionally conserved across bacterial species (Blocker et al., 2003; Rosqvist et al., 1995). In spite of advances in our understanding of the function of specific type III effectors and gross structural characteristics of the TTSS (a detailed description is provided in chapter 9), the mechanism of type III protein secretion and translocation remains poorly understood. Several conclusions, however, can be reached regarding type III secretion in general (Galan and Collmer, 1999). First, type III secreted proteins do not contain a recognizable signal sequence and, unlike proteins secreted by the classical *sec*-dependent pathway, are not processed by signal peptidase. Second, the secretion process occurs in a single step without the presence of periplasmic intermediates. Finally, the secretion and translocation process is often mediated by type III chaperones, a specialized family of small, acidic proteins capable of stabilizing their cognate effectors, targeting them to the secretion apparatus, establishing a secretion hierarchy, and, in a few cases, regulating the transcription of effector genes (Parsot et al., 2003).

    When grown in tissue culture medium, EPEC secretes large quantities of proteins into the supernatant via its TTSS as a result of upregulated expression of LEE genes (Kenny and Finlay, 1995). Some of the secreted proteins, namely, EspA, EspB, and EspD, are not effectors but structural proteins that assemble into an extracellular translocon extending from the basal type III secretion apparatus at the bacterial membrane (Ide et al., 2001; Knutton et al., 1998). The EPEC type III translocon, which consists of a hollow sheath-like EspA filament and a hetero-oligomeric pore on the host plasma membrane formed by EspB and EspD, is responsible for mediating protein translocation across the host cell membrane (Fig. 3). In the presence of cultured cells, EPEC translocates at least five different effectors into its host, including EspF, EspG, EspH, Map, and Tir (Clarke et al., 2003). Tir, the receptor for the outer membrane adhesin intimin, is the best characterized of the EPEC effectors and is widely considered the most central to subsequent EPEC pathogenicity (Kenny, 2002; Kenny et al., 1997).

The detailed mechanism of Tir secretion and translocation into the host cell remains largely uncharacterized. CesT, the type III chaperone of Tir, binds within the first 100 residues of Tir and maintains the stability of Tir in the bacterial cytoplasm prior to secretion and translocation (Abe et al., 1999; Elliott et al., 1999b). Despite the lack of sequence identity between CesT and other type III chaperones, the crystal structure of EHEC CesT (97% identity to EPEC CesT) reveals that CesT adopts an overall fold and dimeric structure similar to that of other type III chaperones (Fig. 4) (Luo et al., 2001). The CesT dimer contains an extensive hydrophobic surface that is thought to mediate specific interactions with Tir. CesT binding does not appear to globally unfold Tir, since CesT-bound Tir can still bind its natural ligand, intimin, with high affinity (Fig. 5) (Luo et al., 2001). A number

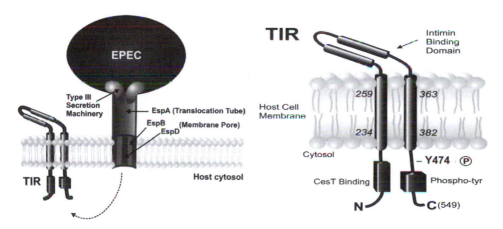

Figure 3. EPEC type III secretion and proposed membrane topology of EPEC Tir. Upregulation of LEE gene expression results in the assembly of a TTSS across the bacterial membrane. Three type III secreted proteins, EspA, EspB, and EspD, then assemble a translocon structure which mediates protein translocation into the host cell. EspA polymerizes into a filament which is thought to be the translocation tube, and EspB and EspD form a hetero-oligomeric pore on the surface of host cells. Type III effectors including Tir are translocated into the host on proper assembly of the TTSS and its translocon. Once translocated into the host cell, Tir is thought to insert into the host cell membrane and adopt a hairpin-like conformation. Tir is predicted to contain two transmembrane segments (TM1, residues 259 to 234; TM2, residues 363 to 382). The N-terminal (residues 1 to 233) and C-terminal (residues 383 to 549) tails of Tir localize to the host cytoplasm. The intervening region between the two putative transmembrane domains makes up the intimin-binding domain.

CesT   **SigE**   **SycE**

Figure 4. Structural similarities among type III secretion chaperones. Shown are ribbon representations of three type III secretion chaperones: EHEC CesT, *Salmonella* SigE, and *Yersinia* SycE. Despite the lack of sequence identities among these three proteins, their three-dimensional structures are very similar. All contain hydrophobic surfaces. Diagrams were generated with Molscript (Kraulis, 1991) and Raster3D (Merritt and Bacon, 1997), using PDB accession codes 1K3E (CesT), 1K3S (SigE), and 1JYA (SycE). Due to domain swapping, two possible homodimers can be chosen for the CesT structure (Luo et al., 2001), only one of which is shown here.

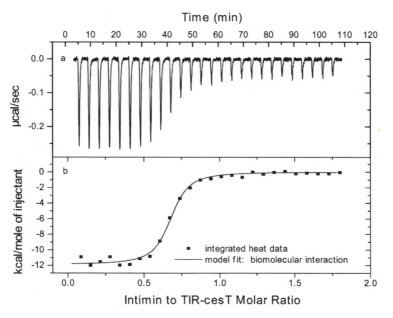

Figure 5. Isothermal titration calorimetric analysis of Tir-intimin in the presence of the Tir chaperone CesT. CesT-bound Tir was titrated against the extracellular fragment of intimin in solution (Luo et al., 2001). Tir retains its ability to bind intimin with high affinity (K_a, $2.2 \times 10^7\ \mathrm{M}^{-1}$) even in the presence of CesT, suggesting that the CesT chaperone does not act to globally unfold Tir.

of type III chaperones locally unfold a limited N-terminal region of their cognate effectors for facilitation of the secretion process at the bacterial inner membrane. Recent work suggests that CesT may assist Tir in docking directly to the type III ATPase at the inner membrane (Gauthier et al., 2003), putatively providing an unfolded N-terminus of CesT-bound Tir to initiate type III secretion (a mechanism reminiscent of that proposed for the SecB chaperone-preprotein complex and the ATPase SecA in the *sec*-dependent translocation system in bacteria). Exactly how Tir passes efficiently through the extended EspA-EspB-EspD translocon structure to be ultimately translocated into the host is also not clearly understood. Low-resolution electron microscopic studies of the EspA filament, the putative translocation tube, have shown that the inner diameter of this hollow filament is approximately 25 Å, indicating at least partial unfolding of the effector during translocation (Daniell et al., 2003). However, without detailed structures of the translocon at atomic resolution, it is not possible to understand the energetics of this molecular event.

Once translocated, Tir inserts itself into the host plasma membrane, where it is thought to adopt a hairpin-like conformation (Fig. 3). A puzzling question is how Tir inserts itself into the host cell membrane after translocation, since CesT, like other type III chaperones, is not known to be translocated into the host. One mechanism proposes that Tir is first delivered into the host cytoplasm, where it would become phosphorylated and ultimately adopt a conformation which facilitates its insertion into the membrane (Kenny, 2002). An alternative mechanism, which is based on the observation that Tir is detected in the host cell membrane shortly after infection and subsequently in the host cytoplasm, suggests that Tir is delivered directly into the host cell membrane by the EPEC TTSS (Gauthier et al., 2000).

Tir AND INTIMIN MEDIATE INTIMATE BACTERIAL ATTACHMENT

Tir and intimin are the two molecules responsible for establishing and maintaining the tight adherence of EPEC to its host during infection. Intimin, the first bacterial factor associated with the A/E phenotype, is a 939-residue outer membrane adhesin which directly mediates host-pathogen interactions. The receptor for intimin, initially thought to be a host protein, is the bacterially derived and type III translocated effector Tir (Kenny et al., 1997). The use of Tir by EPEC represents a novel pathogenic mechanism in which a pathogen injects its own receptor into mammalian cells for binding. The ability to synthesize and deliver its own receptor means that EPEC can avoid the usual dependence on the expression of a host-derived factor to achieve effective binding to its host. Indeed, EPEC can infect a broad range of cell types including macrophages and red blood cells in tissue cultures (Vallance and Finlay, 2000). Intimin binds and focuses Tir beneath the infecting EPEC and clusters it in a manner similar to integrin clustering by the extracellular matrix in focal adhesion formation in mammalian cells (Goosney et al., 2000a). The specific binding and receptor clustering not only allows EPEC to overcome the natural charge-charge repulsion with the host membrane but also may provide sufficient interaction energy to overcome dislodging by the constant peristaltic flushing as well as the ensuing host diarrheal response within the intestinal environment (Vallance and Finlay, 2000). Tir also binds via its N terminus to the host cytoskeletal proteins α-actinin, talin, and vinculin, all of which are involved in focal adhesion (Freeman et al., 2000; Goosney et al., 2000b). These interactions anchor EPEC firmly to the host cell by directly linking the bacterium to the actin cytoskeleton.

Recent structural and biochemical studies have provided further insights into the molecular details of the Tir-intimin interaction with respect to intimate attachment; and we review some of these aspects in the following sections.

STRUCTURAL CHARACTERISTICS OF INTIMIN, THE COGNATE LIGAND OF Tir

Intimin, the cognate ligand of Tir, is an outer membrane adhesin which is composed of two structurally distinct regions: an N-terminal region containing a putative peptidoglycan-binding domain (residues 1 to 155) connected to a membrane-anchoring domain with similarity to the porin β-barrels (residues 155 to 550), followed by a C-terminal surface-exposed extracellular region (residues 551 to 939) which mediates cell binding and contains the Tir-binding domain. The N-terminal region of intimin forms a ring-like structure, with an approximately 7-nm external diameter and an electron-dense core (Touze et al., 2004). This central β-barrel membrane-anchoring domain appears to direct the dimerization of intimin, which could explain the mechanism of Tir clustering by intimin during A/E lesion formation (Fig. 6). The C-terminal extracellular fragment of intimin has been more thoroughly characterized at the atomic level; this has been greatly aided by the availability of a high-resolution crystal structure and a nuclear magnetic resonance spectroscopy structure determined for the C-terminal ~280 residues of intimin (Batchelor et al., 2000; Luo et al., 2000).

The crystal structure of the EPEC intimin C-terminal fragment (residues 658 to 939) shows a rigid rod-shaped molecule with three adjacent domains aligned in tandem (Fig. 7a). The first two domains, designated D1 and D2, are each composed of a β-sheet sandwich with topology resembling that of an immunoglobulin (Ig) type I domain. A third Ig-like domain, D0, is thought to be located directly upstream of D1 and close to the bacterial membrane. The most C-terminal domain, designated D3, is formed by three helices that surround two antiparallel β-sheets, a fold that is analogous to that of C-type lectins such as the well-characterized mannose-binding protein (MBP) (Fig. 7b). Despite the high degree of overall structural similarity to the fold of C-type lectins, the intimin D3 domain does not contain the essential structural features pertaining to calcium and carbohydrate binding and thus is not likely to bind to host cell surface carbohydrate. The D3 domain, however, does contain the specific Tir-binding pocket (see below). The net overall electrostatic potential of intimin (residues 658 to 939) is positive and is localized primarily at the terminal D3 domain. The largely nonpolar surface with a positively charged tip may allow intimin to sit favorably on top of the membrane or anionic sialylated cellular mask of the host cell during infection.

Despite the low sequence identity (21%), the C-terminal fragment of intimin is structurally similar to the extracellular fragment of *Yersinia pseudotuberculosis* invasin, an important adhesin which binds β$_1$ integrins and mediates *Yersinia* adherence and invasion of host cells (Fig. 7a) (Hamburger et al., 1999). Invasin has four Ig-like domains (termed D1 through D4) and a C-type lectin-like domain (D5), which has been shown through mutagenesis studies to contain residues essential for binding integrin. The D2 and D3 domains of intimin form a superdomain with approximately 1,500 Å^2 of buried surface in a similar

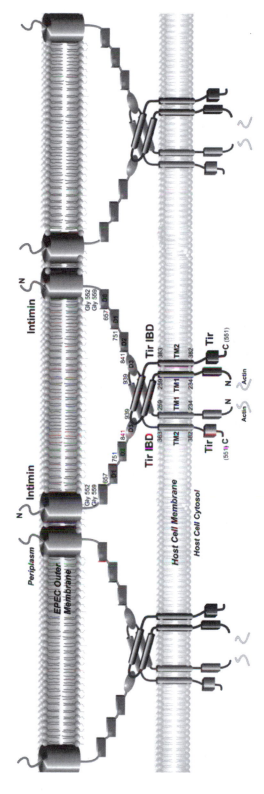

Figure 6. The EPEC-host cell adhesion interface based on the structural model of Tir IBD-intimin. Intimin contains an eight-residue linker with a glycine near each end that serves as a flexible hinge, mediating mechanical movement between the rigid extracellular domains (D0 to D3) and the bacterial outer membrane to which it is anchored. Dimeric Tir IBD, which forms a four-helix bundle, is stabilized by multiple hydrophobic and hydrogen-bonded interactions with intimin. Finally, EPEC is anchored to the host actin cytoskeleton via the N-terminal tails of Tir, which binds focal adhesion proteins.

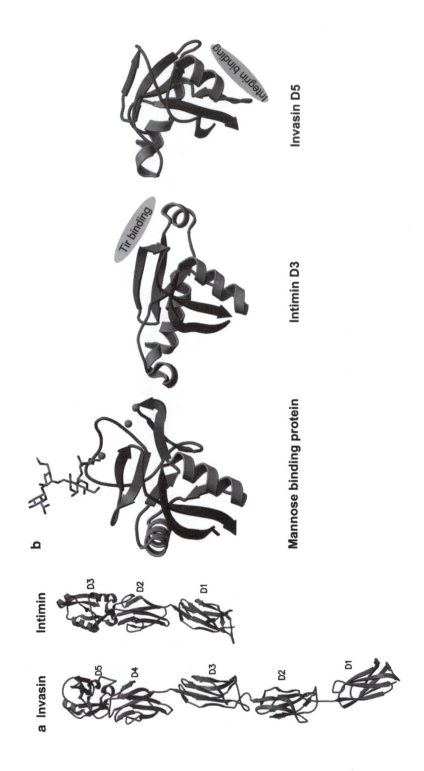

manner to the D4 and D5 domains of invasin. In addition, the intimin D1 domain super-poses well with each of the invasin D1, D2, and D3 domains, indicating similar tandemly organized Ig-like domains. Although the domain organization is very similar between in-timin and invasin, the anchoring of these extracellular receptor-binding fragments to the bacterial membrane may differ somewhat. The first Ig domain of invasin, D1, is linked to its last putative transmembrane β-strand by only a two-residue linker, whereas intimin has a longer (eight-residue) linker. This longer linker, which connects the putative first Ig do-main, D0, to the last residue in the transmembrane anchor, contains a glycine residue at each end (Gly552 and Gly559). The conformational variability of the glycine residues could introduce significant flexibility and may play an important role in mediating me-chanical movement between the rigid extracellular arm (domains D0 to D3) and the bacte-rial outer membrane to which it is anchored during optimization of intimate attachment to the host cell. The entire extracellular fragment of intimin is an elongated and relatively rigid rod, but it is approximately 40 Å shorter than invasin, which contains an additional Ig-like domain. Thus, these adhesins appear to use a similar structural motif that is uniquely adapted to their particular cell-receptor-binding scenario (by varying the numbers of tandem Ig-like domains to create a rod with an appropriate length), different levels of mechanical flexibility at the bacterial membrane interface, and a customized C-terminal lectin-like domain to interact specifically with their cognate receptors.

STRUCTURAL CHARACTERISTICS OF Tir-MEMBRANE TOPOLOGY AND RECEPTOR DIMERIZATION

Sequence analysis of Tir indicates that it possesses two transmembrane domains (TM1 [residues 234 to 259] and TM2 [residues 363 to 382]) and suggests that it would adopt a hairpin-like arrangement in the host cell membrane (Fig. 3). This hypothesis was con-firmed by topological analyses which demonstrated the cytoplasmic localization of the ter-mini and the presence of an extracellular intimin-binding domain spanning the putative transmembrane segments (de Grado et al., 1999; Hartland et al., 1999; Kenny, 1999). This topology was further substantiated by the observation that host cytoskeletal proteins are di-rectly recruited by the N and C termini of Tir (Goosney et al., 2000a). Thus, Tir appears to

Figure 7. Structural similarities between the extracellular fragment of *Yersinia* invasin and EPEC intimin. (a) De-spite low sequence identity (22%), the overall architecture between invasin and intimin is conserved, with 241 of 282 backbone Cα of the intact intimin structure matching invasin, with a 2.9-Å root mean squared deviation. Both molecules are shaped as rigid rods composed of tandemly repeated Ig domains (three for intimin and four for in-vasin) and terminate with a receptor-binding domain which adopts a fold similar to that of C-type lectins (D3 for intimin and D5 for invasin) and which directly mediates the specific and unique interactions with the cognate re-ceptor ligand (Tir for intimin and β₁-integrins for invasin). In both systems, the two most C-terminal domains are intimately associated, with ~1,500 Å² of buried surface between them. Ribbon representations were generated using data in PDB accession codes 1CWV (invasin) and 1F00 (intimin). (b) Comparison of the C-type lectin-like domains. The D3 domain of intimin and the D5 domain of invasin structurally resemble the C-type lectin family, a well-characterized example of which is MBP. However, unlike MBP, which is shown with bound calcium (spheres) and carbohydrate (ball and stick), intimin D3 and invasin D5 lack the necessary structural features re-quired for such interactions with calcium and carbohydrate. The Tir-binding face of intimin and the putative inte-grin-binding surface of invasin are shown in grey. The ribbon representation of the MBP-carbohydrate complex was generated using PDB accession code 2MSB.

be divided evenly into three distinct domains: an N-terminal domain responsible for linking EPEC to the cytoskeleton, an extracellular domain for binding its ligand intimin, and a C-terminal domain for mediating signal transduction (Fig. 8).

The available structural information on Tir centers around the extracellular intimin-binding domain (IBD) (residues 272 to 376) (Luo et al., 2000). The hairpin-shaped Tir IBD monomer is formed by two long α-helices, a three-residue 3_{10} helix, and a β-hairpin. The two long helices (designated HA and HB) mediate Tir homodimerization by forming an antiparallel four-helix bundle. The dimerization interface between the two Tir IBD molecules is significant (total area of 1,971 $\text{\AA}^2$) and relatively hydrophobic, with a rim of polar and charged side chains. These features match the characteristics of other native four-helix bundle proteins such as ColE1 Rop protein (Fig. 9), supporting the hypothesis that there is a physiological role for the observed Tir dimerization. The charged groups on the rim of

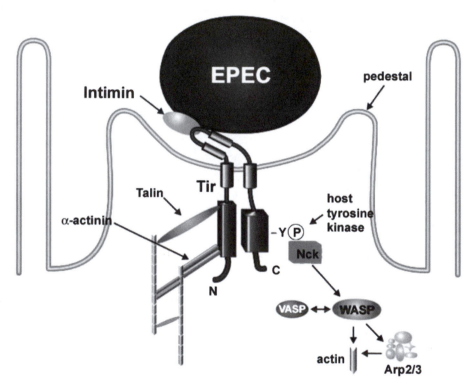

Figure 8. Tir is a multifunctional protein. The three different domains of Tir perform unique biological functions. Tir acts as a receptor of intimin, and the central extracellular intimin-binding domain mediates its interaction with intimin, resulting in intimate attachment. The N-terminal region of Tir recruits focal adhesion proteins, including talin and α-actinin, which anchors EPEC to the host actin cytoskeleton. The C-terminal region of Tir functions as a signaling platform. Phosphorylation of tyrosine 474 by an unknown host kinase results in recruitment of the host adaptor protein Nck and subsequent downstream activation of WASP and the Arp2/3 complex. These events ultimately lead to formation of actin-rich pedestals.

Tir IBD dimer   **ColE Rop**

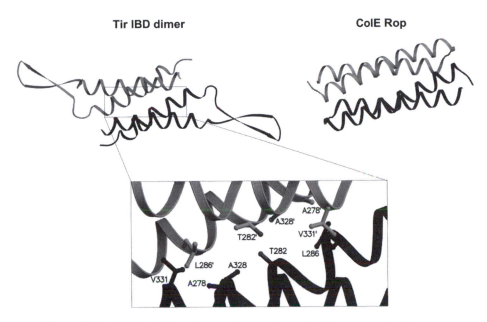

Figure 9. Dimerization of Tir and structural comparison with ColE1 Rop. Each Tir IBD monomer consists of two long α-helices that mediate dimerization by forming an antiparallel four-helix bundle similar to ColE1 Rop (the root mean squared deviation on 80 common backbone Cα atoms is 3.2 Å). As shown, the dimerization interface between the two Tir IBDs is relatively hydrophobic. Data used for this representation were taken from PDB accession codes 1ROP (ColE1 Rop) and 1F02 (Tir IBD).

the dimerization interface give the Tir IBD an overall negative electrostatic potential, complementary to that the positively charged intimin D3 domain. The first transmembrane domain (TM1) and one of the long helices of the Tir IBD are linked by a 12-residue segment, containing a significant number of prolines, and its sequence is highly conserved among the Tir from different A/E pathogens. Although this unusual sequence was not analyzed crystallographically, it is predicted that this proline-rich segment could serve as a conformationally rigid anchor of the dimeric Tir IBD to the transmembrane segment TM1, potentially stabilizing the IBD for interaction with intimin.

STRUCTURAL INSIGHTS INTO THE Tir-INTIMIN INTERACTION

The crystal structure of the EPEC Tir IBD in complex with the intimin C-terminal fragment reveals that binding of Tir and intimin is mediated primarily by the highly specific interaction between the β-hairpin and the N-terminal helix of the Tir IBD and the lectin-like D3 domain of intimin (Color Plate 19a [see color insert]). The specific intimin residues observed to be important in this interaction are consistent with data obtained from nuclear magnetic resonance spectroscopy titration and mutagenesis studies (Batchelor et al., 2000; Liu et al., 2002). The binding of the Tir IBD does not induce significant conformational changes of intimin as judged by superposing the structures of free intimin and the Tir IBD-bound form. The interaction between the Tir IBD and intimin is driven primarily by hydrophobic interactions, and this binding buries 1,335 Å^2 of flat and relatively uncharged surface, which accounts for the favorable enthalpy change and the large heat capacity change observed after formation of the complex in isothermal titration calorimetry (Luo et al., 2000; Luo et al., 2001). Beneath the β-hairpin of the Tir IBD, an unusual main chain conformation of intimin Gln920 enables the side chains of Lys919, Leu923, Asn925, and Ile926 to form a flat interface with the side chains of Arg850, Thr895, Ile897, and Asn932, in which intimin Ile926 and Tir Ile301 are buried (Color Plate 19b). The extensive interactions observed at the Tir-intimin interface indicate that the orientation of the β-hairpin of Tir IBD is fixed.

Charge complementarity between Tir IBD (overall net negative charge) and intimin (positively charged tip) also plays a role in this interaction (Color Plate 19c). The positively charged tip of intimin (formed by a short α-helix) is hydrogen bonded through the main chain carbonyl oxygens of Ser909 and Lys908 to the main chain amides of Asp315 and Asp316 within helix HA of the Tir IBD. In addition to providing additional stability, electrostatic interactions appear to play a role in determining binding specificity. Glu312 of the Tir IBD, which is in close proximity to Lys298, forms a direct salt bridge with Lys927 of intimin at the surface of the interface. In the Tir from the related pathogen EHEC, residue 298 is also a lysine but Glu312 is replaced by a nonpolar valine. The potential electrostatic repulsion between Lys298 of EHEC Tir and Lys927 of EPEC intimin in the absence of the complementary Glu312 explains the 20-fold reduced affinity of EHEC Tir toward EPEC intimin compared with its natural ligand, EHEC intimin, which has an asparagine at position 927. The structure of the EPEC Tir IBD-intimin complex also helps explain why a fortuitous PCR-generated substitution mutant of EHEC intimin V906A (equivalent to V911 of EPEC intimin) is seriously impaired in Tir binding. The side chain of EPEC intimin Val 911 is entirely buried in the binding interface, and mutation of this residue to an alanine would disrupt packing at the interface.

Despite the overall structural similarities between invasin and intimin, examination of the Tir IBD structural interface within the intimin D3 domain indicates that these two related adhesins have evolved distinctive sites for recognizing their respective receptors (Fig. 7b). The predicted integrin-binding motif of invasin and the structurally determined Tir-binding motif of intimin are localized on opposite sides of their respective lectin domains, indicating very specific and distinct binding interfaces. Other structural differences which may dictate binding specificity in the two adhesins include the lack in invasin of the short α-helix involved in Tir binding and, conversely, the lack in intimin of the charge constellation thought to be required for binding integrins.

MORE THAN A RECEPTOR: THE ROLE OF EPEC Tir
IN REORGANIZATION OF THE HOST CYTOSKELETON

In addition to serving as the intimate attachment receptor, Tir plays an important role in subverting the host actin signaling pathway. EPEC is among a number of bacterial pathogens known to alter the host actin cytoskeleton during infection; after type III translocation of effectors, the host cell exhibits localized destruction of microvilli and its underlying actin cytoskeleton is progressively turned into an actin-rich pedestal structure. EPEC resides at the tip of this pedestal, which can extend up to 10 μm in length beneath the pathogen (Rosenshine et al., 1996). After type III translocation of effectors including Tir, intimin binding and clustering of Tir is the final and only bacterial signal needed to trigger the remodeling of the host actin cytoskeleton into pedestals. Evidence supporting this hypothesis includes observations that normally nonpathogenic *E. coli* K-12 strains altered to express intimin or latex beads coated with the C-terminal region of intimin can induce pedestal formation on host cells preinfected with an intimin-deficient EPEC strain (which can translocate type III effectors including Tir into the host but cannot intimately attach) (Liu et al., 1999; Rosenshine et al., 1996).

While the N terminus of Tir functions primarily as an anchor by binding focal adhesion proteins, the initiation of actin reorganization is mediated by its C terminus, which contains a critical tyrosine residue at position 474. Although EPEC Tir is phosphorylated on two serine residues by host protein kinase A on translocation, phosphorylation of Tyr474 by an as yet unidentified host kinase is absolutely required for Tir-based actin signaling (Kenny, 1999; Warawa and Kenny, 2001). EPEC strains deficient in *tir*, when complemented with *tir* containing a Y474F substitution mutation, cannot form pedestals in spite of its ability to efficiently deliver and focus Tir. Earlier immunofluoresence studies have revealed that Tir recruits, in addition to actin, various actin-associated proteins to the pedestal; these proteins include α-actinin, calpactin, cofilin, cortactin, ezrin, gelosin, LIM-containing lipoma-preferred partner, fimbrin, villin, talin, myosin II, tropomyosin, zyxin, vasodilator-stimulated phosphoprotein (VASP), neural-Wiskott-Aldrich syndrome protein (N-WASP), and the actin-related protein 2 and 3 (Arp2/3) complex (Goosney et al., 2001). Some of these cytoskeletal proteins, including α-actinin and talin, are known to bind directly to Tir regardless of the phosphorylation state of Tir. However, the recruitment of a subset of these proteins such as VASP and N-WASP, absolutely requires Tir to be phosphorylated at Tyr474, but none of these proteins appears to bind directly to Tir. The missing link and the recruitment hierarchy were partially resolved when the host adaptor protein Nck was shown to bind directly to the phosphotyrosine motif in the C terminus of Tir (Gruenheid et al., 2001). Nck, an SH2/SH3 adaptor protein consisting of three Src homology 3 (SH3) domains and one Src homology 2 (SH2) domain, serves as an important link between extracellular signals and the cytoskeleton in mammalian cells (McCarty, 1998). The SH2 domain of Nck binds preferentially to the consensus sequence pYDE(P/D/V) (a sequence similar to that observed near phosphorylated tyrosine 474 of EPEC Tir [pYDEV]) of a number of receptor tyrosine kinases and tyrosine-phosphorylated docking proteins. On the other hand, its SH3 domains bind the WASP family of proteins such as N-WASP, which are key regulators of the actin cytoskeleton that promote actin filament nucleation and branching by activating the actin nucleation center of mammalian cells: the heptameric Arp2/3 complex. By mimicking a natural upstream activator of Nck of its host, Tir directly exploits the intracellular actin signaling cascade, specifically N-WASP and the Arp2/3 complex, in remodeling the host cytoskeleton (Campellone and Leong, 2003). Although the signaling downstream of Nck in pedestal formation has not been mapped in detail, it is currently thought that Nck, after its recruitment by tyrosine-phosphorylated Tir, either binds and directly stimulates N-WASP or indirectly recruits N-WASP through WIP (WASP interacting protein) or WIP-like proteins (Fig. 8). Recruitment of N-WASP by Nck leads to stimulation of Arp2/3-based actin nucleation and subsequent formation of actin filaments. It is important to note that while N-WASP could be directly activiated by the Rho family of small GTPases, which play key roles in actin remodeling in mammalian cells, neither Rho, Rac, nor Cdc42 plays a role in EPEC pedestal formation since they are absent in the pedestal, and agents that inhibit these proteins do not inhibit pedestal formation (Ben-Ami et al., 1998; Goosney et al., 2001). In addition to Nck, Tir recruits a number of adaptor proteins, including CrkII, Grb2, and Shc, to the pedestal in a phosphotyrosine-dependent manner, but whether they directly interact with Tir is not known, and further work is required to establish their roles in pedestal formation.

Exploitation of the N-WASP–Arp2/3 pathway by EPEC Tir is remarkably similar to the strategy employed by intracellular microbial pathogens such as *Listeria monocytogenes, Rickettsia conorii,* and *Shigella flexneri* in promoting actin-based motility within host cell (Egile et al., 1999; Gouin et al., 2004; Welch et al., 1997). The *Listeria* ActA protein and the *Rickettsia* RickA protein, which are located in the periphery of their respective bacteria, can bind to and directly activate the Arp2/3 complex to promote actin tail formation. The *Shigella* surface protein IcsA, instead of targeting the Arp2/3 complex, activates N-WASP directly to promote actin nucleation and polymerization.

DIVERGENCE OF Tir-BASED ACTIN SIGNALING

EHEC serotype O157:H7 induces pedestals that are morphologically similar to EPEC pedestals (Goosney et al., 2001). The highly similar structure and composition of their pedestals, together with the high level of sequence conservation observed between the LEE of EHEC and EPEC, led early investigators to conclude that these two closely related pathogens employ the same molecular mechanisms for eliciting the A/E lesion (Perna et al., 1998). It was soon discovered that quite surprisingly fundamental differences exist in the mechanisms of pedestal formation by EHEC and EPEC, specifically in Tir-based signaling (DeVinney et al., 2001).

EHEC Tir cannot complement an EPEC Tir-deficient strain to restore the A/E phoenotype, whereas EPEC Tir can complement an EHEC Tir-deficient strain. In other words, EPEC Tir is functionally interchangeable for that of EHEC O157:H7 but the reverse is not true (Kenny, 2001). The C-terminal regions of EHEC and EPEC Tir, responsible for interacting with the host signaling machinery, are only 40% identical, compared to the overall 60% identity between the two molecules. Most importantly, EHEC Tir lacks the critical Tyr474 (replaced by a serine) that is conserved among Tir proteins of EPEC and other A/E pathogens. As expected, tyrosine phosphorylation of Tir is not required for EHEC pedestal formation. The absence of Tyr474 also suggests that Nck is not involved in EHEC Tir-based actin signaling. Indeed, Nck and other host adaptor molecules, Grb2 and CrkII, present in EPEC pedestals are not found in EHEC pedestals, and EHEC still forms pedestals in a Nck1-Nck2 double-mutant cell line (Goosney et al., 2001; Gruenheid et al., 2001). Since the EHEC LEE does not contain all the factors for forming pedestals (cloned EHEC LEE does not confer the A/E phenotype on laboratory strains of *E. coli*) (Elliott et al., 1999a), the current hypothesis is that an as yet identified type III effector protein outside of the EHEC LEE pathogenicity island substitutes for Nck to recruit N-WASP in the pedestal formation process (Campellone and Leong, 2003).

The uniqueness of EHEC Tir does, however, provide a good system for further investigation of the details of the EPEC Tir signaling mechanism. Through generation and examination of numerous chimeric EHEC and EPEC Tir proteins, it was discovered that a 12-residue sequence of EPEC Tir containing Tyr474 could confer actin-signaling capabilities on EHEC Tir when this particular chimera is expressed in EPEC, indicating that Nck probably only needs to recognize a minimal 12-residue peptide in the C terminus of Tir to be activated (Campellone et al., 2002).

ARE THERE ANY OTHER HOST RECEPTORS OF BACTERIAL ORIGIN?

Since the discovery of EPEC Tir, bacterially derived receptors have been identified only in the A/E family of pathogens, and all of them are Tir orthologues. It is still puzzling why similar types of receptors have not been identified in other bacterial pathogens. We speculate that a number of reasons contribute to the difficulty in identifying other receptors of bacterial origin.

First, the cognate adhesin may bind only transiently to its receptor and thus may not exhibit the high-affinity interaction observed with Tir-intimin. In this situation, experimental protocols such as stringent washing could prevent the observation of these lower-affinity receptor-adhesin interactions. For example, the EspA filament, thought to be the molecular conduit for type III translocation, could not be observed until a gentler washing procedure was used (Knutton et al., 1998). In addition to its essential role in type III protein translocation, the EspA filament has been implicated to play a role in the initial adherence prior to intimate attachment, but its receptor has yet to be identified (Cleary et al., 2004). Second, the need for the receptor to be delivered to the host means that only a subset of pathogens possessing specialized secretion systems could have the potential to evolve a mechanism for a translocated receptor. The type III and type IV secretion systems are the only characterized systems shown to be capable of delivering bacterial proteins directly into the host. Unfortunately, none of the type III effectors characterized so far displayed a similar receptor function to Tir, and very few type IV effectors have been identified to date. Finally, there may not be a need for most pathogens to develop such a sophisticated and multifaceted receptor molecule in the first place. EPEC and related A/E pathogens persist extracellularly in a turbulent environment that conceivably forces them to develop effective means of attaching intimately to their host. While adherence enables most pathogens to get a first grip onto the host, many pathogens invade host cells and reside intracellularly during infection. In other cases where adherence is essential, a particular pathogen may already possess an adhesin that binds effectively and with high affinity to existing host receptors, eliminating the need to develop its own receptor and masking the identification of lower-affinity binding receptor-adhesin pairs.

Recent experimental evidence, however, indicates that *Helicobacter pylori,* the causative agent of chronic gastritis, gastroduodenal ulcer disease, and gastric carcinoma, may possess a novel type of bacterial receptor. *H. pylori,* which resides extracellularly on human gastric epithelial cells for extended periods, possesses a functional type IV secretion system that injects a key effector molecule known as CagA (cytotoxin-associated gene A) into its host (Odenbreit et al., 2000; Stein et al., 2000). Once translocated into the host, CagA localizes to the host membrane and is tyrosine phosphorylated by the Src family of kinases (Stein et al., 2002). Phosphorylation of CagA leads to reorganization of the underlying cytoskeleton into structures which are reminiscent of EPEC pedestals (Segal et al., 1996). While the molecular mechanisms of CagA in remodeling the host cytoskeleton as well as inducing the host inflammatory response have been extensively studied, it remains to be confirmed whether CagA, which has been observed to localize to the regions of bacterial attachment, actually mediates adherence and binds an *H. pylori* surface adhesin.

Acknowledgments. We thank Wanyin Deng, Elizabeth Frey, and Winco Wu for critical reading of the manuscript.

We thank the Canadian Institute of Health Research and the Howard Hughes Medical Institute International Scholar Program for funding. C.Y. acknowledges NSERC and the Michael Smith Foundation for Health Research for fellowships.

REFERENCES

Abe, A., M. de Grado, R. A. Pfuetzner, C. Sanchez-Sanmartin, R. Devinney, J. L. Puente, N. C. Strynadka, and B. B. Finlay. 1999. Enteropathogenic *Escherichia coli* translocated intimin receptor, Tir, requires a specific chaperone for stable secretion. *Mol. Microbiol.* **33:**1162–1175.

Batchelor, M., S. Prasannan, S. Daniell, S. Reece, I. Connerton, G. Bloomberg, G. Dougan, G. Frankel, and S. Matthews. 2000. Structural basis for recognition of the translocated intimin receptor (Tir) by intimin from enteropathogenic *Escherichia coli*. *Embo J.* **19:**2452–2464.

Ben-Ami, G., V. Ozeri, E. Hanski, F. Hofmann, K. Aktories, K. M. Hahn, G. M. Bokoch, and I. Rosenshine. 1998. Agents that inhibit Rho, Rac, and Cdc42 do not block formation of actin pedestals in HeLa cells infected with enteropathogenic *Escherichia coli*. *Infect. Immun.* **66:**1755–1758.

Blocker, A., K. Komoriya, and S. Aizawa. 2003. Type III secretion systems and bacterial flagella: insights into their function from structural similarities. *Proc. Natl. Acad. Sci. USA* **100:**3027–3030.

Campellone, K. G., and J. M. Leong. 2003. Tails of two Tirs: actin pedestal formation by enteropathogenic *E. coli* and enterohemorrhagic *E. coli* O157:H7. *Curr. Opin. Microbiol.* **6:**82–90.

Campellone, K. G., A. Giese, D. J. Tipper, and J. M. Leong. 2002. A tyrosine-phosphorylated 12–amino-acid sequence of enteropathogenic *Escherichia coli* Tir binds the host adaptor protein Nck and is required for Nck localization to actin pedestals. *Mol. Microbiol.* **43:**1227–1241.

Clarke, S. C., R. D. Haigh, P. P. Freestone, and P. H. Williams. 2003. Virulence of enteropathogenic *Escherichia coli*, a global pathogen. *Clin. Microbiol. Rev.* **16:**365–378.

Cleary, J., L. C. Lai, R. K. Shaw, A. Straatman-Iwanowska, M. S. Donnenberg, G. Frankel, and S. Knutton. 2004. Enteropathogenic *Escherichia coli* (EPEC) adhesion to intestinal epithelial cells: role of bundle-forming pili (BFP), EspA filaments and intimin. *Microbiology* **150:**527–538.

Daniell, S. J., E. Kocsis, E. Morris, S. Knutton, F. P. Booy, and G. Frankel. 2003. 3D structure of EspA filaments from enteropathogenic *Escherichia coli*. *Mol. Microbiol.* **49:**301–308.

de Grado, M., A. Abe, A. Gauthier, O. Steele-Mortimer, R. DeVinney, and B. B. Finlay. 1999. Identification of the intimin-binding domain of Tir of enteropathogenic *Escherichia coli*. *Cell. Microbiol.* **1:**7–17.

Deng, W., J. L. Puente, S. Gruenheid, Y. Li, B. A. Vallance, A. Vazquez, J. Barba, J. A. Ibarra, P. O'Donnell, P. Metalnikov, K. Ashman, S. Lee, D. Goode, T. Pawson, and B. B. Finlay. 2004. Dissecting virulence: systematic and functional analyses of a pathogenicity island. *Proc. Natl. Acad. Sci. USA* **101:**3597–3602.

DeVinney, R., J. L. Puente, A. Gauthier, D. Goosney, and B. B. Finlay. 2001. Enterohaemorrhagic and enteropathogenic *Escherichia coli* use a different Tir-based mechanism for pedestal formation. *Mol. Microbiol.* **41:**1445–1458.

Donnenberg, M. S., J. B. Kaper, and B. B. Finlay. 1997. Interactions between enteropathogenic *Escherichia coli* and host epithelial cells. *Trends Microbiol.* **5:**109–114.

Egile, C., T. P. Loisel, V. Laurent, R. Li, D. Pantaloni, P. J. Sansonetti, and M. F. Carlier. 1999. Activation of the CDC42 effector N-WASP by the *Shigella flexneri* IcsA protein promotes actin nucleation by Arp2/3 complex and bacterial actin-based motility. *J. Cell Biol.* **146:**1319–1332.

Elliott, S. J., J. Yu, and J. B. Kaper. 1999a. The cloned locus of enterocyte effacement from enterohemorrhagic *Escherichia coli* O157:H7 is unable to confer the attaching and effacing phenotype upon *E. coli* K-12. *Infect. Immun.* **67:**4260–4263.

Elliott, S. J., L. A. Wainwright, T. K. McDaniel, K. G. Jarvis, Y. K. Deng, L. C. Lai, B. P. McNamara, M. S. Donnenberg, and J. B. Kaper. 1998. The complete sequence of the locus of enterocyte effacement (LEE) from enteropathogenic *Escherichia coli* E2348/69. *Mol. Microbiol.* **28:**1–4.

Elliott, S. J., S. W. Hutcheson, M. S. Dubois, J. L. Mellies, L. A. Wainwright, M. Batchelor, G. Frankel, S. Knutton, and J. B. Kaper. 1999b. Identification of CesT, a chaperone for the type III secretion of Tir in enteropathogenic *Escherichia coli*. *Mol. Microbiol.* **33:**1176–1189.

Elliott, S. J., V. Sperandio, J. A. Giron, S. Shin, J. L. Mellies, L. Wainwright, S. W. Hutcheson, T. K. McDaniel, and J. B. Kaper. 2000. The locus of enterocyte effacement (LEE)-encoded regulator controls expression of both LEE- and non-LEE-encoded virulence factors in enteropathogenic and enterohemorrhagic *Escherichia coli. Infect. Immun.* **68:**6115–6126.

Finlay, B. B., and S. Falkow. 1997. Common themes in microbial pathogenicity revisited. *Microbiol. Mol. Biol. Rev.* **61:**136–169.

Freeman, N. L., D. V. Zurawski, P. Chowrashi, J. C. Ayoob, L. Huang, B. Mittal, J. M. Sanger, and J. W. Sanger. 2000. Interaction of the enteropathogenic *Escherichia coli* protein, translocated intimin receptor (Tir), with focal adhesion proteins. *Cell Motil. Cytoskeleton* **47:**307–318.

Galan, J. E., and A. Collmer. 1999. Type III secretion machines: bacterial devices for protein delivery into host cells. *Science* **284:**1322–1328.

Gauthier, A., M. de Grado, and B. B. Finlay. 2000. Mechanical fractionation reveals structural requirements for enteropathogenic *Escherichia coli* Tir insertion into host membranes. *Infect. Immun.* **68:**4344–4348.

Gauthier, A., J. L. Puente, and B. B. Finlay. 2003. Secretin of the enteropathogenic *Escherichia coli* type III secretion system requires components of the type III apparatus for assembly and localization. *Infect. Immun.* **71:**3310–3319.

Goosney, D. L., S. Gruenheid, and B. B. Finlay. 2000a. Gut feelings: enteropathogenic *E. coli* (EPEC) interactions with the host. *Annu. Rev. Cell Dev. Biol.* **16:**173–189.

Goosney, D. L., R. DeVinney, and B. B. Finlay. 2001. Recruitment of cytoskeletal and signaling proteins to enteropathogenic and enterohemorrhagic *Escherichia coli* pedestals. *Infect. Immun.* **69:**3315–3322.

Goosney, D. L., R. DeVinney, R. A. Pfuetzner, E. A. Frey, N. C. Strynadka, and B. B. Finlay. 2000b. Enteropathogenic *E. coli* translocated intimin receptor, Tir, interacts directly with alpha-actinin. *Curr. Biol.* **10:**735–738.

Gouin, E., C. Egile, P. Dehoux, V. Villiers, J. Adams, F. Gertler, R. Li, and P. Cossart. 2004. The RickA protein of *Rickettsia conorii* activates the Arp2/3 complex. *Nature* **427:**457–461.

Gruenheid, S., R. DeVinney, F. Bladt, D. Goosney, S. Gelkop, G. D. Gish, T. Pawson, and B. B. Finlay. 2001. Enteropathogenic *E. coli* Tir binds Nck to initiate actin pedestal formation in host cells. *Nat. Cell Biol.* **3:**856–859.

Hamburger, Z. A., M. S. Brown, R. R. Isberg, and P. J. Bjorkman. 1999. Crystal structure of invasin: a bacterial integrin-binding protein. *Science* **286:**291–295.

Hartland, E. L., M. Batchelor, R. M. Delahay, C. Hale, S. Matthews, G. Dougan, S. Knutton, I. Connerton, and G. Frankel. 1999. Binding of intimin from enteropathogenic *Escherichia coli* to Tir and to host cells. *Mol. Microbiol.* **32:**151–158.

Hueck, C. J. 1998. Type III protein secretion systems in bacterial pathogens of animals and plants. *Microbiol. Mol. Biol. Rev.* **62:**379–433.

Ide, T., S. Laarmann, L. Greune, H. Schillers, H. Oberleithner, and M. A. Schmidt. 2001. Characterization of translocation pores inserted into plasma membranes by type III-secreted Esp proteins of enteropathogenic *Escherichia coli. Cell. Microbiol.* **3:**669–679.

Kenny, B. 1999. Phosphorylation of tyrosine 474 of the enteropathogenic *Escherichia coli* (EPEC) Tir receptor molecule is essential for actin nucleating activity and is preceded by additional host modifications. *Mol. Microbiol.* **31:**1229–1241.

Kenny, B. 2001. The enterohaemorrhagic *Escherichia coli* (serotype O157:H7) Tir molecule is not functionally interchangeable for its enteropathogenic *E. coli* (serotype O127:H6) homologue. *Cell. Microbiol.* **3:**499–510.

Kenny, B. 2002. Enteropathogenic *Escherichia coli* (EPEC)—a crafty subversive little bug. *Microbiology* **148:**1967–1978.

Kenny, B., and B. B. Finlay. 1995. Protein secretion by enteropathogenic *Escherichia coli* is essential for transducing signals to epithelial cells. *Proc. Natl. Acad. Sci. USA* **92:**7991–7995.

Kenny, B., R. DeVinney, M. Stein, D. J. Reinscheid, E. A. Frey, and B. B. Finlay. 1997. Enteropathogenic *E. coli* (EPEC) transfers its receptor for intimate adherence into mammalian cells. *Cell* **91:**511–520.

Knutton, S., I. Rosenshine, M. J. Pallen, I. Nisan, B. C. Neves, C. Bain, C. Wolff, G. Dougan, and G. Frankel. 1998. A novel EspA-associated surface organelle of enteropathogenic *Escherichia coli* involved in protein translocation into epithelial cells. *Embo J.* **17:**2166–2176.

Kraulis, P. J. 1991. MOLSCRIPT: a program to produce both detailed and schematic plots of protein structures. *J. App. Crystallogr.* **24:**946–950.

Liu, H., L. Magoun, S. Luperchio, D. B. Schauer, and J. M. Leong. 1999. The Tir-binding region of enterohaemorrhagic *Escherichia coli* intimin is sufficient to trigger actin condensation after bacterial-induced host cell signalling. *Mol. Microbiol.* **34:**67–81.

Liu, H., P. Radhakrishnan, L. Magoun, M. Prabu, K. G. Campellone, P. Savage, F. He, C. A. Schiffer, and J. M. Leong. 2002. Point mutants of EHEC intimin that diminish Tir recognition and actin pedestal formation highlight a putative Tir binding pocket. *Mol. Microbiol.* **45**:1557–1573.

Luo, Y., E. A. Frey, R. A. Pfuetzner, A. L. Creagh, D. G. Knoechel, C. A. Haynes, B. B. Finlay, and N. C. Strynadka. 2000. Crystal structure of enteropathogenic *Escherichia coli* intimin-receptor complex. *Nature* **405**:1073–1077.

Luo, Y., M. G. Bertero, E. A. Frey, R. A. Pfuetzner, M. R. Wenk, L. Creagh, S. L. Marcus, D. Lim, F. Sicheri, C. Kay, C. Haynes, B. B. Finlay, and N. C. Strynadka. 2001. Structural and biochemical characterization of the type III secretion chaperones CesT and SigE. *Nat. Struct. Biol.* **8**:1031–1036.

McCarty, J. H. 1998. The Nck SH2/SH3 adaptor protein: a regulator of multiple intracellular signal transduction events. *Bioessays* **20**:913–921.

McDaniel, T. K., and J. B. Kaper. 1997. A cloned pathogenicity island from enteropathogenic *Escherichia coli* confers the attaching and effacing phenotype on *E. coli* K-12. *Mol. Microbiol.* **23**:399–407.

Merritt, E. A., and D. J. Bacon. 1997. Raster3D: photorealistic molecular graphics. *Methods Enzymol.* **277**: 505–524.

Moon, H. W., S. C. Whipp, R. A. Argenzio, M. M. Levine, and R. A. Giannella. 1983. Attaching and effacing activities of rabbit and human enteropathogenic *Escherichia coli* in pig and rabbit intestines. *Infect. Immun.* **41**:1340–1351.

Nicholls, A., K. Sharp, and B. Honig. 1991. Protein folding and association: insights from the interfacial and thermodynamic properties of hydrocarbons. *Proteins Struct., Funct. Gene.* **11**:281–296.

Odenbreit, S., J. Puls, B. Sedlmaier, E. Gerland, W. Fischer, and R. Haas. 2000. Translocation of *Helicobacter pylori* CagA into gastric epithelial cells by type IV secretion. *Science* **287**:1497–1500.

Parsot, C., C. Hamiaux, and A. L. Page. 2003. The various and varying roles of specific chaperones in type III secretion systems. *Curr. Opin. Microbiol.* **6**:7–14.

Perna, N. T., G. F. Mayhew, G. Posfai, S. Elliott, M. S. Donnenberg, J. B. Kaper, and F. R. Blattner. 1998. Molecular evolution of a pathogenicity island from enterohemorrhagic *Escherichia coli* O157:H7. *Infect. Immun.* **66**:3810–3817.

Rosenshine, I., S. Ruschkowski, M. Stein, D. J. Reinscheid, S. D. Mills, and B. B. Finlay. 1996. A pathogenic bacterium triggers epithelial signals to form a functional bacterial receptor that mediates actin pseudopod formation. *Embo J.* **15**:2613–2624.

Rosqvist, R., S. Hakansson, A. Forsberg, and H. Wolf-Watz. 1995. Functional conservation of the secretion and translocation machinery for virulence proteins of yersiniae, salmonellae and shigellae. *Embo J.* **14**:4187–4195.

Segal, E. D., S. Falkow, and L. S. Tompkins. 1996. *Helicobacter pylori* attachment to gastric cells induces cytoskeletal rearrangements and tyrosine phosphorylation of host cell proteins. *Proc. Natl. Acad. Sci. USA* **93**: 1259–1264.

Stebbins, C. E., and J. E. Galan. 2001. Maintenance of an unfolded polypeptide by a cognate chaperone in bacterial type III secretion. *Nature* **414**:77–81.

Stein, M., R. Rappuoli, and A. Covacci. 2000. Tyrosine phosphorylation of the *Helicobacter pylori* CagA antigen after cag-driven host cell translocation. *Proc. Natl. Acad. Sci. USA* **97**:1263–1268.

Stein, M., F. Bagnoli, R. Halenbeck, R. Rappuoli, W. J. Fantl, and A. Covacci. 2002. c-Src/Lyn kinases activate *Helicobacter pylori* CagA through tyrosine phosphorylation of the EPIYA motifs. *Mol. Microbiol.* **43**: 971–980.

Touze, T., R. D. Hayward, J. Eswaran, J. M. Leong, and V. Koronakis. 2004. Self-association of EPEC intimin mediated by the beta-barrel-containing anchor domain: a role in clustering of the Tir receptor. *Mol. Microbiol.* **51**:73–87.

Vallance, B. A., and B. B. Finlay. 2000. Exploitation of host cells by enteropathogenic *Escherichia coli*. *Proc. Natl. Acad. Sci. USA* **97**:8799–8806.

Warawa, J., and B. Kenny. 2001. Phosphoserine modification of the enteropathogenic *Escherichia coli* Tir molecule is required to trigger conformational changes in Tir and efficient pedestal elongation. *Mol. Microbiol.* **42**:1269–1280.

Welch, M. D., A. Iwamatsu, and T. J. Mitchison. 1997. Actin polymerization is induced by Arp2/3 protein complex at the surface of *Listeria monocytogenes*. *Nature* **385**:265–269.

Structural Biology of Bacterial Pathogenesis
Edited by G. Waksman et al.
© 2005 ASM Press, Washington, D.C.

Chapter 5

The Chaperone-Usher Pathway
of Pilus Fiber Biogenesis

Frederic G. Sauer, Scott J. Hultgren, and Gabriel Waksman

Many species of gram-negative bacteria employ a conserved protein secretion system termed the chaperone-usher pathway to assemble a diverse array of multisubunit protein fibers on their surfaces (Hung et al., 1996; Thanassi et al., 1998a; Vallet et al., 2001). These fibers range in morphology from relatively thick pilus rods that radiate outward from the bacteria to more flexible pili to very thin fibrillae. Fibers assembled by the chaperone-usher pathway play critical roles in bacterially mediated disease: they mediate bacterial attachment to host tissues, often an essential early step in pathogenesis; they facilitate the evasion of host defenses; and they promote biofilm formation, a contributing factor both to the establishment of infection and to bacterial resistance to antibiotic treatment.

Fibers assembled by the chaperone-usher pathway are typically encoded in individual gene clusters. Each gene cluster encodes one or more fiber subunit proteins, also known as pilins. Certain pilins interact specifically with receptor molecules and are termed adhesins. Each gene cluster also encodes a periplasmic chaperone and an outer membrane usher, which together orchestrate the assembly of the fiber. Two well-studied pilus fibers expressed by uropathogenic *Escherichia coli*—P pili and type 1 pili—exemplify the assembly mechanisms of the chaperone-usher pathway. P pili are encoded by the (for *pap* "pilus associated with pyelonephritis") gene cluster (*papA–papK*) (Hull et al., 1981). They bind specifically to the globoseries of glycolipids present in human kidneys and are required for the establishment of pyelonephritis, or kidney infection (Lund et al., 1987; Roberts et al., 1994). Each P pilus consists of a relatively rigid rod, up to 2 μm in length, with a thinner, more flexible tip fibrillum, $\sim$42 $\pm$ 16 nm in length, at its distal end (Kuehn et al., 1992) (Color Plate 20 [see color insert]). The rod contains PapA subunits wound in a tight, one-start, right-handed helix with a diameter of 6.8 nm, a pitch of 2.5 nm, and 3.3 subunits per turn (Bullitt and Makowski, 1995; Gong and Makowski, 1992). The interior of the rod

Frederic G. Sauer • Section of Microbial Pathogenesis, Boyer Center for Molecular Medicine, Yale University School of Medicine, 295 Congress Ave., New Haven, CT 06536. *Scott J. Hultgren* • Department of Molecular Microbiology, Washington University School of Medicine, St. Louis, MO 63110. *Gabriel Waksman* • Institute of Structural Molecular Biology, Birkbeck and University College London, Malet St., London WC1 7HX, United Kingdom.

contains a channel, roughly 2.5 by 1.5 nm and with its center displaced ~0.5 nm from the helical axis, that winds through the center of the pilus. This central cavity is connected to the exterior milieu by a set of radial channels. The tip fibrillum of the P pilus contains PapE subunits arranged in an open helical conformation (Kuehn et al., 1992). The PapG adhesin, which binds the digalactoside-containing sugars of the globoseries of glycolipids, is at the distal end of the tip fibrillum (Kuehn et al., 1992; Lund et al., 1987). The PapF and PapK subunits link the adhesin to the tip fibrillum and the tip fibrillum to the rod, respectively (Jacob-Dubuisson et al., 1993). The PapD chaperone and the PapC usher constitute the assembly machinery of the P pilus (Dodson et al., 1993; Lindberg et al., 1989).

Type 1 pili are encoded by the *fim* gene cluster (*fimA* to *fimI*) and bind to mannose residues present in the human bladder (Hull et al., 1981). They are important in the establishment of cystitis (bladder infection) and have also been implicated in attachment to abiotic surfaces in a model biofilm system (Connell et al., 1996; Langermann et al., 1997; Mulvey et al., 1998; Pratt and Kolter, 1998). Each type 1 pilus consists of a rod with a short, stubby tip fibrillum at its distal end (Hahn et al., 2002; Jones et al., 1995). The rod contains FimA subunits, and the tip fibrillum contains FimF and FimG subunits, with the mannose-specific FimH adhesin at its very end (Jones et al., 1995; Saulino et al., 2000). The FimC chaperone and the FimD usher constitute the assembly machinery of the type 1 pilus (Jones et al., 1993; Saulino et al., 1998).

Two other well-studied surface organelles, the capsular F1 antigen of *Yersinia pestis* and the hemagglutinating pilus of the human respiratory pathogen *Haemophilus influenzae,* highlight the structural diversity of fibers assembled by the chaperone-usher pathway. The capsular F1 antigen (Caf1) of *Y. pestis,* the causative agent of bubonic plague, is encoded by the *caf1* genes and expressed at 37°C. It enhances bacterial resistance to uptake by phagocytic cells and consists of thin fibrils of polymerized Caf1 subunits assembled by the Caf1M chaperone and the Caf1A usher (Du et al., 2002; Galyov et al., 1991; MacIntyre et al., 2001; Zavialov et al., 2002). Hemagglutinating pili of *H. influenzae* are encoded by the *hif* gene cluster (*hifA* to *hifE*). These pili promote attachment to human buccal epithelial cells and tracheobronchial mucin and facilitate colonization of the respiratory tract (Sable et al., 1985; Weber et al., 1991). Each hemagglutinating pilus is a composite structure, with a flexible rod that contains HifA subunits and a short, thin distal tip fibrillum that contains the HifD and HifE subunits (McCrea et al., 1994; St. Geme et al., 1996). HifB and HifC are the chaperone and usher, respectively. In three-dimensional reconstructions of electron micrograph images, the pilus rod appears as a three-start, left-handed helix, 7.0 nm in diameter, with three subunits per turn and a central channel, ~2.0 nm in diameter, that runs straight along its axis (Mu et al., 2002).

OVERVIEW OF PILUS FIBER ASSEMBLY

Pilus fiber assembly is a controlled protein subunit polymerization reaction that proceeds by the addition of new subunits to the base of the growing structure. During assembly, subunits enter the bacterial periplasm via the Sec translocation machinery (Manting and Driessen, 2000) (Color Plate 20). Subsequent assembly steps occur in the absence of ATP or electrochemical gradients (Jacob-Dubuisson et al., 1994). In the periplasm, each subunit interacts with a chaperone to form a stable protein-protein complex. Interaction with the chaperone facilitates the folding of the subunit; stabilizes the subunit, preventing its aggregation and/or degradation; caps subunit interactive surfaces, inhibiting premature fiber formation; and maintains the subunit in an activated state, primed for assembly (Barnhart et al., 2000; Bullitt et al., 1996; Jones et al., 1997; Kuehn et al., 1991, 1993; Sauer et al., 2002; Soto et al., 1998). The chaperone is required for assembly; in its absence, subunits misfold and/or aggregate and are proteolytically degraded (Hultgren et al., 1989; Slonim et al., 1992). Thus, in vivo, subunits are not found as soluble monomers but only in a complex with a chaperone or as parts of the mature fiber. Chaperone-subunit complexes in the periplasm are targeted to the outer membrane usher (Dodson et al., 1993). In P and type 1 pili, the chaperone-adhesin complex binds most rapidly and tightly to the usher, forming a ternary complex that initiates the assembly of the fiber and places the adhesin at its tip (Saulino et al., 1998). Additional subunits are then incorporated sequentially into the base of the fiber in a defined order. The usher is thought to facilitate chaperone uncapping to expose the interactive surfaces on the subunits that drive their polymerization. The usher forms a ring with a pore 2 to 3 nm in diameter, wide enough to allow passage to the extracellular milieu of a tip fibrillum and an unwound pilus rod but not of a rod in its final helical structure. Thus, it has been proposed that the pilus winds into its final quaternary structure once it has traversed the usher pore and is outside the cell (Saulino et al., 2000; Thanassi et al., 1998b).

THE CHAPERONE AND DONOR STRAND COMPLEMENTATION

The periplasmic chaperones of the chaperone-usher pathway share conserved structural features first revealed in the crystal structure of PapD (Holmgren and Branden, 1989; Knight et al., 2002; Pellecchia et al., 1998) (Color Plate 21 [see color insert]). Each chaperone consists of two domains that are oriented at an approximate right angle to each other to produce an L-shaped molecule. Each domain has an immunoglobulin-like (Ig) fold with seven primary β-strands (strands A to G). The Ig fold consists of two β-sheets that face each other and pack together such that their seven β-strands form the staves of a β-barrel. Alternating hydrophobic residues in each strand face the interior of the β-barrel and constitute the hydrophobic core of the fold. The first β-sheet includes the B, E, and D strands; the second includes the C, F, and G strands. The A strand is divided into the A1 and A2 segments; the A1 segment interacts with the B strand, but then the A strand switches sheets such that the A2 segment interacts with the G strand. An extended loop connects the F and G strands of the N-terminal domain of the chaperone (the F_1 and G_1 strands, hence the F_1-G_1 loop). This F_1-G_1 loop lies at the end of one arm of the L and contains a conserved motif of solvent-exposed alternating hydrophobic residues that are involved in chaperone-subunit complex formation (Hung et al., 1999). Chaperones can be classified based on the lengths of their F_1-G_1 loops; those with shorter loops (termed FGS for "F_1-G_1 short", chaperones, including PapD and FimC) are typically associated with rod-like pilus fibers, while those with longer loops (termed FGL for "F_1-G_1 long", chaperones, including Caf1M of *Y. pestis*) are typically associated with fibers with a thinner, more flexible morphology (Hung et al., 1996). The chaperone G_1 strand runs from the F_1-G_1 loop toward the cleft between the two domains of the chaperone, at the angle of the L. In the cleft lie two conserved basic residues, a lysine in the G_1 strand and an arginine in the adjacent A_1 strand, that are essential for chaperone-subunit complex formation (Kuehn et al., 1993; Slonim et al., 1992) (Color Plate 21).

Subunits assembled by the chaperone-usher pathway do not fold efficiently in the absence of the chaperone (Barnhart et al., 2000; Vetsch et al., 2002). Subunit features first revealed in the crystal structures of the FimC-FimH and PapD-PapK chaperone-subunit complexes shed light on these observations (Choudhury et al., 1999; Sauer et al., 1999) (Color Plate 22 [see color insert]). Most subunits, such as PapK, consist of a single so-called pilin domain that interacts with the chaperone. Adhesin subunits, such as FimH, generally have both a pilin domain and an N-terminal receptor-binding domain that interacts with a specific receptor. As revealed in the chaperone-subunit complexes, each subunit pilin domain has an incomplete Ig fold. The pilin domain possesses strands A to F of a canonical Ig fold but lacks a C-terminal G strand. The missing strand leaves the subunit β-barrel and hydrophobic core incomplete and results in a groove that runs the length of the subunit. In the chaperone-subunit complex, in an interaction termed donor strand complementation, the chaperone contributes a strand to complete the Ig fold of the subunit. Specifically, residues from the chaperone F_1-G_1 loop, including its alternating hydrophobic residues, shift conformation to become part of the chaperone G_1 strand. This elongated portion of the chaperone G_1 strand occupies the groove of the subunit. The three alternating hydrophobic residues that were formerly exposed as part of the chaperone F_1-G_1 loop now project into the groove and complete the subunit hydrophobic core as part of the chaperone G_1 strand. The G_1 strand lies between the A and F strands on either side of the pilin groove, running parallel to the latter strand. The chaperone G_1 strand thus becomes the G strand of the pilin domain and completes its Ig fold. The fold is slightly atypical, however, since the chaperone G_1 strand runs parallel to the subunit F strand, whereas in a canonical Ig fold the G strand runs in the opposite direction, antiparallel to the F strand. The subunit F strand runs the length of the elongated G_1 strand and terminates at the base of the chaperone cleft. There, the C-terminal negatively charged carboxylate group of the subunit, at the end of the F strand, lies between two conserved basic, positively charged residues in the cleft (Arg8 and Lys112 [Color Plate 21]). This clamping interaction, possible only because the noncanonical parallel orientation of the chaperone G_1 and the subunit F strands places the C terminus of the subunit in the chaperone cleft, is thought to position the subunit F strand properly relative to the chaperone G_1 strand during subunit folding and complex formation. Subunit folding and chaperone-subunit complex formation represent a continuous process. By contributing a critical strand and hydrophobic core residues to the subunit, the chaperone first facilitates the adoption of an Ig fold by the subunit and then remains bound to it, stabilizing it as part of the chaperone-subunit complex (Barnhart et al., 2000; Choudhury et al., 1999; Sauer et al., 2000; Soto et al., 1998).

DONOR STRAND EXCHANGE

Mutagenesis studies indicated that the residues in the subunit F strand that line one side of the groove participate in subunit-subunit interactions, identifying the groove as a critical surface in subunit polymerization (Soto et al., 1998). In the chaperone-subunit complex, the presence of the chaperone G_1 strand in the groove therefore transiently caps this interactive surface. Upstream of the A strand that forms the other side of the groove, non-adhesin subunits have an N-terminal extension (roughly residues 1 to 11 of the mature protein in the PapK subunit, for example) that, in the chaperone-subunit complex, is disordered and does not contribute to the subunit Ig fold (Sauer et al., 1999). Like the chaperone G_1 strand, the N-terminal extension contains a conserved motif of alternating hydrophobic residues. Mutagenesis studies indicated that these alternating hydrophobic residues participate in subunit-subunit interactions, identifying the N-terminal extension as a second critical element in subunit polymerization (Soto et al., 1998). During fiber assembly, then, subunit incorporation occurs via a donor strand exchange mechanism, in which the N-terminal extension of one subunit replaces the chaperone G_1 strand in the groove of the neighboring subunit (Choudhury et al., 1999; Sauer et al., 1999) (Color Plate 23 [see color insert]). The adhesin at the tip of a pilus fiber lacks an N-terminal extension—it has instead its N-terminal receptor-binding domain—but still has a groove that can interact with the N-terminal extension of an adjacent subunit. Donor strand exchange therefore occurs such that the N-terminal extension of the incoming subunit occupies the groove of the most recently incorporated subunit, producing a mature organelle in which each subunit contributes its N-terminal extension to complete the Ig fold of its more distal neighbor.

Subunit-subunit interactions in the fiber are even more stable than chaperone-subunit interactions. For example, subunit-subunit complexes and whole pilus fibers, but not chaperone-subunit complexes, resist dissociation in 2% sodium dodecyl sulfate at room temperature (Soto et al., 1998; Striker et al., 1994). In addition, in vitro studies in the absence of other factors reveal that an excess of the PapK subunit, whose N-terminal extensions bind in the groove of the PapE subunit, displaces the chaperone, but not a peptide corresponding to the N-terminal extension of PapK, from the PapE subunit groove (Sauer et al., 2002). These results indicate that the subunit is in an activated state when bound to the chaperone and attains its "ground" state only after donor strand exchange, when it is bound to the N-terminal extension of its neighbor. The chaperone therefore primes the subunit for donor strand exchange.

Structural studies of the P pilus and of the *Y. pestis* Caf1 fiber systems, by providing views of subunits before and after donor strand exchange, reveal the mechanism of subunit priming by the chaperone. Two crystal structures of the PapE pilus subunit, in complex with the PapD chaperone and in complex with a peptide corresponding to the N-terminal extension of the PapK subunit, respectively, were determined in the P-pilus system (Sauer et al., 2002) (Color Plate 23). In the PapD-PapE and other chaperone-subunit complexes, at the end of the subunit nearest the cleft of the chaperone, the G_1 strand curves away from the centerline of the groove, pulling with it the parallel subunit F strand and holding it away from the A strand on the other side of the groove. The chaperone acts as a wedge between the ends of the A and F strands here, holding this end of the groove in a relatively open conformation and preventing the compact packing of the underlying hydrophobic core residues (Zavialov et al., 2003). In the subunit–N-terminal-extension complex, the N-terminal extension has replaced the chaperone G_1 strand and occupies the subunit groove. The N-terminal extension runs antiparallel to the subunit F strand, in the opposite direction to that of the chaperone G_1 strand in the chaperone-subunit complex, to produce a perfectly canonical Ig fold in the subunit. The alternating hydrophobic residues of the N-terminal extension replace those of the G_1 strand in the subunit groove, but to produce a snug fit they are shifted in register, one position further from the open end of the groove. The displacement of the chaperone and the subsequent orientation and register of the N-terminal extension in the groove trigger a conformation change in the subunit. The removal of the chaperone wedge and the close fit of the alternating hydrophobic residues of the N-terminal extension in the subunit groove allow the entire hydrophobic core of the subunit to pack more tightly. The ends of the subunit A and F strands move together, and the groove closes, sealing the N-terminal extension in place as the subunit adopts a more ordered and compact overall state (Color Plate 23).

The structure of a chaperone-subunit-subunit Caf1M-Caf1-Caf1 ternary complex was determined in the Caf1 system (Zavialov et al., 2003). The Caf1M chaperone, an FGL chaperone with its characteristic long F_1-G_1 loop, is bound to one Caf1 subunit, which in turn donates its N-terminal extension to interact with the second subunit. The structure of the complex thus provides views of both chaperone-subunit and subunit-subunit interactions. The chaperone-subunit and subunit-subunit interactions exhibit the characteristic features of donor strand complementation and exchange, including the maintenance of the groove in an open conformation by the wedging action of the chaperone and the reversal in complementing-strand orientation on donor strand exchange. The groove of the Caf1 subunit is longer than those of the subunits from P pili and type 1 pili, and hence the chaperone G_1 strand, derived in part from the long F_1-G_1 loop, is correspondingly longer, contributing five alternating residues to fill the subunit groove (MacIntyre et al., 2001). Upon donor strand exchange, the N-terminal extension likewise contributes five alternating residues to the groove, but, unlike the P pilus system, these residues are in the same register as those of the chaperone G_1 strand. However, after donor strand exchange the groove is more shallow and narrow than it is in the chaperone-subunit complex, and to fit snugly in the groove the alternating residues from the N-terminal extension donor strand are correspondingly smaller than those of the chaperone G_1 strand. The shallow character of the groove results from the removal of the chaperone wedge, whose relatively bulky alternating G_1 strand residues penetrate more deeply into the subunit hydrophobic core and hold the two sides of the groove away from each other. The displacement of the chaperone allows the A and F strands on either side of the groove to move together and permits the tight packing of the underlying subunit hydrophobic core residues to yield compactly packed molecules.

The chaperone thus primes the subunit for donor strand exchange by stabilizing it in an activated conformation. The chaperone initially facilitates the folding of the subunit, but, by holding the groove open and preventing the compact packing of the subunit hydrophobic core, it traps the subunit in a folding transition state (Sauer et al., 2002; Zavialov et al., 2003). The displacement of the chaperone by the N-terminal extension releases the stored folding energy that drives fiber formation (Zavialov et al., 2003). The subunit then completes its folding process, undergoing a topological transition from a loosely packed, non-canonical Ig fold to its final compact, canonical Ig fold in the context of the fiber, whose characteristic stability then derives from the contribution by each subunit of an integral strand to the fold of its neighbor (Sauer et al., 2002; Zavialov et al., 2003).

DONOR STRAND EXCHANGE AT THE USHER

Donor strand exchange occurs very rapidly in vivo but only relatively slowly and inefficiently in vitro in the absence of the usher (Jacob-Dubuisson et al., 1994; Sauer et al., 2002). This suggests that while the chaperone primes the subunit for donor strand exchange, additional interactions with the usher may facilitate subunit uncapping during fiber formation. Fiber assembly at the usher includes the targeting of chaperone-adhesin and other chaperone-subunit complexes to the usher, the uncapping of subunits and the facilitation of donor strand exchange, and the translocation of the growing fiber through the usher pore. Recent studies have only begun to reveal the mechanisms that drive these processes.

Experimental evidence indicates that the chaperone plays a role in donor strand exchange at the usher. Mutations in two conserved surface-exposed residues of the PapD chaperone—Thr53 and Arg68—reduce pilus assembly without affecting chaperone-adhesin complex formation or the interaction of this complex with the usher (Hung et al., 1999; Hung et al., 1996). These results indicate that the chaperone is involved in a critical step in fiber formation after the formation of the chaperone-adhesin-usher complex. Additional studies point to a role for the chaperone in the incorporation into the fiber of the subunit to which it is bound (Barnhart et al., 2003). Further work will elucidate the details of chaperone function at the usher.

The usher has been modeled as an outer membrane β-barrel, with its larger interstrand loops facing the periplasm. The periplasmic loops are thought to include both a region near the N terminus and the C-terminal region of the protein, both of which contain a single conserved disulfide bond (Thanassi et al., 2002). In P pilus and type 1 pilus assembly, the chaperone-adhesin complex initiates fiber formation by binding to the usher (Dodson et al., 1993; Saulino et al., 1998). It has been shown that both the receptor-binding and pilin domains of the FimH adhesin in the type 1 pilus interact with the FimD usher. It was proposed that the receptor-binding domain of FimH inserts into the pore, while the pilin domain, bound to the chaperone, straddles the periplasmic surface of FimD (Barnhart et al., 2003). The binding of the FimC-FimH chaperone-adhesin complex to the usher induces a conformational change in FimD and yields a stable FimC-FimH-FimD chaperone-adhesin-usher ternary complex, which has been successfully isolated (Saulino et al., 1998). Limited proteolysis of this complex indicated that FimC-FimH interacts with the C-terminal half of FimD. Expression of the subunit FimG with FimC-FimH-FimD permitted the isolation of a FimC-FimG-FimH-FimD chaperone-subunit-adhesin-usher complex in which FimG was present in multiple copies. Electron microscopic visualization of the complex revealed a fiber emerging from one end of the usher pore (Saulino et al., 2000). The presence of the FimC chaperone in these complexes suggests that the chaperone of the most recently incorporated subunit remains bound at the base of the pilus, on the periplasmic side of the usher, until it is displaced by the incoming chaperone-subunit complex during fiber assembly (Jones et al., 1995; Saulino et al., 2000). The expression of the FimF subunit resulted in its incorporation into the fiber and reduced the extent of FimG multimerization, indicating that FimF terminates FimG incorporation. The additional expression of the FimA rod subunit then permitted the isolation of a complex that included the chaperone and an intact type 1 pilus, including tip fibrillum and rod, with an usher at its base.

While the FimC-FimH chaperone-adhesin complex was found bound to a C-terminal fragment of the FimD usher in the FimC-FimH-FimD ternary complex, recent studies indicate that an N-terminal periplasmic region of FimD also binds to chaperone-subunit complexes (Nishiyama et al., 2003). Likewise, deletion studies of the PapC usher from the P pilus system indicate that the PapD-PapG chaperone-adhesin complex binds to PapC truncations that lack significant portions of their C-terminal region (Thanassi et al., 2002). However, the PapC C-terminal region is required for pilus assembly, since these PapC truncations do not assemble fibers in vivo. These results suggest that the usher contains two binding sites for chaperone-subunit complexes, one in its N-terminal region for chaperone-subunit complex targeting and one in its C-terminal region for subunit assembly (Thanassi et al., 2002).

CONCLUSIONS

Although fibers assembled by the chaperone-usher pathway resemble single gigantic proteins composed of repeating Ig domains, their individual component subunits are unable to fold efficiently and are unstable, aggregative, and potentially toxic to the bacterium. Fiber assembly by the chaperone-usher pathway thus requires the rapid and precise coordination of the folding, interaction, and assembly of a great number of subunits in the absence of ATP hydrolysis or other cellular energy sources. The chaperone and usher orchestrate this process by harnessing the problematic inherent instability of single subunits to drive assembly.

Because each subunit lacks its C-terminal G strand, it requires additional steric information to fold, information that is provided by the chaperone. Subunits do not chaperone each other; that is, the N terminus of one subunit is not sufficient to permit the folding of another, and subunits expressed in the absence of the chaperone are degraded. This requirement for the chaperone in subunit folding thus prevents premature fiber assembly in the periplasm. The chaperone provides a template for subunit folding. The clamping interaction between the positively charged residues in the chaperone cleft and the negatively charged terminal carboxylate of the subunit positions the subunit to fold on the preformed chaperone G_1 strand. However, even in the stable chaperone-subunit complex thus formed, the folding of the subunit remains incomplete. The noncanonical orientation of the chaperone G_1 strand and the imperfect fit of its alternating hydrophobic residues in the subunit groove prevent the most energetically favorable packing of the β-strands and hydrophobic core of the subunit. The chaperone thus traps the subunit in an activated folding transition state, storing the subunit folding energy for subsequent use as the driver of fiber assembly. The imperfect fit of the chaperone G_1 strand in the subunit groove also accounts for the relative promiscuity of the donor strand complementation interaction, allowing the PapD chaperone, for example, to interact with the PapG, PapF, PapE, PapK, and Pap A subunits.

Interaction of the initial chaperone-subunit complex (i.e., the chaperone-adhesin) with the usher probably partially destabilizes the chaperone-adhesin interaction, allowing the N-terminal extension of the subsequent incoming subunit (itself in complex with a chaperone, of course) to displace the chaperone from the adhesin groove and form the initial subunit-subunit interaction in the pilus. This process in turn destabilizes the interaction of the most recently incorporated subunit with its chaperone, allowing another round of donor strand exchange to occur as the assembly process continues. In contrast with the chaperone that it replaces, the N-terminal extension of a subunit both completes the fold of the adjacent subunit in a topologically canonical manner and contributes alternating hydrophobic residues that pack well with the subunit core residues. Donor strand exchange thus allows the β-strands and hydrophobic core residues of a subunit to pack in a more energetically favorable way, and the subunit completes its folding process. It is this release of stored folding energy on donor strand exchange that powers pilus fiber assembly. The donor strand exchange reaction is more specific than the relatively promiscuous donor strand complementation interaction; subunits favor interactions with the particular subunits that provide the best-fitting N-terminal extension donor strands. The relative specificity of donor strand exchange hence determines, at least in part, the function of individual subunits and their order of incorporation into the fiber.

Acknowledgments. This work was supported by NIH grant AI49950, MRC grant 58149, and Wellcome Trust grant 065932 to G.W. and by NIH grants AI29549, DK51406, and AI48689 to S.J.H. F.G.S. was supported by a Damon Runyon Cancer Research Fellowship.

REFERENCES

Barnhart, M. M., J. S. Pinkner, G. E. Soto, F. G. Sauer, S. Langermann, G. Waksman, C. Frieden, and S. J. Hultgren. 2000. PapD-like chaperones provide the missing information for folding of pilin proteins. *Proc. Natl. Acad. Sci. USA* **97:**7709–7714.

Barnhart, M. M., F. G. Sauer, J. S. Pinkner, and S. J. Hultgren. 2003. Chaperone-subunit-usher interactions required for donor strand exchange during bacterial pilus assembly. *J. Bacteriol.* **185:**2723–2730.

Bullitt, E., C. H. Jones, R. Striker, G. Soto, F. Jacob-Dubuisson, J. Pinkner, M. J. Wick, L. Makowski, and S. J. Hultgren. 1996. Development of pilus organelle subassemblies in vitro depends on chaperone uncapping of a beta zipper. *Proc. Natl. Acad. Sci. USA* **93:**12890–12895.

Bullitt, E., and L. Makowski. 1995. Structural polymorphism of bacterial adhesion pili. *Nature* **373:**164–167.

Carson, M. 1997. Ribbons. *Methods Enzymol.* **277:**493–505.

Choudhury, D., A. Thompson, V. Stojanoff, S. Langermann, J. Pinkner, S. J. Hultgren, and S. D. Knight. 1999. X-ray structure of the FimC-FimH chaperone-adhesin complex from uropathogenic *Escherichia coli.* *Science* **285:**1061–1066.

Connell, H., W. Agace, P. Klemm, M. Schembri, S. Marild, and C. Svanborg. 1996. Type 1 fimbrial expression enhances *Escherichia coli* virulence for the urinary tract. *Proc. Natl. Acad. Sci. USA* **93:**9827–9832.

Dodson, K. W., F. Jacob-Dubuisson, R. T. Striker, and S. J. Hultgren. 1993. Outer-membrane PapC molecular usher discriminately recognizes periplasmic chaperone-pilus subunit complexes. *Proc. Natl. Acad. Sci. USA* **90:**3670–3674.

Du, Y., R. Rosqvist, and A. Forsberg. 2002. Role of fraction 1 antigen of *Yersinia pestis* in inhibition of phagocytosis. *Infect. Immun.* **70:**1453–1460.

Galyov, E. E., A. V. Karlishev, T. V. Chernovskaya, D. A. Dolgikh, O. Smirnov, K. I. Volkovoy, V. M. Abramov, and V. P. Zav'yalov. 1991. Expression of the envelope antigen F1 of *Yersinia pestis* is mediated by the product of *caf1M* gene having homology with the chaperone protein PapD of *Escherichia coli.* *FEBS Lett.* **286:**79–82.

Gong, M., and L. Makowski. 1992. Helical structure of P pili from *Escherichia coli.* Evidence from X-ray fiber diffraction and scanning transmission electron microscopy. *J. Mol. Biol.* **228:**735–742.

Hahn, E., P. Wild, U. Hermanns, P. Sebbel, R. Glockshuber, M. Haner, N. Taschner, P. Burkhard, U. Aebi, and S. A. Muller. 2002. Exploring the 3D molecular architecture of *Escherichia coli* type 1 pili. *J. Mol. Biol.* **323:**845–857.

Holmgren, A., and C. I. Branden. 1989. Crystal structure of chaperone protein PapD reveals an immunoglobulin fold. *Nature* **342:**248–251.

Hull, R. A., R. E. Gill, P. Hsu, B. H. Minshew, and S. Falkow. 1981. Construction and expression of recombinant plasmids encoding type 1 or D-mannose-resistant pili from a urinary tract infection *Escherichia coli* isolate. *Infect. Immun.* **33:**933–938.

Hultgren, S. J., F. Lindberg, G. Magnusson, J. Kihlberg, J. M. Tennent, and S. Normark. 1989. The PapG adhesin of uropathogenic *Escherichia coli* contains separate regions for receptor binding and for the incorporation into the pilus. *Proc. Natl. Acad. Sci. USA* **86:**4357–4361.

Hung, D. L., S. D. Knight, and S. J. Hultgren. 1999. Probing conserved surfaces on PapD. *Mol. Microbiol.* **31:**773–783.

Hung, D. L., S. D. Knight, R. M. Woods, J. S. Pinkner, and S. J. Hultgren. 1996. Molecular basis of two subfamilies of immunoglobulin-like chaperones. *EMBO J.* **15:**3792–3805.

Jacob-Dubuisson, F., J. Heuser, K. Dodson, S. Normark, and S. J. Hultgren. 1993. Initiation of assembly and association of the structural elements of a bacterial pilus depend on two specialized tip proteins. *EMBO J.* **12:**837–847.

Jacob-Dubuisson, F., R. Striker, and S. J. Hultgren. 1994. Chaperone-assisted self-assembly of pili independent of cellular energy. *J. Biol. Chem.* **269:**12447–12455.

Jones, C. H., P. N. Danese, J. S. Pinkner, T. J. Silhavy, and S. J. Hultgren. 1997. The chaperone-assisted membrane release and folding pathway is sensed by two signal transduction systems. *EMBO J.* **16:**6394–6406.

Jones, C. H., J. S. Pinkner, A. V. Nicholes, L. N. Slonim, S. N. Abraham, and S. J. Hultgren. 1993. FimC is a periplasmic PapD-like chaperone that directs assembly of type 1 pili in bacteria. *Proc. Natl. Acad. Sci. USA* **90:**8397–8401.

Jones, C. H., J. S. Pinkner, R. Roth, J. Heuser, A. V. Nicholes, S. N. Abraham, and S. J. Hultgren. 1995. FimH adhesin of type 1 pili is assembled into a fibrillar tip structure in the *Enterobacteriaceae. Proc. Natl. Acad. Sci. USA* **92:**2081–2085.

Knight, S. D., D. Choudhury, S. Hultgren, J. Pinkner, V. Stojanoff, and A. Thompson. 2002. Structure of the S pilus periplasmic chaperone SfaE at 2.2 Å resolution. *Acta Crystallogr. Ser. D* **58:**1016–1022.

Kuehn, M. J., J. Heuser, S. Normark, and S. J. Hultgren. 1992. P pili in uropathogenic *E. coli* are composite fibres with distinct fibrillar adhesive tips. *Nature* **356:**252–255.

Kuehn, M. J., S. Normark, and S. J. Hultgren. 1991. Immunoglobulin-like PapD chaperone caps and uncaps interactive surfaces of nascently translocated pilus subunits. *Proc. Natl. Acad. Sci. USA* **88:**10586–10590.

Kuehn, M. J., D. J. Ogg, J. Kihlberg, L. N. Slonim, K. Flemmer, T. Bergfors, and S. J. Hultgren. 1993. Structural basis of pilus subunit recognition by the PapD chaperone. *Science* **262:**1234–1241.

Langermann, S., S. Palaszynski, M. Barnhart, G. Auguste, J. S. Pinkner, J. Burlein, P. Barren, S. Koenig, S. Leath, C. H. Jones, and S. J. Hultgren. 1997. Prevention of mucosal *Escherichia coli* infection by FimH-adhesin-based systemic vaccination. *Science* **276:**607–611.

Lindberg, F., J. M. Tennent, S. J. Hultgren, B. Lund, and S. Normark. 1989. PapD, a periplasmic transport protein in P-pilus biogenesis. *J. Bacteriol.* **171:**6052–6058.

Lund, B., F. Lindberg, B. I. Marklund, and S. Normark. 1987. The PapG protein is the alpha-D-galactopyranosyl-(1-4)-beta-D-galactopyranose-binding adhesin of uropathogenic *Escherichia coli. Proc. Natl. Acad. Sci. USA* **84:**5898–5902.

MacIntyre, S., I. M. Zyrianova, T. V. Chernovskaya, M. Leonard, E. G. Rudenko, V. P. Zav'yalov, and D. A. Chapman. 2001. An extended hydrophobic interactive surface of *Yersinia pestis* Caf1M chaperone is essential for subunit binding and F1 capsule assembly. *Mol. Microbiol.* **39:**12–25.

Manting, E. H., and A. J. Driessen. 2000. *Escherichia coli* translocase: the unravelling of a molecular machine. *Mol. Microbiol.* **37:**226–238.

McCrea, K. W., W. J. Watson, J. R. Gilsdorf, and C. F. Marrs. 1994. Identification of *hifD* and *hifE* in the pilus gene cluster of *Haemophilus influenzae* type b strain Eagan. *Infect. Immun.* **62:**4922–4928.

Mu, X.-Q., E. H. Egelman, and E. Bullitt. 2002. Structure and function of Hib pili from *Haemophilus influenzae* type b. *J. Bacteriol.* **184:**4868–4874.

Mulvey, M. A., Y. S. Lopez-Boado, C. L. Wilson, R. Roth, W. C. Parks, J. Heuser, and S. J. Hultgren. 1998. Induction and evasion of host defenses by type 1–piliated uropathogenic *Escherichia coli. Science* **282:**1494–1497. (Erratum, **283:**795, 1999.)

Nicholls, A., K. A. Sharp, and B. Honig. 1991. Protein folding and association: insights from the interfacial and thermodynamic properties of hydrocarbons. *Protein Struct. Funct. Genet.* **11:**281–296.

Nishiyama, M., M. Vetsch, C. Puorger, I. Jelesarov, and R. Glockshuber. 2003. Identification and characterization of the chaperone-subunit complex-binding domain from type 1 pilus assembly platform FimD. *J. Mol. Biol.* **330:**513–525.

Pellecchia, M., P. Guntert, R. Glockshuber, and K. Wuthrich. 1998. NMR solution structure of the periplasmic chaperone FimC. *Nat. Struct. Biol.* **5:**885–890.

Pratt, L. A., and R. Kolter. 1998. Genetic analysis of *Escherichia coli* biofilm formation: roles of flagella, motility, chemotaxis and type I pili. *Mol. Microbiol.* **30:**285–293.

Roberts, J. A., B. I. Marklund, D. Ilver, D. Haslam, M. B. Kaack, G. Baskin, M. Louis, R. Mollby, J. Winberg, and S. Normark. 1994. The Gal(alpha 1-4)Gal-specific tip adhesin of *Escherichia coli* P-fimbriae is needed for pyelonephritis to occur in the normal urinary tract. *Proc. Natl. Acad. Sci. USA* **91:**11889–11893.

Sable, N. S., E. M. Connor, C. B. Hall, and M. R. Loeb. 1985. Variable adherence of fimbriated *Haemophilus influenzae* type b to human cells. *Infect. Immun.* **48:**119–123.

Sauer, F. G., K. Futterer, J. S. Pinkner, K. W. Dodson, S. J. Hultgren, and G. Waksman. 1999. Structural basis of chaperone function and pilus biogenesis. *Science* **285:**1058–1061.

Sauer, F. G., S. D. Knight, G. Waksman, and S. J. Hultgren. 2000. PapD-like chaperones and pilus biogenesis. *Semin. Cell Dev. Biol.* **11:**27–34.

Sauer, F. G., J. S. Pinkner, G. Waksman, and S. J. Hultgren. 2002. Chaperone priming of pilus subunits facilitates a topological transition that drives fiber formation. *Cell* **111:**543–551.

Saulino, E. T., E. Bullitt, and S. J. Hultgren. 2000. Snapshots of usher-mediated protein secretion and ordered pilus assembly. *Proc. Natl. Acad. Sci. USA* **97:**9240–9245.

Saulino, E. T., D. G. Thanassi, J. Pinkner, and S. J. Hultgren. 1998. Ramifications of kinetic partitioning on usher-mediated pilus biogenesis. *EMBO J.* **17:**2177–2185.

Slonim, L. N., J. S. Pinkner, C. I. Branden, and S. J. Hultgren. 1992. Interactive surface in the PapD chaperone cleft is conserved in pilus chaperone superfamily and essential in subunit recognition and assembly. *EMBO J.* **11:**4747–4756.

Soto, G. E., K. W. Dodson, D. Ogg, C. Liu, J. Heuser, S. Knight, J. Kihlberg, C. H. Jones, and S. J. Hultgren. 1998. Periplasmic chaperone recognition motif of subunits mediates quaternary interactions in the pilus. *EMBO J* **17:**6155–6167.

St. Geme, J. W., III, J. S. Pinkner, G. P. Krasan, J. Heuser, E. Bullitt, A. L. Smith, and S. J. Hultgren. 1996. *Haemophilus influenzae* pili are composite structures assembled via the HifB chaperone. *Proc. Natl. Acad. Sci. USA* **93:**11913–11918.

Striker, R., F. Jacob-Dubuisson, C. Freiden, and S. J. Hultgren. 1994. Stable fiber-forming and nonfiber-forming chaperone-subunit complexes in pilus biogenesis. *J. Biol. Chem.* **269:**12233–12239.

Thanassi, D. G., E. T. Saulino, and S. J. Hultgren. 1998a. The chaperone/usher pathway: a major terminal branch of the general secretory pathway. *Curr. Opin. Microbiol.* **1:**223–231.

Thanassi, D. G., E. T. Saulino, M. J. Lombardo, R. Roth, J. Heuser, and S. J. Hultgren. 1998b. The PapC usher forms an oligomeric channel: implications for pilus biogenesis across the outer membrane. *Proc. Natl. Acad. Sci. USA* **95:**3146–3151.

Thanassi, D. G., C. Stathopoulos, K. Dodson, D. Geiger, and S. J. Hultgren. 2002. Bacterial outer membrane ushers contain distinct targeting and assembly domains for pilus biogenesis. *J. Bacteriol.* **184:**6260–6269.

Vallet, I., J. W. Olson, S. Lory, A. Lazdunski, and A. Filloux. 2001. The chaperone/usher pathways of *Pseudomonas aeruginosa:* identification of fimbrial gene clusters (*cup*) and their involvement in biofilm formation. *Proc. Natl. Acad. Sci. USA* **98:**6911–6916.

Vetsch, M., P. Sebbel, and R. Glockshuber. 2002. Chaperone-independent folding of type 1 pilus domains. *J. Mol. Biol.* **322:**827–840.

Weber, A., R. Harris, S. Lohrke, L. Forney, and A. L. Smith. 1991. Inability to express fimbriae results in impaired ability of *Haemophilus influenzae* b to colonize the nasopharynx. *Infect. Immun.* **59:**4724–4728.

Zavialov, A. V., J. Berglund, A. F. Pudney, L. J. Fooks, T. M. Ibrahim, S. MacIntyre, and S. D. Knight. 2003. Structure and biogenesis of the capsular F1 antigen from *Yersinia pestis.* Preserved folding energy drives fiber formation. *Cell* **113:**587–596.

Zavialov, A. V., J. Kersley, T. Korpela, and V. P. Zav'yalov. 2002. Donor strand complementation mechanism in the biogenesis of non-pilus systems. *Mol. Microbiol.* **45:**983–995.

Structural Biology of Bacterial Pathogenesis
Edited by G. Waksman et al.
© 2005 ASM Press, Washington, D.C.

Chapter 6

Structure and Assembly of Type IV Pilins

Katrina T. Forest

Type IV pili are ubiquitous surface organelles of gram-negative bacteria—pathogens, symbionts, and free-living organisms alike. They are long, extremely narrow hair-like filaments that are best known for their role in the initial attachment of microbes to host cells. Pili are virtually homopolymers of the pilin structural subunits, which in some cases interact directly with cellular receptors. *Neisseria gonorrhoeae* pilin binds to CD46, while *Pseudomonas aeruginosa* pilin recognizes βGalNAc-β1,4-Gal on asialogangliosides (Källström et al., 1997; Cachia and Hodges, 2003). Pilus-associated proteins have also been implicated in target cell attachment. *N. meningitidis* requires PilC for epithelial cell binding, while *N. gonorrhoeae* requires PilC and the minor pilin PilV (Nassif et al., 1999; Winther-Larsen et al., 1998). Nonpiliated mutants of many human, animal, and plant pathogens, including *P. aeruginosa, N. gonorrhoeae, Moraxella bovis, Fusiformis nodosus,* enteropathogenic and enterohemorrhagic *Escherichia coli, Vibrio cholerae,* and *Ralstonia solanacearum,* are unable to establish infection (Comolli et al., 1999; Bieber et al., 1998; Herrington et al., 1988; Kang et al., 2002). In addition to receptor binding, type IV pili mediate interactions of microbes with one another. For example, *V. cholerae* forms tight microcolonies by bundling toxin-coregulated pilus (TCP) filaments, both in liquid culture and at sites of colonization (Kirn et al., 2000). Pili also promote attachment to abiotic surfaces; *P. aeruginosa* requires type IV pili for establishment of biofilms (O'Toole and Kolter, 1998), and *Myxococcus xanthus* uses pili for attachment to secreted polysaccharide trails (Li et al., 2003).

After initial attachment to target cells, there are many pilus-dependent downstream responses in eukaryotic host cells, depending on the specific infecting species as well as the tissue in question. For example, in *N. gonorrhoeae,* the binding of pili to CD46 on epithelial cells triggers Ca^{2+} release and casein kinase II signal transduction, as well as PilT-dependent invasion of the endothelial layer and trafficking through this layer (Källström et al., 1998; Merz et al., 1996). Piliated *N. meningitidis* transitions from localized to diffuse adherence, causes shedding of microvilli, and loses its pili 4 to 9 h after model infection of T84 epithelial cells (Pujol et al., 1999).

Additional functions mediated by type IV pili include surface motility, natural transformation (for example, in *N. gonorrhoeae, Legionella pneumophila, Pseudomonas stutzeri,*

Katrina T. Forest • Department of Bacteriology, University of Wisconsin—Madison, Madison, WI 53706.

and *Thermus thermophilus* strain HB27), and phage binding (e.g., in *P. aeruginosa* and *V. cholerae*) (Mattick, 2002; Friedrich et al., 2002). Motility is thought to occur by a "grappling-hook" mechanism whereby pili attached to a surface at their distal ends are retracted, pulling the bacterium along the surface in the process. The same retraction mechanism would bring phage attached to the pilus receptor into contact with the bacterial cell surface and could provide the energy for uptake of transforming DNA. These roles of pili beyond attachment appear to be dependent on functional PilT, a hexameric ATPase required for pilus retraction and first identified in a search for phage-resistant mutants of *P. aeruginosa* (Merz and So, 2000; Maier et al., 2002; Skerker and Berg, 2001; Herdendorf et al., 2002; Bradley, 1980).

Electron micrographs of bacterial cells elaborating 60- to 80-Å-wide, 1- to 4-μm-long type IV pili were available decades ago and raised many questions about the functions of these structures. Were the pilin sequences conserved? Were the assembly mechanisms conserved? Could pili be useful vaccines? How could their great tensile strength be explained? How did subunits assemble to form homopolymeric filaments? What was the molecular nature of receptor binding? Was retraction the mechanism of motility and phage uptake?

As peptide sequences of the 140- to 200-amino-acid pilin proteins became available, one could further define the class of type IV pilins on the basis of (i) the short, positively charged leader sequence which targets pilin subunits to an inner membrane protease for cleavage of the leader and methylation of the N-terminal residue prior to secretion; (ii) the conserved first 25 amino acids of the mature proteins, which are hydrophobic except for glutamate at position 5; and (iii) a cysteine pair in the C-terminal one-third of the molecule, which forms a disulfide bond in the mature protein (Fig. 1a).

Pilin monomer sequences vary substantially among species and can be further subclassified into two groups based on sequence homology. The type IVa pilins are typified by sequences from *Neisseria, Pseudomonas,* and *Myxococcus* spp. The type IVb pilins, such as the bundle-forming pilus of enteropathogenic *E. coli,* the longus pilus of enterotoxigenic *E. coli,* and the TCP of *V. cholerae,* are found on enteric pathogens. These molecular data raised additional questions. How could species-specific sequence differences and antigenic variation within species be tolerated without blocking assembly, attachment, and retraction of pili? Could pilus-based vaccine strategies now be guided by molecular data? What is the three-dimensional structure of these molecules? More recently, higher-resolution information from nuclear magnetic resonance spectroscopy (NMR), X-ray crystallography, fiber diffraction, and cryoelectron microscopy has increased our understanding of the biology of these surface structures and has begun to answer these questions.

This chapter focuses on the contribution of structural biology to our understanding of the type IV pili. Recent reviews cover the developments in research on molecular genetics, regulation of expression and assembly, pilus-mediated surface motility, host responses to pili, and pilus-based vaccines carried out by many laboratories around the world (Alm and Mattick, 1997; McBride, 2001; Mattick, 2002; Merz and So, 2000; Nassif et al., 1999; Cachia and Hodges, 2003).

STRUCTURAL BIOLOGY OF FIBER-FORMING PROTEINS

The major challenge to high-resolution analysis of pilin monomers lies within their very structures. Having evolved the biophysical propensity to form long and stable fibers, which are essentially one-dimensional assemblies, pilins are not amenable to the three-dimensional interactions required for crystallization. Protein crystallographers have cir-

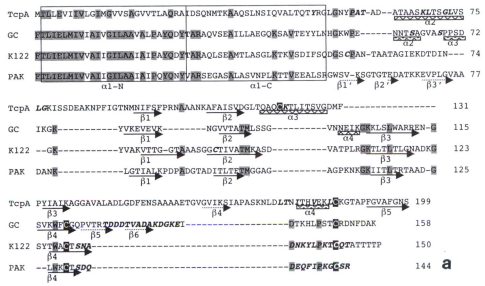

```
TcpA  MTLLEVIIVLGIMGVVSAGVVTLAQRAIDSQNMTKAAQSLNSIQVALTQTYRGLGNYPAT-AD--ATAASKLTSGLVS  75
                                                                        α2
GC    FTLIELMIVIAIVGILAAVALPAYQDYTARAQVSEAILLAEGQKSAVTEYYLNHGKWPE------NNTSAGVASPPSD  72
                                                                        α2    α3
K122  FTLIELMIVVAIIGILAAIAIPAYQDYTARAQLSEAMTLASGLKTKVSDIFSQDGSCPAN-TAATAGIEKDTDIN---  74
PAK   FTLIELMIVVAIIGILAAIAIPQYQNYVARSEGASALASVNPLKTTVEEALSHGWSV-KSGTGTEDATKKEVPLGVAA  77
      α1-N                            α1-C                           β1"  β2"       β3"

TcpA  LGKISSDEAKNPFIGTNMNIFSFPRNAAANKAFAISVDGLTQAQCKTLITSVGDMF----------------  131
        β1              β2            α3
GC    IKGK----------YVKEVEVK--------NGVVTATMLSSG----------VNNEIKGKKLSLWARREN-G  115
        β1              β2                          α4      β3
K122  --GK----------YVAKVTTG-GTAAASGGCTIVATMKASD----------VATPLRGKTLTLTLTLGNADKG  123
        β1              β2                          β3
PAK   DANK----------LGTIALKPDPADGTADITLTETMGGAG----------AGPKNKGKIITLTLTRTAAD-G  125
        β1              β2                          β3

TcpA  PYIAIKAGGAVALADLGDFENSAAAAETGVGVIKSIAPASKNLDLTNITHVEKLCKGTAPFGVAFGNS  199
       β3                        β4          α4     β5
GC    SVKWFCGQPVTRTDDDTVADAKDGKEI-------------------DTKHLPSTCRDNFDAK  158
        β4    β5     β6
K122  SYTWACTSNA--------------------------------DNKYLPKTCQTATTTTP  150
        β4
PAK   --LWKCTSDQ--------------------------------DEQFIPKGCSR  144   a
        β4
```

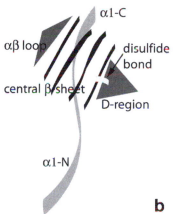

Figure 1. Pilin structure-based sequence alignment. (a) Sequences of *V. cholerae* TcpA, *N. gonorrhoeae* MS11 pilin (GC), *P. aeruginosa* strain K122-4 pilin, and *P. aeruginosa* PAK pilin are aligned based on the three-dimensional structures. Sequence identity (grey shading if three of four residues are the same) is evident in the N-terminal half of the boxed α1-helix. Secondary-structure elements below each sequence highlight the structurally conserved β-sheet (solid arrows) and disulfide bond (reverse video). Species-specific β-strands (dashed arrows) and helices (wavy boxes) occur largely in the αβ loop and D-region. Amino acids contributing to species-specific functionalities are in bold italics. TcpA residues involved in crystal packing and fiber-stabilizing hydrophobic interactions (Y51, P58, A59, T60, K68, L69, G72, L73, L76, G77, K121, L176, T177, I179, V182, and L185); the MS11 glycosylation site (S63), phosphorylation site (S68), and hypervariable region (residues 128 to 141); the PAK and K122-4 receptor-binding loops (residues 131 to 144); and the K122-4 second disulfide bond (C57 to C93) are shown. (b) Conserved α/β-roll pilin fold, with the locations of species-specific αβ loop and D-region shown schematically.

cumvented this problem for other filamentous proteins in several ways including cocrystallization with partner proteins that block polymerization (actin-profilin, actin-DNase I, actin-gelsolin, and type I fimbrial subunits with dedicated chaperones) and limited proteolysis to release soluble crystallizable domains (myosin subfragment 1 head domain and flagellin F41 fragment) (Schutt et al., 1993; Kabsch et al., 1990; Vorobiev et al., 2003; Soto and Hultgren, 1999; Rayment et al., 1993; Samatey et al., 2000). Other biological filaments have been probed by high-resolution fiber diffraction (e.g., tobacco mosaic virus and filamentous phage) (Namba et al., 1989; Gonzalez et al., 1995). Much has been learned from cryoelectron microscopy (e.g., actomyosin complexes, halobacterial flagellin, and type I pili), and the upper resolution of this technique reached 4 Å recently, with the stunning structure of the intact flagellar filament (Cohen-Krausz and Trachtenberg, 2002; Bullitt and Makowski, 1998; Yonekura et al., 2003). Crossing resolution boundaries by combining atomic resolution structures of subunits determined by X-ray crystallography or NMR with moderate-resolution information for intact assemblies based on fiber diffraction, electron microscopy, or atomic force microscopy now allows us to understand the atomic details of extremely large macromolecular complexes and dynamic assemblies (Leiman et al., 2003; Hopfner et al., 2002; VanLoock et al., 2003). Such approaches are beginning to be successfully applied to the filamentous type IV pili.

PILIN SUBUNIT STRUCTURES

The Overall Pilin Fold Is Shared

High-resolution structures of four type IV pilins have been published (see the addendum in proof). Amazingly, despite very low sequence homology beyond the N-terminal 25 to 30 amino acids, these structures have a distinctive protein fold in common (Fig. 1b). The ∼80-Å-long, 52-amino-acid α1-helix forms the backbone of the highly asymmetric pilin subunits. The first 28 to 30 residues of this helix (α1-N) are hydrophobic yet are exposed in the monomer, suggesting that they are buried within the filament. Cradling the C-terminal half of the helix (α1-C) and forming an extensive tightly packed hydrophobic interface is a four-strand antiparallel β-sheet. Each pilin also has unique structural features which enhance its specific functions and dictate its assembly mode as expected, given the low sequence homology among pilins. These sequence and structural differences are observed in two sections of the primary sequence: the $\alpha\beta$-loop, which falls between the C-terminal tip of the long α1-helix and the first strand of the central β-sheet, and the D-region, which lies between the C-terminal disulfide-bonded residues. Connectivity differences for the central sheet are observed between type IVa and type IVb pilins. Loops connecting β-strands also vary, to a lesser extent. In this chapter, the structural details of each pilin subunit are elaborated and compared. Models for pilus filament assembly based on these monomer structures are also discussed. Finally, the limited structural information available for pilus assembly proteins is summarized.

N. gonorrhoeae Pilin

The first diffraction-quality crystals of a type IV pilin were obtained following dissociation of native pilin monomers from intact *N. gonorrhoeae* MS11 pili in the nondenaturing detergent, *n*-octyl-β-D-glucopyranoside (βOG) (Parge et al., 1990). Circular dichroism showed that the secondary structure remains unchanged on dissociation, as was also true of *P. aeruginosa* pilin (Watts et al., 1983). This property of type IV pili is probably critical to the retraction and subsequent disassembly which mediate motility, and it distinguishes type IV from type I pili, whose constituent subunits must be denatured to be separated. Soluble MS11 pilin monomers were purified by extensive filtration through a membrane with a restrictive molecular weight cutoff, removing oligomers and leading to a substantial improvement in crystal quality (Parge et al., 1995).

The 2.6-Å-resolution structure of the gonococcal *pilE* gene product first revealed the pilin fold (Color Plate 24a [see color insert]). The long α-helix curves at Pro22 and Gly42 to form the 85-Å S-shaped spine of pilin, in which the first 28 amino acids do not interact with the rest of the protein. Within the crystal, a pilin dimer is formed by an antiparallel coiled-coil interaction of these hydrophobic amino acids with the same amino acids in a symmetry-related molecule (Parge et al., 1995). The α1-helix is followed by an extended posttranslationally modified "sugar loop" (residues 55 to 77), which forms the $\alpha\beta$ loop in MS11 pilin. Residues 78 to 122 form the central β-sheet, four-strands with +1 connections, which packs via a hydrophobic interface against α1-C. The sequence-variable D-region delimited by cysteines 121 and 151 comprises an additional β-hairpin and well-structured C-terminal tail (Color Plate 24a).

The amino acid sequence of the *N. gonorrhoeae* pilin subunit is subject to antigenic variation at a frequency of 10^{-2} (Serkin and Seifert, 1998), especially within a hypervariable stretch near the C terminus (Fig. 1a; Color Plate 24a). The sequence changes can lead to a nonpiliated state, presumably due to assembly defects, but a wide range of amino acid changes, insertions, and deletions can be accommodated in the subunit structure without preventing folding, assembly of functional pili, receptor binding, transformability, or motility. Antigenic variation of pilin and other surface proteins is part of the reason why gonococcal infection does not lead to lasting immunity. This region of the pilin sequence is surface exposed, since antibodies raised against peptides from this region can bind to intact purified pili. It is also immunodominant, since serum specificity from animals immunized with pili maps frequently to these amino acids (Forest et al., 1996). Although immunization with pili can lead to protection against challenge with the homologous strain in controlled experiments, it is not effective as a vaccine (Boslego et al., 1991). For this reason, understanding the plasticity of pilin has potential import for vaccine design.

The crystal structure of MS11 pilin provided a three-dimensional explanation for the structural accommodation of antigenic variation. Somewhat surprisingly, the D-region was seen to adopt a 14-amino-acid regular β-hairpin followed by a 14-residue loop and an 8-residue C-terminal tail (Color Plate 24a). Arg127 in the center of β5 and Ile142 in the C-terminal loop delimit the variable region. The spacing between these two residues varies among gonococcal pilins from 12 to 21 amino acids and, although considered hypervariable, is dominated by the small and/or polar side chains of Gly, Ala, Thr, Asp, Glu, Gln, and Lys. The invariant Arg127 side chain tacks the β-hairpin to main chain atoms of residues 85, 110, and 129 in the central β-sheet. β2 Val87, β3 Trp109, and β5 Val125 contribute the invariant side chains of a small, hydrophobic stabilization interface for the hypervariable region, which contributes only Val133 to the packing (Color Plate 25 [see color insert]). The disulfide bond itself connects the C-terminal tail to the tip of β4 in the central sheet. Thus, conserved amino acids provide links from the D-region hairpin and loop to the rest of the pilin structure, allowing the length and sequence variability observed in *N. gonorrhoeae* pilins.

Although no additional *Neisseria* pilin structure has been solved, it is likely that the variable amino acids in other gonococcal pilins adopt the same fold in the D-region. The β-hairpin is not an integral part of the pilin fold, and so its length may vary to accommodate more or fewer amino acids within the variable region. The stabilizing interactions which tack the hairpin and loop to the pilin α/β-roll fold are invariant. The residues within the hairpin maintain their largely polar nature and are likely to remain surface exposed. Every example of the hypervariable region includes at least one valine, to play the role of Val133 in MS11 and provide a hydrophobic contact on the inward face of the D-region hairpin (Color Plate 25). Also, each variable region includes at least one aspartate, potentially available to stabilize the hairpin turn and hydrogen bond with the β3-β4 turn, as Asp130 does in the MS11 structure. This hypothesis remains to be experimentally tested.

The electron density map of MS11 pilin contains two well-ordered regions not explained by the polypeptide chain alone, suggesting that pilin is posttranslationally modified. A covalently attached saccharide at Ser63 was originally fitted and refined as O-linked GlcNAc-α1,3-Gal (Parge et al., 1995) and has been clarified recently by mass spectrometry as 2,4-diacetamido-2,4,6-trideoxyhexose (Hegge et al., 2004). Glycosylation of meningococcal pilin by the trisaccharide diagalactosyl-2,4-diacetamido-2,4,6-trideoxy-hexose was identified within the peptide at positions 60 to 63 by tryptic digests and mass spectrometry (Stimson et al., 1995), and its O linkage to Ser63 was confirmed by site-directed mutagenesis (Marceau et al., 1998). Expression of several genes in the enzymatic pathway for pilin glycosylation in *Neisseria* species is subject to phase variation (Power et al., 2003). In particular, the glycosyl transferase PglE which adds the distal β1-4-linked galactose contains a heptanucleotide repeat which leads to phase variation in *N. meningitidis* C311#3. *N. gonorrhoeae* MS11 carries these repeats also, but they are out of frame (Power et al., 2003), explaining the presence of the di- rather than trisaccharide on the crystallized pilin monomer. The biological significance of the difference in the size of the saccharide between the two species is not understood, although it was suggested that this difference could contribute to antigenic variation or host tropism (Parge et al., 1995; Marceau et al., 1998).

To assess the biological role of glycosylation of the two extremely closely related *Neisseria* pilins (which are no more different from one another than are pilins from different strains of *N. gonorrhoeae*), Ser63 was changed to alanine in the *N. meningitidis* and *N. gonorrhoeae* pilins. As predicted, the Ser63Ala mutation did prevent pilin glycosylation in both species (Marceau et al., 1998; Marceau and Nassif, 1999). A small increase in the adhesion of *N. meningitidis* to cultured epithelial cells was observed (Marceau et al., 1998); this increase is probably attributable to a corresponding slight increase in piliation. Production of a normal secreted soluble pilin, cleaved from mature pilin near the α1-N/α1-C boundary, was completely blocked in the meningicoccal mutant and decreased in the gonococcal mutant. These effects on S-pilin production may be due to the general solubilization of pilin by the modification (Marceau and Nassif, 1999). Further in vivo experiments are required to unravel the biological roles of pilin glycosylation, which may be subtle and involve signaling within the host or communication with the immune system.

The second posttranslational modification discovered in *N. gonorrhoeae* MS11 pilin by crystallographic analysis was phosphoserine at residue Ser68 (Forest et al., 1999). Site-specific mutation of this residue to alanine did not negatively impact pilus assembly, adhesion to cultured epithelial cells, twitching motility, or natural transformation frequencies but did result in a loss of phosphorylation as determined by mass spectrometry of tryptic peptides (Forest et al., 1999). The mass spectrometry data indicated a mass loss too large for phosphoserine alone and were interpreted as either the concomitant loss of glycosylation and phosphorylation or the coincidental purification of nonglycosylated peptides (Forest et al., 1999). It is now clear from an exhaustive mass spectrometry analysis that the modification can be either phosphoethanolamine or phosphocholine (Hegge et al., 2004). This highlights a limitation of moderate-resolution crystallography in identifying post-translational modifications when mobile atoms cannot be modeled. Ser68Ala pilus filaments did appear more curly by electron microscopy, presumably due to the loss of repulsive negatively charged phosphates along the length of the pilus filament. As with glycosylation, phosphorylation may exert an effect at downstream sites in infection, such as interaction with signaling cascades or with the innate or humoral immune system.

Thus, the atomic structure of gonococcal pilin revealed the α/β-roll fold of a type IV pilin monomer, explained the biophysical nature of antigenic variation, revealed two post-translational modifications and allowed the design of soluble pilin monomers for the high-resolution determination of the structures of pilins from other species via deletion of the exposed hydrophobic α1-N portion of the assembly helix.

P. aeruginosa Strain K Pilin

Crystal structures of *P. aeruginosa* strain K pilin (PAK pilin) revealed a strong structural similarity to the gonococcal pilin and helped explain the binding mode for PAK pilin to its receptor (Color Plate 24b) (Hazes et al., 2000; Craig et al., 2003). The PAK pilin structure was solved for two crystal forms of the full-length monomer dissociated in β-octyl glucoside and showed almost perfect structural correspondence between the N-terminal α-helices of the PAK and MS11 pilins (Craig et al., 2003). The two crystal forms of full-length PAK pilin contain a total of three independently refined full-length pilin monomers and allow a direct comparison of the conformations of three independent α1-helices. Pro22, conserved as glycine or proline among type IVa pilins, forms a hinge. The propagated positional differences are as much as 10 Å for the splayed N termini, while α1-C residues remain superimposable.

An engineered truncation of the first 28 amino acids of PAK pilin based on the structure of full-length gonococcal pilin was readily expressed with a periplasmic targeting signal and purified from *E. coli,* leading to high-quality crystals that diffracted to better than 1.0-Å resolution (Hazes et al., 2000). The resulting 1.6-Å structure of the head domain is essentially indistinguishable from the lower-resolution full-length structures, confirming the hypothesis that loss of the N-terminal helix does not affect the structure of the pilin head domain. This result can be used to justify future structural biology experiments on soluble truncated pilins. It has already been applied to K122-4 and TcpA pilins (see below), and the technique will undoubtedly yield additional type IV pilin structures.

The PAK pilin monomer also displays the overall α/β pilin fold (Fig. 1b; Color Plate 24b). Despite a mere 17.2% sequence identity beyond the first 29 amino acids, the structure of PAK pilin can be superimposed on the structure of MS11 pilin over 83 Cα atoms with a root mean square deviation of only 1.9 Å. More surprising than the structural similarity within the α1-helix and four β-strands is the further structural conservation of two loops (Color Plate 24a and b) (Hazes et al., 2000). The first of these, from amino acids 107 to 114 (97 to 104 in MS11), connects β2 and β3. The second is within the disulfide-bonded D-region at the C terminus, from amino acids 132 to 143 (140 to 151 in MS11), encompassing the receptor-binding loop of PAK pilin.

Two major differences between MS11 and PAK pilins occur in the αβ loop and in the D-region. PAK pilin has a small β-sheet in the αβ loop; the two central strands are only 3 residues each, and the flanking strands (one of which is in the loop between β1 and β2 in the central sheet) are formed by only a single main chain H bond (Color Plate 24b). This minor β-sheet is probably a structural feature in only a subset of *P. aeruginosa* type IV pilins (including PAK, PAO, PA103, CD, and T2A), whereas the members of a second group (including K122-4 and P1) have sequence homology to the sugar loop of the MS11 pilin in the αβ loop (Hazes et al., 2000). PAK pilin is not known to be posttranslationally modified, but this minor sheet occupies a similar position and volume to the saccharide in the MS11 pilin (Color Plate 24a and b).

The D-region of PAK pilin contains the proposed binding site for the host epithelial cell receptor βGalNAc-β1,4-Gal on asialo-GM$_1$. NMR studies of peptides encompassing this C-terminal receptor binding loop of pilins from four *P. aeruginosa* strains (PAK, PAO, KB7, and P1) revealed a type I β-turn (residues 134 to 137) followed by a type II β-turn (residues 139 to 142) (McInnes et al., 1993; Cachia and Hodges, 2003). The turn conformations are maintained in the PAK pilin crystal structures, although the orientation of the two turns with respect to one another is different from that in the peptides alone. This result underscores the effect of context on peptide structure and may have implications for peptide vaccine trials and a chimeric pilin/exotoxinA vaccine which focuses on this loop from residues 128 to 144 (Cachia and Hodges, 2003; Hertle et al., 2001). Peptide NMR studies on K122-4 pilin show that its receptor-binding loop also contains two β-turns (Cachia and Hodges, 2003). The second is predicted to be a type I turn for two reasons. First, Gly141 is replaced by threonine in K122-4 pilin, restricting the accessible Φ/Ψ angles. Second, MS11 pilin has an equivalent threonine and is observed to have two type I β-turns in this loop (Hazes et al., 2000; Cachia and Hodges, 2003).

Despite the small predicted change in turn conformation, each of these *P. aeruginosa* type IVa pilins is able to bind the same βGalNAc-β1,4-Gal on asialoglycolipid receptors (Hazes et al., 2000). Hazes et al. (2000) show the potential for a main chain mode of saccharide binding, explaining how several pilins with different sequences might bind the same receptor. Analysis by alanine substitution through the receptor-binding loop indicated that Ser131, Gln136, Ile138, Pro139, Gly141, and Lys/Arg144 were required to maintain the ability to bind to A549 cells (Wong et al., 1995). Side chains of some of these most highly conserved residues are involved in stabilizing the interactions of the receptor-binding loop with the rest of the protein, rather than being solvent exposed (Color Plate 26 [see color insert]). The solvent-accessible surface of the two D-region β-turns is composed largely of main chain atoms, with a groove between the two hairpins forming a potential binding pocket (Color Plate 26) (Hazes et al., 2000). This proposed binding pocket in the high-resolution PAK structure suggests hydrophobic interactions for a sugar ring with Ile138 and H-bond donors and acceptors in Ile138, Gly141, Cys142, and Arg144 main chain atoms (Hazes et al., 2000). Despite efforts by multiple groups, no cocrystal structure of PAK pilin with its saccharide receptor is available, and there is not a single obvious orientation for the bound disaccharide predicted from the structure (Hazes et al., 2000). This is a clear next step for structural analysis.

Pseudomonas K122-4 Pilin

Pilin from *P. aeruginosa* K122-4 has also been expressed and purified as a truncated soluble protein. At high concentrations, this protein blocks the binding of purified PAK pilus filaments to immobilized asialo-GM$_1$ and the engineered pilin subunit mitigates the effects of murine infection with PAK if administered before the challenge (Keizer et al., 2001), arguing that the truncation has not severely affected the overall fold or binding surfaces of the pilin monomer.

The NMR structure of K122-4 pilin is well defined over the amino acids in the truncated α1-C helix as well as the central four-strand β-sheet. Not surprisingly, given the 20% sequence identity, the structure of K122-4 pilin is similar to that of PAK pilin, although the angle of the α1-helix with respect to the sheet has shifted by approximately 45° relative to the nearly parallel alignment in PAK pilin (Color Plate 24c) (Keizer et al., 2001). This conformation may be stabilized in part via a second disulfide bond in K122-4 pilin (Cys57-Cys93) that ties α1-C to β2. A second difference between the two *P. aeruginosa* pilins appears to be the structure of the αβ loop. Although it is not well ordered in the K122-4 pilin NMR structure, there is no evidence for the minor β-sheet observed in PAK pilin. Rather, the αβ loop appears to adopt a more extended structure, as seen in MS11 pilin and suggested by sequence alignments (Hazes et al., 2000).

An obvious question is whether the main chain receptor-binding mode predicted by the PAK pilin structures is found in K122-4 pilin. However, the lack of order in the receptor-binding domain makes it difficult to draw any conclusions. Well-diffracting crystals of K122-4 pilin have recently been reported (Audette, 2003) and should help to address the conformation and sugar-binding mode in this C-terminal loop.

V. cholerae Type IVb Pilin

Sequence and functional comparisons of type IVa and IVb pilins provided little reason to suspect structural homology. Beyond the first 20 amino acids, the only identifiable sequence homology for the type IVb pilin monomers from enteric pathogens (enterotoxigenic *E. coli* longus, enteropathogenic *E. coli* bundle forming pilus, and *V. cholerae* TCP) with the type IVa pilins is the presence of two disulfide-bonded cysteines (Fig. 1a). The protein subunit of the *V. cholerae* TCP, TcpA, is significantly longer than that of *N. gonorrhoeae* or *P. aeruginosa* pilins: 199 amino acids for TcpA as opposed to 158 for MS11 and 144 for PAK. These proteins are not known to mediate natural transformation, and their motility proteins are either absent (TCP) or more distantly related to the PilT and PilU retraction proteins (bundle-forming pilus).

To address questions regarding the shared nature of the structural fold and assembly of type IV pilins and to be able to fully interpret data from a large collection of assembly and interaction mutants and protective epitopes in early vaccine trials, the crystal structure of TcpA was investigated. A soluble N-terminally truncated TcpA protein (in which the first 28 residues of the mature pilin were replaced by a 21-amino-acid expression tag) was crystallized, and the structure was solved and refined at 1.3-Å resolution (Craig et al., 2003). Remarkably, the crystal structure of soluble TcpA could be superimposed on that of PAK pilin over 30 Cα atoms within the α1-C helix, β1, and β2 with a root mean square deviation of 1.8 Å, emphasizing again the structural conservation of type IV pilins in the absence of sequence conservation (Color Plate 24d). The hydrophobic core of TcpA is composed of α1-C, three of the strands in the central β-sheet, and an additional shorter helix (α3) in the connection between β2 and β3 (Color Plate 24d).

The TcpA structure has substantial differences in connectivity from the type IVa proteins, with the third strand in a five-strand β-sheet formed by the final 9 amino acids of the protein rather than the nearest-neighbor connectivity observed in the three type IVa pilins (Fig. 1a; Color Plate 24d). The αβ loop of TcpA contains a 13-residue α-helix perpendicular to α1 and stretches of random coil (Color Plate 24d). The D-region between Cys120 and Cys186 is a large, complex structure encompassing two strands of the β-sheet (β3 and β4), two additional short α-helices (α3 and α4), and random coil for a total of 65 amino acids (Fig. 1a; Color Plate 24d). The substantially larger D-region is the site of the largest insertion compared to the type IVa pilins, contributing to an overall increase in the size of the monomer from 26 by 34 by 85 Å to 30 by 40 by 85 Å (Color Plate 24). The longer length of the D-region is shared by other type IVb pilins and suggests that they may share the TcpA topology.

Monomer Conclusions

The high-resolution X-ray crystallographic structures of *N. gonorrhoeae* MS11, *P. aeruginosa* PAK, and *V. cholerae* classical strain pilins and the NMR structure of *P. aeruginosa* K122-4 pilin lead to several general conclusions. The overall scaffold of each type IV pilin is an α/β-roll fold consisting of a highly conserved hydrophobic ~52-residue α-helix, whose N-terminal half is exposed and whose C-terminal half is stabilized by a hydrophobic interface with a central β-sheet. The connectivity of this sheet differs between type IVa and type IVb pilins. The greatest variations among species occur in the αβ loop between the α1-helix and β1 and in the D-region between the invariant, disulfide-bonded cysteines. These two regions form the "edges" of the conserved α/β-roll fold of type IV pilins, varying in size and structure and imparting unique features to each pilin.

PILUS FILAMENT MODELS

The biological functions of pilin require the in vivo assembly of thousands of monomers into a filament. The results of X-ray crystallography analyses of monomers leave central questions regarding the structures of these pili unanswered. How do monomers interact with one another to form the filament? To what extent are these interactions conserved across species? To what extent is the overall packing symmetry conserved across species? What is the role of the sequence-conserved α1-helix in assembly? How is tensile strength achieved? Computational assembly models for type IVa and IVb pili can help address these questions and have been calculated by combining high-resolution crystal structure information from pilin monomers with low-resolution information about filament symmetry from fiber diffraction or cryoelectron microscopy.

Type IVa Pili

The extreme asymmetry of the MS11 pilin monomer placed severe constraints on a filament model and made it possible to generate and screen computationally nearly 100,000 models after initial computer graphic analysis to limit reasonable starting orientations. Six parameters define the helical assembly. The first three are the angles that orient the monomer with respect to the *x*, *y*, and *z* axes of the filament. Holding a given monomer orientation and radial displacement of the subunits from the center axis fixed, the pilus assembly is created by repeatedly applying a rise (longitudinal translation) and twist (azimuthal rotation) before placing the next subunit.

Prior to the crystal structure determination of *N. gonorrhoeae* pilin, fiber diffraction data were available for *P. aeruginosa* PAK pilin, indicating a 1-start helical assembly with four monomers per turn and a per-subunit rise of 8.1 Å (Color Plate 27a [see color insert]), with α-helices running approximately parallel to the filament axis and either a narrow channel or a hydrophobic interior (Folkhard et al., 1981). In the absence of similar data for the *N. gonorrhoeae* filaments, a composite model was created, taking into account the diameter of MS11 pili measured by electron microscopy, the MS11 pilin structure, sequence conservation among *N. gonorrhoeae* pilins and type IVa pilins in general, presumed solvent exposure of the saccharide and the hypervariable region, chemically reasonable burial of the conserved hydrophobic α-helix, and an approximate match to PAK pilus filament helical parameters. Potential solutions were computationally scored to assess burial of hydrophobic residues, potential for formation of salt bridges between conserved residues, and minimal atomic clashes.

The resulting gonococcal pilus model is a right-handed one-start helix with five subunits per turn. Three functional layers mirror the sequence conservation among pilins (Forest and Tainer, 1997b). In the interior of the filament, the hydrophobic α1-helices pack together, providing a resilient interaction that may allow fibers to bend without breaking (Color Plate 27b). Given the conservation of this helix across type IVa and IVb pilins, its hydrophobic nature, and the observation that mutations within this region abolish filament formation, it is clear that the core of all the type IV pili will be similarly formed. At the time this model was generated, only a single conformation of the helix had been observed crystallographically, and it was this conformation that was chosen for model building. Data now exist demonstrating the flexibility of this helix in pilin, and future modeling should take this conformational adaptability into account. It could allow for proteins of the pilus assembly-disassembly machinery to interact with the helical tail of a single pilin at the base of the filament. The assembly model suggests a role for the invariant Glu5 side chain in neutralizing the positively charged N terminus of a neighboring monomer, keeping the interior hydrophobic packing environment neutral and biochemically setting the longitudinal registration for pilin subunits (Parge et al., 1995).

The middle layer of the gonococcal filament model is formed by the four-strand β-sheet and α/β sugar loop. The main chain hydrogen-bonding pattern of the sheet is proposed to extend from β4 of subunit n to the sugar loop of subunit $n + 1$, thus forming a continuous 25-strand β-wrap around the core (referral to Fig. 1b may aid in conceptualization). Conserved residues 144 to 152 surrounding Cys151 also fall structurally in this middle layer and are not exposed on the surface of the filament (Forest and Tainer, 1997b). This β-sheet layer may contribute to the mechanical stability to withstand 100-pN forces on retraction (Merz et al., 2000; Maier et al., 2002). Given the structural differences in the αβ loop and D-region of other pilins, these interactions cannot be identical for all type IV pilus structures.

Finally, the most widely exposed outer layer of the gonococcal pilus filament includes the posttranslational modifications and the hypervariable region. These are fully surface accessible as would be predicted for antigenically varying, immunodominant epitopes (Color Plate 27b). An interesting confirmation of the overall assembly model is that amino acids 48 to 56 at the tip of α1-C and 96 to 102 in the β2-β3 connection are predicted to form a longitudinal contact within the filament (Forest and Tainer, 1997b), and antipeptide antibodies to these regions indeed recognize only the ends of purified pili (Color Plate 27c) (Forest et al., 1996).

A recent PAK filament model is based directly on helical symmetries from fiber diffraction data (Folkhard et al., 1981) and built up from the most strongly kinked of the three available full-length PAK pilin monomer structures (Craig et al., 2003; Craig and Tainer, 2004). This one-start right-handed helix has four subunits per turn and a pitch of 41 Å. The hydrophobic α1-helices are buried as expected on the interior of the filament. Along the filament surface, the minor β-sheet packs against the receptor-binding loop of subunit $n + 3$. (The filament can be equivalently described as a three-start left-handed helix with a 123-Å pitch, in which case the packing is from subunit n to subunit $n + 1$ within one of the three-start strands.) This packing arrangement leaves the receptor-binding loop partially exposed at the tips and along the sides of each filament.

The involvement of type IV pili in competence is intriguing. There is certainly no structural evidence for a hypothesis in which nucleic acids are brought into the cell by transiting through the pilus; no model for an assembled pilus has a wide enough channel to accommodate even single-stranded DNA or RNA. Moreover, the hydrophobic character within the interior would not provide appropriate electrostatic attraction for the negatively charged phosphate backbone. This is a clear difference between type IV pili and some other structurally similar protein filaments whose assembly is based on coiled-coil interactions of α-helices. Filamentous phage do carry nucleic acids within, and they do have interior volumes and charge distributions that allow this packing (Gonzalez et al., 1995; Marvin, 1998). Flagellar assembly precedes from the distal tip outward, as subunits are exported through a narrow channel. While the channel is smaller than previously suspected and does suggest that proteins may be substantially unfolded during transport, there nonetheless is an opening with a diameter of 20 Å (Yonekura et al., 2003), in contrast to the situation for type IV pili. Perhaps pilus-mediated transformation is actually the clever coopting of an assembled functional pore in the membrane, which takes advantage of PilT-mediated retraction and does not require specific binding of DNA to the pilus (Wolfgang et al., 1999). Nonspecific electrostatic interactions with the pilus surface may play a role in transformation or phage DNA uptake, and this hypothesis should be directly testable (Keizer et al., 2001).

Type IVb Pili

Explicit knowledge of the helical symmetry underlying the structure of TCP was obtained by Fourier transform analysis of cryopreserved filaments and negatively stained bundles of filaments observed by electron microscopy (Craig et al., 2003). This analysis indicated a 3-start helical assembly, with a pitch of 45 Å (the distance from one strand to the next [Color Plate 26d]). Remarkably, crystal-packing interactions in the $P6_3$ TcpA crystal lattice also created a symmetric 3-start helical assembly with six subunits per turn of helix and a pitch of 37.5 Å. Although the 28 N-terminal residues of the native mature TcpA subunit were not present, 21 amino acids from the expression and purification tag were apparently filling the same space in the core of the filament, yielding crystal packing which provided strong clues to the biological interactions in the TCP filament. Given the symmetry of the crystal packing, both left- and right-handed helices were present. However, only subunits in the left-handed twist had a consistent conserved buried interface for all three monomers within the asymmetric unit, suggesting that these specific contacts between neighboring monomers are biologically relevant. An additional longitudinal contact along the crystallographic fiber axis was maintained in all of the subunits.

Using a similar computational approach to that pioneered for the MS11 pilus filament model (although with 23×10^6 parameter combinations tested), a model for the intact assembled TCP filament was derived by using a subunit structure whose N-terminal helix conformation was based on the PAK pilin structure (Color Plate 27e). The final TCP model maintains the right-handed 3-start helix predicted by electron microscopy and crystal symmetry. The energetically favorable and relatively nonspecific burial of the hydrophobic α1-helices on the interior of the filament is conserved as expected, although, in type IVb pilins, α1 does not have the proline or glycine residues that impart the S-shaped spine to type IVa pilins. Echoing the crystal-packing interactions, side chains of Pro58, Leu69,

Leu73, and Leu76 and the Gly72 main chain carbon from within or adjacent to α2 in the αβ loop of one monomer interact with Lys121 from α3 and with hydrophobic side chains of Leu176, Thr177, Ile179, Val182, and Leu185 from α4 in the D-region of the adjacent subunit (Color Plate 27f).

The TCP filament model can be used to explain mutational and immunological results. A collection of 16 single-amino-acid substitutions of the cysteines and the charged residues within the TcpA monomer D-region was created to explore the roles of individual residues in folding of TcpA, assembly of TCP filaments, and bundling interactions among multiple filaments (Kirn et al., 2000). A group of "structural" mutations which preclude pilus assembly are found within the first two-thirds of the D-region. These residues make critical side chain–main chain contacts within the subunit and are likely to prevent proper folding of the TcpA monomer. The TCP model shows that a second group, the "interaction" mutants that lower autoagglutination, are found near to but not buried in the subunit-subunit interface. These are largely positively charged side chains which may alter the surface properties of the pilus when changed to alanine. Taylor and colleagues have demonstrated the efficacy of a peptide vaccine for cholera in an infant-mouse model (Meeks et al., 2001; Wu et al., 2001). The most protective epitopes cover amino acids 145 to 168 and 174 to 199 within the D-region, and these localize to a patch on the surface of the assembled filament model.

Filament Conclusions

Conserved hydrophobic N-terminal α-helices on the interior of type IV pilus filaments form a strong, flexible, and well-packed core for all type IV pilus assemblies. Species-specific differences in pilin structure and size dictate unique molecular packing interactions, helical symmetries, and surface properties for each type IV pilus system. The gonococcal pilus filament model has been a helpful springboard for experimental designs to further probe type IV pilus structure, e.g., soluble expression of other pilins and pilin-like proteins. It has suggested preliminary assembly models for the pilins of other species in the absence of more detailed information. The TCP pilus filament model is based on the high-resolution crystal structure of a subunit combined with helical symmetry information from the same filament and may be a truer representation of that biological filament. However, this chapter is an appropriate venue in which to clarify that type IV pilus filament models are not refined atomic resolution structures of pilus quaternary assemblies. To increase the accuracy and resolution of any type IV pilus assembly model, the model must be rigorously refined against moderate-resolution data for intact pili, such as cryoelectron microscopy reconstructions or complete high-resolution fiber diffraction data sets. These have been challenging to obtain for the very smooth and thin type IV pili.

PILUS ASSEMBLY PROTEINS

Outer Membrane Secretin

Assembly of type IV pili involves secretion of pilin monomers through the inner membrane, across the periplasm through the peptidoglycan layer, and across the bacterial outer membrane. The outer membrane secretin PilQ forms a required channel for the exiting pilus filament. PilQ purified from *N. meningitidis* has been studied by negative-stain transmission electron microscopy by Derrick and colleagues (Collins et al., 2003). Single-particle averaging of detergent-solubilized PilQ showed strong 12-fold molecular symmetry, as predicted by earlier studies and as seen in related outer membrane secretins from filamentous phage assembly systems (Collins et al., 2001; Russel et al., 1997). Powerful 12-fold symmetry averaging was thus enforced on the final model, which has a nominal resolution of 25 Å.

The PilQ secretin is a cone-shaped dodecameric assembly with an overall form suggesting a gated outer membrane channel (Collins et al., 2003). Viewed from the top, the PilQ secretin is a ring with an outer diameter of 155 Å and an inner diameter of 65 Å, sufficient to accommodate an assembled *Neisseria* pilus filament (Fig. 2A). However, the opening is not continuous. The upper cavity begins to taper at a depth of ~6 nm, narrowing to an inner diameter of about 30 Å at a depth of 10 nm and forming a closed plug below this depth. Viewed from the side, the 8-nm-high upper chamber is seen to constrict to an outer diameter of 110 Å at the junction of the ring region with the plug (Fig. 2B). The plug tapers to a 45-Å-diameter blunt tip over a 4-nm height change, for a total height of the secretin assembly of 12 nm. The position of the larger-diameter ring region is assumed to coincide with the outer membrane in vivo, and the plug is presumed to reside within the periplasm, although there are no experimental results to support this orientation.

Two possibilities for the pilus assembly mechanism present themselves in light of the PilQ structure (Collins et al., 2003). In the first, fully assembled pilus filaments with an approximate diameter of 6.5 nm pass intact through the PilQ channel after a major conformational change in the plug. In the second, monomers are added to the base of a growing pilus from within PilQ, requiring a smaller opening and smaller conformational change within the plug. In either model, intriguing questions remain regarding the molecular details of assembly and also of reversal for retraction.

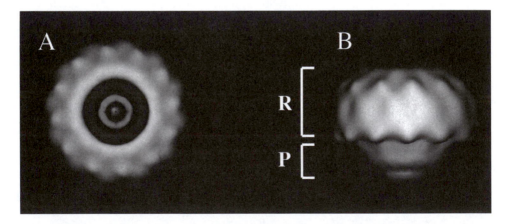

Figure 2. Structure of the *N. meningitidis* outer membrane secretin PilQ as determined by negative-stain electron microscopy reconstruction with 12–fold averaging. (A) Top-down view showing the central cavity; overall width, 155 Å. (B) Side-on view, rotated 90° about the *x* axis from panel A; overall height, 120 Å. R and P indicate the ring and plug regions, respectively. Reprinted from Collins et al. (2003) with permission.

Pilin-Like Proteins

Pilin monomers are added to the base of the growing pilus, and the entire filament remains anchored in the bacterial envelope, while the pilus filament, often several times the length of the bacterium, is found outside the cell. Very little is known about the structural biology of the proteins required for pilus assembly. Pilin monomers direct their own secretion to the inner membrane, probably by virtue of the hydrophobic conserved N-terminal helix. A polytopic inner membrane protease, PilD, cleaves a conserved leader peptide and N methylates the resulting N-terminal amino acid. Many "pilin-like" proteins have been identified and play diverse roles, including pilus assembly, anchoring, twitching motility, and competence (Mattick, 2002). Two examples from *N. gonorrhoeae* are ComP, in which a mutation abolishes natural transformation but does not otherwise affect pilus structure, and PilV, which is required for host cell attachment (Aas et al., 2002; Winther-Larsen et al., 1998). Pilin-like proteins have 66% identity and nearly 100% sequence similarity to pilins across the first 30 amino acids (Aas et al., 2002) and are synthesized as prepilins and cleaved by PilD. In the absence of crystal structure information, these are convincing arguments that pilin-like proteins share the long α1-helix structure with pilin. This allows us to consider interactions with the major subunit proteins through the shared N-terminal hydrophobic helix that mimic filament assembly interactions. It has been proposed that pilin-like proteins are components of a pseudopilus that forms a scaffold or tunnel across the periplasmic space, joining the inner and outer membrane (Alm and Mattick, 1997). Such a model is supported by electron microscopy data revealing pilus-like filaments on the surface of *E. coli* cells engineered to overexpress a pilin-like protein from the *Klebsiella* type II secretion system (Dupuy and Pugsley, 1994), which has many similarities to the type IV pilus assembly system.

Assembly and Retraction ATPases

Each type IV pilus assembly pathway requires the action of a conserved putative ATPase associated with the inner membrane, which presumably provides the energy for secretion of pilin subunits (Alm and Mattick, 1997). These proteins (named PilB in *P. aeruginosa,* for example) are a subset of a larger family of conserved type II and type IV secretion ATPases (Planet et al., 2001). Although no high-resolution information is available for a PilB family member, general conclusions from high-resolution structures of type II (Robien et al., 2003) and type IV (Yeo et al., 2000) secretion ATPases are likely to be applicable to PilB, in particular within a core region of homology that includes the Walker A and B boxes for nucleoside triphosphate binding and hydrolysis as well as highly conserved aspartate and histidine boxes. Each of these secretion systems is addressed in a separate chapter within this book, to which the reader is referred.

Unique to some type IV pilus systems is a second class of ATPase, PilT/PilU. As described in the introduction to this chapter, PilT is dispensable for pilus assembly and initial attachment but required for other pilus-mediated functions including motility, natural transformation, phage sensitivity, and downstream signaling events in host cells. PilT is a ~330-amino-acid member of its own group within the type II secretion ATPase family (Planet et al., 2001). It is a hexameric ATPase whose enzymatic activity is required for its biological functions (Herdendorf et al., 2002; Okamoto and Ohmori, 2002; Bieber et al., 1998). Elegant biophysical experiments estimated the force per retraction event to be 100 pN and thus established PilT as one of the strongest biological motors known. However, the mechanism of retraction and the structure of PilT remain unknown. Proposals for the PilT retraction mechanism include an ATP-driven conformational change in PilT coupled to disassembly of pili, or a rotary motor akin to the F_1ATPase (Merz and Forest, 2002; Herdendorf et al., 2002; Kaiser, 2000). Gonococcal PilT has been shown by freeze-etch electron microscopy to form regular rings with an outer diameter of approximately 12 nm in the presence of ATP, and diffraction quality crystals of PilT have recently become available (Forest et al., 2004).

CONCLUSION

Crystallography of membrane proteins and multiprotein complexes is rapidly advancing to include larger and less stable biological assemblies, while at the same time the results of cryoelectron microscopy approach atomic resolution. High-resolution X-ray crystal structures of several pilin subunits are now available. They have revealed a conserved α/β-roll fold for widely sequence-divergent pilins, with unique functionalities contained in the $\alpha\beta$ loop and D-region of each pilin. Models have been derived for type IVa and IVb pilus filaments that match most available biochemical, immunological, and genetic data and permit the design of further experiments. These models share a conserved interaction of hydrophobic α-helices within an outer layer whose structure is dictated by species-specific interactions which determine unique helical parameters for each pilin. The surface of each filament is dominated by nonconserved structures and permits these type IV pili to adapt to a wide range of environmental niches and microbe-specific functions. Structure-based vaccine development in the future will probably include structurally stabilized PAK pilin receptor-binding loops to mimic the conformation seen in the intact protein (Hazes et al., 2000), gonococcal pilin subunits engineered to remove the hypervariable region (Forest and Tainer, 1997a) and predicted surface-exposed peptides for TCP (Craig et al., 2003). X-ray crystallography on additional pilin subunits, high-resolution electron microscopy reconstructions of intact pilus filaments, and structural biology of additional proteins in the type IV pilus pathway are clearly important next steps for understanding the mechanisms of pilus assembly, retraction, and function.

The PDB codes for structures discussed in this review are 2PIL for *N. gonorrhoeae* MS11 pilin and fiber model, 1DZO for truncated *P. aeruginosa* PAK pilin, 1OQW for full-length *P. aeruginosa* PAK pilin, 1HPW for truncated *P. aeruginosa* K122-4 pilin, 1OQV for *V. cholerae* TcpA, and 1OR9 for the TCP fiber model.

Acknowledgments. I am grateful to Lisa Craig for continuing valuable discussions on type IV pilus structure and to Lisa Craig and Kelly Aukema for insightful comments on the manuscript.

Funds from the NIH (GM59721) and the W. M. Keck Foundation supported the work in my laboratory during the preparation of this chapter.

Addendum in Proof. Since this chapter was submitted, additional structures of type IV pilins have become available: *Salmonella enterica* serovar Typhi PilS (X. F. Xu, Y. W. Tan, L. Lam, J. Hackett, M. Zhang, and Y. K. Mok, *J. Biol. Chem.* **279:**31599–31605, 2004), *P. aeruginosa* K122-4 pilin at 1.54-Å resolution (G. F. Audette, R. T. Irvin, and B. Hazes, *Biochemistry* **43:**11427–11435, 2004), and *Klebsiella oxytoca* pseudopilin PulG (R. Köhler, K. Schafer, S. Muller, G. Vignon, K. Diederichs, A. Philippsen, P. Ringler, A. P. Pugsley, A. Engel, and W. Welte, *Mol. Microbiol.* **54:**647–664, 2004).

REFERENCES

Aas, F. E., M. Wolfgang, S. Frye, S. Dunham, C. Lovold, and M. Koomey. 2002. Competence for natural transformation in *Neisseria gonorrhoeae:* components of DNA binding and uptake linked to type IV pilus expression. *Mol. Microbiol.* **46:**749–760.

Alm, R. A., and J. S. Mattick. 1997. Genes involved in the biogenesis and function of type-4 fimbriae in *Pseudomonas aeruginosa. Gene* **192:**89–98.

Audette, G. F. 2003. Purification, crystallization and preliminary diffraction studies of the *Pseudomonas aeruginosa* strain K122-4 monomeric pilin. *Acta Crystallogr. Ser. D* **59:**1665–1667.

Bieber, D., S. W. Ramer, C.-Y. Wu, W. J. Murray, T. Tobe, R. Fernandez, and G. Schoolnik. 1998. Type IV pili, transient bacterial aggregates, and virulence of enteropathogenic *Escherichia coli. Science* **280:**2114–2118.

Boslego, J. W., E. C. Tramont, R. C. Chung, D. G. McChesney, J. Ciak, J. C. Sadoff, M. V. Piziak, J. D. Brown, C. C. Brinton, Jr., S. W. Wood, et al. 1991. Efficacy trial of a parenteral gonococcal pilus vaccine in man. *Vaccine* **9:**154–162.

Bradley, D. E. 1980. A function of *Pseudomonas aeruginosa* PAO polar pili: twitching motility. *Can. J. Microbiol.* **26:**146–154.

Bullitt, E., and L. Makowski. 1998. Bacterial adhesion pili are heterologous assemblies of similar subunits. *Biophys. J.* **74:**623–632.

Cachia, P. J., and R. S. Hodges. 2003. Synthetic peptide vaccine and antibody therapeutic development: prevention and treatment of *Pseudomonas aeruginosa. Biopolymers* **71:**141–168.

Cohen-Krausz, S., and S. Trachtenberg. 2002. The structure of the archeabacterial flagellar filament of the extreme halophile *Halobacterium salinarum* R1M1 and its relation to eubacterial flagellar filaments and type IV pili. *J. Mol. Biol.* **321:**383–395.

Collins, R. F., L. Davidsen, J. P. Derrick, R. C. Ford, and T. Tonjum. 2001. Analysis of the PilQ secretin from *Neisseria meningitidis* by transmission electron microscopy reveals a dodecameric quaternary structure. *J. Bacteriol.* **183:**3825–3832.

Collins, R. F., R. C. Ford, A. Kitmitto, R. O. Olsen, T. Tonjum, and J. P. Derrick. 2003. Three-dimensional structure of the *Neisseria meningitidis* secretin PilQ determined from negative-stain transmission electron microscopy. *J. Bacteriol.* **185:**2611–2617.

Comolli, J. C., L. L. Waite, K. E. Mostov, and J. N. Engel. 1999. Pili binding to asialo-GM$_1$ on epithelial cells can mediate cytotoxicity or bacterial internalizatin by *Pseudomonas aeruginosa. Infect. Immun.* **67:**3207–3214.

Craig, L., M. E. Pique, and J. A. Tainer. 2004. Type IV pilus structure and bacterial pathogenicity. *Nat. Rev. Microbiol.* **2:**363–378.

Craig, L., R. K. Taylor, M. E. Pique, B. D. Adair, A. S. Arvai, M. Singh, S. J. Lloyd, D. S. Shin, E. D. Getzoff, M. Yeager, K. T. Forest, and J. A. Tainer. 2003. Type IV pilin structure and assembly: X-ray and EM analyses of *Vibrio cholerae* toxin-coregulated pilus and *Pseudomonas aeruginosa* PAK pilin. *Mol. Cell* **11:**1139–1150.

Dupuy, B., and A. P. Pugsley. 1994. Type IV prepilin peptidase gene of *Neisseria gonorrhoeae* MS11: presence of a related gene in other piliated and nonpiliated *Neisseria* strains. *J. Bacteriol.* **176:**1323–1331.

Folkhard, W., D. A. Marvin, T. H. Watts, and W. Paranchych. 1981. Structure of polar pili from *Pseudomonas aeruginosa* strains K and O. *J. Mol. Biol.* **149:**79–93.

Forest, K. T., S. L. Bernstein, E. D. Getzoff, M. So, G. Tribbick, H. M. Geysen, C. D. Deal, and J. A. Tainer. 1996. Assembly and antigenicity of the *N. gonorrhoeae* pilus mapped with antibodies. *Infect. Immun.* **64:**644–652.

Forest, K. T., S. J. Dunham, M. Koomey, and J. A. Tainer. 1999. Crystallographic structure of phosphorylated pilin from *Neisseria:* phosphoserine sites modify type IV pilus surface chemistry and morphology. *Mol. Microbiol.* **31:**743–752.

Forest, K. T., K. A. Satyshur, G. A. Worzalla, J. K. Hansen, and T. J. Herdendorf. 2004. The pilus-retraction protein PilT: ultrastructure of the biological assembly. *Acta Crystallogr.* **D60:**978–982.

Forest, K. T., and J. A. Tainer. 1997a. Type IV pilin structure, assembly and immunodominance: applications to vaccine design, p. 167–173. *In* F. Brown, D. Burton, P. Doherty, J. J. Mekalanos, and E. Norrby (ed.), *Vaccines 97: Molecular Approaches to the Control of Infectious Diseases.* Cold Spring Harbor Laboratory Press, Cold Spring Harbor, N.Y.

Forest, K. T., and J. A. Tainer. 1997b. Type-4 pilus structure: outside to inside and top to bottom—a minireview. *Gene* **192**:165–169.

Friedrich, A., C. Prust, T. Hartsch, A. Henne, and B. Averhoff. 2002. Molecular analyses of the natural transformation machinery and identification of pilus structures in the extremely thermophilic bacterium *Thermus thermophilus* strain HB27. *Appl. Environ. Microbiol.* **68**:745–755.

Gonzalez, A., C. Nave, and D. A. Marvin. 1995. Pf1 filamentous bacteriophage: refinement of a molecular model by simulated annealing using 3.3 Å resolution X-ray fibre diffraction data. *Acta Crystallogr. Ser. D* **51**: 792–804.

Hazes, B., P. A. Sastry, K. Hayakawa, R. J. Read, and R. T. Irwin. 2000. Crystal structure of *Pseudomonas aeruginosa* PAK pilin suggests a main-chain-dominated mode of receptor binding. *J. Mol. Biol.* **299**:1005–1017.

Hegge, F. T., P. G. Hitchen, F. E. Aas, H. Kristiansen, C. Lovold, W. Egge-Jacobsen, M. Panico, W. Y. Leong, V. Bull, M. Virji, H. R. Morris, A. Dell, and M. Koomey. 2004. Unique modifications with phosphocholine and phosphoethanolamine define alternate antigenic forms of *Neisseria gonorrhoeae* type IV pili. *Proc. Natl. Acad. Sci. USA* **101**:10798–10803.

Herdendorf, T. J., D. McCaslin, and K. T. Forest. 2002. *A. aeolicus* PilT, homologue of a twitching motility protein, is an oligomeric ATPase. *J. Bacteriol.* **184**:6465–6471.

Herrington, D. A., R. H. Hall, G. Losonsky, J. J. Mekalanos, R. K. Taylor, and M. M. Levine. 1988. Toxin, toxin-coregulated pilus, and the toxR regulon are essential for *Vibrio cholerae* pathogenesis in humans. *J. Exp. Med.* **168**:1487–1492.

Hertle, R., R. Mrsny, and D. J. Fitzgerald. 2001. Dual-function vaccine for *Pseudomonas aeruginosa:* characterization of chimeric exotoxin A-pilin protein. *Infect. Immun.* **69**:6962–6969.

Hopfner, K. P., L. Craig, G. Moncalian, R. A. Zinkel, T. Usui, B. A. Owen, A. Karcher, B. Henderson, J. L. Bodmer, C. T. McMuray, J. H. Petrini, and J. A. Tainer. 2002. The Rad50 zinc-hook is a structure joining Mre11 complexes in DNA recombination and repair. *Nature* **418**:562–566.

Kabsch, W., H. G. Mannherz, D. Suck, E. F. Pai, and K. C. Holmes. 1990. Atomic structure of the actin:DNAse I complex. *Nature* **347**:37–43.

Kaiser, D. 2000. Bacterial motility: how do pili pull? *Curr. Biol.* **10**:R777–R780.

Källström, H., M. S. Islam, P. O. Berggren, and A. B. Jonsson. 1998. Cell signaling by the type IV pili of pathogenic *Neisseria. J. Biol. Chem.* **273**:21777–21782.

Källström, H., M. K. Liszewski, J. P. Atkinson, and A. B. Jonsson. 1997. Membrane cofactor protein is a cellular pilus receptor for pathogenic *Neisseria. Mol. Microbiol.* **25**:639–647.

Kang, Y., H. Liu, S. Genin, M. A. Schell, and T. P. Denny. 2002. *Ralstonia solanacearum* requires type 4 pili to adhere to multiple surfaces and for natural transformation and virulence. *Mol. Microbiol.* **46**:427–437.

Keizer, D. W., C. M. Slupsky, M. Kalisiak, A. P. Campbell, M. P. Crump, P. A. Sastry, B. Hazes, R. T. Irvin, and B. D. Sykes. 2001. Structure of a pilin monomer from *Pseudomonas aeruginosa*—implications for the assembly of pili. *J. Biol. Chem.* **276**:24186–24193.

Kirn, T. J., M. J. Lafferty, C. M. P. Sandoe, and R. K. Taylor. 2000. Delineation of pilin domains required for bacterial association into microcolonies and intestinal colonization by *Vibrio cholerae. Mol. Microbiol.* **35**:896–910.

Leiman, P. G., M. M. Shneider, V. A. Kostyuchenko, P. R. Chipman, V. V. Mesyanzhinov, and M. G. Rossmann. 2003. Structure and location of gene product 8 in the bacteriophage T4 baseplate. *J. Mol. Biol.* **328**:821–833.

Li, Y., H. Sun, X. Ma, A. Lu, R. Lux, D. Zusman, and W. Shi. 2003. Extracellular polysaccharides mediate pilus retraction during social motility of *Myxococcus xanthus. Proc. Natl. Acad. Sci. USA* **100**:5443–5448.

Maier, B., L. Potter, M. So, H. S. Seifert, and M. P. Sheetz. 2002. Single pilus motor forces exceed 100 pN. *Proc. Natl. Acad. Sci. USA* **99**:16012–16017.

Marceau, M., K. T. Forest, J. A. Tainer, and X. Nassif. 1998. Role of O-linked glycosylation of meningococcal type IV pilin for piliation and pilus-mediated adhesion. *Mol. Microbiol.* **27**:705–715.

Marceau, M., and X. Nassif. 1999. Role of glycosylation at Ser63 in production of soluble pilin in pathogenic *Neisseria. J. Bacteriol.* **181**:656–661.

Marvin, D. A. 1998. Filamentous phage structure, infection and asembly. *Curr. Opin. Struct. Biol.* **8**:150–158.

Mattick, J. S. 2002. Type IV pili and twitching motility. *Annu. Rev. Microbiol.* **56**:289–314.

McBride, M. J. 2001. Bacterial gliding motility: multiple mechanisms for cell movement over surfaces. *Annu. Rev. Microbiol.* **55**:49–75.

McInnes, C., F. D. Soennichsen, C. M. Kay, R. S. Hodges, and B. D. Sykes. 1993. NMR solution structure and flexibility of a peptide antigen representing the receptor binding domain of *Pseudomonas aeruginosa. Biochemistry* **32:**13432–13440.

Meeks, M. D., T. K. Wade, R. K. Taylor, and W. F. Wade. 2001. Immune response genes modulate serologic responses to *Vibrio cholerae* TcpA pilin peptides. *Infect. Immun.* **69:**7687–7694.

Merz, A. J., and K. T. Forest. 2002. Bacterial surface motility: slime trails, grappling hooks and nozzles. *Curr. Biol.* **12:**R297–R303.

Merz, A. J., D. B. Rifenbery, C. G. Arvidson, and M. So. 1996. Traversal of a polarized epithelium by pathogenic neisseriae: facilitation by type IV pili and maintenance of epithelial barrier function. *Mol. Med.* **2:**745–754.

Merz, A. J., and M. So. 2000. Interactions of pathogenic neisseriae with epithelial cell membranes. *Annu. Rev. Cell Dev. Biol.* **16:**423–457.

Merz, A. J., M. So, and M. P. Sheetz. 2000. Pilus retraction powers bacterial twitching motility. *Nature* **407:**98–102.

Namba, K., R. Pattanayek, and G. Stubbs. 1989. Visualization of protein-nucleic acid interactions in a virus. Refined structure of intact tobacco mosaic virus at 2.9 Å resolution by X-ray fiber diffraction. *J. Mol. Biol.* **208:**307–325.

Nassif, X., C. Pujol, P. Mornad, and E. Eugene. 1999. Interactions of pathogenic *Neisseria* with host cells. Is it possible to assemble the puzzle? *Mol. Microbiol.* **32:**1124–1132.

Okamoto, S., and M. Ohmori. 2002. The cyanobacterial PilT protein responsible for cell motility and transformation hydrolyzes ATP. *Plant Cell Physiol.* **43:**1127–1136.

O'Toole, G. A., and R. Kolter. 1998. Flagellar and twitching motility are necessary for *Pseudomonas aeruginosa* biofilm development. *Mol. Microbiol.* **30:**295–304.

Parge, H. E., S. L. Bernstein, C. D. Deal, D. E. McRee, D. Christensen, M. A. Capozza, B. W. Kays, T. M. Fieser, D. Draper, M. So, E. D. Getzoff, and J. A. Tainer. 1990. Biochemical purification and crystallographic characterization of the fiber-forming protein pilin from *Neisseria gonorrhoeae. J. Biol. Chem.* **265:**2278–2285.

Parge, H. E., K. T. Forest, M. J. Hickey, D. A. Christensen, E. D. Getzoff, and J. A. Tainer. 1995. Structure of the fibre-forming protein pilin at 2.6 Å resolution. *Nature* **378:**32–38.

Planet, P. J., S. C. Kachlany, R. DeSalle, and D. H. Figurski. 2001. Phylogeny of genes encoded for secretion NTPases: identification of the widespread tadA subfamily and development of a diagnostic key for gene classification. *Proc. Natl. Acad. Sci. USA* **98:**2503–2508.

Power, P. M., L. F. Roddam, K. Rutter, S. Z. Fitzpatrick, Y. N. Srikhanta, and M. P. Jennings. 2003. Genetic characterization of pilin glycosylation and phase variation in *Neisseria meningitidis. Mol. Microbiol.* **49:**833–847.

Pujol, C., E. Eugene, M. Marceau, and X. Nassif. 1999. The meningococcal PilT protein is required for induction of intimate attachment to epithelial cells following pilus-mediated adhesion. *Proc. Natl. Acad. Sci. USA* **96:**4017–4022.

Rayment, I., W. R. Rypniewski, K. Schmidt-Base, R. Smith, D. R. Tomchick, M. M. Benning, D. A. Winkelmann, G. Wesenberg, and H. M. Holden. 1993. Three-dimensional structure of myosin subfragment-1: a molecular motor. *Science* **261:**50–58.

Robien, M. A., B. E. Krumm, M. Sandkvist, and W. G. J. Hol. 2003. Crystal structure of the extracellular protein secretion NTPase EpsE of *Vibrio cholerae. J. Mol. Biol.* **333:**657–674.

Russel, M., N. A. Linderoth, and A. Sali. 1997. Filamentous phage assembly: variation on a protein export theme. *Gene* **192:**23–32.

Samatey, F. A., K. Imada, F. Vonderviszt, Y. Shirakihara, and K. Namba. 2000. Crystallization of the F41 fragment of flagellin and data collection from extremely thin crystals. *J. Struct. Biol.* **132:**106–111.

Schutt, C. E., J. C. Myslik, M. D. Rozycki, N. C. W. Goonesekere, and U. Lindberg. 1993. The structure of crystalline profilin:β-actin. *Nature* **365:**810–816.

Serkin, C. D., and H. S. Seifert. 1998. Frequency of pilin antigenic variation in *Neisseria gonorrhoeae. J. Bacteriol.* **180:**1955–1959.

Skerker, J. M., and H. C. Berg. 2001. Direct obervation of extension and retraction of type IV pili. *Proc. Natl. Acad. Sci. USA* **98:**6901–6904.

Soto, G., and S. J. Hultgren. 1999. Bacterial adhesins: common themes and variations in architecture and assembly. *J. Bacteriol.* **181:**1059–1071.

Stimson, E., M. Virji, S. Barker, M. Panico, I. Blench, J. Saunders, G. Payne, E. R. Moxon, A. Dell, and H. R. Morris. 1996. Discovery of a novel protein modification: α-glycerophosphate is a substituent of meningococcal pilin. *Biochem. J.* **316:**29–33.

Stimson, E., M. Virji, K. Makepeace, A. Dell, H. R. Morris, G. Payne, J. R. Saunders, M. P. Jennings, S. Barker, M. Panico, I. Blench, and E. R. Moxon. 1995. Meningococcal pilin: a glycoprotein substituted with digalactosyl-2,4–diacetamido-2,4,6-trideoxyhexose. *Mol. Microbiol.* **17:**1201–1214.

VanLoock, M. S., X. Yu, S. Yang, A. L. Lai, C. Low, M. J. Campbell, and E. H. Egelman. 2003. ATP-mediated conformational changes in the RecA filament. *Structure* **11:**187–196.

Vorobiev, S., B. Strolopytov, D. G. Drubin, C. Frieden, S. Ono, J. Condeelis, P. A. Rubenstein, and S. C. Almo. 2003. The structure of nonvertebrate actin: implications for the ATP hydrolytic mechanism. *Proc. Natl. Acad. Sci. USA* **100:**5760–5765.

Watts, T. H., C. M. Kay, and W. Paranchych. 1983. Spectral properties of three quaternary arrangements of *Pseudomonas* pilin. *Biochemistry* **22:**3640–3646.

Winther-Larsen, H. C., F. T. Hegge, M. Wolfgang, S. F. Hayes, J. P. van Putten, and M. Koomey. 1998. *Neisseria gonorrhoeae* PilV, a type IV pilus-associated protein essential to human epithelial cell adherence. *Proc. Natl. Acad. Sci. USA* **98:**15276–15281.

Wolfgang, M., J. P. van Putten, S. F. Hayes, and J. M. Koomey. 1999. The comP locus of *Neisseria gonorrhoeae* encodes a type IV prepilin that is dispensable for pilus biogenesis but essential for natural transformation. *Mol. Microbiol.* **31:**1345–1357.

Wong, W. Y., A. P. Campbell, C. McInnes, B. D. Sykes, W. Paranchych, R. T. Irvin, and R. S. Hodges. 1995. Structure-function analysis of the adherence-binding domain on the pilin of *Pseudomonas aeruginosa* strains PAK and KB7. *Biochemistry* **34:**12963–12972.

Wu, J. Y., W. F. Wade, and R. K. Taylor. 2001. Evaluation of cholera vaccines formulated with toxin-coregulated pilin peptide plus polymer adjuvant in mice. *Infect. Immun.* **69:**7695–7702.

Yeo, H. J., S. N. Savvides, A. B. Herr, E. Lanka, and G. Waksman. 2000. Crystal structure of the hexameric traffic ATPase of the *Helicobacter pylori* type IV secretion system. *Mol. Cell* **6:**1461–1472.

Yonekura, K., S. Maki-Yonekura, and K. Namba. 2003. Complete atomic model of the bacterial flagellar filament by electron cryomicroscopy. *Nature* **424:**643–650.

Structural Biology of Bacterial Pathogenesis
Edited by G. Waksman et al.
© 2005 ASM Press, Washington, D.C.

Chapter 7

Sortase Pathways in Gram-Positive Bacteria

Kevin M. Connolly and Robert T. Clubb

Surface proteins in gram-positive pathogens promote bacterial adhesion to specific organ tissues, resistance to phagocytic killing, and host cell invasion. Research within the past decade has revealed that a large fraction of bacterial surface proteins are covalently anchored to the cell wall by the action of sortase enzymes, in a universally conserved process that is important for infectivity. Many excellent reviews on the subject of sortase enzymes and surface proteins have been written (Navarre and Schneewind, 1999; Cossart and Jonquieres, 2000; Novick, 2000; Mazmanian et al., 2001). This chapter presents a review of the structural basis of sortase-mediated cell wall anchoring, drawing on recent structural, biochemical, and bioinformatic studies of this enzyme family.

BACKGROUND

Protein Display on the Cell Wall

The cell wall of gram-positive bacteria is a critical organelle for survival, both protecting the bacterium from osmotic and mechanical stress and providing a scaffold for a diverse array of proteins that interact with the cell's environment (Sara, 2001). The peptidyl-glycan and secondary cell wall polymers (teichuronic acid [TUA], teichoic acid [TA], and lipoteichoic acid [LTA]) serve as attachment points for a myriad of surface proteins charged with roles in bacterial adhesion, resistance to phagocytic killing, and invasion of nonprofessional phagocytic cells (Navarre and Schneewind, 1999). In addition to embedding proteins into the underlying membrane (e.g., membrane proteins and lipoproteins), gram-positive bacteria have developed several methods to display surface proteins, each with its own distinctive structural features. Several surface proteins are associated with the cell wall through noncovalent interactions (Fig. 1). For example, the LytA amidase of *Streptococcus pneumoniae* utilizes six choline-binding repeats that interact with choline-substituted TA or LTA polymers (Sanchez-Puelles et al., 1990; Varea et al., 2000). The repeats form short β-hairpins that stack into a left-handed superhelix, with the choline-binding pockets located between successive hairpins (Fernandez-Tornero et al., 2001). A

Kevin M. Connolly and Robert T. Clubb • Department of Chemistry and Biochemistry, Molecular Biology Institute, and UCLA-DOE Institute for Genomics and Proteomics, University of California, Los Angeles, Los Angeles, CA 90095-1570.

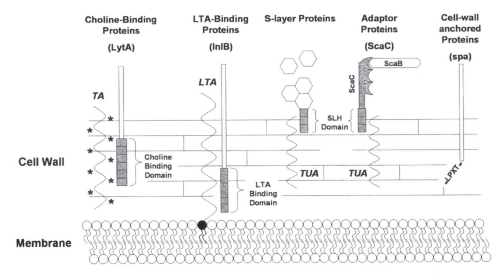

Figure 1. Mechanisms of cell wall attachment. Choline-binding proteins recognize choline-substituted TA and LTA (asterisks) through 20-residue C-terminal repeats, which form stacked β-hairpins. LTA can be directly bound by the GW domains of InlB. Crystalline S-layer proteins self-assemble at the cell surface; they are tethered to TUA polymers through SLH domains. The SLH domain of ScaC affixes it to the membrane, and its three cohesin domains serve as attachment points for the adaptor proteins and enzymes of the cellulosome. Covalent linkage of Cws-containing proteins by a sortase enzyme produces an amide linkage between the threonine of the LPXTG motif of the sorting signal and the peptide cell wall crossbridge. Adapted from Cossart and Jonquieres (2000).

distinct structural solution is used by the InlB protein of *Listeria monocytogenes,* which directly associates with LTA via GW-type domains that adopt an SH3-type protein fold (Braun et al., 1997; Marino et al., 1999, 2002). A third commonly used noncovalent attachment mechanism is exhibited by S-layer proteins, which employ a 50- to 60-amino-acid S-layer-homologous (SLH) motif that interacts with TUA (Lupas et al., 1994; Engelhardt and Peters, 1998; Sara, 2001). Cell wall-attached proteins can also act as adaptors, serving as the attachment points for additional surface proteins (Navarre and Schneewind, 1999). The scaffoldin protein ScaC of the *Acetivibrio* cellulosome is one such example; it uses a C-terminal SLH domain to attach to the cell wall, and associates with additional proteins through its three cohesin domains (Xu et al., 2003).

In contrast to the structural diversity exhibited in noncovalent attachment mechanisms, a highly conserved process appears to be used by nearly all gram-positive bacteria to covalently tether proteins to their cell wall. This process, called sorting, is exemplified by the anchoring of *Staphylococcus aureus* protein A, which is attached (sorted) to the cell wall through the action of a novel cysteine transpeptidase called sortase. The sortase enzyme recognizes a C-terminal LPXTG sorting signal in protein A and performs a transpeptidation reaction that links the protein to the cell wall pentaglycine crossbridge. Numerous studies (detailed below) have shown that this sorting reaction plays an important role in the virulence of a variety of human pathogens. A recent analysis of the completely sequenced genome of *L. monocytogenes* highlights the prevalence of cell wall sorting: of the 133 putative surface proteins in this pathogen, 41 contain LPXTG motifs (Cabanes et al., 2002).

Medical Importance of Sortase Enzymes

Sortase enzymes may be excellent targets for new broad-spectrum anti-infective agents, since sortase-like enzymes and surface proteins containing a cell wall sorting signal are present in nearly all gram-positive bacteria (Janulczyk and Rasmussen, 2001; Pallen et al., 2001). Most importantly, several recent studies have shown that these enzymes play a critical role in pathogenesis. For example, sortase-mediated cell wall anchoring is essential for the pathogenesis of acute *S. aureus* infections, since sortase-deficient strains display defects in their ability to form renal abscesses in mice (Mazmanian et al., 2000) and in their ability to cause septic arthritis (Jonsson et al., 2002; Jonsson et al., 2003). The cariogenic microbes *Streptococcus gordonii* and *Streptococcus mutans* require a sortase to adhere to and colonize the oral mucosa (Bolken et al., 2001; Igarashi et al., 2003; Lee and Boran, 2003). *Actinomyces* spp. have also been implicated in several types of oral disease and may require sortase enzymes for fimbria-mediated adhesion to salivary acidic proline-rich proteins during infection (Li et al., 2001). *S. pneumoniae,* which naturally resides in the human nasopharyngeal mucosa, is able to colonize and invade human hosts through a large number of surface-located virulence factors (Jedrzejas, 2001; Hava and Camilli, 2002). Both noncovalently and covalently attached surface proteins are involved in pneumococcal infections; however, the elimination of sortase from pneumococci drastically reduces their ability to adhere to and invade human pharyngeal cells in vitro (Kharat and Tomasz, 2003). Finally, the elimination of the sortase enzyme from the food-borne pathogen *L. monocytogenes* significantly attenuates its ability to invade human hepatocytes in vitro and enables mice to recover from otherwise lethal doses of this pathogen (Bierne et al., 2002; Garandeau et al., 2002).

Brief History of Protein Sorting

It has long been known that some proteins in gram-positive bacteria are covalently linked to the cell wall, but the enzymes that place them there have only recently been identified. In the early 1970s, Sjoquist and colleagues demonstrated that protein A was covalently attached to the cell of *S. aureus* by showing that it could be released only after treatment with the peptidoglycan-digesting enzymes lysostaphin (a *Staphylococcus simulans* enzyme that cleaves at the glycine within the cell wall crossbridge) or lysozyme (an *N*-acetylmuramidase that cleaves the polysaccharide) (Sjoquist et al., 1972a; Sjoquist et al., 1972b). Nearly 20 years later, Fischetti's group discovered the signal that predestines proteins for attachment. Building on studies of the M protein from *Streptococcus pyogenes* that had implicated its C terminus in cell wall attachment (Fischetti et al., 1985), Fischetti compared the amino acid sequences of several surface-anchored proteins and identified a conserved sequence motif, dubbed the cell wall-sorting signal (Cws) (Fischetti et al., 1990, 1991; Schneewind et al., 1991). As shown in Fig. 2A, the Cws consists of an LPXTG sequence motif, where X is any amino acid, followed by a C-terminal hydrophobic domain and a tail of mostly positively charged residues (Schneewind et al., 1991). The importance of the Cws in protein anchoring was rapidly demonstrated, and it was soon found in a wide range of bacterial proteins, indicating that it was involved in a general mechanism of protein attachment used by many species of gram-positive bacteria (Schneewind et al., 1993).

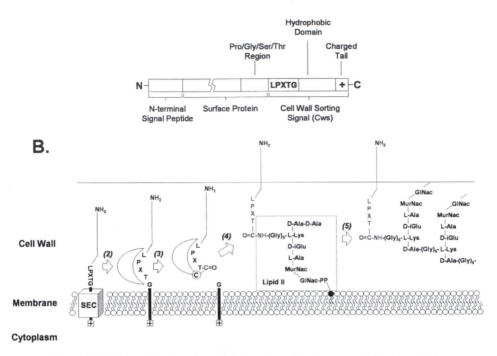

Figure 2. (A) Schematic of a sortase substrate. The protein is composed of an N-terminal signal peptide and a C-terminal Cws. The Cws contains a conserved LPXTG motif followed by a hydrophobic stretch of amino acids and positively charged residues at the C terminus. (B) Proposed model for the cell wall sorting reaction. The full-length surface protein precursor is secreted through the membrane via an N-terminal signal sequence. A charged tail at the C terminus of the protein may serve as a stop transfer signal. Following cleavage of this secretion signal, a sortase enzyme cleaves the protein between the threonine and glycine residues of the LPXTG motif, forming a thioacyl-enzyme intermediate. The free amine of the cell wall crossbridge of lipid II is deprotonated in the SrtA active site and serves as the acceptor for the transpeptidation reaction. The covalently linked protein-lipid II intermediate may then serve as a substrate for the transglycosylation reaction of cell wall synthesis.

Although the Cws was clearly important, the way in which the signal was used to tether proteins to the cell surface remained unknown. Work in Schneewind's laboratory over the past decade has now revealed the enzymatic machinery and general mechanism of attachment. The conserved LPXTG motif within surface proteins had been postulated to be the site of attachment to the cell wall, since protein A produced from lysostaphin-treated cell walls was shorter than the full-length protein (Schneewind et al., 1992). The exact structure of the anchor was elucidated by a mass spectrometry analysis of muramidase-digested *S. aureus* cell wall containing a reporter construct of protein A (Navarre et al., 1998). This work revealed that attached protein A was C-terminally truncated and that it contained three additional glycine residues amide-linked to the threonine of the LPXTG motif. These results, taken together, led to the model in which the LPXTG motif was the recognition sequence for a yet to be identified enzyme, which presumably cleaved between the threonine and glycine residues of the Cws and catalyzed the amide linkage of the threonine to the pentaglycine bridge of the *S. aureus* cell wall (Navarre and Schneewind, 1994; Schneewind et al., 1995). It was some time later that the enzyme responsible for performing this transpeptidation reaction was identified in a screen of temperature-sensitive *S. aureus* mutants that showed the aberrant accumulation of missorted surface proteins (Mazmanian et al., 1999). This enzyme was named sortase, and the gene encoding it was called *srtA*. SrtA-related proteins were found in nearly all gram-positive bacteria with sequenced genomes and, in several cases, were demonstrated to be key determinants of infectivity.

SORTASE PROTEIN SORTING

Overview of Protein Sorting

The protein-anchoring pathway mediated by the *S. aureus* SrtA protein has been characterized in detail, and an overview of this process is presented in Fig. 2B. A full-length precursor protein destined for sorting contains an amino terminal leader peptide that enables it to be exported from the cytoplasm through the secretory (Sec) pathway (Lofdahl et al., 1983; Uhlen et al., 1984). A C-terminal Cws in the precursor surface protein prevents it from being released into the extracellular milieu and contains the processing site for the sortase enzyme. SrtA recognizes the amino acid sequence LPXTG within the Cws (Mazmanian et al., 1999; Ton-That et al., 1999) and cleaves it between the threonine and glycine residues (Navarre and Schneewind, 1994), presumably by forming a thioester link to the threonine carbonyl carbon. The sortase enzyme then catalyzes the formation of a new peptide bond between the carbonyl carbon of the threonine and the free amine of the branched cell wall precursor lipid II [undecaprenyl-pyrophosphate-MurNAc(-L-Ala–D-iGln–L-Lys $(NH_2\text{-}Gly_5)$–D-Ala–D-Ala)-$\beta 1{\rightarrow}4$-GlcNAc)] (Higashi et al., 1967, 1970; Perry et al., 2002; Ruzin et al., 2002). Surface proteins tethered to lipid II are then incorporated into the peptidoglycan via the transglycosylation and transpeptidation reactions of bacterial cell wall synthesis, resulting in covalent attachment of the protein to the cell surface (Schneewind et al., 1995; Ton-That et al., 1997).

Components of the Sorting Reaction

The Cws

All of the components of the Cws (the LPXTG motif, the hydrophobic domain, and the charged tail) are required for proper sorting of proteins to the cell surface by SrtA (Schneewind et al., 1992). The LPXTG motif is processed by SrtA. At present, little is known about the relative importance of amino acids within the LPXTG motif, but limited mutagenesis data have demonstrated that mutation of the proline results in protein missorting while the threonine is tolerant of a T-to-N mutation (Schneewind et al., 1992). The latter finding is consistent with a recent bioinformatics study of *S. aureus,* which identified two potential SrtA substrates in which the threonine residue was replaced by an alanine (Roche et al., 2003). Several multigenome analyses have revealed that a large number of proteins contain a Cws in which the sequence of the LPXTG motif is varied (Janulczyk and Rasmussen, 2001; Pallen et al., 2001; Comfort and Clubb, 2004) (see below). It thus seems likely that different sortase enzymes may have evolved to specifically anchor proteins containing variations of the LPXTG motif. Recent experimental evidence for this conclusion comes from studies of the *S. aureus* SrtB sortase, which exclusively anchors a Cws-bearing protein with the sequence NPQTN instead of LPXTG (Mazmanian et al., 2002). Although the LPXTG motif is the recognition element of SrtA, its positioning relative to the charged tail is critical, since SrtA can process sorting signals only in which these motifs are separated by 31 to 33 amino acids (Schneewind et al., 1993). These results suggest that the sorting reaction is performed in a confined environment on the cell surface, requiring the precise presentation of the LPXTG sorting signal to SrtA.

Resection studies of protein A reveal that only two arginines in the charged tail (positioned 31 and 33 residues from the LPXTG motif) are necessary and sufficient for proper sorting; histidine and lysines may work in concert with these residues in other SrtA-anchored proteins (Schneewind et al., 1993). In addition, mutation of the glutamic acid and cysteine following these basic residues results in a missorted phenotype, but the significance of this finding remains unclear (Schneewind et al., 1993). It is possible that the charged tail causes the Cws to be retained within the secretion channel rather than being inserted into the membrane (Navarre and Schneewind, 1999). This hypothesis is consistent with studies in which fusions of the Cws hydrophobic domain to other secreted proteins (PhoA and Seb) were unable to trigger their stable membrane insertion (Schneewind et al., 1993). Unlike the Cws, polypeptide segments positioned N-terminal to the LPXTG motif are highly variable. However, most contain a 50- to 125-residue region abundant in proline, glycine, serine, and threonine that may be suited for weaving through the tightly packed peptidylglycan (V. A. Fischetti, R. Horstmann, and V. Pancholi, *Program Abstr. 31st Intersci. Conf. Antimicrob. Agents Chemother.,* abstr. S-56, p. 367, 1991).

Cell Wall

The third component of the SrtA-catalyzed transpeptidation reaction is the free amine of the cell wall crossbridge, which is ultimately joined to the threonine of the Cws. The cell wall in *S. aureus* is synthesized from a peptidylglycan precursor, called lipid II, which contains a $\beta(1\rightarrow4)$-linked *N*-acetylglucosamine (MurNAc)–*N*-acetylmuramic acid (GlcNAc) disaccharide that is tethered to the membrane by an undecaprenyl-pyrophosphate molecule (Fig. 3A). The lactic acid group of MurNAc is joined via a peptide bond to the pentapeptide sequence L-Ala-D-iGln-(DA)-D-Ala-D-Ala, where DA is a diamino acid (L-Lys in *S. aureus*). This diamino acid at the third position can serve as a branch point for peptides of the cell wall crossbridge; in *S. aureus,* a pentaglycine peptide extends from the lysine ε-amine. The chemical identity of the diamino acid and peptide branch differs among gram-positive species (reviewed by Goffin and Ghuysen [2002]). During cell wall synthesis, a transglycosylation reaction incorporates lipid II into the growing peptidoglycan and a transpeptidation reaction joins the terminal amine of the crossbridge to the D-Ala at the fourth position of the neighboring glycan strand, forming a $4\rightarrow3$ peptide linkage and releasing its terminal D-Ala (Fig. 3B). In *S. aureus,* nearly all of the glycine crossbridges participate in interchain cross-links (Schneewind et al., 1995).

SrtA of *S. aureus* specifically recognizes the free amine of the branch peptide. Studies of *fem* mutants of *S. aureus* (which can contain glycyl, triglycyl, or unbranched crossbridges) revealed that the wild-type pentaglycine crossbridge is the preferred substrate for SrtA in vivo and only peptides that contain an N-terminal glycine could serve as Srt substrates

(Ton-That et al., 1998). These findings are substantiated by in vitro studies of *S. aureus* SrtA that measured its ability to catalyze the transpeptidation of LPXTG peptides and a series of nucleophiles that mimic the cell wall component of the reaction. This work revealed that maximum transpeptidation activity was obtained when the cell wall substrate contained at least two glycine residues and that replacement of the N-terminal glycine with alanine or valine or acylation of its amine completely abolished activity (Huang et al., 2003).

The demonstrated specificity of *S. aureus* SrtA for the amine of the branch peptide suggests that sortases in other bacteria also exhibit distinct substrate specificities, since the amino acid sequences of their branches differ (Fig. 3C). Recent experimental evidence for branch peptide specificity comes from studies of *Streptococcus pyogenes*. This microbe encodes a SrtA-like enzyme that cleaves the LPXTG motif but is unable to catalyze the transpeptidation of the LPXTG substrate with a lipid II molecule containing a glycine branch. This finding is consistent with the cell wall structure of *S. pyogenes,* which contains branch peptides consisting of the sequence L-Ala–L-Ala and L-Ala–L-Ser. Studies of *S. pneumoniae* further support the notion of branch peptide specificity, since this microbe possesses a similar crossbridge to that of *S. pyogenes* and its sortase is equally adept at recognizing both L-Ala–L-Ala and L-Ala–L-Ser branches (Severin et al., 1996; Filipe et al., 2001; Kharat and Tomasz, 2003). Bacilli have unbranched cell walls, in which a *meso*-diaminopimelic acid (*m*-Dpm) molecule at the third position of the MurNAc peptide is directly linked to the D-Ala of a neighboring glycan strand. The SrtA orthologs of bacilli (Pallen et al., 2001; Comfort and Clubb, 2004) must therefore have evolved to attach proteins to the free amine of *m*-Dpm, a reaction that *S. aureus* SrtA is unable to catalyze (Ton-That et al., 1998; Ruzin et al., 2002). Taken together, these data suggest that sortases may have evolved to specifically recognize the amino acid sequence present in the branch peptide.

Several lines of evidence suggest that sorting occurs prior to cell wall synthesis. First, SrtA cleaves surface protein precursors and forms putative transpeptidation products in *S. aureus* protoplasts (which lack a mature cell wall) (Ton-That and Schneewind, 1999). Second, SrtA is able to selectively use lipid II as a substrate, even when it is present in complex mixtures (Ruzin et al., 2002). Third, lipid II is the only cell surface peptidylglycan precursor that contains a phosphorous atom (Strominger et al., 1967), and cells grown in the presence of [^{32}P]phosphoric acid label an intermediate in the protein sorting pathway (Perry et al., 2002). Finally, antibiotics that inhibit the incorporation of lipid II precursors into the cell wall also interfere with protein sorting (Strominger, 1965; Strominger and Tipper, 1965; Tipper and Strominger, 1965; Ton-That and Schneewind, 1999). Although lipid II is an attachment point for surface proteins in vivo, it is still unclear whether this site is used exclusively, since the available data do not rule out protein attachment to free amines within the growing immature peptidylglycan (Ruzin et al., 2002).

STRUCTURE AND CHEMISTRY OF SORTASE TRANSPEPTIDATION

Sortase Enzyme Structure

Elucidation of the *S. aureus* SrtA structure and recent enzyme kinetic studies have substantially increased our understanding of how sortase enzymes anchor proteins to the cell wall. The three-dimensional structure of truncated SrtA (SrtA$_{\Delta 59}$; residues 60 to 206) was solved by nuclear magnetic resonance (NMR) spectroscopy (PDB ID: 1IJA) (Ilangovan et al., 2001) and possesses a novel eight-strand β-barrel fold that is decorated with two

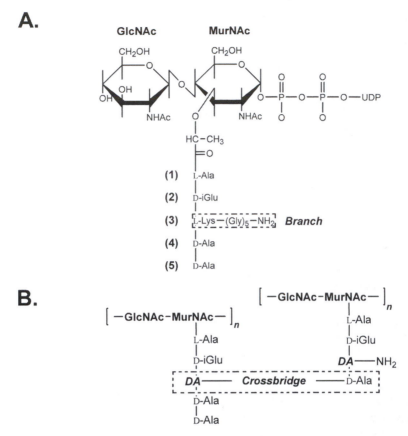

Figure 3. (A) Diagram showing the structure of the second substrate of *S. aureus* SrtA, lipid II [undecaprenyl-pyrophosphate-MurNAc(-L-Ala–D-iGln–L-Lys(NH₂-Gly₅)–D-Ala-D-Ala)-β1→4-GlcNAc)]. The peptide of MurNAc is "branched" by forming a peptide bond between the ε-amine of lysine and a pentaglycine peptide. The terminal glycine α-amine serves as a nucleophile and forms a peptide bond to the carbonyl carbon of the threonine residue within the LPXTG motif. (B) Structure of the cell wall crossbridge. In a 4→3 linkage, a diamino acid (DA; either L-Lys or *m*-Dpm) at position 3 is linked through a peptide bond to the carboxyl of the D-Ala in the position 4 of a neighboring glycan chain, releasing the terminal D-Ala (position 5). These reactions are performed by cell wall transpeptidases (PBPs). The chemical nature of the diamino acid and crossbridge peptides can vary among different bacteria. (C) Cell wall crossbridges of different bacterial strains. In *L. monocytogenes*, bacilli, and corynebacteria, the cell wall peptides are unbranched and the diamino acid *m*-Dpm is directly linked to the D-Ala of the neighboring glycan chain.

short helices and several loops (Fig. 4). The active site of sortase is organized around the catalytically critical C184 and H120 side chains, which are located together at the end of a hydrophobic depression whose floor is formed by strands β7 and β8 and whose walls are constructed by amino acids located in loops connecting strands β3-β4, β2-β3, β6-β7 and β7-β8 (Fig. 5). These two catalytic surface residues (Cys and His) are completely conserved in sortase-related proteins (Pallen et al., 2001; Comfort and Clubb, 2004), and their

C.

S. aureus:

L-Lys —— (Gly)$_5$ —— D-Ala;

L-Lys − (Gly)$_3$(L-Ser)$_2$ − D-Ala

S. pyogenes:

L-Lys — (L-Ala)$_2$ — D-Ala;

L-Lys — L-Ala-L-Ser — D-Ala

Streptomyces, Clostridium:

m-Dpm —— Gly —————— D-Ala

E. faecium, E. hirae:

L-Lys — L-Asp —— D-Ala

L. monocytogenes, Bacilli, Corynebacteria:

m-Dpm ——————— D-Ala

Figure 3. *Continued*

mutation abolishes in vivo and in vitro sortase function (Ton-That et al., 2002). Moreover, methyl methanethiosulfonates (reagents that form a disulfide with the sulfhydryl) prevent LPXTG cleavage by SrtA, but activity can be restored on addition of dithiothreitol, a reductant that regenerates the active site sulfhydryl (Ton-That and Schneewind, 1999). A tryptophan (W194) is positioned on an ordered loop adjacent to the active-site cysteine, and its mutation to alanine reduces activity fourfold (Ton-That et al., 2002). The indole ring partially shields the cysteine thiol group from the solvent, but it is poorly conserved in other sortase enzymes and its exact role in catalysis is unclear.

SrtA presumably examines newly translocated polypeptides, searching for the LPXTG motif. Once the recognition sequence is found and cleaved, the polypeptide is transferred to the α-amine of the terminal glycine residue on lipid II (Schneewind et al., 1993). At present, the binding sites for either of these substrates have not been localized. However, the dimensions of the groove formed by strands β7 and β8 suggest that this surface may interact with the LPXTG motif. This groove is lined with several hydrophobic side chains (A92, L94, A118, I182, and L169) that may interact with the hydrocarbon side chains of the leucine and proline residues within the signal. Interestingly, a large surface loop formed by amino acids connecting strands β6 and β7 rests above this potential binding site. This loop exhibits a modest degree of mobility, suggesting that it may adapt its structure to recognize the sorting signal. The binding site for the crossbridge amine is presumably much smaller, since a single glycine residue can effectively mimic this substrate (Ton-That et al., 2000; Connolly et al., 2003). The surface used to recognize this crossbridge amine is not expected to coincide with the LPXTG-binding surface, since sortase uses a ping-pong type mechanism of catalysis (Huang et al., 2003).

In vitro, SrtA activity is stimulated eightfold when 2 mM calcium (Ca^{2+}) is present in the reaction buffer (Ilangovan et al., 2001). Mn^{2+} and Mg^{2+} can substitute in part for calcium, but other divalent (Fe^{2+}, Zn^{2+}, Cd^{2+}, and Co^{2+}) or monovalent (K^+) cations cannot. The role of calcium activation in vivo remains to be elucidated, but *S. aureus* is expected to encounter extracellular calcium concentrations of 1.5 mM or higher on entry into human tissues, which may facilitate the infection process by increasing the rate of surface protein anchoring. On the addition of calcium to SrtA, large changes occur in the chemical shifts of backbone atoms within residues located in the β3-β4 and β6-β7 loops, suggesting that this region of the protein is responsible for binding the ion (Ilangovan et al., 2001).

A.

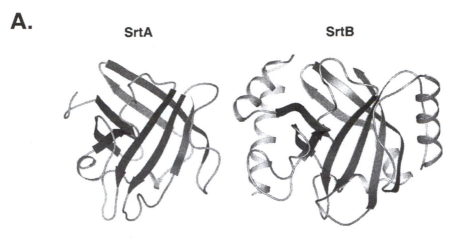

B.

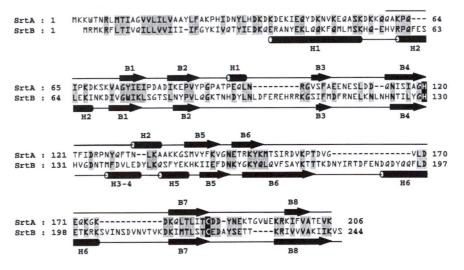

Figure 4. Comparison of the primary and tertiary structures of SrtA and SrtB from *S. aureus*. (A) Ribbon diagrams of the folds of SrtA (PDB ID, 1IJA; residues 60 to 206) (Ilangovan et al., 2001) and SrtB (PDB ID, 1NG5; residues 35 to 244) (Zhang et al., submitted). Both are presented in a similar orientation. The β-barrels of both structures align with a root mean square deviation of 3.2 Å. (B) Sequence alignment of SrtA and SrtB. Similar residues are shaded gray, and the catalytic cysteine and histidine residues are shaded black. Secondary-structure elements are indicated by arrows and cylinders for β-strands and α-helices, respectively. The N termini of both proteins (residues 1 to 59 in SrtA and 1 to 34 in SrtB) are absent in their respective structures and most probably contain a membrane anchor.

A.

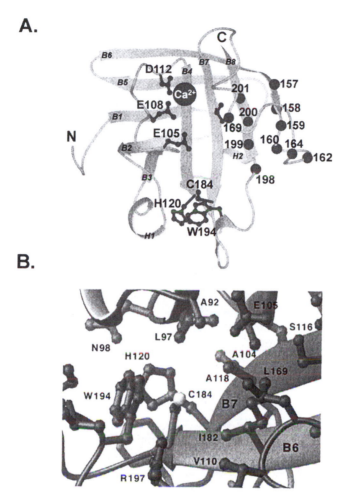

B.

Figure 5. (A) Ribbon drawing of the structure of SrtA$_{\Delta N59}$ (residues 60 to 206 of the *S. aureus* SrtA protein). Beginning at the N terminus, the β1 strand (G74 to I78) is followed by a short hairpin and lies antiparallel to the β2 strand (I83 to Y88). The β2 and β3 strands (V101 to A104) are positioned in parallel and are tethered by a 3$_{10}$ helix (P94 to L97), which crosses over the surface of the enzyme to form the lateral wall of the active site. The β3 and β4 strands (Q113 to G119) lie antiparallel with respect to one another and are followed by a long loop and a second α-helix, which assume a circuitous path to position the β5 strand (S140 to V146) for antiparallel alignment with the β1 strand. The β6 strand (E149 to K155) is then connected by a short hairpin turn for antiparallel pairing and followed by a long loop structure to connect it to the β7 strand (K177 to T183), which is aligned in parallel with β4. The active site sulfhydryl, C184, is positioned at the end of the β7 strand, which includes the LITC signature sequence of sortase enzymes. The structure is completed by a loop (D185 to W194), which connects the β7 and β8 strands (E195 to F200) for antiparallel pairing. Three acidic side chains (E105, E108, and D112) are poised to bind calcium, as judged by localized large-amplitude calcium-dependent changes in their chemical shifts and in surrounding amino acids. Additional residues in the β6-β7 loop displayed calcium-dependent chemical shift changes and may constitute a weaker Ca^{2+}-binding site (spheres). The active-site side chains of H120, C184, and W194 are shown for reference. (B) Expanded view of the active site. H120 and C184 in sortase form a catalytic dyad that mediates the transpeptidation reaction and are positioned within a large hydrophobic pocket suitable for sorting-signal binding. A tryptophan (W194) is positioned in the active site and may play a limited role in catalysis (Ton-That et al., 2002).

Inspection of the electrostatic surface potential of SrtA reveals a continuous acidic surface near the active site. A structurally well-ordered calcium-binding site appears to be formed by the side chains of E105, E108, and D112 within the β3-β4 loop and possibly by E171 from the β6-β7 loop (Fig. 5A). A second interaction surface may be formed by several amino acids in the β6-β7 loop, which may constitute another, weaker ion-binding site. The proximity of this loop to the presumed LPXTG motif-binding surface implies that cations may activate sortase by facilitating substrate binding.

Very recently the structure of the *S. aureus* SrtB sortase was determined by X-ray crystallography (PDB ID: 1NGB) (R. Zhang, G. Joachimiak, and A. Joachimiak, submitted for publication) (Fig. 4A). Although SrtB processes a unique sorting motif NPQTN (instead of LPXTG) and has only 22% sequence identity to SrtA, it adopts a very similar structure and contains an analogous cysteine and histidine catalytic dyad (C223 and H130 in SrtB). Comparison of the SrtA and SrtB structures highlights two significant differences. (i) The SrtB structure contains two N-terminal helices (H1, residues 40 to 57; H2, residues 58 to 67) that are absent in the SrtA structure (Fig. 4A). It is not known whether these helices are present in the full-length SrtA; however, the high sequence similarity between the two proteins in this region (43%) (Fig. 4B) suggests this possibility. These helices are not involved in SrtA catalysis in vitro, since the SrtA$_{\Delta 59}$ mutant and full-length protein have similar activities (Ilangovan et al., 2001). (ii) SrtB contains a large (27-residue) insertion between strands β6 and β7, which adopts a helical fold. This corresponds to the aforementioned β6-β7 loop in SrtA, which may be involved in Ca^{2+} and/or substrate binding.

Chemical Mechanism

Overview

Our understanding of the mechanism of transpeptidation is still in its infancy. However, broad features of this process can be extrapolated from numerous studies of the papain cysteine proteases (Drenth et al., 1968; Storer and Menard, 1994; Somoza et al., 2000). Similar to SrtA, these enzymes utilize a cysteine-histidine catalytic dyad, but are otherwise structurally unrelated. Figure 6 shows an overview of the sortase mediated transpeptidation reaction (Ton-That et al., 2002). In step 1, the thiolate of C184 presumably attacks the carbonyl carbon of the threonine residue within the LPXTG sorting signal, resulting in the formation of a transient tetrahedral intermediate (step 2). This intermediate would then rearrange into a more stable thioester enzyme-substrate linkage, with the concomitant breakage of the threonine-glycine peptide bond (step 3). Several lines of evidence support the existence of a thioacyl-enzyme intermediate, including (i) the demonstrated catalytic importance of C184 (Ton-That et al., 1999; Ton-That et al., 2002), (ii) the sensitivity of catalysis to thiol-modifying compounds (Ton-That and Schneewind, 1999; Ton-That et al., 2000; Ton-That et al., 2002), (iii) the finding that whole cells treated with hydroxylamine release surface proteins with a C-terminal threonine hydroxamate (Ton-That et al., 1999), and (iv) the recent direct observation of this intermediate by electrospray ionization mass spectrometry (Huang et al., 2003). In step 4 of the reaction, the carbonyl carbon in the thioacyl-enzyme intermediate is attacked by the cell wall crossbridge (presumably lipid II), forming a second enzyme-substrate tetrahedral intermediate (step 5), which rearranges to form the new peptide bond (step 6).

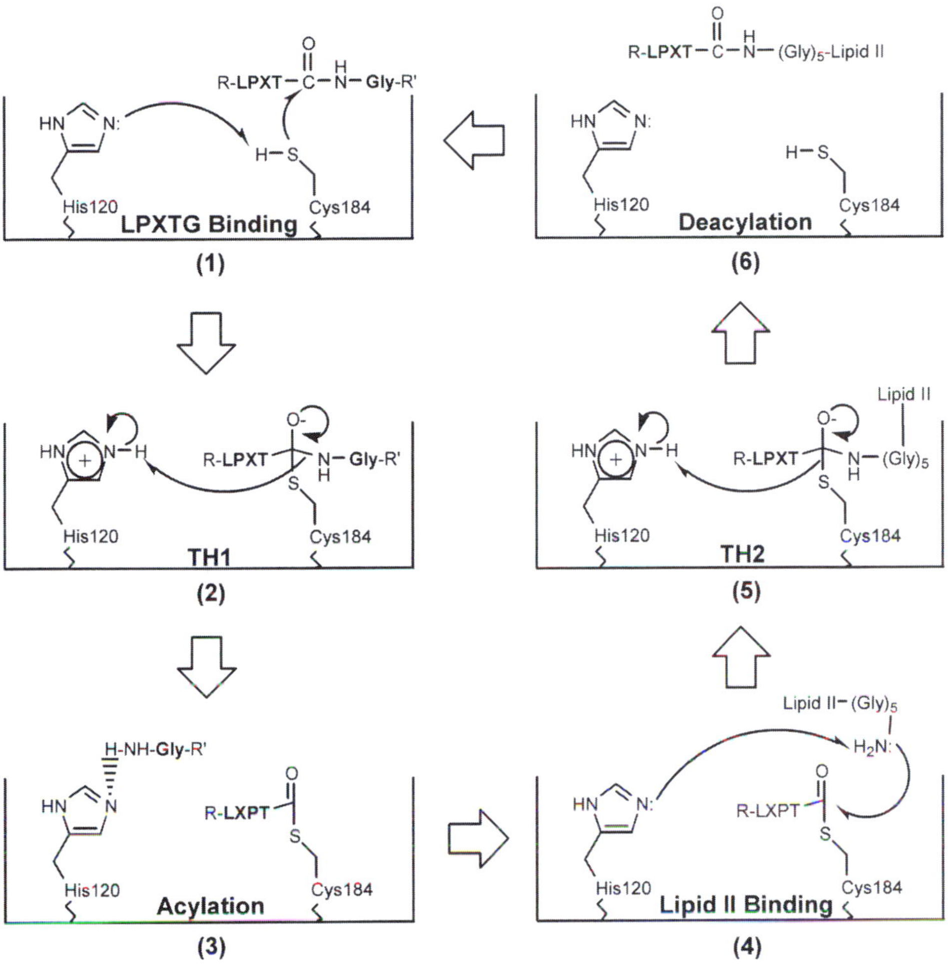

Figure 6. Proposed chemical mechanism of the SrtA sortase. Like the cysteine protease papain, catalysis presumably proceeds through the formation of a thioacyl intermediate. As the LPXTG substrate enters the active site-pocket, H120 functions as a general base, withdrawing a proton from C184 (step 1). The C184 nucleophile attacks the carbonyl carbon of threonine, and, proceeding through a tetrahedral intermediate (TH1) (step 2), results in a thioacyl-enzyme intermediate (step 3). The N-terminal amine of the pentaglycine crossbridge of lipid II then enters the active-site pocket (step 4), serving as an acceptor for the acyl-enzyme intermediate. Deacylation proceeds through a second tetrahedral intermediate (TH2) (step 5), resulting in the formation of a new peptide bond between the threonine of the LPXTG substrate and the terminal glycine of lipid II (step 6).

Kinetics

The transpeptidase activity of sortase has been demonstrated in in vitro reactions that join an LPETG-containing peptide to triglycine (Ton-That et al., 1999; Ton-That et al., 2000). Interestingly, when the triglycine nucleophile is not available, SrtA catalyzes the hydrolysis of the threonine-glycine peptide bond (Ton-That et al., 2000). Hydrolysis presumably occurs through a similar mechanism to transpeptidation, with a water molecule acting as the nucleophile during deacylation. Recent work by Huang and colleagues have supported the double-displacement mechanism outlined in Fig. 6, and revealed that the hydrolysis and transpeptidation reactions have distinct rate limiting steps (Huang et al., 2003). The formation of the acyl intermediate is rate limiting in transpeptidation (Fig. 6, step 2), while the resolution of the thioacyl-enzyme intermediate by a water molecule (step 5) is rate limiting in the hydrolysis reaction (Huang et al., 2003). These results would appear to be consistent with the in vivo function of sortase, since the efficient hydrolysis of the sorting signal would result in the wasteful release of surface proteins and may therefore be slowed unless lipid II, the appropriate cell wall receptor nucleophile, is available. At present it is unclear why the isolated SrtA protein catalyzes the transpeptidation at such a low rate in vitro ($k_{cat} = 10^{-4}$ to 10^{-5} s^{-1}) (Ton-That et al., 2000; Huang et al., 2003), since in vivo it must attach proteins to the cell surface within the doubling time of the bacterium.

Active Site

Although sortases and papain-type cysteine proteases employ a common catalytic dyad, they use a distinct mechanism to activate the cysteine thiol for nucleophilic attack on the carbonyl carbon of the scissile peptide bond. In papain-type cysteine proteases, the cysteine side chain forms a thiolate-imidazolium ion pair with the histidine side chain (C25 and H159 in papain) (Lewis et al., 1981), thereby enhancing the reactivity of the thiol toward electrophiles at weakly acidic and neutral pHs (Otto and Schirmeister, 1997). In this scenario, the imidazole group polarizes the thiol, causing the cysteine thiol to be more acidic and the histidine imidazole to be more basic. For example, the C25 and H159 side chains in papain have anomalous pK_a values of 3.3 and 8.5, respectively (Whittaker et al., 1996; Pinitglang et al., 1997), and methylthiolation of C25 disrupts the ion pair and lowers the pK_a of H159 by 4.5 pH units (Lewis et al., 1981). In contrast, NMR and fluorescence measurements of H120 and C184 in the SrtA active site have revealed nominal pK_a values of 7.0 and 9.4, respectively (Connolly et al., 2003). In addition, mutational exchange of C184 with alanine does not significantly affect the pK_a of H120. These data argue against the existence of an ion pair in SrtA and are consistent with the NMR structure, which shows that the side chain of C184 projects away from H120 (Ilangovan et al., 2001).

The absence of an ion pair in SrtA suggests that its catalytic mechanism may be more similar to the viral 3C proteases (picornains), a structurally and mechanistically distinct group of cysteine proteases related to the trypsin/chymotrypsin serine protease family (reviewed by Otto and Schirmeister [1997] and Barrett and Rawlings [2001]). The pH-dependent alkylation of the active-site cysteine of poliovirus protease 3C with iodoacetamide has resulted in a pK_a of 8.9 (Sarkany et al., 2001), consistent with a reaction mechanism in which the cysteine nucleophile is uncharged at physiological pH and the histidine functions as a general base. A similar activation mechanism may be at work in sortase, with substrate binding initiating a subtle conformational rearrangement in the $\chi 1$ and $\chi 2$ torsion angles of the C184 and H120 side chains, enabling sulfhydryl proton abstraction and subsequent nucleophilic attack (Ilangovan et al., 2001). Substrate-induced activation of sortase may be advantageous, preventing spurious proteolysis reactions without the need for more elaborate inactivation mechanisms, such as prosegment occlusion of the active site. At present it is unclear whether SrtB also uses a general base mechanism of catalysis, since structural information alone is not sufficient to resolve this issue. However, in SrtB, the distance between the cysteine sulfur atom and the closest histidine imidazole nitrogen atom is 4.6 Å, which would appear to preclude the formation of an energetically significant electrostatic interaction (Kumar and Nussinov, 1999).

Relationship to Other Transpeptidases in the Cell Wall

Sortases function in the cell wall milieu, sharing targets with a separate superfamily of transpeptidases, the SxxK acyltransferases (commonly referred to as penicillin-binding proteins). These enzymes are involved in a diverse array of cellular processes including cell wall synthesis, resistance to β-lactams, and signal-transducing penicillin sensing. They can be divided into three groups (reviewed by Goffin and Ghuysen [2002]): group I acyltransferases synthesize (4→3)peptidylglycan and contain the D,D-acyltransferase-PBPs, group II enzymes are diverse in function and contain the β-lactamases that often aid group I enzymes in escaping penicillin susceptibility, and group III enzymes are penicillin-resistant L,D-peptidases that are implicated in (3→3)peptidylglycan synthesis. Although diverse in function, all SxxK acyltransferases use a common mechanism to catalyze the cleavage of D-Ala–D-Ala peptides or its β-lactam mimics and can exist as freestanding PBPs or PBP fusions. Interestingly, they adopt a three-dimensional structure distinct from that of sortase and use a serine residue to mediate the acylation and deacylation steps of catalysis in a manner similar to the serine proteases of the trypsin/chymotrypsin family. Given the great breadth and diversity of SxxK acyltransferases, it will be interesting to see whether members of the sortase family have a wide range of functions or if they are restricted to attaching proteins to the cell wall.

PHYLOGENETIC DISTRIBUTION OF SORTASES

Sortases Are Widely Distributed

An in silico analysis of sequenced genomes has revealed 176 sequence homologs of SrtA and almost 900 potential substrates bearing a Cws (Table 1) (Ton-That et al., 1999; Janulczyk and Rasmussen, 2001; Pallen et al., 2001; Comfort and Clubb, 2004). Sortases are found almost exclusively in gram-positive bacteria and appear to be restricted to species that contain a conventional cell wall, since they are absent in mycobacteria (which have complex, highly crosslinked cell walls) and mycoplasmas (which lack a cell wall) (Pallen et al., 2001). Four genera of gram-negative bacteria from the phylum *Proteobacteria* (*Shewanella, Bradyrhizobium, Colwellia,* and *Microbulbifer*) contain a single sortase and one gene encoding a protein that is its substrate (it contains an amino acid sequence at its C terminus that has all the features of a Cws) (Pallen et al., 2001; Comfort and Clubb, 2004). It will be interesting to see how the gram-negative bacterial enzymes have adapted to attach proteins to the lipopolysaccharide cell wall if these enzymes function in an analogous manner to their gram-positive counterparts. Interestingly, sortases are also present in the archaeal methanobacteria, which possess a pseudo-peptidylglycan (Pallen et al., 2001; Goffin and Ghuysen, 2002). However, substrates for these proteins have yet to be identified, and, given the highly distinct nature of the methanobacterial cell wall, it is possible that they have diverged too greatly to be revealed by a pattern-based genomic search (Pallen et al., 2001).

Too Many Sortases?

Bacteria that encode at least one sortase-like protein more often than not encode multiple sortase-like proteins (up to seven sortases have been found in *Bacillus cereus, Enterococcus faecium,* and *Streptomyces coelicolor*) (Pallen et al., 2001). It seems likely that

Table 1. Examples of sortase family distribution in gram-positive bacteria

Species	No. of sortase homologs in sortase subfamily[a]:						No. of Cws substrates[b]
	A	B	3	4	5	Unclassified[c]	
Actinobacteria (high-G+C gram-positive bacteria)							
Corynebacterium diphtheriae		5		1			13
Corynebacterium efficiens		4		1			8
Corynebacterium glutamicum				1			1
Tropheryma whipplei Twist				1			1
Streptomyces avermitilis				4			13
Streptomyces coelicolor				2		5	13
Thermobifida fusca				1			1
Bifidobacterium longum DJ010A			2	1			9
Chloroflexi (green nonsulfur bacteria)							
Chloroflexus aurantiacus						4	3
Firmicutes (gram-positive bacteria)							
Bacillus anthracis Ames	1	1		1			9
Bacillus cereus ATCC 10987	1	1	4	1			17
Bacillus halodurans		1	2			3	7
Bacillus subtilis			1			1	2
Geobacillus stearothermophilus			1				1
Oceanobacillus iheyensis			2				1
Listeria innocua	1	1					35
Listeria monocytogenes EGD-e	1	1					37
Staphylococcus aureus subsp. *aureus* COL	1	1					21
Staphylococcus epidermidis ATCC 12228	1					1	11
Enterococcus faecalis	1		1			1	27
Enterococcus faecium		5				2	13
Lactobacillus gasseri	1						9
Lactobacillus plantarum	1						10
Pedicoccus pentosaceus	1						3
Leuconostoc mesenteroides		2					3
Oenococcus oeni MCW						2	1
Lactococcus lactis subsp. *lactis*	1	1					6
Streptococcus agalactiae 2603V/R	1	5					22
Streptococcus equi	1	1					22
Streptococcus gordonii	1						21
Streptococcus mitis	1						12
Streptococcus mutans	1						5
Streptococcus pneumoniae TIGR4	1	3					15
Streptococcus pyogenes MI GAS	1	1				1	13
Streptococcus sobrinus	1						12
Streptococcus suis	1	4					19
Streptococcus uberis	1						6
Clostridium acetobutylicum						1	2
Clostridium botulinum						1	1
Clostridium difficile						1	6
Clostridium perfringens ATCC 13124		1				3	11
Clostridium tetani						1	3
Ruminococcus albus		1					2

[a]Sortase homologs are clustered into subfamilies according to sequence homology using BLAST profiles and hidden Markov models.
[b]The total number of Cws-containing proteins identified in each respective genome.
[c]The sortase homolog is not readily classified into a subfamily based on sequence homology.

most of these homologs serve as extracellular transpeptidases or peptidases, since they all contain appropriately positioned active-site residues equivalent to those found in SrtA (SrtA residues H120 and C184) and stretches of hydrophobic amino acids at their polypeptide termini that presumably enable them to associate with the membrane. However, it is less clear whether they all recognize and anchor proteins bearing a Cws, and it is conceivable that many SrtA homologs may function as proteases. Although their functions remain unknown, recent bioinformatics studies suggest that most sortase-like proteins process proteins bearing a Cws, since more than half of the identifiable sortase-like genes are clustered with genes encoding potential substrates (Cws-containing proteins) (Pallen et al., 2001; Comfort and Clubb, 2004). This would seem to functionally link these proteins, since in several instances, biochemical experiments have confirmed that the sortases in these gene clusters are coexpressed with and/or required for the display of an adjacent Cws-containing protein (Pallen et al., 2001; Mazmanian et al., 2002; Osaki et al., 2002; Hava et al., 2003).

The *S. aureus* SrtA and SrtB Sortases Nonredundantly Sort Proteins to the Cell Surface

A total of 46 species of bacteria have been found to encode multiple sortases, but only a limited number of studies have investigated their functions. At present, the results of all studies are consistent with the notion that these bacteria employ their multiple sortases in nonredundant pathways that display surface proteins. The SrtA and SrtB pathways of *S. aureus* are well defined, and the respective sortases selectively sort distinct proteins to the cell surface by recognizing either LPXTG or NPQTN sequence motifs within the Cws. Based on sequence homology, analogous SrtA and SrtB pathways appear in *Listeria* and *Bacillus* spp., and recent work has experimentally verified their presence in *L. monocytogenes* (Bierne et al., 2002). Evidence also suggests that sortase enzymes in streptococci function nonredundantly, since the elimination of a SrtA-like gene in *S. pyogenes* prevents the display of a specific set of proteins, even though this organism encodes additional sortase homologs (Barnett and Scott, 2002).

Twenty-one species of gram-positive bacteria possess two or more sortase-like proteins whose functions cannot be inferred from studies of the *S. aureus* SrtA or SrtB proteins. It is therefore not known whether surface proteins in these organisms are selectively sorted to the cell surface by a specific cognate sortase or whether their sortases have degenerate functions. An answer to this question is of extreme importance, since if bacteria encode a number of functionally redundant sortases, antimicrobial compounds targeted toward a particular sortase could prove ineffective. Moreover, if the sortases in these organisms act nonredundantly, the development of an effective antimicrobial compound will require the identification and inhibition of the specific sortase responsible for anchoring virulence factors.

There Are Six Subfamilies of Sortases

As a step toward defining sortase function and ultimately elucidating the sorting pathways in which they may participate, we have used pairwise BLAST matrices (Altschul et al., 1997) and hidden Markov models (reviewed by Eddy [1998]) to cluster sortases based on their primary sequences (Comfort and Clubb, 2004). Of 176 identifiable homologs, 145 were clustered into six subfamilies (one subfamily of sortases from gram-negative bacteria and five subfamilies from gram-positive bacteria). Following the convention established by Schneewind (Mazmanian et al., 1999, 2002), two of the gram-positive bacterial subfamilies are called SrtA and SrtB, since their members have primary sequences that are most closely related to the well-characterized SrtA and SrtB proteins from *S. aureus*, while the remaining gram-positive bacterial subfamilies have not been extensively characterized and are numbered 3 through 5. The SrtA and subfamily 3 homologs contain the largest number of homologs, and the remaining subfamilies are of nearly equal size (Fig. 7B). The majority of completely sequenced genomes contain genes encoding sortase homologs from at least two of the five subfamilies. These bacteria never encode more than a single copy of a SrtA- or SrtB-type sortase, but they frequently contain multiple copies of subfamily 3, 4, and 5 sortases (Comfort and Clubb, 2004).

Defining Sorting Pathways: Do Other Sortases Selectively Sort Proteins?

Do members of subfamilies 3, 4, and 5 process unique sorting signals? At present, there is no answer to this question, but recent work suggests that, in some cases, this may be true. Working under the assumption that most sortases act on Cws-bearing substrates, a comparative genome analysis predicted the cognate sortase used to process 77% of 892 identifiable Cws-bearing proteins (Comfort and Clubb, 2004). Interestingly, nearly all sortases appear to be involved in surface protein anchoring, since they can be linked to a Cws-bearing substrate. Figure 7A shows the consensus sequence motifs present in the predicted substrates of each subfamily, revealing subfamily specific preferences. As expected, the predicted SrtA subfamily substrates contain the LPXTG motif and the predicted SrtB subfamily substrates contain the distinct sequence NPXT(N/G). However, the subfamily 4 and 5 sortases are predicted to process novel sorting signals, LPXTA(S/T) and LAXTG, respectively. Interestingly, the subfamily 3 sortases are predicted to process a signal similar to that recognized by the SrtA subfamily but are distinguished by the prevalence of a glycine residue immediately following the LPXTG motif (in 83.3% of the 54 sorting signals) and by their membrane topology (discussed below). Figure 7C shows the distribution of substrates classified by the subfamily type that is predicted to be responsible for their processing. Homologs most closely related to the *S. aureus* SrtA protein (members of the SrtA subfamily) are predicted to anchor the largest number of Cws-containing proteins to the cell surface, while members of the remaining subfamilies appear to play a more specialized role, anchoring on average far fewer proteins.

SrtA Subfamily

The majority of surface proteins are predicted to be anchored by sortase enzymes whose primary sequences are most closely related to the *S. aureus* SrtA protein. Members of this subfamily are found in 42 of the 67 gram-positive bacterial genomes analyzed, including the bacterial genera *Bacillus, Enterococcus, Lactobacillus, Lactococcus, Listeria, Staphylococcus,* and *Streptococcus*. Several lines of evidence suggest that members of this subfamily play a housekeeping role in the cell, anchoring a large number of diverse proteins to the cell wall. First, bacteria always encode only a single SrtA-type homolog, which is predicted to anchor a large number of Cws containing proteins (Comfort and Clubb, 2004). Second, the predicted substrates of the SrtA-type proteins are functionally diverse. Third, nearly all of the genes encoding predicted SrtA substrates are isolated, i.e., are not clus-

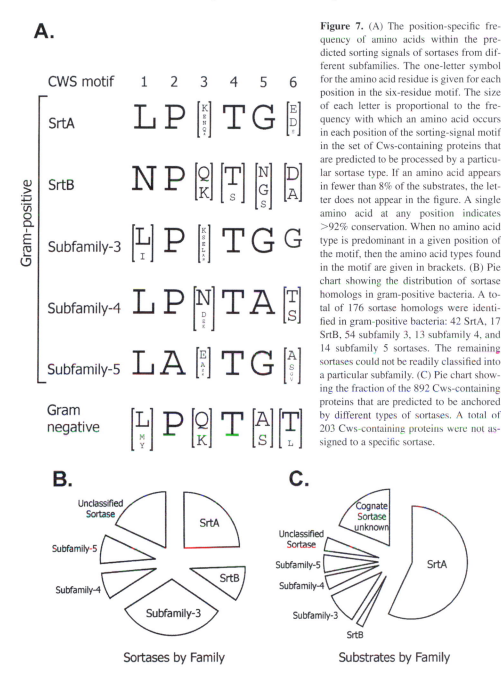

A.

B. Sortases by Family

C. Substrates by Family

Figure 7. (A) The position-specific frequency of amino acids within the predicted sorting signals of sortases from different subfamilies. The one-letter symbol for the amino acid residue is given for each position in the six-residue motif. The size of each letter is proportional to the frequency with which an amino acid occurs in each position of the sorting-signal motif in the set of Cws-containing proteins that are predicted to be processed by a particular sortase type. If an amino acid appears in fewer than 8% of the substrates, the letter does not appear in the figure. A single amino acid at any position indicates >92% conservation. When no amino acid type is predominant in a given position of the motif, then the amino acid types found in the motif are given in brackets. (B) Pie chart showing the distribution of sortase homologs in gram-positive bacteria. A total of 176 sortase homologs were identified in gram-positive bacteria: 42 SrtA, 17 SrtB, 54 subfamily 3, 13 subfamily 4, and 14 subfamily 5 sortases. The remaining sortases could not be readily classified into a particular subfamily. (C) Pie chart showing the fraction of the 892 Cws-containing proteins that are predicted to be anchored by different types of sortases. A total of 203 Cws-containing proteins were not assigned to a specific sortase.

tered with genes encoding other substrates or sortases. This contrasts with the genomic positioning of other types of sortases, which are often clustered with a limited number of substrates whose expression may be coordinately regulated. Although it has been noted that there is a great degree of variation in the sequence motif of the Cws (Navarre and Schneewind, 1999; Janulczyk and Rasmussen, 2001; Pallen et al., 2001), an analysis of the 495 predicted substrates for the SrtA-type sortases suggests that members of this subfamily are highly specific for the sequence LPXTG (Comfort and Clubb, 2004). The N-terminal YSIRK motif recently found to be important in SrtA-mediated processing in *S. aureus* (Bae and Schneewind, 2003) is found in only 20% of SrtA substrates and therefore does not appear to be a major determinant of specificity (Roche et al., 2003; Comfort and Clubb, 2004). A subset of SrtA subfamily sortases found in the genera *Lactococcus* and *Streptococcus* are distinguished by their genomic proximity to the gene encoding DNA gyrase subunit A (Osaki et al., 2002; Kharat and Tomasz, 2003), strongly indicating that they have orthologous functions.

Subfamily 3 Sortases

The subfamily 3 enzymes are the largest (54 homologs) and most widespread group (found in nine genera). The majority of subfamily 3 enzymes (49 of 54) are positioned adjacent to potential substrates that contain an LPXTG sequence motif. Interestingly, many organisms encode both SrtA and subfamily 3 homologs. Do these enzymes serve redundant functions in the cell, or can they differentially process proteins bearing related Cws's? Studies of *S. pyogenes* suggests that the latter is true, because its SrtA and subfamily 3 homologs (called SrtB in their study) each selectively sort proteins containing related LPXTG motifs to the cell surface (Barnett and Scott, 2002). This has led to the hypothesis that SrtA proteins recognize an expanded motif, consisting of an acidic residue immediately following the canonical LPXTG motif (Barnett and Scott, 2002). The SrtA and subfamily 3 enzymes may also be positioned in the membrane differently, since subfamily 3 enzymes contain hydrophobic amino acids at both their N and C termini, suggesting that they are type I membrane proteins (C terminus embedded into the membrane) (Comfort and Clubb, 2004). This contrasts with members of the SrtA subfamily (and all other sortase subfamily types), which contain only an N-terminal stretch of hydrophobic amino acids and are thus predicted to be type II membrane proteins (N terminus embedded into the membrane). The distinct membrane topology of subfamily 3 proteins may enable them to recognize other, as yet undetermined, features of their substrates.

Specialized Sorting Pathways? The Subfamily 4 and 5 and SrtB Sortases

Subfamily 4 sortases appear to mediate a specialized sorting pathway found in bacilli which supplements the SrtA, SrtB, and subfamily 3 pathways. In contrast, several high-G+C gram-positive bacteria have replaced SrtA enzymes with subfamily 5-type homologs, since no bacteria encode both an SrtA- and a subfamily 5-type sortase (Comfort and Clubb, 2004). Homologs most closely related to the well-characterized *S. aureus* SrtB protein (the SrtB subfamily) appear to mediate a minor pathway involved in heme-iron acquisition (Mazmanian et al., 2002). In addition to *S. aureus,* a single SrtB homolog is found in bacteria from the genera *Bacillus* and *Listeria,* and in all cases the SrtB gene is adjacent to one predicted substrate containing an unusual sequence motif within its Cws (NPQTN in all strains of *S. aureus;* NPKSS in *Listeria;* and NPKTG, NPKTD, and NPQTG in *Bacillus*). In *S. aureus,* SrtB anchors an iron-regulated surface determinant (the IsdC protein) involved in heme-iron acquisition (Mazmanian et al., 2002). All SrtB proteins appear to have a conserved function, since, like IsdC, their prospective substrates contain the NEAT domain, which is implicated in iron transport (Andrade et al., 2002).

Sortase Gene Clusters

The predicted sortase-substrate functional linkages reveal several mixed gene clusters, in which surface proteins anchored by sortases from distinct subfamilies are presumably coordinately regulated. Schneewind and colleagues were the first to identify a mixed cluster in *S. aureus,* which contains a single SrtB-encoding gene, its substrate, and two genes that encode proteins processed by the distantly located SrtA homolog (Mazmanian et al., 2002). This cluster is iron regulated and is involved in heme-iron acquisition. It is completely conserved in all strains of *S. aureus* and present in *Bacillus anthracis, L. monocytogenes,* and *L. innocua* (Comfort and Clubb, 2004). The Sekizaki group has described a mixed-gene cluster in *Streptococcus suis* that encodes subfamily 3 homologs, two predicted subfamily 3 substrates, and one substrate that is part of the SrtA pathway (Osaki et al., 2002). In addition to these well-characterized cases, bioinformatic analysis reveals several other mixed clusters that have yet to be demonstrated experimentally. For example, *Corynebacterium diphtheriae* and *C. efficiens* each contain two mixed clusters that pair subfamily 3 sortases with their own substrates and substrates of a distally located subfamily 5 homolog (Comfort and Clubb, 2004). As in the aforementioned SrtA-SrtB mixed clusters, the ability of the subfamily 3 and 5 enzymes to discriminate between these substrates can be readily explained by the predicted distinct Cws specificities of these enzymes (Table 1).

Why have mixed gene clusters? Recent results suggest that the SrtA- and SrtB-type sortases may attach proteins to distinct sites on the cell wall, since enzymatic digestion of the *S. aureus* showed that the SrtA and SrtB attached proteins have distinct electrophoretic properties (Mazmanian et al., 2002). This may explain the purpose of mixed-gene clusters, which may enable the coordinated expression of several proteins that would be attached to distinct surface positions by the action of sortases from different subfamilies.

SORTASE INHIBITION

The broad distribution of sortases in pathogenic bacteria and the essential roles of sortase-anchored surface proteins in the establishment of infection suggest that sortase inhibitors may prove to be effective anti-infective agents. Sortase inhibitors may have a distinctive advantage over conventional antibiotics, in that their anti-infective properties may not induce strong selective pressure that leads to resistance (Mazmanian et al., 1999; Cossart and Jonquieres, 2000). In addition to their therapeutic significance, sortase inhibitors have led, and will continue to lead, to insights into the enzymology of this novel class of enzymes (Connolly et al., 2003). The design of effective sortase inhibitors will require an understanding of the structural biology of sortase-substrate recognition, as well as intimate knowledge of the potentially diverse sortase anchoring pathways and their role in the establishment of infection.

Both natural (Kim et al., 2002; Kim et al., 2003) and synthetic (Kruger et al., 2001; Scott et al., 2002; Connolly et al., 2003) SrtA inhibitors have been described. Kim et al. (2002) have analyzed the inhibition properties of 80 different medicinal plant extracts on sortase transpeptidation in vitro, revealing a 50% inhibitory concentration (IC_{50}) as low as 1.5 μg/ml. The same group has isolated a SrtA inhibitor (β-sitosterol-3-*O*-glucopyransoside) from bulbs of *Fitrillaria verticillata.* This glucosylsterol demonstrated potent SrtA inhibition (IC_{50} = 18 μg/ml), as well as antibacterial activity against *S. aureus, B. subtilis,*

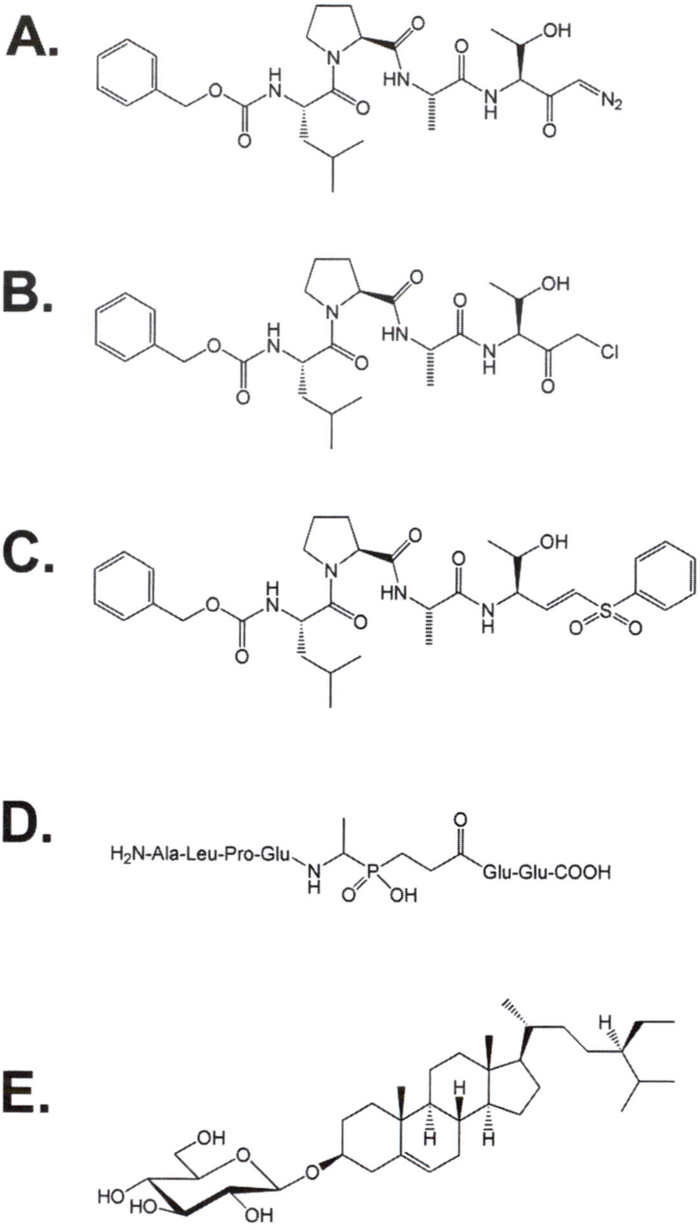

Figure 8. Molecular structures of sortase inhibitors. (A to C) structures that contain the LPXTG substrate mimic but replace the labile T-G peptide bond with a diazomethane (A), chloromethane (B), or vinyl sulfone (C) group for irreversible modification of the C184 nucleophile. (D) The phosphinate octapeptide NH$_2$-ALPEAΨ(PO$_2$HCH$_2$)GEE-OH is a transition state mimic of the T-G scissile bond. (E) The natural SrtA inhibitor β-sitosterol-3-*O*-glucopyransoside from bulbs of *F. verticillata.*

and *Micrococcus luteus* (Kim et al., 2003). Synthetic SrtA inhibitors are transition state mimics (Kruger et al., 2001) or have paired irreversible sulfhydryl modification agents with SrtA recognition sequence mimics (Fig. 8). In this second approach, the irreversible inhibitor consists of the substrate recognition motif of SrtA (LPXTG) but replaces the scissile Thr—Gly amide bond with a reactive "warhead" of vinyl sulfone ($C{=}C{-}SO_2Ph$) (Connolly et al., 2003), diazomethane ($-CH{=}N_2$), or chloromethane ($-CH_2Cl$) (Scott et al., 2002). All three reactive groups have previously been shown to covalently modify the active site thiol in cysteine proteases (Palmer et al., 1995; Bromme et al., 1996; Somoza et al., 2002; Pauly et al., 2003). This is relevant because mutagenesis studies have demonstrated the catalytic importance of SrtA residue C184 (Ton-That et al., 1999; Ton-That et al., 2002), and several sulfhydryl-directed reagents block the activity of SrtA in vitro (Ton-That et al., 2000; Ton-That et al., 2002). The efficacy of both natural and synthetic inhibitors has been tested in vitro by a high-performance liquid chromatography-based assay (Kruger et al., 2001) or by determining how they alter the SrtA-catalyzed hydrolysis of an internally quenched fluorescent substrate analogue (*d*-QALPETGEE-*e*, where *d* is Dabsyl and *e* is Edans) (Ton-That et al., 1999).

The first-order rate constant of inactivation (k_i) and the dissociation constant of inhibitor binding (K_i) for the vinyl sulfone inhibitors are 9×10^{-6} M and 4×10^{-4} min^{-1}, respectively, at physiological pH; the first-order rate constants of the peptidyl-diazomethane (5.8×10^{-3} min^{-1}) and peptidyl-chloromethane (1.1×10^{-2} min^{-1}) SrtA inhibitors are larger than the vinyl sulfone k_i by approximately 10- and 20-fold, respectively (Scott et al., 2002). This difference in reactivity is consistent with the higher electrophilicity of chloro- and diazomethane reactive groups and has been observed in inhibition studies of other cysteine proteases (Otto and Schirmeister, 1997). The development of a quantitative in vivo assay for inhibition will be necessary before we can evaluate the efficacy of peptidyl inhibitors of sortase as anti-infective agents.

SUMMARY

We are just beginning to understand the significance of sortases and the sorting reactions that they catalyze. The rapid elucidation of sortase enzymology, their phylogenetic distribution, and the impact of these enzymes on virulence may have generated as many questions as they have answers. Future areas of sortase research will need to address the functions of the enormous number of sortase homologs in gram-positive and -negative bacteria, the biological pathways in which they are involved, as well as the cellular signals that regulate their activity. Also, since SrtA has already been linked to virulence in many gram-positive pathogens, the design of promising anti-infective therapeutics will require a greater understanding of sortase function.

Acknowledgments. We thank Chu Kong Liew for useful discussions and David Comfort for sharing data prior to publication.

This work was supported by grant AI52217 from the National Institutes of Health to R.T.C.

REFERENCES

Altschul, S. F., T. L. Madden, A. A. Schäffer, J. Zhang, Z. Zhang, W. Miller, and D. J. Lipman. 1997. Gapped BLAST and PSI-BLAST: a new generation of protein database search programs. *Nucleic Acids Res.* **25:** 3389–3402.

Andrade, M. A., F. D. Ciccarelli, C. Perez-Iratxeta, and P. Bork. 2002. NEAT: a domain duplicated in genes near the components of a putative Fe^{3+} siderophore transporter from Gram-positive pathogenic bacteria. *Genome Biol.* **3:**research0047.1–0047.5.

Bae, T., and O. Schneewind. 2003. The YSIRK-G/S motif of staphylococcal protein A and its role in efficiency of signal peptide processing. *J. Bacteriol.* **185:**2910–2919.

Barnett, T. C., and J. R. Scott. 2002. Differential recognition of surface proteins in *Streptococcus pyogenes* by two sortase gene homologs. *J. Bacteriol.* **184:**2181–2191.

Barrett, A. J., and N. D. Rawlings. 2001. Evolutionary lines of cysteine peptidases. *Biol. Chem. Hoppe-Seyler* **382:**727–733.

Bierne, H., S. K. Mazmanian, M. Trost, M. G. Pucciarelli, G. Liu, P. Dehoux, L. Jansch, F. Garcia-del Portillo, O. Schneewind, and P. Cossart. 2002. Inactivation of the *srtA* gene in *Listeria monocytogenes* inhibits anchoring of surface proteins and affects virulence. *Mol. Microbiol.* **43:**869–881.

Bolken, T. C., C. A. Franke, K. F. Jones, G. O. Zeller, C. H. Jones, E. K. Dutton, and D. E. Hruby. 2001. Inactivation of the *srtA* gene in *Streptococcus gordonii* inhibits cell wall anchoring of surface proteins and decreases in vitro and in vivo adhesion. *Infect. Immun.* **69:**75–80.

Braun, L., S. Dramsi, P. Dehoux, H. Bierne, G. Lindahl, and P. Cossart. 1997. InlB: an invasion protein of *Listeria monocytogenes* with a novel type of surface association. *Mol. Microbiol.* **25:**285–294.

Bromme, D., J. L. Klaus, K. Okamoto, D. Rasnick, and J. T. Palmer. 1996. Peptidyl vinyl sulphones—a new class of potent and selective cysteine protease inhibitors—S2p2 specificity of human cathepsin O2 in comparison with cathepsins S and L. *Biochem. J.* **315:**85–89.

Cabanes, D., P. Dehoux, O. Dussurget, L. Frangeul, and P. Cossart. 2002. Surface proteins and the pathogenic potential of *Listeria monocytogenes*. *Trends Microbiol.* **10:**238–245.

Comfort, D., and R. T. Clubb. 2004. A comparative genome analysis identifies distinct sorting pathways in gram-positive bacteria. *Infect. Immun.* **72:**2710–2722.

Connolly, K. M., B. T. Smith, R. Pilpa, U. Ilangovan, M. E. Jung, and R. T. Clubb. 2003. Sortase from *Staphylococcus aureus* does not contain a thiolate-imidazolium ion pair in its active site. *J. Biol. Chem.* **278:**34061–34065.

Cossart, P., and R. Jonquieres. 2000. Sortase, a universal target for therapeutic agents against Gram-positive bacteria? *Proc. Nat. Acad. Sci. USA* **97:**5013–5015.

Drenth, J., J. N. Jansonius, R. Koekoek, H. M. Swen, and B. G. Wolthers. 1968. Structure of papain. *Nature* **218:**929–932.

Eddy, S. R. 1998. Profile hidden Markov models. *Bioinformatics* **14:**755–763.

Engelhardt, H., and J. Peters. 1998. Structural research on surface layers: a focus on stability, surface layer homology domains, and surface layer-cell wall interactions. *J. Struct. Biol.* **124:**276–302.

Fernandez-Tornero, C., R. Lopez, E. Garcia, G. Gimenez-Gallego, and A. Romero. 2001. A novel solenoid fold in the cell wall anchoring domain of the pneumococcal virulence factor LytA. *Nat. Struct. Biol.* **8:**1020–1024.

Filipe, S. R., E. Severina, and A. Tomasz. 2001. Functional analysis of *Streptococcus pneumoniae* MurM reveals the region responsible for its specificity in the synthesis of branched cell wall peptides. *J. Biol. Chem.* **276:**39618–39628.

Fischetti, V. A., K. F. Jones, and J. R. Scott. 1985. Size variation of the M protein in group A streptococci. *J. Exp. Med.* **161:**1384–1401.

Fischetti, V. A., V. Pancholi, and O. Schneewind. 1990. Conservation of a hexapeptide sequence in the anchor region of surface proteins from gram-positive cocci. *Mol. Microbiol.* **4:**1603–1605.

Fischetti, V. A., V. Pancholi, and O. Schneewind. 1991. Common characteristics of the surface proteins from gram-positive cocci, p. 290–294. *In* G. M. Dunny, P. P. Cleary, and L. L. Mckay (ed.), *Genetics and Molecular Biology of Streptococci, Lactococci, and Enterococci: Third International ASM Conference.* American Society for Microbiology, Washington, D.C.

Garandeau, C., H. Reglier-Poupet, I. Dubail, J. L. Beretti, P. Berche, and A. Charbit. 2002. The sortase SrtA of *Listeria monocytogenes* is involved in processing of internalin and in virulence. *Infect. Immun.* **70:**1382–1390.

Goffin, C., and J. M. Ghuysen. 2002. Biochemistry and comparative genomics of SxxK superfamily acyltransferases offer a clue to the mycobacterial paradox: presence of penicillin-susceptible target proteins versus lack of efficiency of penicillin as therapeutic agent. *Microbiol. Mol. Biol. Rev.* **66:**702–738.

Hava, D. L., and A. Camilli. 2002. Large-scale identification of serotype 4 *Streptococcus pneumoniae* virulence factors. *Mol. Microbiol.* **45:**1389–1406.

Hava, D. L., C. J. Hemsley, and A. Camilli. 2003. Transcriptional regulation in the *Streptococcus pneumoniae rlrA* pathogenicity islet by RlrA. *J. Bacteriol.* **185:**413–421.

Higashi, Y., J. L. Strominger, and C. C. Sweeley. 1967. Structure of a lipid intermediate in cell wall peptidoglycan synthesis: a derivative of a C55 isoprenoid alcohol. *Proc. Nat. Acad. Sci. USA* **57:**1878–1884.

Higashi, Y., J. L. Strominger, and C. C. Sweeley. 1970. Biosynthesis of the peptidoglycan of bacterial cell walls. XXI. Isolation of free C55–isoprenoid alcohol and of lipid intermediates in peptidoglycan synthesis from *Staphylococcus aureus. J. Biol. Chem.* **245:**3697–3702.

Huang, X., A. Aulabaugh, W. Ding, B. Kapoor, L. Alksne, K. Tabei, and G. Ellestad. 2003. Kinetic mechanism of *Staphylococcus aureus* sortase SrtA. *Biochemistry* **42:**11307–11315.

Igarashi, T., E. Asaga, and N. Goto. 2003. The sortase of *Streptococcus mutans* mediates cell wall anchoring of a surface protein antigen. *Oral Microbiol. Immunol.* **18:**266–269.

Ilangovan, U., H. Ton-That, J. Iwahara, O. Schneewind, and R. T. Clubb. 2001. Structure of sortase, the transpeptidase that anchors proteins to the cell wall of *Staphylococcus aureus. Proc. Nat. Acad. Sci. USA* **98:**6056–6061.

Janulczyk, R., and M. Rasmussen. 2001. Improved pattern for genome-based screening identifies novel cell wall-attached proteins in gram-positive bacteria. *Infect. Immun.* **69:**4019–4026.

Jedrzejas, M. J. 2001. Pneumococcal virulence factors: structure and function. *Microbiol. Mol. Biol. Rev.* **65:**187–207.

Jonsson, I. M., S. K. Mazmanian, O. Schneewind, T. Bremell, and A. Tarkowski. 2003. The role of *Staphylococcus aureus* sortase A and sortase B in murine arthritis. *Microbes Infect.* **5:**775–780.

Jonsson, I. M., S. K. Mazmanian, O. Schneewind, M. Verdrengh, T. Bremell, and A. Tarkowski. 2002. On the role of *Staphylococcus aureus* sortase and sortase-catalyzed surface protein anchoring in murine septic arthritis. *J. Infect. Dis.* **185:**1417–1424.

Kharat, A. S., and A. Tomasz. 2003. Inactivation of the *srtA* gene affects localization of surface proteins and decreases adhesion of *Streptococcus pneumoniae* to human pharyngeal cells in vitro. *Infect. Immun.* **71:**2758–2765.

Kim, S. H., D. S. Shin, M. N. Oh, S. C. Chung, J. S. Lee, I. M. Chang, and K. B. Oh. 2003. Inhibition of sortase, a bacterial surface protein anchoring transpeptidase, by beta-sitosterol-3-*O*-glucopyranoside from *Fritillaria verticillata. Biosci. Biotechnol. Biochem.* **67:**2477–2479.

Kim, S. W., I. M. Chang, and K. B. Oh. 2002. Inhibition of the bacterial surface protein anchoring transpeptidase sortase by medicinal plants. *Biosci. Biotechnol. Biochem.* **66:**2751–2754.

Kruger, R., S. Pesiridis, and D. G. McCafferty. 2001. Characterization of the *Staphylococcus aureus* sortase transpeptidase: a novel target for the development of chemotherapeutics against gram positive bacteria, p. 565–566. *In* M. Lebl and R. Houghten (ed.), *Peptides: The Wave of the Future.* American Peptide Society, San Diego, Calif.

Kumar, S., and R. Nussinov. 1999. Salt bridge stability in monomeric proteins. *J. Mol. Biol.* **293:**1241–1255.

Lee, S. F., and T. L. Boran. 2003. Roles of sortase in surface expression of the major protein adhesin P1, saliva-induced aggregation and adherence, and cariogenicity of *Streptococcus mutans. Infect. Immun.* **71:**676–681.

Lewis, S. D., F. A. Johnson, and J. A. Shafer. 1981. Effect of cysteine-25 on the ionization of histidine-159 in papain as determined by proton nuclear magnetic resonance spectroscopy. Evidence for a His-159–Cys-25 ion pair and its possible role in catalysis. *Biochemistry* **20:**48–51.

Li, T., M. K. Khah, S. Slavnic, I. Johansson, and N. Stromberg. 2001. Different type 1 fimbrial genes and tropisms of commensal and potentially pathogenic *Actinomyces* spp. with different salivary acidic proline-rich protein and statherin ligand specificities. *Infect. Immun.* **69:**7224–7233.

Lofdahl, S., B. Guss, M. Uhlen, L. Philipson, and M. Lindberg. 1983. Gene for staphylococcal protein A. *Proc. Nat. Acad. Sci. USA* **80:**697–701.

Lupas, A., H. Engelhardt, J. Peters, U. Santarius, S. Volker, and W. Baumeister. 1994. Domain structure of the *Acetogenium kivui* surface layer revealed by electron crystallography and sequence analysis. *J. Bacteriol.* **176:**1224–1233.

Marino, M., M. Banerjee, R. Jonquieres, P. Cossart, and P. Ghosh. 2002. GW domains of the *Listeria monocytogenes* invasion protein InlB are SH3–like and mediate binding to host ligands. *EMBO J.* **21:**5623–5634.

Marino, M., L. Braun, P. Cossart, and P. Ghosh. 1999. Structure of the InlB leucine-rich repeats, a domain that triggers host cell invasion by the bacterial pathogen *L. monocytogenes. Mol. Cell* **4:**1063–1072.

Mazmanian, S. K., G. Liu, T. T. Hung, and O. Schneewind. 1999. *Staphylococcus aureus* sortase, an enzyme that anchors surface proteins to the cell wall. *Science* **285:**760–763.

Mazmanian, S. K., G. Liu, E. R. Jensen, E. Lenoy, and O. Schneewind. 2000. *Staphylococcus aureus* sortase mutants defective in the display of surface proteins and in the pathogenesis of animal infections. *Proc. Nat. Acad. Sci. USA* **97:**5510–5515.

Mazmanian, S. K., H. Ton-That, and O. Schneewind. 2001. Sortase-catalyzed anchoring of surface proteins to the cell wall of *Staphylococcus aureus. Mol. Microbiol.* **40:**1049–1057.

Mazmanian, S. K., H. Ton-That, K. Su, and O. Schneewind. 2002. An iron-regulated sortase anchors a class of surface protein during *Staphylococcus aureus* pathogenesis. *Proc. Nat. Acad. Sci. USA* **99:**2293–2298.

Navarre, W. W., and O. Schneewind. 1994. Proteolytic cleavage and cell wall anchoring at the LPXTG motif of surface proteins in gram-positive bacteria. *Mol. Microbiol.* **14:**115–121.

Navarre, W. W., and O. Schneewind. 1999. Surface proteins of gram-positive bacteria and mechanisms of their targeting to the cell wall envelope. *Microbiol. Mol. Biol. Rev.* **63:**174–229.

Navarre, W. W., H. Ton-That, K. F. Faull, and O. Schneewind. 1998. Anchor structure of staphylococcal surface proteins. II. COOH-terminal structure of muramidase and amidase-solubilized surface protein. *J. Biol. Chem.* **273:**29135–29142.

Novick, R. P. 2000. Sortase: the surface protein anchoring transpeptidase and the LPXTG motif. *Trends Microbiol.* **8:**148–151.

Osaki, M., D. Takamatsu, Y. Shimoji, and T. Sekizaki. 2002. Characterization of *Streptococcus suis* genes encoding proteins homologous to sortase of gram-positive bacteria. *J. Bacteriol.* **184:**971–982.

Otto, H. H., and T. Schirmeister. 1997. Cysteine proteases and their inhibitors. *Chem. Rev.* **97:**133–171.

Pallen, M. J., A. C. Lam, M. Antonio, and K. Dunbar. 2001. An embarrassment of sortases: a richness of substrates? *Trends Microbiol.* **9:**97–101.

Palmer, J. T., D. Rasnick, J. L. Klaus, and D. Bromme. 1995. Vinyl sulfones as mechanism-based cysteine protease inhibitors. *J. Med. Chem.* **38:**3193–3196.

Pauly, T. A., T. Sulea, M. Ammirati, J. Sivaraman, D. E. Danley, M. C. Griffor, A. V. Kamath, I. K. Wang, E. R. Laird, A. P. Seddon, R. Menard, M. Cygler, and V. L. Rath. 2003. Specificity determinants of human cathepsin S revealed by crystal structures of complexes. *Biochemistry* **42:**3203–3213.

Perry, A. M., H. Ton-That, S. K. Mazmanian, and O. Schneewind. 2002. Anchoring of surface proteins to the cell wall of *Staphylococcus aureus*. III. Lipid II is an in vivo peptidoglycan substrate for sortase-catalyzed surface protein anchoring. *J. Biol. Chem.* **277:**16241–16248.

Pinitglang, S., A. B. Watts, M. Patel, J. D. Reid, M. A. Noble, S. Gul, A. Bokth, A. Naeem, H. Patel, E. W. Thomas, S. K. Sreedharan, C. Verma, and K. Brocklehurst. 1997. A classical enzyme active center motif lacks catalytic competence until modulated electrostatically. *Biochemistry* **36:**9968–9982.

Roche, F. M., R. Massey, S. J. Peacock, N. P. Day, L. Visai, P. Speziale, A. Lam, M. Pallen, and T. J. Foster. 2003. Characterization of novel LPXTG-containing proteins of *Staphylococcus aureus* identified from genome sequences. *Microbiology* **149:**643–654.

Ruzin, A., A. Severin, F. Ritacco, K. Tabei, G. Singh, P. A. Bradford, M. M. Siegel, S. J. Projan, and D. M. Shlaes. 2002. Further evidence that a cell wall precursor [C(55)-MurNAc-(peptide)-GlcNAc] serves as an acceptor in a sorting reaction. *J. Bacteriol.* **184:**2141–2147.

Sanchez-Puelles, J. M., J. M. Sanz, J. L. Garcia, and E. Garcia. 1990. Cloning and expression of gene fragments encoding the choline-binding domain of pneumococcal murein hydrolases. *Gene* **89:**69–75.

Sara, M. 2001. Conserved anchoring mechanisms between crystalline cell surface S-layer proteins and secondary cell wall polymers in Gram-positive bacteria? *Trends Microbiology* **9:**47–49.

Sarkany, Z., Z. Szeltner, and L. Polgar. 2001. Thiolate-imidazolium ion pair is not an obligatory catalytic entity of cysteine peptidases: the active site of picornain 3C. *Biochemistry* **40:**10601–10606.

Schneewind, O., A. Fowler, and K. F. Faull. 1995. Structure of the cell wall anchor of surface proteins in *Staphylococcus aureus. Science* **268:**103–106.

Schneewind, O., D. Mihaylova-Petkov, and P. Model. 1993. Cell wall sorting signals in surface proteins of gram-positive bacteria. *EMBO J.* **12:**4803–4811.

Schneewind, O., P. Model, and V. A. Fischetti. 1992. Sorting of protein-a to the staphylococcal cell wall. *Cell* **70:**267–281.

Schneewind, O., V. Pancholi, and V. A. Fischetti. 1991. Surface proteins from gram-positive cocci have a common motif for membrane anchoring, p. 152–154. *In* G. M. Dunny, P. P. Cleary, and L. L. McKay (ed.), *Genetics and Molecular Biology of Streptococci, Lactococci, and Enterococci.* American Society for Microbiology, Washington, D.C.

Scott, C. J., A. McDowell, S. L. Martin, J. F. Lynas, K. Vandenbroeck, and B. Walker. 2002. Irreversible inhibition of the bacterial cysteine protease-transpeptidase sortase (SrtA) by substrate-derived affinity labels. *Biochem. J.* **366:**953–958.

Severin, A., A. Figueiredo, and A. Tomasz. 1996. Separation of abnormal cell wall composition from penicillin resistance through genetic transformation of *Streptococcus pneumoniae. J. Bacteriol.* **178:**1788–1792.

Sjoquist, J., B. Meloun, and H. Hjelm. 1972a. Protein A isolated from *Staphylococcus aureus* after digestion with lysostaphin. *Eur. J. Biochem.* **29:**572–578.

Sjoquist, J., J. Movitz, I. B. Johansson, and H. Hjelm. 1972b. Localization of protein A in the bacteria. *Eur. J. Biochem.* **30:**190–194.

Somoza, J. R., J. T. Palmer, and J. D. Ho. 2002. The crystal structure of human cathepsin F and its implications for the development of novel immunomodulators. *J. Mol. Biol.* **322:**559–568.

Somoza, J. R., H. Zhan, K. K. Bowman, L. Yu, K. D. Mortara, J. T. Palmer, J. M. Clark, and M. E. McGrath. 2000. Crystal structure of human cathepsin V. *Biochemistry* **39:**12543–12551.

Storer, A. C., and R. Menard. 1994. Catalytic mechanism in papain family of cysteine peptidases. *Methods Enzymol.* **244:**486–500.

Strominger, J. L. 1965. Biochemistry of the cell wall of *Staphylococcus aureus. Ann. N.Y. Acad. Sci.* **128:**59–61.

Strominger, J. L., K. Izaki, M. Matsuhashi, and D. J. Tipper. 1967. Peptidoglycan transpeptidase and D-alanine carboxypeptidase: penicillin-sensitive enzymatic reactions. *Fed. Proc.* **26:**9–22.

Strominger, J. L., and D. J. Tipper. 1965. Bacterial cell wall synthesis and structure in relation to the mechanism of action of penicillins and other antibacterial agents. *Am. J. Med.* **39:**708–721.

Tipper, D. J., and J. L. Strominger. 1965. Mechanism of action of penicillins: a proposal based on their structural similarity to acyl-D-alanyl-D-alanine. *Proc. Nat. Acad. Sci. USA* **54:**1133–1141.

Ton-That, H., K. F. Faull, and O. Schneewind. 1997. Anchor structure of staphylococcal surface proteins—a branched peptide that links the carboxyl terminus of proteins to the cell wall. *J. Biol. Chem.* **272:**22285–22292.

Ton-That, H., H. Labischinski, B. Berger-Bachi, and O. Schneewind. 1998. Anchor structure of staphylococcal surface proteins. III. Role of the FemA, FemB, and FemX factors in anchoring surface proteins to the bacterial cell wall. *J. Biol. Chem.* **273:**29143–29149.

Ton-That, H., G. Liu, S. K. Mazmanian, K. F. Faull, and O. Schneewind. 1999. Purification and characterization of sortase, the transpeptidase that cleaves surface proteins of *Staphylococcus aureus* at the LPXTG motif. *Proc. Nat. Acad. Sci. USA* **96:**12424–12429.

Ton-That, H., S. K. Mazmanian, L. Alksne, and O. Schneewind. 2002. Anchoring of surface proteins to the cell wall of *Staphylococcus aureus*—cysteine 184 and histidine 120 of sortase form a thiolate-imidazolium ion pair for catalysis. *J. Biol. Chem.* **277:**7447–7452.

Ton-That, H., S. K. Mazmanian, K. F. Faull, and O. Schneewind. 2000. Anchoring of surface proteins to the cell wall of *Staphylococcus aureus*—sortase catalyzed in vitro transpeptidation reaction using LPXTG peptide and NH2–Gly(3) substrates. *J. Biol. Chem.* **275:**9876–9881.

Ton-That, H., and O. Schneewind. 1999. Anchor structure of staphylococcal surface proteins IV. Inhibitors of the cell wall sorting reaction. *J. Biol. Chem.* **274:**24316–24320.

Uhlen, M., B. Guss, B. Nilsson, S. Gatenbeck, L. Philipson, and M. Lindberg. 1984. Complete sequence of the staphylococcal gene encoding protein A. *J. Biol. Chem.* **259:**1695–1702.

Varea, J., J. L. Saiz, C. Lopez-Zumel, B. Monterroso, F. J. Medrano, J. L. Arrondo, I. Iloro, J. Laynez, J. L. Garcia, and M. Menendez. 2000. Do sequence repeats play an equivalent role in the choline-binding module of pneumococcal LytA amidase? *J. Biol. Chem.* **275:**26842–26855.

Whittaker, C. J., D. L. Clemans, and P. E. Kolenbrander. 1996. Insertional inactivation of an intrageneric co-aggregation-relevant adhesion locus from *Streptococcus gordonii* DL1 (challis). *Infect. Immun.* **64:**4137–4142.

Xu, Q., W. Gao, S.-Y. Ding, R. Kenig, Y. Shoham, E. A. Bayer, and R. Lamed. 2003. The cellulosome system of *Acetivibrio cellulolyticus* includes a novel type of adaptor protein and a cell surface anchoring protein. *J. Bacteriol.* **185:**4548–4557.

Chapter 8

Structural Determinants of *Haemophilus influenzae* Adherence to Host Epithelia: Variations on Type V Secretion

Neeraj K. Surana, Shane E. Cotter, Hye-Jeong Yeo, Gabriel Waksman, and Joseph W. St. Geme III

Haemophilus influenzae is a gram-negative bacterium that causes naturally acquired disease exclusively in humans. Encapsulated strains express one of six structurally and antigenically distinct polysaccharide capsules, designated serotypes (types) a to f. Nonencapsulated strains are defined by their failure to react with antisera against the recognized *H. influenzae* capsular polysaccharides and are referred to as nontypeable.

Type b strains and to a lesser extent other encapsulated *H. influenzae* strains represent an important etiology of invasive (bacteremic) disease, including septicemia, meningitis, epiglottitis, arthritis, and cellulitis. Since the late 1980s, the incidence of *H. influenzae* type b disease has dropped precipitously in the United States, reflecting the routine use of *H. influenzae* conjugate vaccines, which contain type b capsular polysaccharide conjugated to one of several non-*H. influenzae* immunogenic carrier proteins (Centers for Disease Control and Prevention, 2002; Wenger and Ward, 2004). However, *H. influenzae* type b disease remains prevalent in underdeveloped countries, where immunization rates are low (Peltola, 2000).

Nontypeable *H. influenzae* is a common cause of middle ear infection, sinusitis, and conjunctivitis (Turk, 1984; Rao et al., 1999). In addition, this organism is a frequent cause of exacerbations of underlying lung disease (e.g., chronic bronchitis, bronchiectasis, and cystic fibrosis) and is an important etiology of community-acquired pneumonia, especially in elderly adults and among children in developing countries (Turk, 1984; Rao et al.,

Neeraj K. Surana, Shane E. Cotter, and Hye-Jeong Yeo • Edward Mallinckrodt Department of Pediatrics, Washington University School of Medicine, 660 South Euclid Ave., St. Louis, MO 63110. *Gabriel Waksman* • Institute of Structural Molecular Biology, Birkbeck and University College London, Malet St., London, United Kingdom. *Joseph W. St. Geme III* • Edward Mallinckrodt Department of Pediatrics and Department of Molecular Microbiology, Washington University School of Medicine, 660 South Euclid Ave., St. Louis, MO 63110.

1999). Nontypeable strains are sometimes recovered from systemic sites, usually in neonates or in the setting of compromised host immunity (Krasan and St. Geme, 1997; Rao et al., 1999).

The pathogenesis of disease caused by *H. influenzae* begins with colonization of the nasopharynx (Murphy et al., 1987). Successful colonization requires that the organism overcome mechanical forces, including the mucociliary escalator, coughing, and sneezing. Adherence to respiratory epithelium represents an important strategy to circumvent these forces and is influenced by a number of *H. influenzae* factors. In this chapter, we focus on Hap, the HMW1/HMW2 adhesins, and the Hia/Hsf adhesins. All of these adhesive proteins are grouped in the type V secretion pathway, which includes the autotransporter pathway and the two-partner secretion (TPS) pathway.

Hap is ubiquitous among *H. influenzae* isolates, including both typeable and nontypeable strains (Rodriguez et al., 2003). HMW1/HMW2-like proteins are expressed by the majority of nontypeable clinical isolates and are absent from typeable strains (Barenkamp and Leininger, 1992; Rodriguez et al., 2003). Hia is expressed by nearly all nontypeable strains that lack HMW adhesins (Barenkamp and St. Geme, 1996; St. Geme et al., 1998), and Hsf is a homolog of Hia that is universally present among typeable *H. influenzae* strains, including both type b and non-type b strains (St. Geme et al., 1998; Rodriguez et al., 2003). Hap is an example of a conventional autotransporter, Hia/Hsf represent variant autotransporters, and HMW1/HMW2 belong to the TPS pathway.

THE *H. INFLUENZAE* Hap ADHESIN

Hap was first discovered based on its ability to facilitate adherence and low-level invasion in assays with cultured human epithelial cells (St. Geme et al., 1994). More recent observations have established that Hap mediates adherence to selected extracellular matrix proteins and promotes the formation of bacterial microcolonies (Hendrixson and St. Geme, 1998; Fink et al., 2002). In addition, Hap has serine protease activity, resulting in autoproteolysis at four distinct sites (Fink et al., 2001, 2003).

To mediate adhesion and invasion, the Hap adhesive domains must be localized on the bacterial surface. In considering the structural determinants of Hap secretion, it is informative to examine the structural elements that are common to Hap and other conventional autotransporters.

Autotransporter proteins contain three distinct functional regions, including an N-terminal signal peptide, an internal passenger domain, and a C-terminal translocator domain (also referred to as a helper domain or β-domain) (Color Plate 28 [see color insert]). The passenger domain harbors the effector functions of the protein, with examples including adherence, invasion, protease activity, cytotoxicity, serum resistance, and cell-to-cell spread (Henderson and Nataro, 2001). In Hap the passenger domain is called Hap_S and the translocator domain is called Hap_β (Color Plate 28).

The first step in secretion of a protein from the cytoplasm to the bacterial surface involves traversal of the cytoplasmic membrane (Henderson et al., 1998). In autotransporters this task is accomplished at least in part via interaction between the signal peptide of the protein and the Sec machinery. In Hap, the signal peptide is 25 amino acids in length and has a characteristic sequence, with a positively charged N domain, a hydrophobic H domain, and a signal peptidase I cleavage site (Stathopoulos et al., 2000).

Following Sec-dependent delivery of the autotransporter to the periplasm, the translocator domain inserts into the outer membrane, ultimately facilitating translocation of the passenger domain across the outer membrane. The Hap$_\beta$ translocator domain is ~360 amino acids in length and migrates at an apparent molecular mass of ~45 kDa on denaturing sodium dodecyl sulfate-polyacrylamide gels, with the size varying slightly from strain to strain (Hendrixson et al., 1997; Cutter et al., 2002). Hap$_\beta$ has significant sequence homology to the C-terminal translocator domains of other autotransporters and is predicted to form a transmembrane α-helix and a β-barrel with 14 transmembrane β-strands, analogous to most conventional autotransporters (Schirmer and Cowan, 1993; Loveless and Saier, 1997). In other autotransporters, deletion of the α-helix (the linker region) abrogates surface localization of the passenger domain, suggesting that the α-helix and the β-barrel regions function as the miminal translocation unit (Suzuki et al., 1995; Maurer et al., 1999; Oliver et al., 2003).

Consistent with secondary-structure analysis predicting that the translocator domains of autotransporters form a β-barrel, black lipid bilayer and liposomal swelling assays using purified protein have demonstrated that *Bordetella pertussis* BrkA, *Neisseria gonorrhoeae* immunoglobulin A1 (IgA1) protease, *Pseudomonas aeruginosa* PalA, and *N. meningitidis* NalP translocators have pore-forming activity, with a pore size estimated at 0.84 to 2.4 nm in diameter (Shannon and Fernandez, 1999; Veiga et al., 2002; Lee and Byun, 2003; Oomen et al., 2004). Recently, the crystal structure of the NalP translocator domain was solved, revealing a monomer with an N-terminal α-helix and 12 antiparallel β-sheets (Color Plate 29 [see color insert]). The 12 β-sheets form a β-barrel with a hydrophilic pore measuring 1 by 1.25 nm, and the α-helix spans the pore, presumably facilitating extrusion of the passenger domain across the bacterial outer membrane (Oomen et al., 2004). Interestingly, analysis of the NalP structure indicates that autotransporter translocator domains obey the same structural constraints that apply to other bacterial outer membrane β-barrel proteins (Schulz, 2000).

Review of the crystal structures available for bacterial outer membrane β-barrel proteins reveals a list of parameters that are critical for β-barrel formation. In all cases, the β-barrel is composed of an even number of antiparallel amphipathic β-strands that are connected to their next neighbors along the peptide chain. All hydrogen-bonding potential of the main chain is saturated between neighboring strands, while alternating hydrophobic and hydrophilic side chains face the lipid bilayer and the aqueous pore, respectively. Furthermore, the first and last β-strands interact to complete the formation of a cylinder with an enclosed, hydrophilic interior and a hydrophobic, hydrogen bond-saturated exterior. Finally, both the N terminus and the C terminus of the barrel face the periplasm. All of these constraints are satisfied in the predicted structure of Hap$_\beta$.

Examination of the amino acid sequences of bacterial outer membrane β-barrel-forming proteins reveals a consensus motif at the immediate C terminus, corresponding to the final amphipathic transmembrane β-strand (Loveless and Saier, 1997). The final amino acid is always aromatic, most commonly phenylalanine and occasionally tryptophan but never tyrosine. Furthermore, the preceding residues alternate between aromatic (or hydrophobic) and polar (or charged) (Jose et al., 1995; Loveless and Saier, 1997). It has been postulated that this consensus motif targets the protein to the outer membrane (Henning and Koebnik, 1994). Indeed, in studies of the PhoE porin, deletion of the final phenylalanine alone results in a marked decrease in both protein targeting to the outer membrane and protein stability in the outer membrane, and deletion of the final transmembrane β strand abrogates outer membrane localization altogether (Struyve et al., 1991; de Cock et al., 1997). Consistent with these observations, deletion or substitution of the final three residues of Hap results in loss of detectable protein in the outer membrane (Hendrixson et al., 1997).

Among crystallized bacterial outer membrane β-barrel proteins, some are monomers but most are oligomers (Schulz, 2002). For example, the porins exist as homotrimers, with each monomer generating a β-barrel with an independent pore (Weiss et al., 1990; Dutzler et al., 1999). In these proteins, multimerization occurs at interface sites on the external faces of the barrels and is thought to increase the stability of the individual subunits in both the periplasm (where multimers are formed) and the outer membrane (Koebnik et al., 2000). In *Escherichia coli* TolC, three distinct subunits contribute four β-strands each to produce a single functional β-barrel (Koronakis et al., 2000).

Interestingly, while the crystal structure of the NalP translocator domain reveals a monomer, recent work with the *N. gonorrhoeae* IgA1 protease translocator domain raises the possibility that in at least some conventional autotransporters the translocator domain undergoes multimerization. After purification from *E. coli,* the 50-kDa Iga$_\beta$ protein was found to migrate at ~250 kDa in nondenaturing gels and at ~500 kDa on size exclusion columns, leading to the conclusion that the IgA1 protease translocator domain is a stable multimer containing 6 to 10 subunits (Veiga et al., 2002). This oligomeric structure is thought to consist of separate, monomeric β-barrels that together form a central pore. Biochemical studies and analysis by electron microscopy have determined the pore diameter to be 2 nm.

Based on information about NalP, IgA1 protease, and other bacterial β-barrel proteins, at least four different architectures are possible for Hap$_\beta$. Perhaps most likely is that Hap$_\beta$ may reside in the outer membrane as a monomer with a single central pore, similar to NalP (Color Plate 30A [see color insert]). Alternatively, Hap$_\beta$ may form a multimer with a single common pore, similar to the current model for Iga$_\beta$ (Color Plate 30B). As another option, Hap$_\beta$ may form a multimer with each subunit forming a separate pore, analogous to porins (Color Plate 30C). Finally, Hap$_\beta$ may form a single multimeric β-barrel containing β-strands from multiple individual subunits, akin to *E. coli* TolC. With each of these architectures, the Hap β-barrel facilitates translocation of the passenger domain from the periplasm to the bacterial surface.

In recent work, investigation of the *B. pertussis* BrkA autotransporter established that deletion of a ~100-residue region at the C-terminal end of the passenger domain resulted in incomplete folding of the passenger domain on the bacterial surface (Oliver et al., 2003b). Interestingly, when the deleted region was expressed in *trans,* the mutant passenger domain was able to assume a wild-type conformation, suggesting intramolecular chaperone activity. A homology search with this ~100-residue region identified a homologous region in Hap and in many other autotransporter proteins. In all cases, the homologous region is located near the C terminus of the passenger domain, at a variable distance N-terminal to the translocator domain. One hypothesis is that the proposed intramolecular chaperone region allows for correct folding of the passenger domain as it traverses the β-domain pore and is presented on the bacterial surface. To date, pertactin is the only autotransporter for which a crystal structure of the full-length passenger domain (referred to as P.69) has been solved (Emsley et al., 1996). The proposed intramolecular chaperone region can be visual-

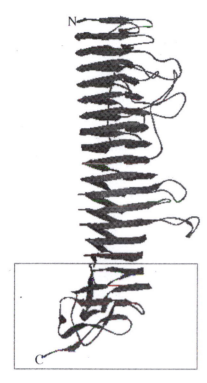

Figure 1. Structure of pertactin passenger domain. Ribbon diagram of the full passenger domain of pertactin (PDB code 1dab). The protein fold consists of a 16-strand parallel β-helix with a V-shaped cross section. The structure appears as a helix with several protruding loops. The proposed intramolecular chaperone region is enclosed in a box.

ized within this structure as a β-helical loop at the base of the right-handed cylindrical β-helix that makes up the majority of the protein (Fig. 1). Autotransporter proteins that belong to the subtilisin subfamily contain a motif called the junction domain at the C-terminal end of the passenger domain. This motif is required for correct folding and function of subtilisin-like autotransporters, implying functional homology to the intramolecular chaperone of BrkA (Ohnishi et al., 1994). Interestingly, despite the functional homology, no sequence homology exists.

Once on the bacterial surface, most conventional autotransporters undergo a cleavage event that separates the passenger domain from the translocator domain (Henderson et al., 1998). This cleavage is mediated via either autoproteolysis or a distinct membrane-bound protease (Pohlner et al., 1987; Egile et al., 1997; Shere et al., 1997). In some cases the passenger domain is liberated into the extracellular space (Pohlner et al., 1987), and in others it remains noncovalently associated with the cell surface (Benz and Schmidt, 1992a; Oliver et al., 2003a). Early studies demonstrated that Hap undergoes cleavage via autoproteolysis, resulting in extracellular release of Hap$_S$. More recent work has established that Hap is a member of the SA (chymotrypsin) clan of serine proteases, with a catalytic triad that consists of S243, D140, and H98, with S243 falling within a consensus serine protease motif, GD$\underline{S}$GS. Mutation of any of the three active-site residues to alanine results in complete loss of autoproteolysis (Hendrixson et al., 1997; Fink et al., 2001).

The GD$\underline{S}$GS serine protease motif and the arrangement of the catalytic residues in Hap are typical of a number of other autotransporters with protease activity and with roles in virulence. Examples include the IgA1 proteases expressed by *N. gonorrhoeae*, *N. meningitidis*, and *H. influenzae* and a group of autotransporters expressed by members of the *Enterobacteriaceae* and referred to as SPATEs. The *H. influenzae* IgA1 protease cleaves the hinge region of human IgA1 (Mulks et al., 1980), while the *Neisseria* IgA1 proteases cleave both human IgA1 and human LAMP1, an endosomal protein (Lin et al., 1997). The SPATEs cleave a variety of host factors, such as hemoglobin, fodrin, mucin, and complement proteins (Otto et al., 1998; Henderson et al., 1999; Villaseca et al., 2000). To date, the only known substrate for Hap is Hap itself, although cleavage of host factors has not been fully studied. Hap autoproteolysis occurs predominantly via intermolecular cleavage, with one molecule cleaving neighboring molecules. The primary site of cleavage is the peptide bond between Leu1036 and Asn1037; other sites include Leu1046-Thr1047, Phe1077-Ala1078, and Phe1067-Ser1068, defining a consensus motif [(Q/R)(S/A)X(L/F)] in the P1 to P4 positions (Fink et al., 2001).

Given that the Hap_S passenger domain harbors adhesive activity, at first glance cleavage of Hap_S from the bacterial surface seems counterintuitive. However, it is notable that Hap_S is a strong immunogen, as evidenced by the fact that immunization of BALB/c mice with this protein results in a decreased density of *H. influenzae* nasopharyngeal colonization (Cutter et al., 2002). Consequently, liberation of Hap_S may prevent deposition of Hap-specific antibodies on the bacterial surface and may result in binding of circulating antibody before surface-associated Hap can be engaged. Alternatively, cleavage of Hap_S from the bacterial surface may play an important role in allowing the separation of bacteria from epithelial cells, extracellular matrix proteins, and bacterial aggregates, facilitating the seeding of other niches in the human host. In support of this point, the mutant expressing proteolytically inactive HapS243A demonstrates both increased adherence to epithelial cells and enhanced bacterial autoaggregation (Hendrixson and St. Geme, 1998). As another possibility, following liberation from the bacterial surface, Hap may cleave unidentified host substrates, facilitating subsequent steps in the pathogenesis of disease. Of note, the host may play a role in Hap adhesive potential, since physiologic levels of host secretory leukocyte protease inhibitor inhibit Hap proteolytic activity, leading to increased adhesive capacity in vitro (Hendrixson and St. Geme, 1998).

The region of Hap responsible for adherence to epithelial cells maps to the final 311 residues of the passenger domain (Fink et al., 2003). Interestingly, Hap-Hap-mediated microcolony formation also maps to the final 311 residues of Hap_S. It remains unknown whether these two activities are attributed to the same binding pocket or instead to separate modules that are located nearby each other.

In addition to mediating adherence to epithelial cells, Hap promotes binding to fibronectin, laminin, and collagen IV (Fink et al., 2003). The ability to bind maximally to these extracellular matrix proteins requires the final 511 residues of Hap_S, arguing that the binding pockets for the extracellular matrix are distinct from those involved in adherence to epithelial cells and formation of bacterial microcolonies.

To summarize, the Hap precursor is exported out of the cytoplasm via a typical prokaryotic N-terminal signal peptide. Subsequently, Hap is targeted to the outer membrane, where the Hap_β domain is predicted to form a channel, allowing for surface localization of Hap_S. Once on the bacterial surface, Hap_S gains serine protease activity and cleaves neighboring Hap precursor molecules, releasing the Hap_S domain from Hap_β and the cell wall. In the presence of SLPI and potentially other factors, Hap_S protease activity is blocked, resulting in retention of cell-associated Hap_S and potentiating Hap adhesive activity.

THE *H. INFLUENZAE* HMW1 AND HMW2 ADHESINS
AND THE TWO-PARTNER SECRETION PATHWAY

In contrast to Hap and other conventional autotransporters, proteins that belong to the TPS family are characterized by a secreted "passenger domain" (referred to as a TpsA protein) and an outer membrane "translocator domain" (referred to as a TpsB protein) that are expressed as separate proteins (Jacob-Dubuisson et al., 2000; Jacob-Dubuisson et al., 2001). TpsA proteins secreted by the TPS pathway are found in a wide variety of bacterial species and include *B. pertussis* filamentous hemagglutinin (FHA), *Serratia marcescens* ShlA, *Proteus mirabilis* HpmA, and the *H. influenzae* HMW1 and HMW2 proteins (Domenighini et al., 1990; Poole et al., 1988; Uphoff and Welch, 1990; Barenkamp and Leininger, 1992).

The HMW1 and HMW2 proteins are high-molecular-weight, nonpilus adhesins that were originally identified as major targets of the human serum antibody response during acute otitis media (Barenkamp and Leininger, 1992). They are present in 70 to 80% of non-typeable *H. influenzae* strains and mediate adherence to macrophages and a variety of epithelial cell types (Barenkamp and Leininger, 1992; St. Geme et al., 1993; Noel et al., 1994). HMW1 is encoded by the *hmw1A* gene, while HMW2 is encoded by the *hmw2A* gene. *hmw1A* and *hmw2A* are located at physically distinct positions on the chromosome and are flanked downstream by accessory genes designated *hmw1B/hmw1C* and *hmw2B/hmw2C*, respectively. These accessory genes are required for proper expression and localization of the adhesins (St. Geme and Grass, 1998). The HMW1B and HMW2B proteins are 99% identical, and the HMW1C and HMW2C proteins are 97% identical. Consistent with these findings, the HMW1B and HMW1C proteins are functionally interchangeable with the HMW2B and HMW2C proteins (St. Geme and Grass, 1998). Given the similarity between the HMW1 and HMW2 systems, in this section we focus primarily on HMW1 (Color Plate 31 [see color insert]).

Initial examination of the *hmw1* locus revealed that the predicted amino acid sequence of HMW1C lacks a signal sequence, suggesting a cytoplasmic location. Elimination of this protein affected the apparent size and stability of HMW1 (Grass et al., 2003). Further analysis demonstrated that HMW1 is glycosylated, with a carbohydrate structure that contains galactose, glucose, and mannose. Glycosylation appears to be important for tethering of HMW1 to the bacterial surface, a prerequisite for HMW1-mediated adherence. Subsequent analyses established that glycosylation of HMW1 occurs in the cytoplasm in an HMW1C-dependent manner, suggesting the possibility that HMW1C is a glycosyltransferase. Consistent with this possibility, HMW1C has homology to a family of eukaryotic *O*-GlcNAc transferases. Alternatively, HMW1C may serve to stabilize HMW1 in the cytoplasm and allow another enzyme to mediate glycosylation.

In experiments addressing the mechanism of HMW1 export from the cytoplasm, both chemical and genetic disruption of Sec apparatus components resulted in diminished quantities of periplasmic HMW1 and elimination of surface-localized HMW1, demonstrating that HMW1 is translocated to the periplasm via the Sec system (Grass and St. Geme, 2000). Further study revealed that HMW1 has an atypical signal sequence. In particular, the immediate N terminus bears little resemblance to a signal sequence while residues 48 to 68 contain features of a typical prokaryotic signal sequence, including a stretch of hydrophobic residues and a predicted signal peptidase I cleavage site. Examination of chimeras consisting of the HMW1 N terminus fused to PhoA and N-terminal sequencing of the periplasmic form of HMW1 confirmed that residues 1 to 68 represent the signal sequence (Grass and St. Geme, 2000).

HMW1-like atypical signal sequences have been found in a number of autotransporter and TpsA proteins, including the diarrheagenic *E. coli* AIDA-I adhesin, *Enterobacteriaceae* proteins within the SPATE family, and *B. pertussis* FHA (Benz and Schmidt, 1992b; Guyer et al., 2000; Lambert-Buisine et al., 1998). Closer examination of the signal sequences of these proteins reveals that the majority have an N-terminal extension with the sequence $M_1N(R/K)X(Y/F)X(I/L/V/T)X(W/Y/F/K)(N/S/C)(L/W/F)(V/N/I)$ (A/V) (A/V/C) SE(L/F/G)(A/T/S)(R/K), where X represents any amino acid. Thus far, HMW1, *Shigella* IcsA, and *E. coli* Hbp are the only examples of these proteins that have been demonstrated to utilize components of the Sec apparatus (Grass and St. Geme, 2000; Brandon et al., 2003; Sijbrandi et al., 2003). Given the recent evidence that some Sec components are also used by the signal recognition particle-dependent pathway, it is also possible that proteins with an HMW1-like N-terminal extension undergo transport across the inner membrane via the signal recognition particle pathway in addition to the Sec apparatus (Valent et al., 1998; Qi and Bernstein, 1999; Sijbrandi et al., 2003). It is also possible that the N-terminal extension influences the kinetics of transport across the inner membrane, perhaps by interacting with a cytoplasmic factor that impedes translocation. The sequence conservation of this N-terminal extension probably underlies some common function.

In considering the mechanism by which HMW1 is transported from the periplasm to the bacterial surface, it is noteworthy that HMW1B has significant sequence homology to outer membrane translocator proteins belonging to the TpsB family, including *B. pertussis* FhaC, *S. marcescens* ShlB, and *P. mirabilis* HpmB (Willems et al., 1994; Poole et al., 1988; Uphoff and Welch, 1990). HMW1B also has limited homology to outer membrane usher proteins involved in pilus biogenesis, including *H. influenzae* HifC, *Salmonella enterica* serovar Enteritidis SefC, and *E. coli* CssD (Watson et al., 1994; Clouthier et al., 1993; Willshaw et al., 1988). Consistent with this homology, deletional analysis established that expression of HMW1B is required for HMW1 surface localization and for HMW1-mediated adherence (St. Geme and Grass, 1998). HMW1B exhibits heat modifiability, a hallmark of integral outer membrane proteins that form β-barrels (Nakamura and Mizushima, 1976; Surana et al., 2004). Secondary-structure predictions suggest that HMW1B has 22 transmembrane β-strands, slightly more than the 20 and 19 β-strands described for ShlB and FhaC, respectively (St. Geme and Grass, 1998; Konninger et al., 1999; Guédin et al., 2000). Recent work has established that HMW1B exists as a tetramer (Surana et al., 2004). Furthermore, expression of HMW1B by *E. coli* results in increased sensitivity to relatively large (600- to 900-Da) antibiotics and compensation for the absence of LamB, allowing growth on larger maltosaccharides as the only carbon source (G. G. Hardy and J. W. St. Geme III, unpublished observations). To determine whether HMW1B has translocator activity, chimeric proteins consisting of either HMW1 or Hia_{50-779} fused to HMW1B were constructed. In both cases, the adhesive epitopes were surface localized and were able to mediate adherence to epithelial cells (Hardy and St. Geme, unpublished). Taken together, these findings provide strong evidence that HMW1B forms a β-barrel in the outer membrane and has both pore-forming and translocating activities.

The HMW1 adhesin contains a peptide sequence defined by amino acids 69 to 441 that is lacking in conventional autotransporters. This peptide has homology at the N-terminal end to a highly conserved 110-residue region at the N terminus of other TpsA proteins (St. Geme and Grass, 1998; Jacob-Dubuisson et al., 2001). Based on C-terminal deletion experiments with *S. marcescens* ShlA, *P. mirabilis* HpmA, and FHA and examination of an in-frame deletion of residues 72 to 441 in HMW1, the conserved 110-residue region has been implicated in secretion (Schonherr et al., 1993; Uphoff and Welch, 1994; Renauld-Mongenie et al., 1996; Grass and St. Geme, 2000). In further studies involving ShlA, FHA, and HMW1, site-directed mutagenesis of conserved N(P/S)(N/H)L and N(P/T)NG motifs in the 110-residue region resulted in reduced translocation across the outer membrane, raising the hypothesis that these motifs participate in interactions with the cognate outer membrane translocator (Schonherr et al., 1993; Jacob-Dubuisson et al., 1997; St. Geme and Grass, 1998; Grass and St. Geme, 2000). With this information in mind, Jacob-Dubuisson et al. proposed that the 110-residue region is a "secretion domain" (Jacob-Dubuisson et al., 2000; Jacob-Dubuisson et al., 2001). Consistent with this proposal, in recent work we generated a chimeric protein containing HMW1 amino acids 1 to 441 (HMW1$_{1-441}$) and a portion of the passenger domain of the *H. influenzae* Hia adhesin (Hia$_{50-779}$) and found that the chimera was secreted into the supernatant in an HMW1B-dependent manner (Surana et al., 2004). In contrast, this chimera was not secreted when coexpressed with the Hia translocator domain, demonstrating specificity between the N-terminal fragment of HMW1 and the cognate translocator. Additionally, we have demonstrated via far-Western analysis that HMW1B interacts directly with HMW1$_{69-441}$, providing a mechanism for how the secretion domain mediates the secretion of TpsA proteins (Surana et al., 2004). Sometime after interaction with HMW1B, HMW1$_{69-441}$ is cleaved and released into the supernatant, leaving the mature adhesin on the bacterial surface (S. Grass and J. W. St. Geme III, unpublished observations).

Recently, Clantin et al. (2004) solved the crystal structure of the secretion domain of FHA (Fha30), revealing a right-handed parallel β-helix (Color Plate 32 [see color insert]). While most of the sequence of Fha30 is well conserved with other TpsA proteins, there are two less highly conserved regions that form extrahelical motifs. One of these regions forms three β-strands and serves to cap the N terminus of Fha30, and the other forms a four-strand β-sheet and extends from the core β-helix. The conserved N(P/S)(N/H)L and N(P/T)NG motifs form type I β-turns and appear to stabilize the overall fold. Consistent with this possibility, we have demonstrated via far-Western analysis that HMW1$_{1-441}$ with mutations in the NPNGI and NTNG motifs are still able to interact with HMW1B (Surana et al., unpublished). It has been speculated that the less highly conserved regions may be the domains that mediate interaction with the translocator domain, providing a basis for the observed specificity between the TpsA protein and the cognate TpsB translocator (Jacob-Dubuisson et al., 1997; Clantin et al., 2004).

Once on the surface of the organism, both HMW1 and HMW2 function as adhesins (St. Geme et al., 1993). Despite the sequence similarity between these proteins, they have differing cellular binding specificities (Barenkamp and Leininger, 1992; Hultgren et al., 1993). Using chimeras containing part of HMW1 and part of HMW2, the binding domains were narrowed to a ~360-residue region near the N terminus of the mature HMW1 and HMW2 proteins (Dawid et al., 2000). Interestingly, HMW1 and HMW2 have 71% identity over their entire protein sequences, but the binding domains correspond to the regions of highest dissimilarity, with only ~35% identity, perhaps explaining the difference in binding specificities. Biochemical studies have demonstrated that HMW1 recognizes an α2-3-linked sialylated glycoprotein on human epithelial cells (St. Geme, 1994). The nature of the HMW2 receptor remains unknown.

Although little is known about the structure of either HMW1 or HMW2, structural information on other related proteins may provide some insights. In early work, Barenkamp and Leininger (1992) noted that the HMW proteins share antigenic relatedness to FHA. In particular, an antibody derived against recombinant HMW1 reacted with purified FHA in both an enzyme-linked immunosorbent assay and a Western immunoblot analysis and a monoclonal antibody against FHA reacted with HMW1 and HMW2 from a diverse set of clinical isolates. The sequence homology, similar functions, shared secretion pathway, and shared immunoreactive epitopes between the HMW proteins and FHA may indicate that these adhesins have similar structures. Electron microscopy studies coupled with secondary-structure predictions suggested that FHA adopts a β-helix structure, with a length of ~500 Å and a diameter of ~40 Å (Makhov et al., 1994; Kajava et al., 2001). With this information in mind, it is interesting that both HMW1 and HMW2 are predicted to be rich in β-strands and β-turns, the principal components of a β-helix (A. Z. Buscher and J. W. St. Geme III, unpublished observations). Furthermore, carbohydrate binding is a property common to many β-helical proteins, and it is known that HMW1 binds α2-3-linked sialic acid (Kajava et al., 2001; St. Geme, 1994). For these reasons, we suggest that HMW1 and HMW2 may form β-helical structures. Additional investigation of the structure of these proteins is currently underway.

To summarize, following synthesis, HMW1 is glycosylated in the cytoplasm in an HMWC-dependent manner. Subsequently, the atypical signal sequence (HMW1$_{1-68}$) targets the protein to the Sec apparatus and is then cleaved by signal peptidase I. Within the periplasm, the ~110-residue secretion domain mediates interaction with HMWB, allowing translocation of the adhesin to the bacterial surface. At some time during this translocation process, HMW1$_{69-441}$ is cleaved and released into the supernatant. Once on the surface, the mature adhesin is loosely tethered to the bacterial surface via glycosylation. Finally, given the sequence and functional homology to FHA, we suggest that the HMW1 and HMW2 adhesins adopt β-helical structures.

THE *H. INFLUENZAE* Hia AND Hsf ADHESINS: EXAMPLES OF TRIMERIC AUTOTRANSPORTERS

Hia was first identified in nontypeable *H. influenzae* strain 11, which was isolated from a child with acute otitis media and lacks HMW1 and HMW2 (Barenkamp and St. Geme, 1996). In a series of studies, Hia was discovered to have adhesive capabilities and was found to be present in 15 to 30% of nontypeable *H. influenzae* isolates (Barenkamp and St. Geme, 1996; St. Geme et al., 1998). Hsf is a homolog of Hia that is much larger and is ubiquitous among encapsulated strains of *H. influenzae* (St. Geme et al., 1996).

Immunoelectron microscopy revealed that Hia is surface localized (St. Geme and Cutter, 2000). Sequence analysis indicated that the C-terminal 319 residues of Hia (Hia$_{780-1098}$) were similar to the C termini of many autotransporter proteins, suggesting that Hia may belong to the autotransporter family. Consistent with this possibility, secondary-structure predictions suggested that the C-terminal 319 amino acids form an α-helix followed by 14 transmembrane β-strands, identical to predictions for the translocator domains of conventional autotransporters (Color Plate 33 [see color insert]). When the passenger domain from Hap (Hap$_S$) was fused to Hia$_{780-1098}$, Hap$_S$ was localized on the bacterial surface and was able to promote adherence to epithelial cells, demonstrating that the Hia C terminus has translocator activity and that Hia is an autotransporter (St. Geme and Cutter, 2000). Similar to a number of autotransporters (and selected TpsA proteins such as HMW1 and HMW2), Hia has an atypical signal sequence with an N-terminal extension (St. Geme and Cutter, 2000). However, distinct from the vast majority of autotransporters, Hia remains covalently linked to its translocator domain on the bacterial surface and fails to undergo a cleavage event (St. Geme and Cutter, 2000).

On comparing the Hia sequences from 10 different clinical isolates, we found significant similarity throughout the entire protein sequence. However, Hia expressed by strain 32 (Hia$_{32}$) was exceptional in that it lacked a sequence corresponding to residues 590 to 976 of Hia from the prototypic strain 11 (Hia$_{11}$). Despite the fact that this sequence omission covers the N-terminal half of the proposed Hia translocator domain, Hia$_{32}$ is still present on the bacterial surface and is able to mediate high-level adherence (Surana et al., 2004). With this information in mind, we speculated that the functional translocator unit from other Hia isolates may also be smaller than originally recognized. A series of in-frame deletions within Hia$_{11}$ and a set of chimeric proteins containing a heterologous passenger domain fused to varying regions of the Hia C terminus revealed that the final 76 residues were necessary and sufficient for translocation (Surana et al., 2004).

This region of 76 residues is predicted to form a maximum of 5 β-strands, significantly fewer than the 8 that are present in *E. coli* OmpA and OmpX and the 10 that are present in *E. coli* OmpT, the smallest known β-barrels (Pautsch and Schulz, 1998; Vogt and Schulz, 1999; Vandeputte-Rutten et al., 2001; Schulz, 2000). We speculated that the Hia β-domain may oligomerize to increase the number of β-strands and form a more typical β-barrel. Indeed, examination of Hia$_{1023-1098}$ present in outer membrane fractions revealed that this peptide migrates at a molecular mass consistent with a trimer (Surana et al., 2004). Denaturation of these complexes with formic acid resolved the trimers to their predicted monomeric molecular mass. Considering the constraints governing β-barrel formation (Schulz, 2000), it is likely that Hia$_{1023-1098}$ contains four, rather than the predicted five, β-strands, resulting in an even number of total β-strands in the trimeric structure. Each Hia$_{1023-1098}$ monomer probably contributes its β-strands to a common β-barrel, similar to observations with TolC (Koronakis et al., 2000). While Hia$_{1023-1098}$ has properties of a trimer, it remains possible that a higher-order multimer is formed in vivo, perhaps consisting of a multiple of three subunits.

To begin to address the generality of these findings with the Hia C terminus, we examined *Yersinia* YadA and *N. meningitidis* NhhA and found that the C-terminal 98 residues of YadA and the C-terminal 119 residues of NhhA have translocator activity, establishing that both of these proteins are autotransporters that make use of a short translocator domain, analogous to Hia (Surana et al., 2004). Searching the database for other members of this novel subfamily of autotransporter proteins revealed a total of 28 proteins at the time of this writing, including a few proteins that have been previously predicted to be autotransporters. All 28 of these proteins appear to be distinct from conventional autotransporters in that they uniformly lack a consensus sequence that defines the final 20 residues in almost all other known autotransporters (Loveless and Saier, 1997). In addition, this subfamily of autotransporter proteins lacks the intramolecular chaperone region described for *B. pertussis* BrkA (Oliver et al., 2003b). Furthermore, among the few proteins within this subfamily that have been examined, the passenger domain remains covalently attached to the β-domain in all cases (Hoiczyk et al., 2000; St. Geme and Cutter, 2000). Finally, the C terminus of Hia and the C terminus of YadA form trimers (Roggenkamp et al., 2003; Surana et al., 2004). Taken together, these observations support the notion that Hia-like autotransporters represent a distinct subfamily of autotransporter proteins, characterized by a short, trimeric translocator domain.

Studies of Hia adhesive activity revealed that Hia mediates high-level adherence to a variety of epithelial cell types (Barenkamp and St. Geme, 1996). As a first step toward defining Hia adhesive activity in more detail, Hia_{1-779} was fused to Hap_β. The resulting chimera was introduced into a nonadherent laboratory strain of *H. influenzae* and was found to promote bacterial attachment to epithelial cells (Laarmann et al., 2002). This finding localized Hia adhesive activity to the passenger domain, thus mimicking observations with other autotransporters (Henderson et al., 1998).

To further map the adhesive activity of the Hia passenger domain, glutathione *S*-transferase (GST) fusion proteins containing smaller regions of the Hia passenger domain were generated. Interestingly, two distinct regions of the Hia passenger domain exhibited adherence activity in multiple assays. A GST fusion protein containing residues 541 to 714 of Hia showed robust adherence, with a K_d of 0.05 to 0.1 nM. A GST fusion protein containing residues 50 to 374 of Hia showed slightly reduced adherence, with a K_d of 1 to 2 nM. Further experimentation has narrowed this adhesive domain to residues 50 to 175. $Hia_{541-714}$ is now called the primary binding domain (HiaBD1), and Hia_{50-175} is called the secondary binding domain (HiaBD2). Competition experiments demonstrated that the primary and secondary binding domains target the same receptor structure. To complement these in vitro assays, the primary and secondary binding domains of the native adhesin were disrupted by small deletions. As expected, disruption of either the primary or the secondary binding domain alone still allowed Hia-mediated bacterial attachment, while disruption of both domains together completely eliminated adhesive activity (Laarmann et al., 2002).

Hia is the first example of a bacterial adhesin containing multiple binding pockets that target the same receptor structure, a property that is reminiscent of the function of several eukaryotic proteins involved in cell-cell adhesion (Ranheim et al., 1996; Chappuis-Flament et al., 2001). This architecture allows Hia-mediated adherence to epithelial cells via a multivalent interaction, resulting in an increase in the avidity and stability of the bacterial-epithelial cell interaction. This increased stability may be critical to colonization of the human upper respiratory tract, considering the shear forces induced by coughing and sneezing and the expulsive actions of the mucociliary escalator.

STRUCTURAL BASIS FOR HOST CELL
RECOGNITION BY THE Hia ADHESIN

Following characterization of the general architecture of the Hia protein, attention turned to the structural basis for host recognition by Hia. As a first step, the Hia primary binding domain ($Hia_{541-714}$) was overexpressed and purified. Examination of the purified protein by size exclusion chromatography revealed stable oligomers. On a denaturing sodium dodecyl sulfate-polyacrylamide gel, unboiled HiaBD1 migrated as a trimer. Ulti-mately, the structure of HiaBD1 was solved by X-ray crystallography (Yeo et al., 2004).

As shown in Color Plate 34A (see color insert), HiaBD1 is a trimer with three identical subunits and threefold symmetry. The trimer has a mushroom shape with a broad stem at the N-terminal end and an elongated cap at the C-terminal end; it is 80Å long and 45 Å wide. Each subunit contains three well-defined structural domains, designated domain 1 (residues 548 to 585), domain 2 (residues 586 to 653), and domain 3 (residues 654 to 705) (Color Plate 34A, left). Domain 1 is a four-strand antiparallel β-sheet (β1 to β4), which forms a slightly concave sheet structure. One side of the sheet is composed of hydrophobic residues, while the other side consists primarily of polar residues. Domain 2 is a globular domain with mixed structure, containing several loops, four short β-strands (β5 to β8) and four helices (αA, αB, 3_{10}a, and 3_{10}b). This domain protrudes laterally and forms a knob. Domain 3 is an all-β domain and consists of five long β-strands (β9 to β13) and one short β-strand (β9′). These strands are twisted and segregate into three distinct subdomains, in-cluding a β-hairpin formed by the antiparallel β9/β9′ and β10 strands, a connector formed by the β11 strand, and a β-hairpin formed by the antiparallel β12 and β13 strands.

Interestingly, there are very few macromolecular contacts within a given subunit, and each subunit fold appears inherently unstable by itself. The subunit fold is stabilized by trimerization, with each domain in each subunit lending complementary secondary struc-tures to the other subunits (Color Plate 34A, middle). Overall, an unusually large amount of surface area ($\sim$15,240 Å^2) is buried on trimerization. Of the available solvent-accessible surface area, about 25% of domain 1, 24% of domain 2, and 49% of domain 3 are used for trimer formation. The domain 1 β-strands in one subunit interact with the corresponding β-strands in both neighboring subunits to form an almost perfect triangle (the stem of the mushroom). The domain 2 β5 strand in one subunit forms a small β-sheet together with the domain 1 β4 strand in a second subunit and the domain 2 β8 strand in a third subunit, while the αB helix in one subunit interacts with the αB helix in the other subunits to form a tight hydrophobic core. The domain 3 β-strands in one subunit interdigitate with the do-main 3 β-strands in the other two subunits, resulting in extensive subunit-subunit interac-tions and a very compact structure. Each of the domain 3 subdomains from one subunit contributes to a separate side of the three-sided tapering mushroom cap. In addition, the domain 3 region holds the trimer together through an extensive hydrophobic/aromatic core consisting primarily of phenylalanine and tryptophan residues.

The surface of the HiaBD1 trimer reveals several interesting structural features (Color Plate 34A, right). The concave β-sheet of domain 1 forms a shallow groove that runs par-allel to the threefold axis of the trimer. Domain 2 forms a knob that extends laterally from the structure, into the solvent in vitro and into the extracellular milieu in vivo. This knob consists of the αA helix, the 3_{10}a helix, and the loop structures between β6 and β8. As re-vealed by comparison of the Hia amino acid sequences from diverse strains, this knob is the most variable portion of the HiaBD1 structure, suggesting that the knob may be a ma-jor target of the host immune response, perhaps in part because of its overt accessibility to the immune system. Domains 2 and 3 form a major groove that runs diagonal to the three-fold axis of the trimer, just underneath the knob.

The regular arrangement of domain 1 β-strands along a triangle in the HiaBD1 trimer is reminiscent of the passenger domain of the *B. pertussis* pertactin autotransporter, which is a classical right-handed β-helical structure with a V-shaped or triangular cross-section

(Fig. 1) (Emsley et al., 1996). However, in pertactin the three sides of the triangle are formed by β-strands contributed by a single polypeptide, reflecting the fact that pertactin is a conventional autotransporter and has a monomeric passenger domain. In contrast, in Hia the three sides of the triangle are contributed by three separate subunits, each related to the others by a perfect threefold axis coincident with the crystallographic threefold axis. The arrangement of domain 1 in the HiaBD1 trimer also bears some resemblance to the YadA collagen-binding domain (YadA$_{26-241}$) (Nummelin et al., 2004). YadA is also trimeric, and the YadA collagen-binding domain forms a nine-coil left-handed parallel β-roll that is held together at the C-terminal end by a safety pin-like fold that results in extensive subunit-subunit interactions (Color Plate 34B) (Nummelin et al., 2004).

To identify the receptor-binding region of the HiaBD1 structure, site-directed mutagenesis was performed and over 40 site-specific mutations were generated. Examination of the resulting mutants established that residues N617, D618, A620, V656, E668, and E678 contribute to the receptor-binding pocket of HiaBD1 (Color Plate 34A, right). Mutations of D618, A620, and V656 abrogated adhesive activity in purified HiaBD1 and also in HiaΔBD2 (harboring a small deletion that eliminates HiaBD2 activity) expressed on the surface of bacteria. In contrast, mutations of N617, E668, and E678 abolished adhesive activity in purified HiaBD1 but had little effect on Hia expressed by bacteria. Residues D618, A620, and V656 form the rim of the binding pocket, while residues N617, E668, and E678 lie behind the rim or at the base of the pocket. One possibility is that N617, E668, and E678 are sites of secondary contact with the Hia receptor; as a consequence, mutations of these residues may result in only a modest decrease in HiaBD1 binding affinity, an effect that can be overcome by the cooperative effect of multiple molecules of Hia on the surface of an organism, at least under the conditions of an in vitro adherence assay.

The observation that a single HiaBD1 trimer contains three identical receptor-binding pockets is reminiscent of the tumor necrosis factor (TNF) surperfamily of proteins (Aggarwal, 2003). TNFα, TNFβ, and all other members of this family are trimeric proteins with three spatially distinct but equivalent receptor-binding sites that recognize members of the TNF receptor superfamily (Locksley et al., 2001). In general, each trimer interacts with three receptor monomers in a final 3:3 complex (Banner et al., 1993; Locksley et al., 2001). It is possible that a single HiaBD1 trimer in Hia interacts with three separate receptor molecules or with three related domains on one receptor, resulting in a multivalent interaction with the host cell surface and more stable adherence and potentially initiating host cell signaling events.

Based on alignment of the amino acid sequences of Hia proteins from nine different strains of *H. influenzae*, it appears that the HiaBD1 receptor-binding pocket is highly conserved, with absolute conservation of D618, A620, V656, E668, and E678. Comparison of the HiaBD1 and HiaBD2 primary and secondary binding domains reveals absolute conservation of A620, V656, E668, and E678. In contrast, HiaBD1 has an aspartic acid at position 618 while HiaBD2 has a glutamine at the corresponding position, perhaps accounting in part for the differences in HiaBD1 and HiaBD2 binding affinity.

In considering the structure of full-length Hia, it is notable that HiaBD1 has significant homology to other parts of the Hia passenger domain (Yeo et al., 2004). In particular, residues 585 to 705 (HiaBD1 domains 2 and 3) have 44% identity and 76% similarity to residues 50 to 166 (within HiaBD2), and residues 641 to 705 (encompassing the β8 strand and the αB helix of domain 2 and all of domain 3 in HiaBD1) have 27% identity and 63% similarity to residues 250 to 316 and 19% identity and 65% similarity to residues 359 to 422. These observations, combined with information from alignment of Hia sequences from multiple strains, suggest that the Hia passenger domain adopts a modular structure.

CONCLUSION

To conclude, *H. influenzae* is an important human pathogen that initiates infection by colonizing the upper respiratory tract. The process of colonization involves adherence to host epithelium and is influenced by bacterial adhesins that engage in lock-in-key interactions with complementary host cell receptors. The major *H. influenzae* adhesins include Hap, the HMW1 and HMW2 proteins, and the Hia and Hsf proteins, all of which belong to the type V secretion pathway. Hap is a conventional autotransporter and contains a ~300-amino-acid C-terminal translocator domain. Following surface localization of the passenger domain, Hap undergoes autoproteolysis, separating the passenger domain from the translocator domain. Hia and Hsf are examples of trimeric autotransporters and contain a very short translocator domain that trimerizes. These proteins remain uncleaved at the C terminus and fully cell associated. HMW1 and HMW2 are secreted by the two-partner secretion pathway and contain an N-terminal secretion domain that targets them to a cognate tetrameric outer membrane translocator, which is expressed as a separate polypeptide. As highlighted in this chapter, recent crystal structures have provided important insights into the mechanisms by which Hap, Hia/Hsf, HMW1/HMW2, and other type V secretion pathway proteins are presented on the bacterial surface and are able to interact with host cells.

Acknowledgments. Neeraj K. Surana and Shane E. Cotter contributed equally to the preparation of this chapter.

The work described in this chapter was supported in part by Public Health Service grants R01 AI44167, R01 DC02873, and R01 AI49322 to J.W.S.G.

REFERENCES

Aggarwal, B. B. 2003. Signalling pathways of the TNF superfamily: a double-edged sword. *Nat. Rev. Immunol.* **3:**745–756.

Banner, D. W., A. D'Arcy, W. Janes, R. Gentz, H. J. Schoenfeld, C. Broger, H. Loetscher, and W. Lesslauer. 1993. Crystal structure of the soluble human 55 kd TNF receptor-human TNF beta complex: implications for TNF receptor activation. *Cell* **73:**731–745.

Barenkamp, S. J., and E. Leininger. 1992. Cloning, expression, and DNA sequence analysis of genes encoding nontypeable *Haemophilus influenzae* high-molecular-weight proteins related to filamentous hemagglutinin of *Bordetella pertussis. Infect. Immun.* **60:**1302–1313.

Barenkamp, S. J., and J. W. St. Geme III. 1996. Identification of a second family of high-molecular-weight adhesion proteins expressed by non-typable *Haemophilus influenzae. Mol. Microbiol.* **19:**1215–1223.

Benz, I., and M. A. Schmidt. 1992a. AIDA-I, the adhesin involved in diffuse adherence of the diarrhoeagenic *Escherichia coli* strain 2787 (O126:H27), is synthesized via a precursor molecule. *Mol. Microbiol.* **6:**1539–1546.

Benz, I., and M. A. Schmidt. 1992b. Isolation and serologic characterization of AIDA-I, the adhesin mediating the diffuse adherence phenotype of the diarrhea-associated *Escherichia coli* strain 2787 (O126:H27). *Infect. Immun.* **60:**13–18.

Brandon, L. D., N. Goehring, A. Janakiraman, A. W. Yan, T. Wu, J. Beckwith, and M. B. Goldberg. 2003. IcsA, a polarly localized autotransporter with an atypical signal peptide, uses the Sec apparatus for secretion, although the Sec apparatus is circumferentially distributed. *Mol. Microbiol.* **50:**45–60.

Centers for Disease Control and Prevention. 2002. Progress toward elimination of *Haemophilus influenzae* type b disease among infants and children: United States, 1998–2000. *Morb. Mortal. Wkly. Rev.* **51:**234–237.

Chappuis-Flament, S., E. Wong, L. D. Hicks, C. M. Kay, and B. M. Gumbiner. 2001. Multiple cadherin extracellular repeats mediate homophilic binding and adhesion. *J. Cell Biol.* **154:**231–243.

Clantin, B., H. Hodak, E. Willery, C. Locht, F. Jacob-Dubuisson, and V. Villeret. 2004. The crystal structure of filamentous hemagglutinin secretion domain and its implications for the two-partner secretion pathway. *Proc. Natl. Acad. Sci. USA* **101:**6194–6199.

Clouthier, S. C., K. H. Muller, J. L. Doran, S. K. Collinson, and W. W. Kay. 1993. Characterization of three fimbrial genes, *sefABC,* of *Salmonella enteritidis. J. Bacteriol.* **175:**2523–2533.

Cutter, D., K. W. Mason, A. P. Howell, D. L. Fink, B. A. Green, and J. W. St. Geme III. 2002. Immunization with *Haemophilus influenzae* Hap adhesin protects against nasopharyngeal colonization in experimental mice. *J. Infect. Dis.* **186:**1115–1121.

Dawid, S., S. Grass, and J. W. St. Geme III. 2001. Mapping of the binding domains of nontypeable *Haemophilus influenzae* HMW1 and HMW2 adhesins. *Infect. Immun.* **69:**307–314.

de Cock, H., M. Struyve, M. Kleerebezem, T. van der Krift, and J. Tommassen. 1997. Role of the carboxy-terminal phenylalanine in the biogenesis of outer membrane protein PhoE of *Escherichia coli* K-12. *J. Mol. Biol.* **269:**473–478.

Domenighini, M., D. Relman, C. Capiau, S. Falkow, A. Prugnola, V. Scarlato, and R. Rappuoli. 1990. Genetic characterization of *Bordetella pertussis* filamentous haemagglutinin: a protein processed from an unusually large precursor. *Mol. Microbiol.* **4:**787–800.

Dutzler, R., G. Rummel, S. Alberti, S. Hernandez-Alles, P. Phale, J. Rosenbusch, V. Benedi, and T. Schirmer. 1999. Crystal structure and functional characterization of OmpK36, the osmoporin of *Klebsiella pneumoniae. Struct. Fold. Des.* **7:**425–434.

Egile, C., H. d'Hauteville, C. Parsot, and P. J. Sansonetti. 1997. SopA, the outer membrane protease responsible for polar localization of IcsA in *Shigella flexneri. Mol. Microbiol.* **23:**1063–1073.

Emsley, P., I. G. Charles, N. F. Fairweather, and N. W. Isaacs. 1996. Structure of *Bordetella pertussis* virulence factor P.69 pertactin. *Nature* **381:**90–92.

Fink, D. L., A. Z. Buscher, B. Green, P. Fernsten, and J. W. St. Geme III. 2003. The *Haemophilus influenzae* Hap autotransporter mediates microcolony formation and adherence to epithelial cells and extracellular matrix via binding regions in the C-terminal end of the passenger domain. *Cell. Microbiol.* **5:**175–186.

Fink, D. L., L. D. Cope, E. J. Hansen, and J. W. St. Geme III. 2001. The *Haemophilus influenzae* Hap autotransporter is a chymotrypsin clan serine protease and undergoes autoproteolysis via an intermolecular mechanism. *J. Biol. Chem.* **276:**39492–39500.

Fink, D. L., B. A. Green, and J. W. St. Geme III. 2002. The *Haemophilus influenzae* Hap autotransporter binds to fibronectin, laminin, and collagen IV. *Infect. Immun.* **70:**4902–4907.

Grass, S., A. Z. Buscher, W. E. Swords, M. A. Apicella, S. J. Barenkamp, N. Ozchlewski, and J. W. St. Geme III. 2003. The *Haemophilus influenzae* HMW1 adhesin is glycosylated in a process that requires HMW1C and phosphoglucomutase, an enzyme involved in lipooligosaccharide biosynthesis. *Mol. Microbiol.* **48:**737–751.

Grass, S., and J. W. St. Geme III. 2000. Maturation and secretion of the non-typable *Haemophilus influenzae* HMW1 adhesin: roles of the N-terminal and C-terminal domains. *Mol. Microbiol.* **36:**55–67.

Guédin, S., E. Willery, J. Tommassen, E. Fort, H. Drobecq, C. Locht, and F. Jacob-Dubuisson. 2000. Novel topological features of FhaC, the outer membrane transporter involved in secretion of the *Bordetella pertussis* filamentous hemagglutinin. *J. Biol. Chem.* **275:**30202–30210.

Guyer, D. M., I. R. Henderson, J. P. Nataro, and H. L. Mobley. 2000. Identification of Sat, an autotransporter toxin produced by uropathogenic *Escherichia coli. Mol. Microbiol.* **38:**53–66.

Henderson, I. R., J. Czeczulin, C. Eslava, F. Noriega, and J. P. Nataro. 1999. Characterization of Pic, a secreted protease of *Shigella flexneri* and enteroaggregative *Escherichia coli. Infect. Immun.* **67:**5587–5596.

Henderson, I. R., and J. P. Nataro. 2001. Virulence functions of autotransporter proteins. *Infect. Immun.* **69:**1231–1243.

Henderson, I. R., F. Navarro-Garcia, and J. P. Nataro. 1998. The great escape: structure and function of the autotransporter proteins. *Trends Microbiol.* **6:**370–378.

Hendrixson, D. R., M. L. de la Morena, C. Stathopoulos, and J. W. St. Geme III. 1997. Structural determinates of processing and secretion of the *Haemophilus influenzae* Hap protein. *Mol. Microbiol.* **26:**505–518.

Hendrixson, D. R., and J. W. St. Geme III. 1998. The *Haemophilus influenzae* Hap serine protease promotes adherence and microcolony formation, potentiated by a soluble host protein. *Mol. Cell* **2:**841–850.

Henning, U., and R. Koebnik. 1994. Outer membrane proteins of *Escherichia coli:* mechanism of sorting and regulation of synthesis, p. 381–395. *In* J.-M. Ghuysen and R. Hukenbeck (ed.), *Bacterial Cell Wall.* Elsevier Science, Amsterdam, The Netherlands.

Hoiczyk, E., A. Roggenkamp, M. Reichenbecher, A. Lupas, and J. Heesemann. 2000. Structure and sequence analysis of *Yersinia* YadA and *Moraxella* UspAs reveal a novel class of adhesins. *EMBO J.* **19:**5989–5999.

Hultgren, S. J., S. Abraham, M. Caparon, P. Falk, J. W. St. Geme III, and S. Normark. 1993. Pilus and non-pilus bacterial adhesins: assembly and function in cell recognition. *Cell* **73:**887–901.

Jacob-Dubuisson, F., R. Antoine, and C. Locht. 2000. Autotransporter proteins, evolution and redefining protein secretion: response. *Trends Microbiol.* **8:**533–534.

Jacob-Dubuisson, F., C. Buisine, E. Willery, G. Renauld-Mongenie, and C. Locht. 1997. Lack of functional complementation between *Bordetella pertussis* filamentous hemagglutinin and *Proteus mirabilis* HpmA hemolysin secretion machineries. *J. Bacteriol.* **179:**775–783.

Jacob-Dubuisson, F., C. Locht, and R. Antoine. 2001. Two-partner secretion in Gram-negative bacteria: a thrifty, specific pathway for large virulence proteins. *Mol. Microbiol.* **40:**306–313.

Jose, J., F. Jahnig, and T. F. Meyer. 1995. Common structural features of IgA1 protease-like outer membrane protein autotransporters. *Mol. Microbiol.* **18:**378–380.

Kajava, A. V., N. Cheng, R. Cleaver, M. Kessel, M. N. Simon, E. Willery, F. Jacob-Dubuisson, C. Locht, and A. C. Steven. 2001. Beta-helix model for the filamentous haemagglutinin adhesin of *Bordetella pertussis* and related bacterial secretory proteins. *Mol. Microbiol.* **42:**279–292.

Koebnik, R., K. P. Locher, and P. Van Gelder. 2000. Structure and function of bacterial outer membrane proteins: barrels in a nutshell. *Mol. Microbiol.* **37:**239–253.

Konninger, U. W., S. Hobbie, and V. Braun. 1999. The haemolysin-secreting ShlB protein of the outer membrane of *Serratia marcescens:* determination of surface-exposed residues and formation of ion-permeable pores by ShlB mutants in artificial lipid bilayer membranes. *Mol. Microbiol.* **32:**1212–1225.

Koronakis, V., A. Sharff, E. Koronakis, and B. Luisi. 2000. Crystal structure of the bacterial membrane protein TolC central to multidrug efflux and protein export. *Nature* **405:**914–919.

Krasan, G. P., and J. W. St. Geme III. 1997. Invasive disease due to nontypeable *Haemophilus infuenzae. Rep. Pediatr. Infect. Dis.* **7:**11–12.

Laarmann, S., D. Cutter, T. Juehne, S. J. Barenkamp, and J. W. St. Geme III. 2002. The *Haemophilus influenzae* Hia autotransporter harbours two adhesive pockets that reside in the passenger domain and recognize the same host cell receptor. *Mol. Microbiol.* **46:**731–743.

Lambert-Buisine, C., E. Willery, C. Locht, and F. Jacob-Dubuisson. 1998. N-terminal characterization of the *Bordetella pertussis* filamentous haemagglutinin. *Mol. Microbiol.* **28:**1283–1293.

Lee, H. W., and S. M. Byun. 2003. The pore size of the autotransporter domain is critical for the active translocation of the passenger domain. *Biochem. Biophys. Res. Commun.* **307:**820–825.

Lin, L., P. Ayala, J. Larson, M. Mulks, M. Fukuda, S. R. Carlsson, C. Enns, and M. So. 1997. The *Neisseria* type 2 IgA1 protease cleaves LAMP1 and promotes survival of bacteria within epithelial cells. *Mol. Microbiol.* **24:**1083–1094.

Locksley, R. M., N. Killeen, and M. J. Lenardo. 2001. The TNF and TNF receptor superfamilies: integrating mammalian biology. *Cell* **104:**487–501.

Loveless, B. J., and M. H. Saier, Jr. 1997. A novel family of channel-forming, autotransporting, bacterial virulence factors. *Mol. Membr. Biol.* **14:**113–123.

Makhov, A. M., J. H. Hannah, M. J. Brennan, B. L. Trus, E. Kocsis, J. F. Conway, P. T. Wingfield, M. N. Simon, and A. C. Steven. 1994. Filamentous hemagglutinin of *Bordetella pertussis.* A bacterial adhesin formed as a 50-nm monomeric rigid rod based on a 19-residue repeat motif rich in beta strands and turns. *J. Mol. Biol.* **241:**110–124.

Maurer, J., J. Jose, and T. F. Meyer. 1999. Characterization of the essential transport function of the AIDA-I autotransporter and evidence supporting structural predictions. *J. Bacteriol.* **181:**7014–7020.

Mulks, M. H., S. J. Kornfeld, and A. G. Plaut. 1980. Specific proteolysis of human IgA by *Streptococcus pneumoniae* and *Haemophilus influenzae. J. Infect. Dis.* **141:**450–456.

Murphy, T. F., J. M. Bernstein, D. M. Dryja, A. A. Campagnari, and M. A. Apicella. 1987. Outer membrane protein and lipooligosaccharide analysis of paired nasopharyngeal and middle ear isolates in otitis media due to nontypeable *Haemophilus influenzae:* pathogenic and epidemiologic observations. *J. Infect. Dis.* **5:**723–731.

Nakamura, K., and S. Mizushima. 1976. Effects of heating in dodecyl sulfate solution on the conformation and electrophoretic mobility of isolated major outer membrane proteins from *Escherichia coli* K-12. *J. Biochem.* (Tokyo) **80**:1411–1422.

Noel, G. J., S. J. Barenkamp, J. W. St. Geme III, W. N. Haining, and D. M. Mosser. 1994. High-molecular-weight surface-exposed proteins of *Haemophilus influenzae* mediate binding to macrophages. *J. Infect. Dis.* **169**:425–429.

Nummelin, H., M. C. Merckel, J. C. Leo, H. Lankinen, M. Skurnik, and A. Goldman. 2004. The *Yersinia* adhesin YadA collagen-binding domain structure is a novel left-handed parallel β-roll. *EMBO J.* **23**:701–711.

Ohnishi, Y., M. Nishiyama, S. Horinouchi, and T. Beppu. 1994. Involvement of the COOH-terminal pro-sequence of *Serratia marcescens* serine protease in the folding of the mature enzyme. *J. Biol. Chem.* **269**:32800–32806.

Oliver, D. C., G. Huang, and R. C. Fernandez. 2003a. Identification of secretion determinants of the *Bordetella pertussis* BrkA autotransporter. *J. Bacteriol.* **185**:489–495.

Oliver, D. C., G. Huang, E. Nodel, S. Pleasance, and R. C. Fernandez. 2003b. A conserved region within the *Bordetella pertussis* autotransporter BrkA is necessary for folding of its passenger domain. *Mol. Microbiol.* **47**:1367–1383.

Oomen, C. J., P. Van Ulsen, P. Van Gelder, M. Feijen, J. Tommassen, and P. Gros. 2004. Structure of the translocator domain of a bacterial autotransporter. *EMBO J.* **23**:1257–1266.

Otto, B. R., S. J. van Dooren, J. H. Nuijens, J. Luirink, and B. Oudega. 1998. Characterization of a hemoglobin protease secreted by the pathogenic *Escherichia coli* strain EB1. *J. Exp. Med.* **188**:1091–1103.

Pautsch, A., and G. E. Schulz. 1998. Structure of the outer membrane protein A transmembrane protein. *Nat. Struct. Biol.* **5**:1013–1017.

Peltola, H. 2000. Worldwide *Haemophilus influenzae* type b disease at the beginning of the 21st century: global analysis of the disease burden 25 years after the use of the polysaccharide vaccine and a decade after the advent of conjugates. *Clin. Microbiol. Rev.* **13**:302–317.

Pohlner, J., R. Halter, K. Beyreut, and T. F. Meyer. 1987. Gene structure and extracellular secretion of *Neisseria gonorrhoeae* Iga protease. *Nature* **325**:458–462.

Poole, K., E. Schiebel, and V. Braun. 1988. Molecular characterization of the hemolysin determinant of *Serratia marcescens*. *J. Bacteriol.* **170**:3177–3188.

Qi, H. Y., and H. D. Bernstein. 1999. SecA is required for the insertion of inner membrane proteins targeted by the *Escherichia coli* signal recognition particle. *J. Biol. Chem.* **274**:8993–8997.

Ranheim, T. S., G. M. Edelman, and B. A. Cunningham. 1996. Homophilic adhesion mediated by the neural cell adhesion molecule involves multiple immunoglobulin domains. *Proc. Natl. Acad. Sci. USA* **93**:4071–4075.

Rao, V. K., G. P. Karsan, D. R. Hendrixson, S. Dawid, and J. W. St. Geme III. 1999. Molecular determinants of the pathogenesis of disease due to non-typable *Haemophilus influenzae*. *FEMS Microbiol. Rev.* **23**:99–129.

Renauld-Mongenie, G., J. Cornette, N. Mielcarek, F. D. Menozzi, and C. Locht. 1996. Distinct roles of the N-terminal and C-terminal precursor domains in the biogenesis of the *Bordetella pertussis* filamentous hemagglutinin. *J. Bacteriol.* **178**:1053–1060.

Rodriguez, C. A., V. Avadhanula, A. Buscher, A. L. Smith, J. W. St. Geme III, and E. E. Adderson. 2003. Prevalence and distribution of adhesins in invasive non-type b encapsulated *Haemophilus influenzae*. *Infect. Immun.* **71**:1635–1642.

Roggenkamp, A., N. Ackermann, C. A. Jacobi, K. Truelzsch, H. Hoffmann, and J. Heesemann. 2003. Molecular analysis of transport and oligomerization of the *Yersinia enterocolitica* adhesin YadA. *J. Bacteriol.* **185**:3735–3744.

Schirmer, T., and S. W. Cowan. 1993. Prediction of membrane-spanning β-strands and its application to maltoporin. *Protein Sci.* **2**:1361–1363.

Schonherr, R., R. Tsolis, T. Focareta, and V. Braun. 1993. Amino acid replacements in the *Serratia marcescens* haemolysin ShlA define sites involved in activation and secretion. *Mol. Microbiol.* **9**:1229–1237.

Schulz, G. E. 2000. Beta-barrel membrane proteins. *Curr. Opin. Struct. Biol.* **10**:443–447.

Schulz, G. E. 2002. The structure of bacterial outer membrane proteins. *Biochim. Biophys. Acta* **1565**:308–317.

Shannon, J. L., and R. C. Fernandez. 1999. The C-terminal domain of the *Bordetella pertussis* autotransporter BrkA forms a pore in lipid bilayer membranes. *J. Bacteriol.* **181**:5838–5842.

Shere, K. D., S. Sallustio, A. Manessis, T. G. D'Aversa, and M. B. Goldberg. 1997. Disruption of IcsP, the major *Shigella* protease that cleaves IcsA, accelerates actin-based motility. *Mol. Microbiol.* **25**:451–462.

Sijbrandi, R., M. L. Urbanus, C. M. ten Hagen-Jongman, H. D. Bernstein, B. Oudega, B. R. Otto, and J. Luirink. 2003. Signal recognition particle (SRP)-mediated targeting and Sec-dependent translocation of an extracellular *Escherichia coli* protein. *J. Biol. Chem.* **278:**4654–4659.

Stathopoulos, C., D. R. Hendrixson, D. G. Thanassi, S. J. Hultgren, J. W. St. Geme III, and R. Curtiss III. 2000. Secretion of virulence determinants by the general secretory pathway in gram-negative pathogens: an evolving story. *Microbes Infect.* **2:**1061–1072.

St. Geme, J. W., III, M. L. de la Morena, and S. Falkow. 1994. A *Haemophilus influenzae* IgA protease-like protein promotes intimate interaction with human epithelial cells. *Mol. Microbiol.* **14:**217–233.

St. Geme, J. W., III, and D. Cutter. 2000. The *Haemophilus influenzae* Hia adhesin is an autotransporter protein that remains uncleaved at the C terminus and fully cell associated. *J. Bacteriol.* **182:**6005–6013.

St. Geme, J. W., III, D. Cutter, and S. J. Barenkamp. 1996. Characterization of the genetic locus encoding *Haemophilus influenzae* type b surface fibrils. *J. Bacteriol.* **178:**6281–6287.

St. Geme, J. W., III, S. Falkow, and S. J. Barenkamp. 1993. High-molecular-weight proteins of nontypable *Haemophilus influenzae* mediate attachment to human epithelial cells. *Proc. Natl. Acad. Sci. USA* **90:**2875–2879.

St. Geme, J. W., III, and S. Grass. 1998. Secretion of the *Haemophilus influenzae* HMW1 and HMW2 adhesins involves a periplasmic intermediate and requires the HMWB and HMWC proteins. *Mol. Microbiol.* **27:**617–630.

St. Geme, J. W., III, V. V. Kumar, D. Cutter, and S. J. Barenkamp. 1998. Prevalence and distribution of the *hmw* and *hia* genes and the HMW and Hia adhesins among genetically diverse strains of nontypeable *Haemophilus influenzae*. *Infect. Immun.* **66:**364–368.

St. Geme, J. W., III. 1994. The HMW1 adhesin of nontypeable *Haemophilus influenzae* recognizes sialylated glycoprotein receptors on cultured human epithelial cells. *Infect. Immun.* **62:**3881–3889.

Struyve, M., M. Moons, and J. Tommassen. 1991. Carboxy-terminal phenylalanine is essential for the correct assembly of a bacterial outer membrane protein. *J. Mol. Biol.* **218:**141–148.

Surana, N. K., D. Cutter, S. J. Barenkamp, and J. W. St. Geme III. 2004. The *Haemophilus influenzae* Hia autotransporter contains an unusually short trimeric translocator domain. *J. Biol. Chem.* **279:**14679–14685.

Surana, N. K., S. Grass, G. G. Hardy, H. Li, D. G. Thanassi, and J. W. St. Geme III. 2004. Evidence for conservation of architecture and physical properties of Omp85-like proteins throughout evolution. *Proc. Natl. Acad. Sci. USA* **101:**14497–14502.

Suzuki, T., M. C. Lett, and C. Sasakawa. 1995. Extracellular transport of VirG protein in *Shigella. J. Biol. Chem.* **270:**30874–30880.

Turk, D. C. 1984. The pathogenicity of *Haemophilus influenzae. J. Med. Microbiol.* **18:**1–16.

Uphoff, T. S., and R. A. Welch. 1990. Nucleotide sequencing of the *Proteus mirabilis* calcium-independent hemolysin genes (*hpmA* and *hpmB*) reveals sequence similarity with the *Serratia marcescens* hemolysin genes (*shlA* and *shlB*). *J. Bacteriol.* **172:**1206–1216.

Uphoff, T. S., and R. A. Welch. 1994. Structural and functional analysis of HpmA hemolysin of *Proteus mirabilis,* p. 283–292. *In* C. I. Cado and J. H. Crosa (ed.), *Molecular Mechanisms of Bacterial Virulence.* Kluwer Academic Publishers, Dordrecht, The Netherlands.

Valent, Q. A., P. A. Scotti, S. High, J. W. de Gier, G. von Heijne, G. Lentzen, W. Wintermeyer, B. Oudega, and J. Luirink. 1998. The *Escherichia coli* SRP and SecB targeting pathways converge at the translocon. *EMBO J.* **17:**2504–2512.

Vandeputte-Rutten, L., R. A. Kramer, J. Kroon, N. Dekker, M. R. Egmond, and P. Gros. 2001. Crystal structure of the outer membrane protease OmpT from *Escherichia coli* suggests a novel catalytic site. *EMBO J.* **20:**5033–5039.

Veiga, E., E. Sugawara, H. Nikaido, V. de Lorenzo, and L. A. Fernandez. 2002. Export of autotransported proteins proceeds through an oligomeric ring shaped by C-terminal domains. *EMBO J.* **21:**2122–2131.

Villaseca, J. M., F. Navarro-Garcia, G. Mendoza-Hernandez, J. P. Nataro, A. Cravioto, and C. Eslava. 2000. Pet toxin from enteroaggregative *Escherichia coli* produces cellular damage associated with fodrin disruption. *Infect. Immun.* **68:**5920–5927.

Vogt, J., and G. E. Schulz. 1999. The structure of the outer membrane protein OmpX from *Escherichia coli* reveals possible mechanisms of virulence. *Struct. Fold Des.* **7:**1301–1309.

Watson, W. J., J. R. Gilsdorf, M. A. Tucci, K. W. McCrea, L. J. Forney, and C. F. Marrs. 1994. Identification of a gene essential for piliation in *Haemophilus influenzae* type b with homology to the pilus assembly platform genes of gram-negative bacteria. *Infect. Immun.* **62:**468–475.

Weiss, M. S., T. Wacker, J. Weckesser, W. Welte, and G. E. Schulz. 1990. The three-dimensional structure of porin from *Rhodobacter capsulatus* at 3 Å resolution. *FEBS Lett.* **267:**268–272.

Wenger, J. D., and J. J. Ward. 2004. *Haemophilus influenzae* vaccine, p. 229–268. *In* S. A. Plotkin and W. A. Orenstein (ed.), *Vaccines,* 4th ed. The W. B. Saunders Co., Philadelphia, Pa.

Willems, R. J., C. Geuijen, H. G. van der Heide, G. Renauld, P. Bertin, W. M. van den Akker, C. Locht, and F. R. Mooi. 1994. Mutational analysis of the *Bordetella pertussis fim/fha* gene cluster: identification of a gene with sequence similarities to haemolysin accessory genes involved in export of FHA. *Mol. Microbiol.* **11:** 337–347.

Willshaw, G. A., H. R. Smith, M. M. McConnell, and B. Rowe. 1988. Cloning of genes encoding coli-surface (CS) antigens in enterotoxigenic *Escherichia coli. FEMS Microbiol. Lett.* **49:**473–478.

Yeo, H. J., S. E. Cotter, S. Laarmann, T. Juehne, J. W. St. Geme III, and G. Waksman. 2004. Structural basis for host recognition by the *Haemophilus influenzae* Hia autotransporter. *EMBO J.* **23:**1245–1256.

Structural Biology of Bacterial Pathogenesis
Edited by G. Waksman et al.
© 2005 ASM Press, Washington, D.C.

Chapter 9

Type III Secretion Machinery and Effectors

C. Erec Stebbins

Historically, bacterial infection has represented one of the greatest human health hazards. Plague alone exterminated over one-third of the European population in the Middle Ages and has been estimated to have taken the lives of over 200 million people worldwide (Perry and Fetherston, 1997). Each year, there are over one billion new human cases of infections due to *Salmonella* spp. (and nearly four million deaths), and millions of dollars are spent to control *Salmonella* infection in the dairy and poultry industries (Pang et al., 1995). Significant investment is also made to control plant diseases due to bacterial pathogens such as *Pseudomonas aeruginosa* PA14, *Erwinia* spp., *Pseudomonas syringae, Ralstonia solanacearum,* and *Xanthomonas* spp., and opportunistic pathogens such as *P. aeruginosa* prey on immunocompromised persons.

The ravages of bacterial diseases have been kept in check through the use of potent antibiotics since the 1930s. However, the development of widespread antibiotic resistance has raised the specter of a "postantibiotic" world, where even simple infections may become life-threatening (Cohen, 1992, 1994, 2000). The double-edged sword of modern molecular biology also provides the opportunity for scientists to participate in "black biology," that is, the engineering of pathogens of enhanced virulence and drug resistance for use in war or terrorism (Fraser and Dando, 2001; Nixdorff et al., 2000).

All the gram-negative bacterial pathogens mentioned above are linked in that they utilize a highly specialized virulence related secretion system, termed "type III" or "contact dependent," to achieve a remarkable translocation of bacterial proteins directly into cells of the host organism. These virulence proteins, often called effectors, hijack eukaryotic biochemical processes in sophisticated ways for the benefit of the pathogen. While the secretion machinery itself appears to be highly conserved among different bacteria (although more divergent between animal and plant pathogens), the effectors themselves are a highly variable component of the virulence repertoire in these miocroorganisms (Buttner and Bonas, 2002a; Cornelis and Van Gijsegem, 2000; Galan and Collmer, 1999; Hueck, 1998; Zaharik et al., 2002).

C. Erec Stebbins • Laboratory of Structural Microbiology, The Rockefeller University, New York, NY 10021.

The translocated effectors harbor many different activities and can be used in different combinations by various bacteria to exert highly specific and unique effects on the host cell. Additionally, even homologous effectors with identical enzymatic activity can vary enough in substrate specificity and delivery to make their effect on the host cell tailored for a given pathogen. In understanding the biology of these systems, moderate- and high-resolution structural information has often played a key role and, furthermore, revealed aspects and themes in the pathogenesis of these systems that were much less clear from data generated from more indirect experimental techniques. This chapter represents only a skimming of the surface in examining these systems from a structural point of view, but it is nonetheless a fascinating and informative tour.

THE TYPE III SECRETION SYSTEM: VIRULENCE THROUGH A MOLECULAR SYRINGE

Composed of more than 20 proteins and related to the flagellar assembly apparatus, type III secretion systems are one of the most complex protein secretion systems to be discovered. Although the basic components of these systems have been identified, the actual mechanisms of protein secretion are poorly understood. At the heart of the type III secretion system is a supermolecular organelle-sized entity that has been called variably the injectosome, the molecular syringe, and the needle complex. First identified as associated with the type III secretion system of *Salmonella enterica* serovar Typhimurium (Kubori et al., 1998), this structure does indeed resemble a syringe with needle, which, given the proposed translocation activity of this system, provides a nearly literal interpretation for a structure-function relationship (Fig. 1). The genes encoding the proteins of this apparatus are contained in the so-called pathogenicity islands on the chromosome or on virulence plasmids of these bacteria and were shown to be essential virulence determinants in most gram-negative pathogens with type III secretion systems.

Needle Complex

Moderate-resolution electron microscopy (EM) studies of osmotically shocked *Salmonella, Shigella,* and *Escherichia coli* organisms revealed startling structures embedded in the inner and outer membranes of these pathogenic bacteria (Fig. 1) (Blocker et al., 2001; Cordes et al., 2003; Kimbrough and Miller, 2000; Kubori et al., 1998; Kubori et al., 2000; Sekiya et al., 2001; Tamano et al., 2000; Journet et al., 2003). These structures bear a substantial resemblance to the flagellar basal body, but they appear even in flagellar knockout bacterial mutants (Kubori et al., 1998). Further distinguishing them from the flagellar system is the extracellular portion of the structure. Instead of a hook and flagellum, the type III needle complexes have a single long filamentous structure (the needle). While the needle complexes from other gram-negative bacteria have not been isolated and imaged as well as those from *Salmonella* and *Shigella,* the proteins that comprise the complex are highly conserved and are typically essential virulence determinants.

Structurally, the needle complex can be subdivided into three main regions: the extracelluar needle, the outer membrane rings, and the base (which spans the inner membrane and the periplasmic space). Overall, the complex is characterized by radial symmetry that

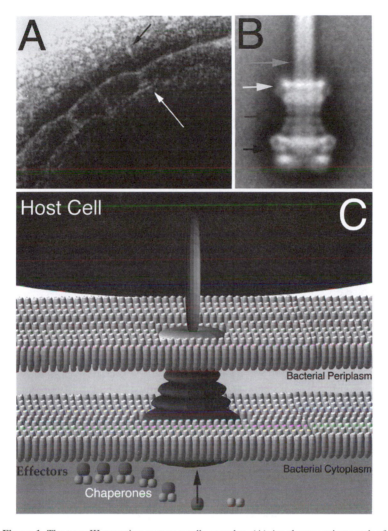

Figure 1. The type III secretion system needle complex. (A) An electron micrograph of osmotically shocked bacterial cells from *S. enterica* serovar Typhimurium reveals a structure with an inner membrane-associated base (white arrow), outer membrane ring, and filamentous extension (the "needle"; black arrow). Reprinted from Kubori et al. (1998) with permission. (B) Higher-resolution EM reconstructions of isolated needle complexes from *Shigella* reveal the details of the substructures of the secretion apparatus. The needle filament (light gray arrow), the outer membrane secretin rings (white arrow), the periplasmic rings (dark gray arrow), and the inner membrane rings (black arrow) are shown. Reprinted from Blocker et al. (2001) with permission. (C) Schematic of the needle complex, illustrating the docking of the filament with the host cell pore, the inner and outer membrane elements such as the secretin rings, periplasmic rings, and inner membrane rings. The bacterial cytoplasm contains homodimeric secretion chaperones binding to effector molecules prior to translocation through the needle complex.

extends from the needle through the outer rings and base. The needle itself is a long and very straight filamentous structure, with lengths varying from about 50 to 80 nm in length and from 7 to 12 nm in width (Blocker et al., 2001; Cordes et al., 2003; Kimbrough and Miller, 2000; Kubori et al., 1998; Kubori et al., 2000; Tamano et al., 2000); it consists primarily of a single, small protein that oligomerizes (Kubori et al., 2000; Kimbrough and Miller, 2000; Hoiczyk and Blobel, 2001). The assembly of the needle filament takes place after the formation of the basal body and is dependent on a functional type III secretion system (Kimbrough and Miller, 2002; Sukhan et al., 2001). In other words, the needle is probably secreted through the nearly completed system, after which it assembles into a filamentous form by direct analogy to flagellin in the related flagellar system. Mutations in other, presumably regulatory, genes can cause the loss or reduced assembly of these structures or, in some cases, even dramatic elongation of the needle filament, so much so that it can induce aggregation of the bacteria (Kubori et al., 2000). The needle filament contains a channel through its center with a diameter of approximately 2 to 3 nm (20 to 30 Å). High-resolution structural information of the flagellum shows that this related system has a channel with diameter of 20 Å, which is lined with acidic residues (Yonekura et al., 2003). Should the interior of the needle filament of the pathogenic type III secretion systems resemble that of the flagellar system, then, unless the channel can increase its diameter in a fashion not yet observed, it probably represents a space too confined for most globular proteins to traverse in their folded state. EM of the needle filament from *Shigella* (Cordes et al., 2003) has shown that it forms a helical structure, with approximately 5.6 subunits per turn and a helical pitch of 24 Å, very similar to the properties of the flagellum (despite the very different subunit makeup of these filaments).

The outer membrane rings are formed from an oligomer of a single protein that is a member of the large secretin family of outer membrane channels (Crago and Koronakis, 1998; Cornelis, 2002; Burghout et al., 2004). The rings are secreted into the periplasmic space via the Sec-dependent pathway and assemble and insert into the outer membrane independently of the rest of the needle complex. It is thought that they then dock with the lower portions of the base once they have correctly formed. The outer rings together span about 15 to 20 nm in a direction parallel to the needle and are approximately 15 nm wide. When expressed in nonpathogenic *E. coli* strains, they can assemble into outer membrane pores with a diameter of about 5 to 10 nm (Crago and Koronakis, 1998). Recent EM work with the viral secretin homolog pIV revealed at 22-Å resolution a structure with a barrel-like shape, 24 nm in length and 13.5 nm in diameter, composed of three separate, cylindrical domains (Opalka et al., 2003). The pore was measured at 6 to 9 nm.

The base of the structure is composed primarily of two proteins, which together form an assembly that resembles the base of the flagellum. Two rings, about 10 to 20 nm high by about 20 to 25 nm wide, are observed in the inner membrane and contain a large channel through their centers. Like the outer ring structures, the base can be expressed heterologously and assembled by coexpressing these two essential and highly conserved genes from the pathogenicity islands (Kimbrough and Miller, 2000, 2002; Schuch and Maurelli, 2001).

Other structural and functional components are likely to be part of the needle complex, perhaps aiding the insertion and stability of the rings within membranes as well as the interaction with cytoplasmic elements in the bacterium. Examples might include capping proteins at the tip of the filament or other factors that may allow the needle to dock with the translocon pore (discussed below) in the host cell membrane. Another example within the bacterium would be the ATPase known to be required for the function of the system (Galan and Collmer, 1999; Hueck, 1998).

Type III Secretion and Other Filaments

In addition to the needle complex, other filamentous structures have been associated with the type III secretion systems of pathogenic bacteria (Ginocchio et al., 1994; Knutton et al., 1998; Wei et al., 2000). This is especially true for plant pathogens, where the large Hrp pilus not only has been shown genetically to be required for virulence, but also has provided to date the only direct evidence of effector translocation through any of these type III systems (He and Jin, 2003). In a set of elegant experiments, immunolabeling of putative translocated proteins revealed that the pilus served directly as a translocation channel (Brown et al., 2001; Jin and He, 2001; Jin et al., 2001). The Hrp pilus is considerably longer than the needle filament (on the order of micrometers), although its internal diameter appears to be approximately the same (Buttner and Bonas, 2002a; Hu et al., 2001; Roine et al., 1997; Van Gijsegem et al., 2002). The greater length in the Hrp pilus has been hypothesized to be needed to aid the bacterium in traversing the cell wall of the plant (Buttner and Bonas, 2002a; He and Jin, 2003). The Hrp pilus appears to be devoted mostly to a translocation function, since bacterial mutants lacking the pilus are not impaired in attachment (Van Gijsegem et al., 2000).

In addition, filamentous appendages have been linked with type III secretion in animal pathogens (Ginocchio et al., 1994; Knutton et al., 1998). In particular, enteropathogenic *E. coli* strains secrete a protein, called EspA, which assembles into a filamentous and elongated structure that interacts with the needle protein and appears to extend the channel formed by the type III secretion system (Daniell et al., 2001; Knutton et al., 1998; Neves et al., 1998; Sekiya et al., 2001). EM has revealed that this filament interacts with the host cell, forms a helical tube with an inner diameter of 25 Å, and shows remarkable helical structural similarity to the flagellar filament, although it is about half the external diameter of the flagellar filament (Daniell et al., 2003). Since EspA homologues have not been identified in most of the other bacterial pathogens, the generality of the results in enteropathogenic *E. coli* in not clear.

The Translocon

Several protein substrates of the type III secretion system associate with the eukaryotic host membrane (Blocker et al., 1999; Buttner et al., 2002; Hayward et al., 2000; Osiecki et al., 2001; Wachter et al., 1999). Several of these have in vitro pore-forming activity and are predicted to contain transmembrane regions (Hakansson et al., 1996b; Neyt and Cornelis, 1999; Cornelis, 2002; Tardy et al., 1999). The deletion of these proteins prevents entry into the host cell of the remaining effectors, although secretion into the surrounding media still occurs through the needle complex (Buttner and Bonas, 2002b). Together, these data have led to models in which these proteins function to form a pore in the host cell that docks with the needle complex, essentially opening a door into the host cell for the effectors to travel through at the end of their journey through the needle complex (Buttner and Bonas, 2002b). Their specific effect on translocation as opposed to secretion has led to their classification as translocases. Several of these putative pore-forming proteins also have effector function within the host cell (Hayward and Koronakis, 2002; Zhou and Galan, 2001). An example is SipC of *Salmonella,* which has, in addition to its transmembrane predicted regions, domains at its NH_2 and COOH termini that function within the host cell to nucleate actin filaments and bundle the fibers formed, leading to cytoskeletal reorganizations that are pivotal to the internalization of this pathogen (Hayward and Koronakis, 1999; McGhie et al., 2001). No structural data are currently available for any of these elements of the type III secretion system, however.

Machinery in the Bacterial Cytoplasm

An analysis of the type III secretion system is not complete without a consideration of the elements that function within the bacterial cytoplasm to achieve the secretion and translocation of effector molecules. While many factors have been implicated on the bacterial side in the function of the system, such as an F_oF_1 family ATPase (Woestyn et al., 1994), very little structural information is available. The one exception to this is a set of proteins known as chaperones that are important for the efficient secretion of the effector substrates.

An important element in the effective translocation of most virulence factors through type III secretion system is a set of specialized secretion chaperones (Page and Parsot, 2002; Wattiau et al., 1996). These are small, often highly acidic proteins, which form tight and highly specific complexes by binding to the NH_2-terminal domain of their cognate effector molecule (Woestyn et al., 1996; Abe et al., 1999; Bronstein et al., 2000; Darwin et al., 2001; Elliott et al., 1999; Fu and Galan, 1998). Unlike standard housekeeping chaperones such as the GroEL and DnaK/Hsp70 family of proteins, which aid in the folding of a variety of proteins (Hartl and Hayer-Hartl, 2002), type III secretion-associated chaperones lack nucleotide-binding or hydrolysis activities and are rather specific for their cognate substrates. Strains lacking a given chaperone are usually impaired in the secretion of the corresponding virulence factor, which prematurely degrades or accumulates within the bacterium (Page and Parsot, 2002; Wattiau et al., 1996).

Important insights into the role of these chaperones were gained by studying the cocrystal structures of the chaperone-binding domains of the *S. enterica* serovar Typhimurium effector SptP and the *Yersinia* effector YopE, with their cognate chaperones SicP and SycE, respectively (Birtalan et al., 2002; Stebbins and Galan, 2001a). These structures showed that these type III secretion system-associated chaperones bind their cognate effectors in a manner that, remarkably, keeps the entire chaperone-binding domain in a nonglobular state (Color Plate 35 [see color insert]). In each of these complexes, an extended region of the NH_2-terminal domain of each molecule of SptP and YopE is wrapped around chaperone homodimer pairs, so that roughly 80 to 100 amino acids of the effector are bound. These regions of the effector have been shown through deletion analysis to be critical to the binding and functioning of the chaperones (Page and Parsot, 2002; Wattiau et al., 1996).

The effector polypeptides run along hydrophobic patches, making extensive hydrophobic contacts, which bind the complexes together. Although the chaperone-binding domains of SptP and YopE in these structure have no hydrophobic core and are completely nonglobular, they contain significant amounts of secondary structure (primarily helical), also forming two intermolecular β-sheets with their chaperones. The polypeptide main chains of SptP and YopE in their chaperone complexes can be roughly superimposed (Color Plate 35), showing that these proteins take the same general three-dimensional route across their respective cognate chaperones.

The details of the molecular interactions between the chaperones and their substrates are very interesting. To give an example from the *Salmonella* structure, SptP and SicP interact extensively in four discrete pockets characterized by the insertion of one or more predominantly hydrophobic residues of the translocation domain of SptP into binding crevices of SicP. Two of these four key interaction sites involve a large hydrophobic groove in the chaperone, which is seen to bind to two different helices of SptP, one from the NH_2 terminus, which interacts with one of the SicP molecules in the chaperone homodimer, and the other from the COOH terminus, which binds the groove in the other monomer of SicP in the chaperone pair. Thus, SicP uses the same structural element to bind two distinct portions of SptP. These helices from SptP run through the groove such that hydrophobic amino acids from the two proteins form an extensive nonpolar interface centered on the deep insertion of a bulky aromatic side chain into the helix-binding groove, where it stacks on Phe36 of SicP (Color Plate 35). The groove wraps around the SptP helix so extensively that nearly every side chain in this helix is able to contact residues in SicP. The ends of the helix are locked down to SicP by side chains making hydrogen bonds.

The structures of two other type III chaperones, CesT from enterohemorrhagic *E. coli* and SigE from *Salmonella,* were solved at nearly the same time, but without their cognate effectors (Birtalan and Ghosh, 2001; Luo et al., 2001; Evdokimov et al., 2002b). Together, these structures showed that these proteins are structurally homologous (despite extremely low primary amino acid sequence similarity), indicating that each of these chaperones is a member of a common protein family. The structural similarity extends beyond the significant conservation of the three-dimensional fold to the molecular surfaces as well. For example, extended hydrophobic patches, and especially a large hydrophobic groove (Color Plate 35), which in SicP and SycE are twice used to bind different helical regions of their effectors, occur in all these other chaperone molecules.

An interaction as observed between the type III effector chaperones and their effector molecules is extremely unusual. Typical protein-protein interactions involve a complementary interface between two globular molecules that are each folded with a well-packed hydrophobic core. How, then, might one explain why these pathogens have evolved this family of customized chaperones that maintain their cognate virulence factors in such an unusual, nonglobular state? The answer to this question probably lies in the state of the proteins as they are secreted. The 25-Å diameter estimate for the central channel within the needle complex (from cryo-EM studies) places severe constraints on the size of the globular molecules which could travel through this apparatus (see above). In fact, previous structural work indicates that several of these effectors are too large to fit through this channel (Evdokimov et al., 2001a; Stebbins and Galan, 2000; Stuckey et al., 1994). Therefore, it is likely that the virulence factors travel through the type III secretion pathway in a nonglobular or partially folded state by analogy to what is thought to occur in the evolutionarily related flagellar system (Samatey et al., 2001; Yonekura et al., 2003). The structural insights into chaperone-virulence factor complexes indirectly argue that the nonglobular nature of the interactions and secretion might be closely coupled. The intimate complex of a chaperone with an extended, nonglobular region of the virulence factor could be seen as an efficient mechanism for priming these factors for secretion by presenting to the export apparatus a protein that is lacking a folded structure and is therefore secretion competent. Since this nonglobular polypeptide segment is threaded through the needle, the energy of ATP hydrolysis that is coupled to type III secretion through an associated ATPase present in all type III secretion systems might unravel the remainder of the molecule and allow it to fit through the needle complex.

The mechanisms by which the chaperone-effector complexes are recognized by the protein secretion machinery are unclear. A secretion signal present within the first ~20 amino acids of all secreted proteins and not covered by the cognate chaperones probably plays a critical role in this process (Cornelis and Van Gijsegem, 2000). It is also possible that structural features of the chaperone-effector complexes may also serve as signals that are specifically recognized by the secretion apparatus. Those signals must consist of structural features conserved among the different chaperone-effector complexes, given the ability of type III secretion systems to engage heterologous effector proteins provided that they are presented to the secretion apparatus in conjunction with their cognate chaperones. Some clues about the potential identity of those recognition signals come from a close examination of the crystal structures of several chaperones and chaperone-effector complexes. Despite their lack of overall primary sequence similarity, all effector chaperones whose structures are known have the same overall fold and have similar hydrophobic regions that could serve as common targeting elements. Furthermore, comparisons of the structures of SptP and YopE in complex with their respective chaperones reveal that the main-chain route over the chaperones is very similar. This observation has led to the hypothesis that these features may serve as recognition signals for targeting the complexes to the type III secretion system (Birtalan et al., 2002). Other aspects of the complex, such as the electrostatic segregation in the 4:2 SicP-SptP complex, may also serve as signals to the export apparatus.

Recently, the structure of the Spa15 chaperone from *Shigella* was determined; it revealed that there is likely to be more diversity in this large protein family than was evinced in the initial set of chaperone structures determined. Spa15 had been classified (along with other chaperones such as InvB of *Salmonella*) in a separate subcategory because it chaperoned multiple effectors (Page et al., 2002). Sequence-threading analysis clearly indicates that Spa15 and InvB have a common tertiary fold with the other chaperones, but the structure of the Spa15 homodimer revealed that the assembly of the dimer—the relative orientation of the two chaperones—in Spa15 was different from those in all the previously determined structures (van Eerde et al., 2004). Since this result was identical in two crystal forms, the likelihood of a crystal-packing artifact distorting the true interaction is low. Interestingly, the large hydrophobic patches seen to bind the effectors SptP and YopE are still present (e.g., the hydrophobic, helix-binding groove), although located in different three-dimensional positions relative to each other than in the other chaperones (due to the differences in dimer formation). This spatial difference has led the authors of this study to conclude that, if the effectors bind these patches in a manner similar to the two known chaperone-effector complexes, it is likely that they will be forced to take different three-space paths over the Spa15 dimer, thereby perhaps challenging the idea that a universal signal to the translocation apparatus lies in the effector conformation. High-resolution cocrystal structures of Spa15 or InvB with their effectors are required before this question can be answered, however.

TYPE III SECRETION SYSTEM SUBSTRATES: TRANSLOCATED EFFECTORS

Overview

Examining the structures of the different type III secretion substrates—the virulence factor effectors that are translocated and act within the host cell—presents a problem in organization. Rather than presenting the structures based on their fold or biochemical activity or on the species which utilize them, this chapter is organized in biological themes and examines the structures in related functional contexts. While imperfect as an organizational system, such a framework prevents the complete divorce that occurs at times between detailed structural analysis and biological context and seeks to find a middle ground in combining general structural dissection with meaning in a disease context.

Modulating the Host Cytoskeleton

Both bacteria and viruses are known to modulate eukaryotic cells in highly specific manners, but the differences between these types of infectious agents demand different strategies of infection. The size difference between the two, in particular, requires that the much larger bacteria grapple with problems that the small viruses can ignore. A recurring theme in bacterial pathogenesis, and one that extends beyond organisms utilizing type III secretion, is that of manipulation of the host cell structure, in particular by modulating the cytoskeleton (Galan and Zhou, 2000; Lerm et al., 2000; Aktories et al., 2000; Barbieri et al., 2002; Galan, 2001). By changing the structure of cells, bacteria can induce their uptake into normally nonphagocytic host cells or prevent their uptake into professional phagocytes. In some cases, internalized bacteria remodel the actin cytoskeleton and microtubule networks for purposes such as the creation and maintenance of a specialized intercellular vacuole or their propulsion within and between cells (Galan, 2001; Steele-Mortimer et al., 2002). In the least sophisticated instances, bacteria can use toxins to irreversibly alter host cytoskeletal structure, often with rapid death of the eukaryotic cell (Barbieri et al., 2002).

Indirect Actin Cytoskeletal Modulation

The eukaryotic actin cytoskeleton is carefully regulated to precisely maintain cell shape and motility. One important signaling pathway in cells that exerts a large influence on the structure of the host cytoskeleton is that involving the Rho family of low-molecular-weight GTPases (Hall, 1998). These proteins undergo conformational changes in two peptide regions, called switch I and switch II, depending on which form of nucleotide they bind (Color Plate 36 [see color insert]) (Bourne et al., 1990; Gamblin and Smerdon, 1998; Sprang, 1997; Wittinghofer and Pai, 1991). When GTP is bound, the switch I and II regions adopt a conformation that renders the GTPase competent to bind signaling molecules that transmit instructions to reprogram the actin cytoskeletal structure (among other signals transduced). GTP binding is therefore associated with the signal "on" state, leading to diverse cellular morphological alterations such as membrane ruffling, filopodia growth, or stress fiber formation. When GTP is hydrolyzed to GDP, a conformational change ensues in the switch regions, rendering the GTPase unable to bind downstream signaling proteins (the signaling "off" state).

Several bacterial pathogens possess factors that modulate the activity of the Rho GTPases and therefore can indirectly control actin dynamics in the cell (Aktories et al.,

2000; Barbieri et al., 2002). Some do so in an irreversible fashion through covalent modifications of these G-proteins (Barbieri et al., 2002). Perhaps of even greater interest from a structural point of view, however, are the several type III effectors which are able to specifically and reversibly modify the activity of these enzymes.

Eukaryotic cells tightly regulate the nucleotide state of the Rho GTPases in a reversible manner (Hall, 1998). Guanine nucleotide exchange factors (GEFs) interact with GDP-bound forms of the small GTPases and lead to nucleotide exchange (Cherfils and Chardin, 1999; Sprang, 1997). The high cellular concentration of GTP relative to GDP then allows GTP to displace the GEF and enter the active site. GEFs therefore function to activate Rho GTPases and stimulate cytoskeletal rearrangements. GTPase-activating proteins (GAPs), on the other hand, bind to the GTP form of the Rho proteins and dramatically stimulate the low intrinsic rate of GTP hydrolysis that these enzymes possess (Sprang, 1997; Vetter and Wittinghofer, 2001). GAPs therefore drive the GTPases to the inactive form and function to downregulate cytoskeletal rearrangements.

SopE: a GEF for Rho GTPases. The enteric pathogen *Salmonella* delivers into the host cell two highly related bacterial proteins, SopE and SopE2, that function as GEFs that activate Rac1 and Cdc42 (Bakshi et al., 2000; Hardt et al., 1998a, 1998b; Stender et al., 2000; Wood et al., 1996). This GTPase activation potently contributes to actin rearrangements (creating membrane ruffles) and subsequent bacterial internalization into intestinal epithelial cells. SopE shows no sequence similarity to any host proteins and is one of the most active GEFs examined (Rudolph et al., 1999). The cocrystal structure of SopE and Cdc42 reveals the mechanism by which SopE induces the exchange of GTP for GDP, which involves a unique insertion of a highly conserved loop into the nucleotide-binding site of Cdc42 and a rearrangement of the active site to make it unable to bind GDP (Buchwald et al., 2002).

Consistent with its lack of sequence similarity to any known proteins, SopE possesses a unique fold that is, in particular, unrelated to host GEF enzymes. The crystallized, catalytic domain of SopE (residues 78 to 240) consists of a pair of three helix bundles oriented such that each bundle contacts the other in a "V" shape (Color Plate 37 [see color insert]). A small, two strand β-sheet is located at the bottom of the molecule proximal to the end of one bundle, and a loop extending from one stand of this sheet crosses over to the other bundle. This loop, as discussed below, is presented outward to the solvent and is the key element in the activity of SopE.

The structure of Cdc42 in the complex is very similar to the structure of the enzyme alone and the structures of other small GTPases such as Rac1 (Hirshberg et al., 1997; Ihara et al., 1998; Rittinger et al., 1997a). The main difference is in the regulatory switch I and II regions of Cdc42, which, like all Rho family GTPases, serve both to determine the activity of the molecule in signal transduction conformationally and to bind and coordinate the nucleotide (and thereby distinguish the GTP-bound from the GDP-bound state). SopE binds primarily to the switch regions of Cdc42, burying a relatively large surface area of 2,800 Å^2, but it leaves the nucleotide-binding pocket open to the solvent (and thus to GTP). No GDP is bound in the complex.

The loop spanning the bottom of SopE (connecting the small β-sheet with one of the three helix bundles and centered on the conserved sequence GAGA) inserts between the two switch regions of Cdc42 and displaces switch I by pushing out from its normal conformation and pulling switch II closer to the loop. These conformational changes are stabilized by several contacts between the switch regions and both SopE and Cdc42. A peptide flip in the conformation of switch II is particularly important since it causes a positioning of an alanine residue (Ala59 of Cdc42) very near the location of a magnesium-binding site. In nucleotide-bound Cdc42 (and all the small Rho GTPases), the magnesium ion plays a crucial role in the coordination of the nucleotide. By reorienting switch II to sterically prevent magnesium binding, SopE further contributes to the loss of contacts already achieved by the repositioning of nucleotide-binding regions in the switch polypeptides.

SptP, YopE, and ExoS: GAP enzymes for the Rho GTPases. Several bacterial pathogens possess GAPs, which function oppositely to the GEFs and therefore shut down Rho family GTPase signaling (Black and Bliska, 2000; Fu and Galan, 1999; Geiser et al., 2001; Goehring et al., 1999). GAP enzymes bind to Rho GTPases in the GTP bound state and stimulate by severalfold the slow, intrinsic GTPase activity of these enzymes (Sprang, 1997; Stebbins and Galan, 2000; Vetter and Wittinghofer, 2001; Wurtele et al., 2001b). While the activity of these bacterial GAPs at the structural level is virtually identical, their use in different cell types and in the context of different effectors, along with their different stability in the host cell, all combine to make their effect of the host cell quite different for each pathogen.

For example, *S. enterica* serovar Typhimurium uses the GAP enzyme SptP to shut down Rho GTPase activity stimulated by SopE (Fu and Galan, 1999; Galan and Zhou, 2000). This decreases the massive ruffling induced by the bacterium that led to its internalization and helps the infected cell recover from the membrane disruption. *Salmonella* therefore uses a GAP to aid in the recovery of the cell to protect its niche for growth. In contrast, *Yersinia* species use the GAP YopE to shut down Rho GTPase signaling in macrophages. Since this signaling is critical to actin cytoskeletal rearrangements, needed by the macrophage to function, YopE contributes to the paralysis of the cell, effectively declawing this family of lymphocytes (Black and Bliska, 2000). In these two cases, we see the same enzymatic activity used to completely different purpose by different pathogens. Such examples make it clear that to understand the function of these molecules, it is not enough to know their structure and biochemistry and that it is crucial to understand the biological context in which the proteins function.

In what follows, the complex of the protein SptP of *Salmonella* will be used to illustrate the key elements in the GAP-GTPase complexes, although nearly every point made about SptP applies to the complex of the *P. aeruginosa* GAP ExoS in complex with Rac1 (Stebbins and Galan, 2000; Wurtele et al., 2001b). Biochemically, the complexes were formed by "fooling" the proteins with a mixture of aluminum fluoride (AlF_3) and GDP. AlF_3 mimics the γ-phosphate group of GTP in these complexes, and the GTPase and GAP function as though the additional phosphate were present, but they remain bound as no hydrolysis can occur. In addition, because SptP of *Salmonella* possesses a GAP as well as a protein tyrosine phosphatase (PTP) domain (discussed in more detail below, along with the homologous *Yersinia* phosphatase) and because the PTP domain does not interact with Rac1, the presence of the PTP domain is ignored in most of the subsequent discussion.

Structurally, the bacterial GAP fold is a four-helix bundle with an up-down-up-down topology containing heptad repeats and supercoiling in the helices (Color Plate 37). A localized deviation from the standard bundle structure creates a bulge at one end of the GAP domain. The folded-out segments interact with the nucleotide GDP and Rac1 elements in the switch I and II regions and are stabilized by inserting several residues into the hydrophobic core of the GAP domain. These residues (highly conserved in the related proteins ExoS and YopE) create a hydrophobic core at one end of the GAP domain that bulges out from the narrow channel-like interior of more standard four-helix bundles (Color Plate 37). Several helices maintain a canonical bundle architecture, with two of them, along with the bulged out portions, comprising a side of the molecule that interacts with Rac1. The catalytically essential residue for the GAP activity, Arg209 in SptP (Arg144 and Arg146 in YopE and ExoS, respectively), extends from a bundle helix and into the active site of Rac1 (Color Plate 37). The GAP domains of ExoS and YopE have nearly identical folds compared with those of SptP and, in the case of the ExoS-Rac1 complex, position the catalytic arginine like SptP (Color Plate 37).

Rac1 as found in complex with SptP is very similar to its reported monomeric structure, with the exception of a complete ordering of switch I, which was found to be disordered even in a 1.3-Å-resolution crystal structure of the isolated human enzyme (Hirshberg et al., 1997). Switch I makes extensive contacts to the GAP domain of SptP and, in so doing, adopts a conformation very similar to that of switch I in other eukaryotic GAP complexes with Rho family members (Gamblin and Smerdon, 1998; Nassar et al., 1998; Rittinger et al., 1997a, 1997b).

The SptP and Rac1 surfaces are highly complementary. The bulge extending out from the helical bundle inserts into the Rac1 active-site cavity, and the bundle helices lie along the surface of Rac1 formed by switches I and II. There is also a surface charge complementarity among all portions of the interface. The interface residues are almost exclusively polar, with only a few hydrophobic van der Waals interactions. A large network of hydrogen bonds, both direct and water mediated, dominate the contacts between SptP and Rac1 (Color Plate 37).

Amino acids from SptP contact important elements in Rac1 associated with the GTP hydrolysis reaction, including the catalytically crucial Gln61 of switch II, the γ-phosphate mimic AlF_3 the phosphate-binding region (P-loop), and the nucleotide itself (Color Plate 37). Arg209 of SptP inserts deeply into the active site, resulting in strong hydrogen bonds of its guanidinium group to the (Pγ-bridging) β-phosphate oxygen of GDP and a fluoride atom of AlF_3. The interactions of Arg209 of SptP with the β-phosphate oxygen and fluoride atom are similar to those in the functionally equivalent arginine of Cdc42GAP in its transition state complex with Cdc42 (Nassar et al., 1998). Arg209 of SptP therefore complements the active site of Rac1 by stabilizing the negative charge that develops on the leaving group during the phosphoryl transfer reaction. An attacking nucleophilic water molecule is effectively positioned for the phosphoryl transfer reaction through a strong hydrogen bond to Gln61. SptP devotes several residues to the positioning of switch II (Color Plate 37C) and especially Gln61 in a catalytically favorable conformation (Color Plate 37). SptP also makes significant contacts to the ribose and guanine groups of GDP (Color Plate 37).

A comparison of the crystal structures of the ExoS/SptP-Rac1 complexes with monomeric ExoS and SptP reveals important conformational changes (Stebbins and Galan, 2000; Wurtele et al., 2001a; Wurtele et al., 2001b) (although this does not seem to be the case for crystallized monomeric YopE [Evdokimov et al., 2002a]). The isolated GAP domain is significantly disordered; in particular, most of the Rac1-binding regions (the bulge from the four-helix bundle fold) are not visible in the electron density maps. Notably, the catalytic arginine of SptP shows no clear density. In contrast, the GAP domain in the transition state complex is well ordered, and residues that are completely lacking interpretable electron density in the monomer become well ordered in the complex (e.g., Arg209). The differences between the monomeric and Rac1-bound crystal structures of SptP may reflect a process that occurs in vivo, and may offer insights into the function and regulation of SptP. In particular, the GAP domain of SptP is immediately adjacent to the amino-terminal chaperone-binding domain discussed above. If the NH_2 terminus of SptP is primed for secretion as a nonglobular polypeptide wound around its chaperone, as it is threaded through the export machinery, the energy of ATP hydrolysis, presumably provided by the associated ATPase of the type III secretion system, might drive the unfolding of the already highly disordered GAP domain if it should fold within the bacterium. Indeed, the most disordered regions of the GAP domain are those immediately adjacent to the chaperone-binding domain. The better-ordered portion of the GAP domain stabilizes an extended-loop region at the amino terminus of the adjacent PTP domain. It is possible that, once the GAP domain has been unraveled, the loss of support for this extended-loop structure could induce the unraveling of the entire PTP domain. In this sense, the unfolded nature of the chaperone-binding domain, along with inherent design features of the GAP and PTP domains, could prime SptP for rapid unfolding on engagement by the secretion apparatus.

SptP and YopH PTP domains. Phosphorylation is also known to play an important role in cytoskeletal regulation. Both *Yersinia* and *Salmonella* spp. possess proteins that can modify the phosphorylation of host proteins to dramatic effect. *Yersinia* spp. translocate the YpkA (also called YopO) serine/threonine kinase that causes massive cytoskeletal disruption (Galyov et al., 1993; Hakansson et al., 1996a). In addition, *Salmonella* and *Yersinia* both translocate a PTP into the host cell, but for very different purposes (Bliska et al., 1991; Guan and Dixon, 1990, 1993; Kaniga et al., 1996). The crystal structures of these enzymes have been illuminating both for understanding the PTP mechanism and for understanding bacterial virulence (Stuckey et al., 1994; Fauman et al., 1996; Stebbins and Galan, 2000).

The COOH-terminal domains of YopH and SptP consist of the canonical protein tyrosine phosphatase fold: eight β-strands form a single, highly twisted sheet with a core structure of four parallel strands at the center of the molecule (Color Plate 38 [see color insert]). Four helices cluster on one face of the sheet, with two on the other. Although these bacterial phosphatases have low sequence identity to the eukaryotic enzyme PTP1B, the tertiary folds of these enzymes align over 200 residues (Color Plate 38), with a Cα-Cα root mean square deviation of around 2.0 Å (Barford et al., 1994; Holm and Sander, 1993; Stuckey et al., 1994; Fauman et al., 1996; Jia et al., 1995; Stebbins and Galan, 2000).

The active site of the phosphatase domains in these effectors is highly conserved. SptP and YopH contain the signature P-loop motif, (I/V)HCXXGXGR(S/T)G, of a phosphatase active site with the catalytic cysteine residue, and this segment forms a strand-loop-helix structure that is nearly identical to the P-loop in other PTPs (Color Plate 38). The active sites are completed by a WPD loop, as in other PTPs (which contains a catalytic aspartic acid). As found in substrate complexes of related enzymes (Barford et al., 1994; Jia et al., 1995; Stuckey et al., 1994), oxygen atoms from the solvent and the substrate mimic tungstate are found to occupy positions in and be coordinated by conserved residues in the SptP and YopH active sites (Color Plate 38).

The active site consists of a pocket whose depth (~8 to 10 Å) gives the tyrosine phosphatases specificity for phosphotyrosine, since only this long amino acid side chain can extend deeply enough into the catalytic core. The phosphate moiety, as judged by cocrystal structures of PTPs and phosphate, sulfate, tungstate, and phosphopeptides, are coordinated by the so-called P-loop, a stretch of glycine-rich amino acids in the active site with a conformation such that the amides of the main-chain point inward and form hydrogen bonds with the phosphate oxygens, and the substrate is further stabilized by the positive electrostatic potential created by main-chain nitrogens as well as a helix dipole (Barford et al., 1994; Jia et al., 1995; Stuckey et al., 1994).

Based on early structural work with the eukaryotic PTP1B and the bacterial YopH, the following mechanism has been proposed for PTP activity (Barford et al., 1994; Barford et al., 1995; Bliska, 1995; Dixon, 1995; Fauman and Saper, 1996; Jia et al., 1995; Stuckey et al., 1994; Zhang, 2002). The absolutely conserved cysteine residue near the bottom of the active site acts as an attacking nucleophile, which forms a thiophosphoryl enzyme intermediate (thereby transferring the phosphate group to the cysteine and from the tyrosine). This attack is facilitated by another absolutely conserved residue, an aspartic acid in the WPD loop, that acts as a general acid in neutralizing the development of negative charge on the leaving group, and by an arginine residue that binds to phosphate oxygens and stabilizes the PO_3 moiety formed during transfer to the cysteine. The same aspartic acid residue then acts as a general base in de-protonating a coordinated water molecule in the active site, which then acts through an in-line attack to hydrolyze the phosphocysteine intermediate, producing a free cysteine residue and orthophosphate, regenerating the enzyme. All the residues involved in this mechanism—cysteine, arginine, and aspartic acid—are absolutely conserved in all PTP enzymes and have been shown through mutagenesis to be essential for activity and virulence.

The remaining homology between these enzymes is restricted almost entirely to the hydrophobic core. SptP and YopH are distinguished from each other by their surface properties, which reveal unique shapes and charge distributions (Color Plate 38). The differences between SptP of *Salmonella* and YopH of *Yersinia* are consistent with their different substrates within the host cell. For example, while YopH disrupts focal adhesions by dephosphorylating p130[cas], the focal adhesion kinase (FAK), and other targets (Black and Bliska, 1997; Hamid et al., 1999; Persson et al., 1997), SptP exerts its function on proteins that are tyrosine phosphorylated on bacterial infection without affecting focal adhesions (Murli et al., 2001). Presumably, the molecular surfaces of these two phosphatases are each well adapted to optimize the interactions with their different targets, thereby contributing to the observed surface divergence.

Dual-function YopH NH$_2$-terminal domain. The NH$_2$-terminal domain of YopH, which in the bacterium serves a function in translocation by interacting with the secretion chaperone SycH (Woestyn et al., 1996), has the unusual property of playing a second role by exerting effector activity within the host cell (Montagna et al., 2001). It does so by binding to phosphotyrosine residues of host targets that have the recognition motif pYxxP (Montagna et al., 2001). The structures of this domain alone and in complex have been determined, revealing the molecular basis for phosphotyrosine recognition (Evdokimov et al., 2001b; Khandelwal et al., 2002; Smith et al., 2001). The YopH NH$_2$-terminal domain possesses a unique $\alpha + \beta$ fold that bears no resemblance to eukaryotic phosphotyrosine-binding motifs (Color Plate 38). The four helices in the structure are located in the middle of the molecule, although they do not form a bundle architecture. These four helices are flanked at either side of the molecule by short, two-stranded β-sheets. Most of the linking loops between these elements of secondary structure are located at one end of the molecule.

Of great interest in these structures are the potential insights into how the secretion and translocation (bacterial side function) and the effector function (host cell) can be packaged into a signal domain. The structure alone provides some excellent clues to the effector function, since it was immediately clear that residues which had been shown by mutagenesis to be important for phosphotyrosine recognition were all clustered to a single side of the molecule (Montagna, 2001). The nuclear magnetic resonance spectroscopy (NMR) structure of the complex of this domain with the consensus phosphopeptide DEpYDDPF is consistent with this but demonstrates that the acidic peptide binds in a largely positively charged surface patch of YopH (Color Plate 38), interacting with some residues identified by mutagenesis but also several others with positive charges (Khandelwal et al., 2002). In particular, a lysine residue in the effector in the first β-sheet is observed to make direct contacts with the phosphotyrosine side chain of the peptide.

In contrast to the insight gained into the effector function of this domain, the secretion and translocation function seems poorly correlated with the structure. Residues 1 to 17 (Sory et al., 1995) of YopH constitute a signal sequence for secretion through the type III apparatus. This segment corresponds almost exactly to the first helix in the molecule, and in the crystals and NMR structure, this helix is well packed against the remainder of the folded domain. The reported translocation domain of this effector—residues 20 to 70 identified through deletion analysis—does not constitute an independently folded domain but forms an integral part of the fold (Evdokimov et al., 2001b). Therefore, the structure-function relationship with respect to secretion and translocation is not clear from the crystal structure of the YopH NH$_2$-terminal domain.

This conundrum is resolved when the results of the chaperone-effector domains previously discussed are considered. In the SptP and YopE complexes with their respective chaperones, the NH$_2$-terminal domains of the effectors are found to be nonglobular and wrapped around the chaperone molecules. This complex is biologically relevant within the bacterium and is associated with secretion and translocation, and hypotheses have been put forth that this domain in this context primes the effectors for a nonglobular secretion and serves as a unique recognition signal for the needle complex machinery. If this is generally true beyond YopE of *Yersinia* and SptP of *Salmonella*, then the NH$_2$-terminal domain of YopH is likely to interact with its chaperone, SycH, as a nonglobular polypeptide. Strengthening this possibility is the fact that SycH, both by sequence analysis and by sequence threading to structure, is strongly predicted to have the same fold as that of the other effector chaperones whose structures have been determined. In that case, the folded structure of the YopH NH$_2$-terminal domain observed in the crystals and solution NMR probably does not represent the structure of the molecule in the bacterium, at least not when it is bound to the chaperone. The globular state is probably relevant only for the effector function of the molecule within the host cell and cannot be used to understand its role in secretion and translocation.

Although the YopH NH$_2$-terminal domain seemed like a unique example of a dual functional role for a virulence protein, subsequent results have raised the possibility that many more translocated molecules may be characterized by chaperone-binding domains that have an effector function. The chaperone-binding domains of SopE and SptP, for example, were recently shown to modulate their protein half-lives in a host ubiquitin-mediated proteolysis pathway (Kubori and Galan, 2003). Other virulence factors may also therefore possess domains with separate functions in the bacterial and host cells.

Direct Actin Cytoskeletal Modulation

In addition to molecules that indirectly modulate cytoskeletal function, many bacterial pathogens translocate type III effectors that directly interact and manipulate actin structure within the cell. For example, the SipC protein of *Salmonella,* already discussed as an element of the translocon, also possesses domains that function within the host cytoplasm to nucleate and bundle actin (Hayward and Koronakis, 1999; McGhie et al., 2001). In addition, several strains of *Salmonella* harbor the SPV virulence plasmid, on which lies the *spvB* gene. SpvB is potentially translocated through a type III secretion system in these strains of *Salmonella,* where it functions as an ADP-ribosyltransferase, covalently modifying actin and preventing its polymerization (Lesnick et al., 2001; Tezcan-Merdol et al., 2001). While no structural information is currently available for SipC or SpvB, the clear homology of the latter protein in sequence and function to other actin-targeting ADP-ribosyltransferases (but which are not type III effectors) whose structures are known suggests that its enzymatic mechanism mirrors that in these other proteins (Han et al., 2001; Han et al., 1999; Tsuge et al., 2003).

S. enterica serovar Typhimurium mutants with a disruption in *Salmonella* invasion protein A (encoded by the *sipA* gene) show attenuated virulence in bovine intestinal models (Zhang et al., 2002), impaired cellular invasion (Jepson et al., 2001; Zhou et al., 1999b), and weaker and less localized membrane ruffling (Higashide et al., 2002; Zhou et al., 1999b). Biochemically, the SipA protein of *Salmonella* is able to bind to actin, reduce the critical concentration for the formation of F-actin, stabilize actin filaments (McGhie et al., 2001; Zhou et al., 1999b), and potentiate the actin-nucleating and bundling activity of SipC (McGhie et al., 2001). A carboxy-terminal region of SipA, spanning residues 446 to 684, possesses these biochemical properties in vitro (Hayward and Koronakis, 1999; McGhie et al., 2001; Zhou et al., 1999b) and interacts indirectly with plastin in a yeast two-hybrid assay (Zhou et al., 1999a). This region of SipA shows no sequence or predicted structural homology to any host proteins or known actin-binding motifs (Lilic et al., 2003).

The crystal structure of the COOH-terminal domain of *S. enterica* serovar Typhimurium SipA (residues 497 to 669, henceforth called SipA$^{497-669}$) retains actin binding and polymerization activities and folds into a compact, heart-shaped molecule dominated by helical secondary structure with dimensions of roughly 30 by 40 by 40 Å (Color Plate 39 [see color insert]). In the crystal, the protein is ordered only between residues 513 and 657 (with 16 and 12 residues disordered at the NH$_2$ and COOH termini, respectively), and the termini are located at opposite sides of the molecule. The molecular surface of the protein is remarkable for a large basic patch, which may aid SipA in efficiently orienting its actin-binding sites to the mostly acidic surface of actin (Color Plate 39).

EM studies of actin-SipA$^{497-669}$ complexes reveal that SipA interacts with polymerized actin as a globular structure with two nonglobular extensions that connect actin protomers on opposite strands (Color Plate 39). These "arms" are positioned at opposite ends of the globular density linking subdomain 4 of actin subunit *n* in the filament to subdomain 1 of subunit *n* + 3. The large globular density is located between two adjacent actin protomers (subunits *n* + 2 and *n* + 3). Comparisons of SipA$^{497-669}$ and a larger construct of SipA, SipA$^{446-684}$, reveal that both have a central density into which the globular crystal structure can be positioned but that they differ significantly in the length of the arms. When the crystal structure is fit into the globular region of the EM density, the best placement occurs with the NH$_2$ and COOH termini located near each arm (Color Plate 39).

The combination of the crystallography and EM of SipA on polymerized actin suggests that the tubular density previously observed with the longer SipA constructs is due to the presence of additional polypeptide in nonglobular extensions from a core globular domain. These structural data suggest a model in which SipA would function as a "molecular staple" in polymerizing actin, centered on the globular domain for binding actin and positioning the arms but using the nonglobular extensions to reach out and tether actin molecules on "opposing strands" (the actin filament can be described as two actin polymers or strands wrapped around each other as a parallel helix). Deletion analysis of the tethering arms has supported this model, since "armless" mutants are far less active in inducing polymerization of actin.

Modulation of Host Signal Transduction

Besides the reversible modulation of the Rho GTPases, several type III effector molecules affect signal transduction within the host cell. No structural information about these effectors is currently available, but for several of them predictions based on sequence homology and activity can be made. For example, *Yersinia* spp. translocate the effector YopT, which causes massive cytoskeletal changes in animal cells and induces the hypersensitive defense response in resistant plant cells (Iriarte and Cornelis, 1998; Zumbihl et al., 1999). In addition, the *Yersinia* effector YopJ inhibits the host immune response and induces apoptosis by interfering with the MAPK and NF-κB signal transduction pathways (Meijer et al., 2000; Monack et al., 1997; Palmer et al., 1998; Palmer et al., 1999; Schesser et al., 1998). While the structures of these molecules have not been determined, their sequences thread well against cysteine proteases, and alignments revealed that the classic catalytic triad of the cysteine proteases appears to be conserved in these molecules (Orth et al., 2000; Shao et al., 2002). These results led to an examination of YopT and YopJ for proteolytic activity. It was recently shown that YopT cleaves the COOH terminus of the Rho GTPases (Shao et al., 2002). This is significant because these GTPases are active at the membrane and are attached to the membrane through lipid modifications at their COOH termini. The protease activity of YopT separates the GTPase from its membrane anchor; it detaches from the membrane and can no longer function in signal transduction. This explains the catastrophic effect of YopT on the host cytoskeleton. YopJ was shown to have a very interesting activity in cleaving the COOH termini of the ubiquitin-like molecules such as SUMO-1 (Orth et al., 2000). These proteins are conjugated by isopeptide bonds to lysine side chains of other proteins (analogously to ubiquitin). However, the ubiquitin-like proteins do not necessarily target the modified proteins for degradation but can modify their stability or activity in different ways, much like phosphorylation can affect signal transduction. The overexpression of YopJ in eukaryotic cells results in the decrease of SUMO-lated proteins as well as ubiquitinated proteins, although the physiologically relevant substrate for YopJ is not yet established. Consistent with this theme, the *Salmonella* effector protein AvrA, which has homology to YopJ, also functions to downregulate the NF-κB pathway and stimulates apoptosis (Collier-Hyams et al., 2002).

Another molecule predicted to effect signal transduction in the host is the *Salmonella* protein SopB, which activates the serine-threonine kinase Akt/PKBα pathway (Marcus et al., 2001; Steele-Mortimer et al., 2000), affects the cytoskeleton of host cells (Zhou et al., 2001), and alters membrane lipid content (Terebiznik et al., 2002). SopB possesses the active-site residues required for a lipid phosphatase and has inositol phosphatase activity (Norris et al., 1998), but no structural information is currently available.

YopM: an Effector with Leucine-Rich Repeats

Finally, we consider the *Yersinia* effector YopM, a 42-kDa protein composed primarily of 15 tandem leucine-rich repeats (LRR) of 20 to 22 residues (flanked by 73 and 24 non-LRR residues at its NH_2 and COOH termini, respectively), which is required for full virulence in the mouse model of infection (Grosdent et al., 2002; Leung and Straley, 1989; Skrzypek et al., 1998). LRR are typically protein-protein interaction motifs, and studies have suggested that YopM interacts with and stimulates the activity of two host cell signaling serine/threonine protein kinases (McDonald et al., 2003), although the significance of the interaction is not yet clear.

The crystal structure of YopM (Color Plate 40 [see color insert]) reveals that, like other LRR-containing proteins, the LRR form a similar loop-strand repeat that stacks one on top of the other throughout the LRR region (Evdokimov et al., 2001a). The YopM structure has variations in the LRR fold, such as the lack of 3/10 helix content and, more dramatically, an overall helical twist in the repeats that cause the protein to spiral slightly about the concave inner surface, the region hypothesized to be involved in molecular interactions for most LRR-containing proteins (Kobe and Kajava, 2001). An interesting, although functionally unclear, observation from the YopM crystal structures is the packing in the crystals of the protein to form tetrameric assemblies described by a superhelical arrangement which forms a hollow channel with a diameter of 35 Å in the center. This crystal packing is observed in three different crystal forms and buries most of the hydrophobic residues of the convex surface to create an extensive nonpolar interaction. At present, however, the biological significance of this startling arrangement, if any, is unclear (Evdokimov et al., 2001a).

Interestingly, once again the secretion and translocation regions of an effector were crystallized without a type III secretion chaperone and therefore presumably in a conformation that would be expected to function within the host cell. In YopM, and unlike YopH, the secretion peptide (residues 1 to 40) is disordered in the crystals. The translocation domain of YopM (for which no chaperone has yet been identified) spans residues up to 100 (Boland et al., 1996), and residues 35 to 75 of YopM adopt an antiparallel two-helix fold that caps the first LRR, although the conformation of this region within the bacterium prior to translocation may be quite different from that observed in the crystallized protein.

Themes

Prokaryotic organisms are fantastically successful life-forms. It has been estimated that bacteria roughly equal the carbon mass of plants and that humans probably carry about 10^{11} bacterial cells, on the order of the total number of human cells in the body (Whitman et al., 1998). Most of these bacteria are usually harmless, not only coexisting peacefully with us but also playing an essential role in our normal metabolism (Galan, 2001). A small subset of these bacteria have evolved a much closer association with their hosts, which sometimes leads to pathology, and are therefore considered pathogens. Due to their long coexistence with vertebrate hosts, these bacteria have evolved an amazing panoply of sophisticated adaptations to secure their survival and replication. In particular, recent studies have begun to reveal that many bacterial pathogens mimic the function of host proteins in order to manipulate host physiology for their own benefit. This contrasts with the strategies used by some pathogens, which involve microbial products with activities lacking clear counterparts in eukaryotic cells (Stebbins and Galan, 2001b). In this section we consider structural work that has provided unique insights into the mechanisms of host mimicry by bacterial virulence factors.

The ability to modulate host cellular activities at the molecular level through functional mimicry is a powerful tool for a bacterial pathogen. It allows the bacterium to be precise and limited in its effects in achieving its goals (e.g., internalization into a host cell through limited damage). Obtaining molecules with such activity presents difficulties, however. There are two means by which a pathogen could acquire such effector molecules: by obtaining "foreign" genes through horizontal transfer (such as host protein homologs) or through the process of convergent evolution. Perhaps the most intriguing is mimicry through convergent evolution, whereby material (genes and the proteins they encode) already accessible to the pathogen is "sculpted" to a new function. Such a protein would usually have a distinct three-dimensional architecture from the molecule it mimics but will have evolved to imitate the surface chemical groups of the functional homolog.

Horizontal Acquisition: Same Activity, Different Application

For many of the functional mimics used by bacterial pathogens, the mimicry is achieved in a straightforward manner through homologous enzymes that have been subverted for the benefit of the pathogen (Stebbins and Galan, 2001b). These enzymes are often easily identifiable by sequence alignments, and the bacterial homologs of host proteins often differ from their host counterparts through alterations in substrate specificity, absence of regulatory control domains, and/or modulation of their intrinsic activity.

An example of a well-characterized virulence factor with such properties is the YopH PTP discussed above. YopH disrupts focal adhesions by dephosphorylating p130^{cas} and the focal adhesion kinase, leading to paralysis of macrophage attack. In this case, the pathogen has obtained a virulence determinant with sequence and structural similarities to host enzymes but with substrates and regulation tailored to the needs of the pathogen. Other examples that are clear from strong sequence conservation with eukaryotic enzymes include YpkA/YopO (the serine/threonine kinase of *Yersinia*); the cysteine proteases YopJ, YopT, and AvrA; and the LRR-containing YopM, among others.

Convergent Evolution

Some virulence factors with activities similar to host enzymes do not show any sequence similarity to eukaryotic proteins. Structural and functional similarities can occur even in the absence of sequence similarity, and thus it was unclear how these virulence factors functioned. The solution of several crystal structures of such factors and their complexes with host targets has illuminated this topic. In some cases, while these "bacterial mimics" possess a structural architecture (the fold) that differs markedly from that of their host functional homologs, the molecular surfaces that interact with their targets, the true level at which natural selection ultimately sculpts, are seen as excellent mimics of host proteins.

An interesting example to consider are the bacterial GAPs (see above) (Color Plate 37). For example, like host cell GAPs, the bacterial GAPs extensively interact with the regulatory switch I and II regions of the GTPase, contacting similar residues (especially catalytic residues) but doing so in a very different manner from that used by host enzymes. Despite the different molecular tentacles extended by bacterial GAPs relative to host proteins, the switch I and II regions of SptP/ExoS-bound Rac1 adopt nearly identical conformations to those observed in host GAP-RhoGTPase transition state complexes (Color Plate 41 [see color insert]). The bacterial GAPs, for example, use several side chains and a main-chain contact to constrain the catalytic Gln61, a key functional element that positions a nucleophilic water molecule for attack on the γ-phosphate of GTP. Positioning Gln61 in this way is considered crucial to the hydrolysis reaction. Therefore, the virulence factors have "learned" to mimic host GAPs to achieve the same goal but use their own unique methods to do so.

A second instance of mimicry in this structure occurs with a donated catalytic residue from the bacterial GAPs (Color Plate 41). The small GTPases of the Ras/Rho family lack a crucial catalytic arginine for the GTP hydrolysis reaction. GAPs for the small GTPases invariably insert an arginine side chain, thereby completing the active site. By analogy to host GAPs, which contain such arginines on extended loops (arginine "finger loops"), it was hypothesized that this was the structural foundation for the GAP activity of the bacterial enzymes (Fu and Galan, 1999). The crystal structures of these virulence factors in complex with host GTPases provide an interesting surprise, however. Indeed, arginine, as predicted from mutagenesis, is the catalytic residue and is inserted into the active site of the GTPase in a manner nearly identical to that for functionally equivalent arginines from host enzymes. The surprise is that, unlike host GAPs, the arginine is not in a loop but instead extends from an α-helix (Stebbins and Galan, 2000; Wurtele et al., 2001b). Thus, these foreign enzymes with a different tertiary structure still achieve a precise chemical mimicry of the host proteins.

Another beautiful example of convergent evolution in bacterial pathogenesis involves not a type III effector but, instead, an outer membrane adhesion molecule important for the virulence of bacteria that use a type III system. *Y. pseudotuberculosis* uses the envelope protein invasin to bind host cell β₁-integrin surface receptors, thereby manipulating host signal transduction and contributing to bacterial attachment and internalization (Isberg and Leong, 1988; Isberg et al., 1987; Leong et al., 1991; Van Nhieu and Isberg, 1991; Young et al., 1990). Invasin is so potent that it will outcompete natural host substrates such as fibronectin for β₁-integrin binding. No sequence similarity between invasin and host proteins can be detected, but the crystal structure of invasin reveals the powerful hand of convergent evolution at work in producing a virulence factor that mimics host activities (Hamburger et al., 1999). In this case, what is mimicked is the integrin-binding surface of fibronectin (Color Plate 41).

Invasin and fibronectin are revealed in their respective crystal structures to be long, rod-like molecules composed of repeated domains (Hamburger et al., 1999). Capping the rods in each case are specialized domains with the integrin-binding surface. The tertiary structures of the integrin-binding domains of invasin and fibronectin seem completely unrelated. However, it becomes apparent when their integrin-binding surfaces are examined that, despite the utterly different protein architectures that scaffold the amino acids forming the binding site, the functional aspects of their molecular surfaces are very similar (Color Plate 41). For example, although fibronectin presents a binding surface that contains an extensive cleft absent in invasin, three key residues absolutely required for integrin binding (two aspartic acids and an arginine) are located in nearly identical positions spanning the breadth of the extensive binding surface. Invasin is therefore demonstrated to be an excellent mimic of its eukaryotic cell counterpart, an ability that gives *Yersinia* an important key into host cells.

Finally, the protein SopE represents a third class of bacterial virulence factor: one that achieves a host-like activity (activation of the Rho GTPases through stimulating nucleotide exchange) without being derived from host-like genes or mimicking the activity of host proteins. As revealed by the crystal structure of the complex of SopE with Cdc42, the bacterial GEF possesses a completely novel mechanism for stimulating nucleotide exchange. SopE therefore represents a case of convergent evolution in function but with complete novelty in mechanism. While the enzyme achieves the same results with respect to the GTPase as host factors, even at the level of positioning the switch I and II regions analogously, it does so through mechanisms that heretofore have not been observed in host GEFs and currently do not appear to function in the host based on sequence alignments. As a comparison, Tiam-1, a host GEF for the Cdc42-related family member Rac1, uses the insertion of a lysine residue as a major part of its efforts to displace the switch regions away from their role in nucleotide coordination (Worthylake et al., 2000). A close alignment of SopE with Tiam-1 shows that although SopE does not use a long amino acid side chain like a lysine, the GAGA catalytic loop inserts in such a way that it reaches nearly the same position as that lysine between the switch regions (Color Plate 41).

FUTURE

Detailed structural analyses of type III secretion virulence systems are only in their infancy. The field awaits high-resolution work on the needle complex, the translocon, the bacterial side elements, and the remaining arsenal of effector molecules used by the different gram-negative pathogens utilizing this translocation system in plants and animals. The field is growing quickly in terms of both results and the number of investigators, and future editions of this chapter will probably require substantial rewrites and additions. Nonetheless, the first structural information from these systems has been immensely informative and very exciting from a biological point of view. Over millions of years, the coevolutionary battle and coexistence of hosts and pathogens have engendered highly complex and sophisticated biochemical cross talk between different species. The coming years will surely only deepen this appreciation, with additional examples for which structural insight will often play a key role.

REFERENCES

Abe, A., M. de Grado, R. A. Pfuetzner, C. Sanchez-Sanmartin, R. Devinney, J. L. Puente, N. C. Strynadka, and B. B. Finlay. 1999. Enteropathogenic *Escherichia coli* translocated intimin receptor, Tir, requires a specific chaperone for stable secretion. *Mol. Microbiol.* **33:**1162–1175.

Aktories, K., G. Schmidt, and I. Just. 2000. Rho GTPases as targets of bacterial protein toxins. *Biol. Chem.* **381:**421–426.

Bakshi, C. S., V. P. Singh, M. W. Wood, P. W. Jones, T. S. Wallis, and E. E. Galyov. 2000. Identification of SopE2, a *Salmonella* secreted protein which is highly homologous to SopE and involved in bacterial invasion of epithelial cells. *J. Bacteriol.* **182:**2341–2344.

Barbieri, J. T., M. J. Riese, and K. Aktories. 2002. Bacterial toxins that modify the actin cytoskeleton. *Annu. Rev. Cell Dev. Biol.* **18:**315–344.

Barford, D., A. J. Flint, and N. K. Tonks. 1994. Crystal structure of human protein tyrosine phosphatase 1B. *Science* **263:**1397–1404.

Barford, D., Z. Jia, and N. K. Tonks. 1995. Protein tyrosine phosphatases take off. *Nat. Struct. Biol.* **2:**1043–1053.

Birtalan, S., and P. Ghosh. 2001. Structure of the *Yersinia* type III secretory system chaperone SycE. *Nat. Struct. Biol.* **8:**974–978.

Birtalan, S. C., R. M. Phillips, and P. Ghosh. 2002. Three-dimensional secretion signals in chaperone-effector complexes of bacterial pathogens. *Mol. Cell* **9:**971–980.

Black, D. S., and J. B. Bliska. 1997. Identification of p130Cas as a substrate of *Yersinia* YopH (Yop51), a bacterial protein tyrosine phosphatase that translocates into mammalian cells and targets focal adhesions. *EMBO J.* **16:**2730–2744.

Black, D. S., and J. B. Bliska. 2000. The RhoGAP activity of the *Yersinia pseudotuberculosis* cytotoxin YopE is required for antiphagocytic function and virulence. *Mol. Microbiol.* **37:**515–527.

Bliska, J. B. 1995. Crystal structure of the *Yersinia* tyrosine phosphatase. *Trends Microbiol.* **3:**125–127.

Bliska, J. B., K. L. Guan, J. E. Dixon, and S. Falkow. 1991. Tyrosine phosphate hydrolysis of host proteins by an essential *Yersinia* virulence determinant. *Proc. Natl. Acad. Sci. USA* **88:**1187–1191.

Blocker, A., P. Gounon, E. Larquet, K. Niebuhr, V. Cabiaux, C. Parsot, and P. Sansonetti. 1999. The tripartite type III secreton of Shigella flexneri inserts IpaB and IpaC into host membranes. *J. Cell Biol.* **147:**683–693.

Blocker, A., N. Jouihri, E. Larquet, P. Gounon, F. Ebel, C. Parsot, P. Sansonetti, and A. Allaoui. 2001. Structure and composition of the *Shigella flexneri* "needle complex," a part of its type III secreton. *Mol. Microbiol.* **39:**652–663.

Boland, A., M. P. Sory, M. Iriarte, C. Kerbourch, P. Wattiau, and G. R. Cornelis. 1996. Status of YopM and YopN in the *Yersinia* Yop virulon: YopM of *Y. enterocolitica* is internalized inside the cytosol of PU5–1.8 macrophages by the YopB, D, N delivery apparatus. *EMBO J.* **15:**5191–5201.

Bourne, H. R., D. A. Sanders, and F. McCormick. 1990. The GTPase superfamily: a conserved switch for diverse cell functions. *Nature* **348:**125–132.

Bronstein, P. A., E. A. Miao, and S. I. Miller. 2000. InvB is a type III secretion chaperone specific for SspA. *J. Bacteriol.* **182:**6638–6644.

Brown, I. R., J. W. Mansfield, S. Taira, E. Roine, and M. Romantschuk. 2001. Immunocytochemical localization of HrpA and HrpZ supports a role for the Hrp pilus in the transfer of effector proteins from *Pseudomonas syringae* pv. tomato across the host plant cell wall. *Mol. Plant-Microbe Interact.* **14:**394–404.

Buchwald, G., A. Friebel, J. E. Galan, W. D. Hardt, A. Wittinghofer, and K. Scheffzek. 2002. Structural basis for the reversible activation of a Rho protein by the bacterial toxin SopE. *EMBO J.* **21:**3286–3295.

Burghout, P., R. Van Boxtel, P. Van Gelder, P. Ringler, S. A. Muller, J. Tommassen, M. Koster. 2004. Structure and electrophysiological properties of the YscC secretin from the type III secretion system of *Yersinia enterocolitica*. *J. Bacteriol.* **186:**4645–4654.

Buttner, D., and U. Bonas. 2002a. Getting across—bacterial type III effector proteins on their way to the plant cell. *EMBO J.* **21:**5313–5322.

Buttner, D., and U. Bonas. 2002b. Port of entry—the type III secretion translocon. *Trends Microbiol.* **10:**186–192.

Buttner, D., D. Nennstiel, B. Klusener, and U. Bonas. 2002. Functional analysis of HrpF, a putative type III translocon protein from *Xanthomonas campestris* pv. vesicatoria. *J. Bacteriol.* **184:**2389–2398.

Cherfils, J., and P. Chardin. 1999. GEFs: structural basis for their activation of small GTP-binding proteins. *Trends Biochem. Sci.* **24:**306–311.

Cohen, M. L. 1992. Epidemiology of drug resistance: implications for a post-antimicrobial era. *Science* **257:**1050–1055.

Cohen, M. L. 1994. Antimicrobial resistance: prognosis for public health. *Trends Microbiol.* **2:**422–425.

Cohen, M. L. 2000. Changing patterns of infectious disease. *Nature* **406:**762–767.

Collier-Hyams, L. S., H. Zeng, J. Sun, A. D. Tomlinson, Z. Q. Bao, H. Chen, J. L. Madara, K. Orth, and A. S. Neish. 2002. Cutting edge: *Salmonella* AvrA effector inhibits the key proinflammatory, anti-apoptotic NF-kappa B pathway. *J. Immunol.* **169:**2846–2850.

Cordes, F. S., K. Komoriya, E. Larquet, S. Yang, E. H. Egelman, A. Blocker, and S. M. Lea. 2003. Helical structure of the needle of the type III secretion system of *Shigella flexneri*. *J. Biol. Chem.* **278:**17103–17107.

Cornelis, G. R. 2002. The *Yersinia* Ysc-Yop virulence apparatus. *Int. J. Med. Microbiol.* **291:**455–462.

Cornelis, G. R., and F. Van Gijsegem. 2000. Assembly and function of type III secretory systems. *Annu. Rev. Microbiol.* **54:**735–774.

Crago, A. M., and V. Koronakis. 1998. *Salmonella* InvG forms a ring-like multimer that requires the InvH lipoprotein for outer membrane localization. *Mol. Microbiol.* **30:**47–56.

Daniell, S. J., E. Kocsis, E. Morris, S. Knutton, F. P. Booy, and G. Frankel. 2003. 3D structure of EspA filaments from enteropathogenic *Escherichia coli. Mol. Microbiol.* **49:**301–308.

Daniell, S. J., N. Takahashi, R. Wilson, D. Friedberg, I. Rosenshine, F. P. Booy, R. K. Shaw, S. Knutton, G. Frankel, and S. Aizawa. 2001. The filamentous type III secretion translocon of enteropathogenic *Escherichia coli. Cell Microbiol.* **3:**865–871.

Darwin, K. H., L. S. Robinson, and V. L. Miller. 2001. SigE is a chaperone for the *Salmonella enterica* serovar Typhimurium invasion protein SigD. *J. Bacteriol.* **183:**1452–1454.

Dixon, J. E. 1995. Structure and catalytic properties of protein tyrosine phosphatases. *Ann. N.Y. Acad. Sci.* **766:** 18–22.

Elliott, S. J., S. W. Hutcheson, M. S. Dubois, J. L. Mellies, L. A. Wainwright, M. Batchelor, G. Frankel, S. Knutton, and J. B. Kaper. 1999. Identification of CesT, a chaperone for the type III secretion of Tir in enteropathogenic *Escherichia coli. Mol. Microbiol.* **33:**1176–1189.

Evdokimov, A. G., D. E. Anderson, K. M. Routzahn, and D. S. Waugh. 2001a. Unusual molecular architecture of the *Yersinia pestis* cytotoxin YopM: a leucine-rich repeat protein with the shortest repeating unit. *J. Mol. Biol.* **312:**807–821.

Evdokimov, A. G., J. E. Tropea, K. M. Routzahn, T. D. Copeland, and D. S. Waugh. 2001b. Structure of the N-terminal domain of *Yersinia pestis* YopH at 2.0 Å resolution. *Acta Crystallogr. Ser. D* **57:**793–799.

Evdokimov, A. G., J. E. Tropea, K. M. Routzahn, and D. S. Waugh. 2002a. Crystal structure of the *Yersinia pestis* GTPase activator YopE. *Protein Sci.* **11:**401–408.

Evdokimov, A. G., J. E. Tropea, K. M. Routzahn, and D. S. Waugh. 2002b. Three-dimensional structure of the type III secretion chaperone SycE from *Yersinia pestis. Acta Crystallogr. Ser. D* **58:**398–406.

Fauman, E. B., and M. A. Saper. 1996. Structure and function of the protein tyrosine phosphatases. *Trends Biochem. Sci.* **21:**413–417.

Fauman, E. B., C. Yuvaniyama, H. L. Schubert, J. A. Stuckey, and M. A. Saper. 1996. The X-ray crystal structures of *Yersinia* tyrosine phosphatase with bound tungstate and nitrate. Mechanistic implications. *J. Biol. Chem.* **271:**18780–18788.

Fraser, C. M., and M. R. Dando. 2001. Genomics and future biological weapons: the need for preventive action by the biomedical community. *Nat. Genet.* **29:**253–256.

Fu, Y., and J. E. Galan. 1998. Identification of a specific chaperone for SptP, a substrate of the centisome 63 type III secretion system of *Salmonella typhimurium. J. Bacteriol.* **180:**3393–3399.

Fu, Y., and J. E. Galan. 1999. A salmonella protein antagonizes Rac-1 and Cdc42 to mediate host-cell recovery after bacterial invasion. *Nature* **401:**293–297.

Galan, J. E. 2001. Salmonella interactions with host cells: type III secretion at work. *Annu. Rev. Cell Dev. Biol.* **17:**53–86.

Galan, J. E., and A. Collmer. 1999. Type III secretion machines: bacterial devices for protein delivery into host cells. *Science* **284:**1322–1328.

Galan, J. E., and D. Zhou. 2000. Striking a balance: modulation of the actin cytoskeleton by *Salmonella. Proc. Natl. Acad. Sci. USA* **97:**8754–8761.

Galyov, E. E., S. Hakansson, A. Forsberg, and H. Wolf-Watz. 1993. A secreted protein kinase of *Yersinia pseudotuberculosis* is an indispensable virulence determinant. *Nature* **361:**730–732.

Gamblin, S. J., and S. J. Smerdon. 1998. GTPase-activating proteins and their complexes. *Curr. Opin. Struct. Biol.* **8:**195–201.

Geiser, T. K., B. I. Kazmierczak, L. K. Garrity-Ryan, M. A. Matthay, and J. N. Engel. 2001. *Pseudomonas aeruginosa* ExoT inhibits in vitro lung epithelial wound repair. *Cell. Microbiol.* **3:**223–236.

Ginocchio, C. C., S. B. Olmsted, C. L. Wells, and J. E. Galan. 1994. Contact with epithelial cells induces the formation of surface appendages on *Salmonella typhimurium. Cell* **76:**717–724.

Goehring, U. M., G. Schmidt, K. J. Pederson, K. Aktories, and J. T. Barbieri. 1999. The N-terminal domain of *Pseudomonas aeruginosa* exoenzyme S is a GTPase-activating protein for Rho GTPases. *J. Biol. Chem.* **274:**36369–36372.

Grosdent, N., I. Maridonneau-Parini, M. P. Sory, and G. R. Cornelis. 2002. Role of Yops and adhesins in resistance of *Yersinia enterocolitica* to phagocytosis. *Infect. Immun.* **70:**4165–4176.

Guan, K. L., and J. E. Dixon. 1990. Protein tyrosine phosphatase activity of an essential virulence determinant in *Yersinia. Science* **249:**553–556.

Guan, K. L., and J. E. Dixon. 1993. Bacterial and viral protein tyrosine phosphatases. *Semin. Cell Biol.* **4**:389–396.

Hakansson, S., E. E. Galyov, R. Rosqvist, and H. Wolf-Watz. 1996a. The *Yersinia* YpkA Ser/Thr kinase is translocated and subsequently targeted to the inner surface of the HeLa cell plasma membrane. *Mol. Microbiol.* **20**:593–603.

Hakansson, S., K. Schesser, C. Persson, E. E. Galyov, R. Rosqvist, F. Homble, and H. Wolf-Watz. 1996b. The YopB protein of *Yersinia pseudotuberculosis* is essential for the translocation of Yop effector proteins across the target cell plasma membrane and displays a contact-dependent membrane disrupting activity. *EMBO J.* **15**:5812–5823.

Hall, A. 1998. Rho GTPases and the actin cytoskeleton. *Science* **279**:509–514.

Hamburger, Z. A., M. S. Brown, R. R. Isberg, and P. J. Bjorkman. 1999. Crystal structure of invasin: a bacterial integrin-binding protein. *Science* **286**:291–295.

Hamid, N., A. Gustavsson, K. Andersson, K. McGee, C. Persson, C. E. Rudd, and M. Fallman. 1999. YopH dephosphorylates Cas and Fyn-binding protein in macrophages. *Microb. Pathog.* **27**:231–242.

Han, S., A. S. Arvai, S. B. Clancy, and J. A. Tainer. 2001. Crystal structure and novel recognition motif of rho ADP-ribosylating C3 exoenzyme from *Clostridium botulinum:* structural insights for recognition specificity and catalysis. *J. Mol. Biol.* **305**:95–107.

Han, S., J. A. Craig, C. D. Putnam, N. B. Carozzi, and J. A. Tainer. 1999. Evolution and mechanism from structures of an ADP-ribosylating toxin and NAD complex. *Nat. Struct. Biol.* **6**:932–936.

Hardt, W. D., L. M. Chen, K. E. Schuebel, X. R. Bustelo, and J. E. Galan. 1998a. *S. typhimurium* encodes an activator of Rho GTPases that induces membrane ruffling and nuclear responses in host cells. *Cell* **93**:815–826.

Hardt, W. D., H. Urlaub, and J. E. Galan. 1998b. A substrate of the centisome 63 type III protein secretion system of *Salmonella typhimurium* is encoded by a cryptic bacteriophage. *Proc. Natl. Acad. Sci. USA* **95**:2574–2579.

Hartl, F. U., and M. Hayer-Hartl. 2002. Molecular chaperones in the cytosol: from nascent chain to folded protein. *Science* **295**:1852–1858.

Hayward, R. D., and V. Koronakis. 1999. Direct nucleation and bundling of actin by the SipC protein of invasive *Salmonella. EMBO J.* **18**:4926–4934.

Hayward, R. D., and V. Koronakis. 2002. Direct modulation of the host cell cytoskeleton by Salmonella actin-binding proteins. *Trends Cell Biol.* **12**:15–20.

Hayward, R. D., E. J. McGhie, and V. Koronakis. 2000. Membrane fusion activity of purified SipB, a Salmonella surface protein essential for mammalian cell invasion. *Mol. Microbiol.* **37**:727–739.

He, S. Y., and Q. Jin. 2003. The Hrp pilus: learning from flagella. *Curr. Opin. Microbiol.* **6**:15–19.

Higashide, W., S. Dai, V. P. Hombs, and D. Zhou. 2002. Involvement of SipA in modulating actin dynamics during *Salmonella* invasion into cultured epithelial cells. *Cell. Microbiol.* **4**:357–365.

Hirshberg, M., R. W. Stockley, G. Dodson, and M. R. Webb. 1997. The crystal structure of human rac1, a member of the rho-family complexed with a GTP analogue. *Nat. Struct. Biol.* **4**:147–152.

Hoiczyk, E., and G. Blobel. 2001. Polymerization of a single protein of the pathogen *Yersinia enterocolitica* into needles punctures eukaryotic cells. *Proc. Natl. Acad. Sci. USA* **98**:4669–4674.

Holm, L., and C. Sander. 1993. Protein structure comparison by alignment of distance matrices. *J. Mol. Biol.* **233**:123–138.

Hu, W., J. Yuan, Q. L. Jin, P. Hart, and S. Y. He. 2001. Immunogold labeling of Hrp pili of *Pseudomonas syringae* pv. tomato assembled in minimal medium and in planta. *Mol. Plant-Microbe Interact.* **14**:234–241.

Hueck, C. J. 1998. Type III protein secretion systems in bacterial pathogens of animals and plants. *Microbiol. Mol. Biol. Rev.* **62**:379–433.

Ihara, K., S. Muraguchi, M. Kato, T. Shimizu, M. Shirakawa, S. Kuroda, K. Kaibuchi, and T. Hakoshima. 1998. Crystal structure of human RhoA in a dominantly active form complexed with a GTP analogue. *J. Biol. Chem.* **273**:9656–9666.

Iriarte, M., and G. R. Cornelis. 1998. YopT, a new *Yersinia* Yop effector protein, affects the cytoskeleton of host cells. *Mol. Microbiol.* **29**:915–929.

Isberg, R. R., and J. M. Leong. 1988. Cultured mammalian cells attach to the invasin protein of *Yersinia pseudotuberculosis. Proc. Natl. Acad. Sci. USA* **85**:6682–6686.

Isberg, R. R., D. L. Voorhis, and S. Falkow. 1987. Identification of invasin: a protein that allows enteric bacteria to penetrate cultured mammalian cells. *Cell* **50:**769–778.

Jepson, M. A., B. Kenny, and A. D. Leard. 2001. Role of *sipA* in the early stages of *Salmonella typhimurium* entry into epithelial cells. *Cell. Microbiol.* **3:**417–426.

Jia, Z., D. Barford, A. J. Flint, and N. K. Tonks. 1995. Structural basis for phosphotyrosine peptide recognition by protein tyrosine phosphatase 1B. *Science* **268:**1754–1758.

Jin, Q., and S. Y. He. 2001. Role of the Hrp pilus in type III protein secretion in *Pseudomonas syringae. Science* **294:**2556–2558.

Jin, Q., W. Hu, I. Brown, G. McGhee, P. Hart, A. L. Jones, and S. Y. He. 2001. Visualization of secreted Hrp and Avr proteins along the Hrp pilus during type III secretion in *Erwinia amylovora* and *Pseudomonas syringae. Mol. Microbiol.* **40:**1129–1139.

Journet, L., C. Agrain, P. Broz, and G. R. Cornelis. 2003. The needle length of bacterial injectisomes is determined by a molecular ruler. *Science* **302:**1757–1760.

Kaniga, K., J. Uralil, J. B. Bliska, and J. E. Galan. 1996. A secreted protein tyrosine phosphatase with modular effector domains in the bacterial pathogen *Salmonella typhimurium. Mol. Microbiol.* **21:**633–641.

Khandelwal, P., K. Keliikuli, C. L. Smith, M. A. Saper, and E. R. Zuiderweg. 2002. Solution structure and phosphopeptide binding to the N-terminal domain of *Yersinia* YopH: comparison with a crystal structure. *Biochemistry* **41:**11425–11437.

Kimbrough, T. G., and S. I. Miller. 2000. Contribution of *Salmonella typhimurium* type III secretion components to needle complex formation. *Proc. Natl. Acad. Sci. USA* **97:**11008–11013.

Kimbrough, T. G., and S. I. Miller. 2002. Assembly of the type III secretion needle complex of *Salmonella typhimurium. Microbes Infect.* **4:**75–82.

Knutton, S., I. Rosenshine, M. J. Pallen, I. Nisan, B. C. Neves, C. Bain, C. Wolff, G. Dougan, and G. Frankel. 1998. A novel EspA-associated surface organelle of enteropathogenic *Escherichia coli* involved in protein translocation into epithelial cells. *EMBO J.* **17:**2166–2176.

Kobe, B., and A. V. Kajava. 2001. The leucine-rich repeat as a protein recognition motif. *Curr. Opin. Struct. Biol.* **11:**725–732.

Kubori, T., Y. Matsushima, D. Nakamura, J. Uralil, M. Lara-Tejero, A. Sukhan, J. E. Galan, and S. I. Aizawa. 1998. Supramolecular structure of the *Salmonella typhimurium* type III protein secretion system. *Science* **280:**602–605.

Kubori, T., A. Sukhan, S. I. Aizawa, and J. E. Galan. 2000. Molecular characterization and assembly of the needle complex of the *Salmonella typhimurium* type III protein secretion system. *Proc. Natl. Acad. Sci. USA* **97:**10225–10230.

Kubori, T., and J. E. Galan. 2003. Temporal regulation of salmonella virulence effector function by proteasome-dependent protein degradation. *Cell* **115:**333–342.

Leong, J. M., R. S. Fournier, and R. R. Isberg. 1991. Mapping and topographic localization of epitopes of the *Yersinia pseudotuberculosis* invasin protein. *Infect. Immun.* **59:**3424–3433.

Lerm, M., G. Schmidt, and K. Aktories. 2000. Bacterial protein toxins targeting rho GTPases. *FEMS Microbiol. Lett.* **188:**1–6.

Lesnick, M. L., N. E. Reiner, J. Fierer, and D. G. Guiney. 2001. The *Salmonella spvB* virulence gene encodes an enzyme that ADP-ribosylates actin and destabilizes the cytoskeleton of eukaryotic cells. *Mol. Microbiol.* **39:**1464–1470.

Leung, K. Y., and S. C. Straley. 1989. The yopM gene of Yersinia pestis encodes a released protein having homology with the human platelet surface protein GPIb alpha. *J. Bacteriol.* **171:**4623–4632.

Lilic, M., V. E. Galkin, A. Orlova, M. S. VanLoock, E. H. Egelman, and C. E. Stebbins. 2003. *Salmonella* SipA polymerizes actin by stapling filaments with nonglobular protein arms. *Science* **301:**1918–1921.

Luo, Y., M. G. Bertero, E. A. Frey, R. A. Pfuetzner, M. R. Wenk, L. Creagh, S. L. Marcus, D. Lim, F. Sicheri, C. Kay, C. Haynes, B. B. Finlay, and N. C. Strynadka. 2001. Structural and biochemical characterization of the type III secretion chaperones CesT and SigE. *Nat. Struct. Biol.* **8:**1031–1036.

Marcus, S. L., M. R. Wenk, O. Steele-Mortimer, and B. B. Finlay. 2001. A synaptojanin-homologous region of *Salmonella typhimurium* SigD is essential for inositol phosphatase activity and Akt activation. *FEBS Lett.* **494:**201–207.

McDonald, C., P. O. Vacratsis, J. B. Bliska, and J. E. Dixon. 2003. The *Yersinia* virulence factor YopM forms a novel protein complex with two cellular kinases. *J. Biol. Chem.* **278:**18514–18523.

McGhie, E. J., R. D. Hayward, and V. Koronakis. 2001. Cooperation between actin-binding proteins of invasive *Salmonella:* SipA potentiates SipC nucleation and bundling of actin. *EMBO J.* **20:**2131–2139.

Meijer, L. K., K. Schesser, H. Wolf-Watz, P. Sassone-Corsi, and S. Pettersson. 2000. The bacterial protein YopJ abrogates multiple signal transduction pathways that converge on the transcription factor CREB. *Cell. Microbiol.* **2:**231–238.

Monack, D. M., J. Mecsas, N. Ghori, and S. Falkow. 1997. *Yersinia* signals macrophages to undergo apoptosis and YopJ is necessary for this cell death. *Proc. Natl. Acad. Sci. USA* **94:**10385–10390.

Montagna, L. G., M. I. Ivanov, and J. B. Bliska. 2001. Identification of residues in the N-terminal domain of the *Yersinia* tyrosine phosphatase that are critical for substrate recognition. *J. Biol. Chem.* **276:**5005–5011.

Murli, S., R. O. Watson, and J. E. Galan. 2001. Role of tyrosine kinases and the tyrosine phosphatase SptP in the interaction of *Salmonella* with host cells. *Cell. Microbiol.* **3:**795–810.

Nassar, N., G. R. Hoffman, D. Manor, J. C. Clardy, and R. A. Cerione. 1998. Structures of Cdc42 bound to the active and catalytically compromised forms of Cdc42GAP. *Nat. Struct. Biol.* **5:**1047–1052.

Neves, B. C., S. Knutton, L. R. Trabulsi, V. Sperandio, J. B. Kaper, G. Dougan, and G. Frankel. 1998. Molecular and ultrastructural characterisation of EspA from different enteropathogenic *Escherichia coli* serotypes. *FEMS Microbiol. Lett.* **169:**73–80.

Neyt, C., and G. R. Cornelis. 1999. Insertion of a Yop translocation pore into the macrophage plasma membrane by *Yersinia enterocolitica:* requirement for translocators YopB and YopD, but not LcrG. *Mol. Microbiol.* **33:**971–981.

Nixdorff, K., J. Brauburger, and D. Hahlbohm. 2000. The biotechnology revolution: the science and applications. *NATO ASI Ser.* **32:**77–124.

Norris, F. A., M. P. Wilson, T. S. Wallis, E. E. Galyov, and P. W. Majerus. 1998. SopB, a protein required for virulence of *Salmonella dublin,* is an inositol phosphate phosphatase. *Proc. Natl. Acad. Sci. USA* **95:**14057–14059.

Opalka, N., R. Beckmann, N. Boisset, M. N. Simon, M. Russel, S. A. Darst. 2003. Structure of the filamentous phage pIV multimer by cryo-electron microscopy. *J. Mol. Biol.* **325:**461–470.

Orth, K., Z. Xu, M. B. Mudgett, Z. Q. Bao, L. E. Palmer, J. B. Bliska, W. F. Mangel, B. Staskawicz, and J. E. Dixon. 2000. Disruption of signaling by *Yersinia* effector YopJ, a ubiquitin-like protein protease. *Science* **290:**1594–1597.

Osiecki, J. C., J. Barker, W. L. Picking, A. B. Serfis, E. Berring, S. Shah, A. Harrington, and W. D. Picking. 2001. IpaC from *Shigella* and SipC from *Salmonella* possess similar biochemical properties but are functionally distinct. *Mol. Microbiol.* **42:**469–481.

Page, A. L., P. Sansonetti, and C. Parsot. 2002. Spa15 of *Shigella flexneri,* a third type of chaperone in the type III secretion pathway. *Mol. Microbiol.* **43:**1533–1542.

Page, A. L., and C. Parsot. 2002. Chaperones of the type III secretion pathway: jacks of all trades. *Mol. Microbiol.* **46:**1–11.

Palmer, L. E., S. Hobbie, J. E. Galan, and J. B. Bliska. 1998. YopJ of *Yersinia pseudotuberculosis* is required for the inhibition of macrophage TNF-alpha production and downregulation of the MAP kinases p38 and JNK. *Mol. Microbiol.* **27:**953–965.

Palmer, L. E., A. R. Pancetti, S. Greenberg, and J. B. Bliska. 1999. YopJ of *Yersinia* spp. is sufficient to cause downregulation of multiple mitogen-activated protein kinases in eukaryotic cells. *Infect. Immun.* **67:**708–716.

Pang, T., Z. A. Bhutta, B. B. Finlay, and M. Altwegg. 1995. Typhoid fever and other salmonellosis: a continuing challenge. *Trends Microbiol.* **3:**253–255.

Perry, R. D., and J. D. Fetherston. 1997. *Yersinia pestis*—etiologic agent of plague. *Clin. Microbiol. Rev.* **10:** 35–66.

Persson, C., N. Carballeira, H. Wolf-Watz, and M. Fallman. 1997. The PTPase YopH inhibits uptake of *Yersinia,* tyrosine phosphorylation of p130Cas and FAK, and the associated accumulation of these proteins in peripheral focal adhesions. *EMBO J.* **16:**2307–2318.

Rittinger, K., P. A. Walker, J. F. Eccleston, K. Nurmahomed, D. Owen, E. Laue, S. J. Gamblin, and S. J. Smerdon. 1997a. Crystal structure of a small G protein in complex with the GTPase-activating protein rhoGAP. *Nature* **388:**693–697.

Rittinger, K., P. A. Walker, J. F. Eccleston, S. J. Smerdon, and S. J. Gamblin. 1997b. Structure at 1.65 A of RhoA and its GTPase-activating protein in complex with a transition-state analogue. *Nature* **389:**758–762.

Roine, E., W. Wei, J. Yuan, E. L. Nurmiaho-Lassila, N. Kalkkinen, M. Romantschuk, and S. Y. He. 1997. Hrp pilus: an *hrp*-dependent bacterial surface appendage produced by *Pseudomonas syringae* pv. tomato DC3000. *Proc. Natl. Acad. Sci. USA* **94:**3459–3464.

Rudolph, M. G., C. Weise, S. Mirold, B. Hillenbrand, B. Bader, A. Wittinghofer, and W. D. Hardt. 1999. Biochemical analysis of SopE from *Salmonella typhimurium,* a highly efficient guanosine nucleotide exchange factor for RhoGTPases. *J. Biol. Chem.* **274:**30501–30509.

Samatey, F. A., K. Imada, S. Nagashima, F. Vonderviszt, T. Kumasaka, M. Yamamoto, and K. Namba. 2001. Structure of the bacterial flagellar protofilament and implications for a switch for supercoiling. *Nature* **410:**331–337.

Schesser, K., A.-K. Spiik, J.-M. Dukuzumuremyi, M. F. Neurath, S. Pettersson, and H. Wolf-Watz. 1998. The *yopJ* locus is required for *Yersinia*-mediated inhibition of NF-kappaB activation and cytokine expression: YopJ contains a eukaryotic SH2–like domain that is essential for its repressive activity. *Mol. Microbiol.* **28:** 1067–1079.

Schuch, R., and A. T. Maurelli. 2001. MxiM and MxiJ, base elements of the Mxi-Spa type III secretion system of *Shigella,* interact with and stabilize the MxiD secretin in the cell envelope. *J. Bacteriol.* **183:**6991–6998.

Sekiya, K., M. Ohishi, T. Ogino, K. Tamano, C. Sasakawa, and A. Abe. 2001. Supermolecular structure of the enteropathogenic *Escherichia coli* type III secretion system and its direct interaction with the EspA-sheath-like structure. *Proc. Natl. Acad. Sci. USA* **98:**11638–11643.

Shao, F., P. M. Merritt, Z. Bao, R. W. Innes, and J. E. Dixon. 2002. A *Yersinia* effector and a *Pseudomonas* avirulence protein define a family of cysteine proteases functioning in bacterial pathogenesis. *Cell* **109:**575–588.

Skrzypek, E., C. Cowan, and S. C. Straley. 1998. Targeting of the *Yersinia pestis* YopM protein into HeLa cells and intracellular trafficking to the nucleus. *Mol. Microbiol.* **30:**1051–1065.

Smith, C. L., P. Khandelwal, K. Keliikuli, E. R. Zuiderweg, and M. A. Saper. 2001. Structure of the type III secretion and substrate-binding domain of *Yersinia* YopH phosphatase. *Mol. Microbiol.* **42:**967–979.

Sory, M. P., A. Boland, I. Lambermont, and G. R. Cornelis. 1995. Identification of the YopE and YopH domains required for secretion and internalization into the cytosol of macrophages, using the cyaA gene fusion approach. *Proc. Natl. Acad. Sci. USA* **92:**11998–12002.

Sprang, S. R. 1997. G proteins, effectors and GAPs: structure and mechanism. *Curr. Opin. Struct. Biol.* **7:**849–856.

Stebbins, C. E., and J. E. Galan. 2000. Modulation of host signaling by a bacterial mimic: structure of the *Salmonella* effector SptP bound to Rac1. *Mol. Cell* **6:**1449–1460.

Stebbins, C. E., and J. E. Galan. 2001a. Maintenance of an unfolded polypeptide by a cognate chaperone in bacterial type III secretion. *Nature* **414:**77–81.

Stebbins, C. E., and J. E. Galan. 2001b. Structural mimicry in bacterial virulence. *Nature* **412:**701–705.

Steele-Mortimer, O., J. H. Brumell, L. A. Knodler, S. Meresse, A. Lopez, and B. B. Finlay. 2002. The invasion-associated type III secretion system of *Salmonella enterica* serovar Typhimurium is necessary for intracellular proliferation and vacuole biogenesis in epithelial cells. *Cell. Microbiol.* **4:**43–54.

Steele-Mortimer, O., L. A. Knodler, S. L. Marcus, M. P. Scheid, B. Goh, C. G. Pfeifer, V. Duronio, and B. B. Finlay. 2000. Activation of Akt/protein kinase B in epithelial cells by the *Salmonella typhimurium* effector sigD. *J. Biol. Chem.* **275:**37718–37724.

Stender, S., A. Friebel, S. Linder, M. Rohde, S. Mirold, and W. D. Hardt. 2000. Identification of SopE2 from *Salmonella typhimurium,* a conserved guanine nucleotide exchange factor for Cdc42 of the host cell. *Mol. Microbiol.* **36:**1206–1221.

Stuckey, J. A., H. L. Schubert, E. B. Fauman, Z. Y. Zhang, J. E. Dixon, and M. A. Saper. 1994. Crystal structure of *Yersinia* protein tyrosine phosphatase at 2.5 Å and the complex with tungstate. *Nature* **370:**571–575.

Sukhan, A., T. Kubori, J. Wilson, and J. E. Galan. 2001. Genetic analysis of assembly of the *Salmonella enterica* serovar Typhimurium type III secretion-associated needle complex. *J. Bacteriol.* **183:**1159–1167.

Tamano, K., S. Aizawa, E. Katayama, T. Nonaka, S. Imajoh-Ohmi, A. Kuwae, S. Nagai, and C. Sasakawa. 2000. Supramolecular structure of the *Shigella* type III secretion machinery: the needle part is changeable in length and essential for delivery of effectors. *EMBO J.* **19:**3876–3887.

Tardy, F., F. Homble, C. Neyt, R. Wattiez, G. R. Cornelis, J. M. Ruysschaert, and V. Cabiaux. 1999. *Yersinia enterocolitica* type III secretion-translocation system: channel formation by secreted Yops. *EMBO J.* **18:** 6793–6799.

Terebiznik, M. R., O. V. Vieira, S. L. Marcus, A. Slade, C. M. Yip, W. S. Trimble, T. Meyer, B. B. Finlay, and S. Grinstein. 2002. Elimination of host cell PtdIns(4,5)P(2) by bacterial SigD promotes membrane fission during invasion by *Salmonella. Nat. Cell Biol.* **4:**766–773.

Tezcan-Merdol, D., T. Nyman, U. Lindberg, F. Haag, F. Koch-Nolte, and M. Rhen. 2001. Actin is ADP-ribosylated by the *Salmonella enterica* virulence-associated protein SpvB. *Mol. Microbiol.* **39:**606–619.

Tsuge, H., M. Nagahama, H. Nishimura, J. Hisatsune, Y. Sakaguchi, Y. Itogawa, N. Katunuma, and J. Sakurai. 2003. Crystal structure and site-directed mutagenesis of enzymatic components from *Clostridium perfringens* iota-toxin. *J. Mol. Biol.* **325:**471–483.

van Eerde, A., C. Hamiaux, J. Perez, C. Parsot, and B. W. Dijkstra. 2004. Structure of Spa15, a type III secretion chaperone from *Shigella flexneri* with broad specificity. *EMBO Rep.* **5:**477–483.

Van Gijsegem, F., J. Vasse, J. C. Camus, M. Marenda, and C. Boucher. 2000. *Ralstonia solanacearum* produces hrp-dependent pili that are required for PopA secretion but not for attachment of bacteria to plant cells. *Mol. Microbiol.* **36:**249–260.

Van Gijsegem, F., J. Vasse, R. De Rycke, P. Castello, and C. Boucher. 2002. Genetic dissection of Ralstonia solanacearum hrp gene cluster reveals that the HrpV and HrpX proteins are required for Hrp pilus assembly. *Mol. Microbiol.* **44:**935–946.

Van Nhieu, G. T., and R. R. Isberg. 1991. The *Yersinia pseudotuberculosis* invasin protein and human fibronectin bind to mutually exclusive sites on the alpha 5 beta 1 integrin receptor. *J. Biol. Chem.* **266:**24367–24375.

Vetter, I. R., and A. Wittinghofer. 2001. The guanine nucleotide-binding switch in three dimensions. *Science* **294:**1299–1304.

Wachter, C., C. Beinke, M. Mattes, and M. A. Schmidt. 1999. Insertion of EspD into epithelial target cell membranes by infecting enteropathogenic *Escherichia coli*. *Mol. Microbiol.* **31:**1695–1707.

Wattiau, P., S. Woestyn, and G. R. Cornelis. 1996. Customized secretion chaperones in pathogenic bacteria. *Mol. Microbiol.* **20:**255–262.

Wei, W., A. Plovanich-Jones, W. L. Deng, Q. L. Jin, A. Collmer, H. C. Huang, and S. Y. He. 2000. The gene coding for the Hrp pilus structural protein is required for type III secretion of Hrp and Avr proteins in *Pseudomonas syringae* pv. tomato. *Proc. Natl. Acad. Sci. USA* **97:**2247–2252.

Whitman, W. B., D. C. Coleman, and W. J. Wiebe. 1998. Prokaryotes: the unseen majority. *Proc. Natl. Acad. Sci. USA* **95:**6578–6583.

Wittinghofer, A., and E. F. Pai. 1991. The structure of Ras protein: a model for a universal molecular switch. *Trends Biochem. Sci.* **16:**382–387.

Woestyn, S., A. Allaoui, P. Wattiau, and G. R. Cornelis. 1994. YscN, the putative energizer of the *Yersinia* Yop secretion machinery. *J. Bacteriol.* **176:**1561–1569.

Woestyn, S., M. P. Sory, A. Boland, O. Lequenne, and G. R. Cornelis. 1996. The cytosolic SycE and SycH chaperones of *Yersinia* protect the region of YopE and YopH involved in translocation across eukaryotic cell membranes. *Mol. Microbiol.* **20:**1261–1271.

Wood, M. W., R. Rosqvist, P. B. Mullan, M. H. Edwards, and E. E. Galyov. 1996. SopE, a secreted protein of *Salmonella dublin*, is translocated into the target eukaryotic cell via a *sip*-dependent mechanism and promotes bacterial entry. *Mol. Microbiol.* **22:**327–338.

Worthylake, D. K., K. L. Rossman, and J. Sondek. 2000. Crystal structure of Rac1 in complex with the guanine nucleotide exchange region of Tiam1. *Nature* **408:**682–688.

Wurtele, M., L. Renault, J. T. Barbieri, A. Wittinghofer, and E. Wolf. 2001a. Structure of the ExoS GTPase activating domain. *FEBS Lett.* **491:**26–29.

Wurtele, M., E. Wolf, K. J. Pederson, G. Buchwald, M. R. Ahmadian, J. T. Barbieri, and A. Wittinghofer. 2001b. How the *Pseudomonas aeruginosa* ExoS toxin downregulates Rac. *Nat. Struct. Biol.* **8:**23–26.

Yonekura, K., S. Maki-Yonekura, and K. Namba. 2003. Complete atomic model of the bacterial flagellar filament by electron cryomicroscopy. *Nature* **424:**643–650.

Young, V. B., V. L. Miller, S. Falkow, and G. K. Schoolnik. 1990. Sequence, localization and function of the invasin protein of *Yersinia enterocolitica*. *Mol. Microbiol.* **4:**1119–1128.

Zaharik, M. L., S. Gruenheid, A. J. Perrin, and B. B. Finlay. 2002. Delivery of dangerous goods: type III secretion in enteric pathogens. *Int. J. Med. Microbiol.* **291:**593–603.

Zhang, S., R. L. Santos, R. M. Tsolis, S. Stender, W. D. Hardt, A. J. Baumler, and L. G. Adams. 2002. The *Salmonella enterica* serotype Typhimurium effector proteins SipA, SopA, SopB, SopD, and SopE2 act in concert to induce diarrhea in calves. *Infect. Immun.* **70:**3843–3855.

Zhang, Z. Y. 2002. Protein tyrosine phosphatases: structure and function, substrate specificity, and inhibitor development. *Annu. Rev. Pharmacol. Toxicol.* **42:**209–234.

Zhou, D., L. M. Chen, L. Hernandez, S. B. Shears, and J. E. Galan. 2001. A *Salmonella* inositol polyphosphatase acts in conjunction with other bacterial effectors to promote host cell actin cytoskeleton rearrangements and bacterial internalization. *Mol. Microbiol.* **39:**248–259.

Zhou, D., and J. Galan. 2001. *Salmonella* entry into host cells: the work in concert of type III secreted effector proteins. *Microbes Infect.* **3:**1293–1298.

Zhou, D., M. S. Mooseker, and J. E. Galan. 1999a. An invasion-associated Salmonella protein modulates the actin-bundling activity of plastin. *Proc. Natl. Acad. Sci. USA* **96:**10176–10181.

Zhou, D., M. S. Mooseker, and J. E. Galan. 1999b. Role of the *S. typhimurium* actin-binding protein SipA in bacterial internalization. *Science* **283:**2092–2095.

Zumbihl, R., M. Aepfelbacher, A. Andor, C. A. Jacobi, K. Ruckdeschel, B. Rouot, and J. Heesemann. 1999. The cytotoxin YopT of *Yersinia enterocolitica* induces modification and cellular redistribution of the small GTP-binding protein RhoA. *J. Biol. Chem.* **274:**29289–29293.

Structural Biology of Bacterial Pathogenesis
Edited by G. Waksman et al.
© 2005 ASM Press, Washington, D.C.

Chapter 10

Type IV Secretion Machinery

Gunnar Schröder, Savvas N. Savvides, Gabriel Waksman, and Erich Lanka

Type IV secretion systems (T4SS) are multicomponent transport machineries mediating the intercellular transfer of specific substrates from bacterial donors to prokaryotic or eukaryotic target cells. The implication of T4SS in the pathogenesis of a large number of human, animal, and plant pathogens (Table 1), as well as the major contribution of T4SS to genetic exchange through bacterial conjugation and DNA uptake and release, has attracted growing interest in this transport system for the past few decades (Cascales and Christie, 2003a; Ding et al., 2003).

The definition of T4SS is historically based on the sequence similarity found in the *Agrobacterium tumefaciens* and *Bordetella pertussis* VirB secretion systems (Salmond, 1994). The conserved components of T4SS are ancestrally related to components of bacterial conjugation systems (Lessl et al., 1992a; Lessl and Lanka, 1994), the first of which was discovered as early as 1946 (Lederberg and Tatum, 1946). The similarity between bacterial conjugation systems and the components of other T4SS extends to a set of 9 to 11 conserved components belonging to the mating pair formation (Mpf) complex (Willetts, 1981). This complex consists of a postulated membrane channel structure topped by a surface-exposed pilus (Grahn et al., 2000). An additional component conserved in most T4SS is the type IV secretion-coupling protein (T4CP), also known as the TraG-like or VirD4-like protein (Cabezón et al., 1997).

In this chapter, we provide an overview of the pathogenic involvement of T4SS, with a focus on the structural aspects and the molecular mechanisms of the type IV secretion process. The assembled knowledge gained from studies of individual secretion systems is used to propose a unified view of the architecture of the type IV secretion machinery at the molecular level. Different mechanistic aspects are discussed.

Gunnar Schröder • Max-Planck-Institut für Molekulare Genetik, Ihnestraße 73, Dahlem, D-14195 Berlin, Germany, and Division of Molecular Microbiology, Biozentrum, University of Basel, Klingelbergstraße 50-70, CH-4056 Basel, Switzerland.   ***Savvas N. Savvides*** • Laboratory for Protein Biochemistry, Ghent University, K. L. Ledeganckstraat 35, B-9000 Ghent, Belgium.   ***Gabriel Waksman*** • Institute of Structural Molecular Biology, Birkbeck and University College London, Malet St., London WC1E 7HX, United Kingdom. ***Erich Lanka*** • Max-Planck-Institut für Molekulare Genetik, Ihnestraße 73, Dahlem, D-14195 Berlin, Germany.

Table 1. T4SS involved in pathogenicity

Donor bacterium (plasmid)	T4SS components	Known substrate(s)	Target cells	T4SS-associated disease
Agrobacterium tumefaciens (pTi)	VirB/VirD4	T-DNA, VirD2, VirE2, VirE3, VirF	Plant cells	Crown gall tumor
Bartonella henselae	VirB/VirD4	Bep proteins	Erythrocytes, vascular epithelial cells of mammals	Intraerythrocytic bacteremia, cat scratch disease
Bordetella pertussis	Ptl	Pertussis toxin	Human ciliated respiratory-epithelial cells	Whooping cough
Brucella suis	VirB	?	Macrophages	Brucellosis
Campylobacter jejuni (pVir)	VirB	?	Intestinal epithelial cells	Diarrheal disease
Coxiella burnetii	Dot/Icm[a]	?	Mononuclear phagocytes	Q fever
Helicobacter pylori	Cag	CagA	Gastric epithelial cells	Gastric diseases
Legionella pneumophila	Dot/Icm	LidA, RalF, SidA–SidH, SdbA–SdbD, SdcA–SdcB, SdeA–SdeE, SdhA–SdhC	Macrophages, protozoa	Pneumonia (Legionnaires' disease)
Wolbachia sp.	VirB/VirD4[a]	?	Arthropods	Male feminization (?)

[a]A contribution of this T4SS to pathogenicity has not yet been demonstrated.

OCCURRENCE OF T4SS

T4SS are found predominantly in gram-negative bacteria. Exceptions are the bacterial conjugation systems of gram-positive bacteria and archaea with weak homology to the conjugation systems of gram-negative bacteria (see the following section), which are regarded as the evolutionarily most ancient T4SS. The *A. tumefaciens* T-DNA transfer system is the first-characterized T4SS mediating transport into eukaryotic hosts. It has been the reference system for the nomenclature of T4SS components studied since then.

Bacterial Conjugation Systems

The conjugation systems of gram-negative bacteria are encoded by conjugative plasmids that often carry antibiotic resistance genes and transposable genetic elements. These plasmids are transferred with high frequency between gram-negative bacteria. Bacterial conjugation thus largely contributes to genetic exchange between bacteria and accounts for the rapid spread of antibiotic resistances between pathogenic bacteria (Waters, 1999). Furthermore, conjugative plasmids endow the bacteria with a high adaptability to different environmental conditions (Lilley et al., 2000; Top et al., 2000). Many of the plasmids are able to replicate in a broad range of gram-negative hosts. A small number of conjugative plasmids have the remarkable ability to transfer between and to replicate in both gram-negative and gram-positive bacteria (Gormley and Davis, 1991; Trieu-Cuot et al., 1987; Charpentier et al., 1999; Kurenbach et al., 2003). In some cases, it has been shown that conjugative DNA transfer can even occur between bacterial and eukaryotic cells (Heinemann and

Sprague, 1989; Waters, 2001), thereby contributing to genetic exchange beyond the bacterial kingdom.

The mechanism of DNA transfer by bacterial conjugation was initially studied by using the F plasmid (belonging to incompatibility group F [IncFI]). A series of other conjugative plasmids, such as R1 (IncFII), RP4 (IncPα), R27 (IncH), ColIb (IncI), pKM101 (IncN), R388 (IncW), and R6K (IncX), were used for later studies of conjugation. Despite the differences in sequence, host range, and compatibility, these conjugation systems have high functional similarity to the F-plasmid transfer (F Tra) system. The investigations carried out with these conjugative plasmids contributed to unraveling some of the fundamental mechanisms of bacterial conjugation (Zechner et al., 2000).

Before transfer, the conjugative plasmid is processed into a single-stranded intermediate by the action of DNA-processing enzymes (Dtr system [Willetts, 1981]). Here, the catalytic key component, the relaxase (a homologue of VirD2 of *A. tumefaciens*), becomes covalently attached to the 5′ end of the DNA (Pansegrau et al., 1990a). The relaxase interacts with the TraG-like protein (T4CP, a homologue of VirD4 of *A. tumefaciens*) (Llosa et al., 2003; Pansegrau and Lanka, 1996a; Schröder et al., 2002; Szpirer et al., 2000), which in turn interacts with the membrane-spanning mating-pair formation complex (Mpf complex) (Cabezón et al., 1997; Gilmour et al., 2003; Hamilton et al., 2000; Llosa et al., 2003). The Mpf complex mediates contact with recipient cells via an elongated, surface-exposed sex pilus structure and transports the DNA into the target cell. It is not known whether the actively transported substrate consists of the DNA itself or whether it is the DNA-linked relaxase that is transported during bacterial conjugation. The finding that relaxases are transferred to recipient cells independently of DNA transfer supports the view that the secretion machinery of bacterial conjugation systems transports the relaxase, which in turn trails the DNA and directs it into the target cell (Llosa and de la Cruz, 2005; Luo and Isberg, 2004). Structural features of relaxases are discussed below (see "T4SS Substrates: Diverse Molecules Transported along the Same Route" below).

Bacterial conjugation systems of gram-positive bacteria are distantly related to those of gram-negative bacteria. Based on sequence similarities, several homologues of conserved T4SS components (namely, homologues of the VirB1, VirB4, VirD4, and VirD2 types) are predicted to exist in gram-positive bacterial conjugation systems such as the ones encoded by plasmids pIP501 and pGO1 (reviewed by Grohmann et al. [2003]). Conjugation systems of archaea exist in thermophilic crenarchaea of the genus *Sulfolobus* (She et al., 1998; Stedman et al., 2000). These conjugation systems encode homologues of the VirB4 and the VirD4 types, but are otherwise unrelated to bacterial T4SS.

The causative agent of gonorrhea, *Neisseria gonorrhoeae,* encodes an exceptional and still unique T4SS that mediates the release of DNA into the growth medium, without the requirement for a recipient or target cell (Dillard and Seifert, 2001). The genes encoding this T4SS include homologues of the F factor conjugation genes and are located on a genomic island that is specific for *N. gonorrhoeae* and is frequently found in strains producing the serious complications of disseminated gonococcal infection.

The *A. tumefaciens* T-DNA Transfer System

A. tumefaciens is a phytopathogen that induces tumorous growth (crown gall tumor) of infected plant tissues (Smith, 1907). The primary hosts are dicotyledons such as grapevines, stone and pome fruit trees, and nut trees, to which most economic damage is

done (Kerr, 1992). *A. tumefaciens* uses the VirB-VirD4 secretion system to transport onco-
genic transfer-DNA (T-DNA) and several effector proteins to the nuclei of plant cells (see
the reviews by Binns and Costantino [1998], de la Cruz and Lanka [1998], Deng and Nester
[1998], Johnson and Das [1998], Rossi et al. [1998], and Tzfira and Citovsky [2000]).

The genes encoding the VirB-VirD4 secretion system of *A. tumefaciens* map on the tu-
mor-inducing plasmids (Ti plasmids), which are exemplified by the extensively studied and
fully sequenced plasmid pTiC58 (Goodner et al., 2001). This 194-kb plasmid also harbors
the T-DNA sequence (23 kb), which is excised and transferred as a single strand during plant
transfection (Stachel et al., 1986; Tinland et al., 1994; Yusibov et al., 1994). The effector
proteins that are cotranslocated along with the T-DNA are VirD2, VirE2, VirE3, and VirF
(Citovsky et al., 1992; Howard et al., 1992; Schrammeijer et al., 2003; Vergunst et al., 2000).
VirD2 is a relaxase that, similar to relaxases of bacterial conjugation systems, is essential for
cleavage of the plasmid DNA and becomes transiently attached to the T-DNA single strand
(Howard et al., 1989; Pansegrau et al., 1993). VirE2, VirE3, and VirF share a C-terminal peptide motif (Arg-Pro-Arg) that is thought to direct their transfer by the VirB-VirD4 transport
machinery (Schrammeijer et al., 2003; Vergunst et al., 2000; Vergunst et al., 2005). VirE2 is
a specialized single-stranded DNA-binding protein that has the ability to form polymeric
channel-like structures in membranes (Citovsky et al., 1989; Duckely and Hohn, 2003). After entering the plant cell, VirE2 binds cooperatively to the VirD2–T-DNA conjugate. By
virtue of nuclear localization sequences, VirE2 and VirD2 then guide the T-DNA to the nucleus of the plant cell. Here, the T-DNA becomes integrated into the plant genome by ille-
gitimate recombination. The T-DNA contains plant oncogenes and genes encoding substrates collectively known as opines, which can be specifically consumed by the infecting *A.
tumefaciens* cells. The oncogenes code for enzymes that catalyze the production of plant
growth hormones, which in turn cause the formation of non-self-limiting tumors. The genes
carried by the T-DNA are not essential for the transfection process itself and can be replaced
by other genes of interest. The discovery of the *A. tumefaciens* T-DNA transfer system has
therefore led to the development of effective tools for the genetic transformation of plants
(Gelvin, 2003). Under laboratory conditions, the T-DNA also transfers into a wider range of
plant hosts (Deng and Nester, 1998) and even into yeast (Bundock et al., 1995; Piers et al.,
1996), fungus (de Groot et al., 1998), and human (Kunik et al., 2001) cells.

Apart from the VirB-VirD4 secretion system, *A. tumefaciens* strain C58 encodes two
additional T4SS: the Tra system (*trb* locus) of the Ti plasmid (Li et al., 1998; von Bodman
et al., 1989) and the AvhB system of the cryptic plasmid pAtC58 (Chen et al., 2002). These
two bacterial conjugation systems are not involved in T-DNA transfer but enable conjugative transfer of the respective plasmids encoding them. *Agrobacterium rhizogenes,* a close
relative of *A. tumefaciens,* induces the formation of "hairy roots" on infected plant tissue.
The virulence determinants of this plant pathogen are encoded by the root-inducing plasmid (pRi), which carries a T4SS closely related to the Ti plasmid of *A. tumefaciens* (Binns
and Costantino, 1998; White and Nester, 1980).

The *H. pylori cag* Pathogenicity Island

Helicobacter pylori is a human pathogen causing gastric diseases such as gastritis, pep-
tic ulcer, and gastric cancer. The pathogenicity of *H. pylori* is associated with the presence
of the cancer-associated genes (*cag*) that localize on the pathogenicity island (PAI)

(Censini et al., 1996). The *cag*-PAI consists of 27 putative genes, 10 of which show homologies to the *virB* and *virD* genes of *A. tumefaciens* or other genes encoding type IV secretion components.

The T4SS encoded by *cag*-PAI directs the transfer of the 145-kDa immunodominant CagA protein into gastric epithelial cells (Asahi et al., 2000; Backert et al., 2000; Odenbreit et al., 2000; Segal et al., 1999a; Stein et al., 2000). After transfer, CagA undergoes tyrosine phosphorylation and induces changes in the tyrosine phosphorylation state of distinct cellular proteins (Backert et al., 2001; Higashi et al., 2002a; Higashi et al., 2002b; Püls et al., 2002; Segal et al., 1999a; Stein et al., 2002). CagA translocation induces morphological changes in the host epithelial cell, characterized by cell scattering, elongation, and spreading (Backert et al., 2001; Churin et al., 2001; Higashi et al., 2002a; Mimuro et al., 2002; Segal et al., 1999a). Transfer of CagA from *H. pylori* is probably supported by a *cag*-PAI-dependent, needle-like surface organelle whose formation is induced after contact with epithelial target cells (Rohde et al., 2003; Tanaka et al., 2003). By analogy to the substrates secreted by the VirB-VirD4 system of *A. tumefaciens* (see above), CagA contains a C-terminal peptide signal that promotes translocation (W. Fischer, personal communication). Independently of CagA secretion, *H. pylori* strains carrying the *cag*-PAI furthermore induce an inflammatory response in gastric epithelial cells, characterized by the production and secretion of chemokines such as interleukin-8 (IL-8) (Crabtree et al., 1995). Distinct genes of the *cag*-PAI are required either for translocation of CagA into host cells, for induction of host cellular IL-8 secretion, or for both of these phenotypes (Fischer et al., 2001; Selbach et al., 2002). It remains unknown what mechanisms initiate the chemokine production in infected host cells. The bacterial surface organelle may interact with receptors of the host cell that initiate the signal cascade leading to chemokine production. Alternatively, yet unknown *H. pylori* effector proteins that are cotranslocated along with CagA into the host may account for the inflammatory response.

Apart from *cag*-PAI, *H. pylori* encodes two other T4SS: the *comB* system for DNA uptake (Hofreuter et al., 2001) and the *tra* region, which possibly enables conjugative transfer of chromosomal DNA between *H. pylori* cells (Backert et al., 2002). The three different T4SS found in *H. pylori*, enabling both import and export of DNA or protein substrates, are a remarkable example of the diversity and the range of application of these secretion systems.

The VirB System of *C. jejuni*

Campylobacter jejuni, which belongs to the same order as *Helicobacter* (*Campylobacterales*), is one of the major causes of diarrheal disease worldwide (Oberhelman and Taylor, 2000). It produces a cytotoxin, the cytolethal distending toxin, that arrests eukaryotic cells in the G_2 phase of the cell cycle and induces IL-8 production by intestinal epithelial cells (Hickey et al., 2000; Whitehouse et al., 1998). Apart from a chromosome-encoded type III flagellin export system, a plasmid-encoded type IV secretion system (VirB, encoded by pVir) is involved in pathogenicity of *C. jejuni* (Bacon et al., 2000; Bacon et al., 2002; Konkel et al., 1999; Parkhill et al., 2000). The *C. jejuni* VirB system, being closely related to the *comB* system of *H. pylori* (see above), mediates the uptake of DNA (Bacon et al., 2000). Additionally, the VirB system supports adherence to and invasion of intestinal epithelial cells (Bacon et al., 2002). It remains unknown how these different functions are related.

The Dot-Icm Systems of *L. pneumophila* and *C. burnetii*

Legionella pneumophila is an intracellular pathogen. It is the causative agent of a potentially fatal form of pneumonia known as Legionnaires' disease (Fields et al., 2002). In the environment, *L. pneumophila* survives and replicates within freshwater amoebae and monocytes. When infecting mammals, it can alter the endocytic pathway of macrophages and replicate in the lungs of the host (Horwitz, 1983). *L. pneumophila* encodes the Dot-Icm secretion system (for "defect in organelle trafficking/intracellular multiplication"), which contributes to the intracellular survival of the bacterium (Andrews et al., 1998; Berger et al., 1994; Brand et al., 1994; Purcell and Shuman, 1998; Segal et al., 1998; Segal and Shuman, 1997).

Legionella infection of macrophages proceeds as follows. The bacteria are first internalized by macrophages and reside in plasma membrane-derived phagosomes. These phagosomes containing *L. pneumophila* recruit secretory vesicles exiting the endoplasmatic reticulum (ER) of the host cell and remodel their membrane. The phagosomes thereby become specialized organelles, resembling rough ER, which escape endocytic lysosome fusion (Roy and Tilney, 2002). Trafficking and maturation of the phagosomes is controlled by the Dot-Icm system, which secretes effector molecules that probably regulate these functions. DotA is an essential component that becomes secreted to the extracellular milieu and is thought to form a pore in the host cell membrane, permitting the passage of effector proteins into the cytoplasm of the host cell (Nagai and Roy, 2001).

An increasing number of proteins secreted by the Dot-Icm system (the actual number is 26) have been identified in different studies (Chen et al., 2004; Conover et al., 2003; Luo and Isberg, 2004; Nagai et al., 2002). None of these proteins are, however, essential for organelle trafficking or intracellular replication of *L. pneumophila* under laboratory conditions. This may be due to a high functional redundancy between the different effector proteins (Luo and Isberg, 2004). Also, the wide range of host cells that are entered by the pathogen may require a battery of effectors that are specialized to the different environments. Limited information is available about the function of some of the secreted effector proteins. LidA is probably secreted to the cytoplasmic surface of the host cellular or phagosomal membrane and becomes involved in recruitment of vesicles exiting the ER (Conover et al., 2003). A similar function is attributed to RalF, which recruits a membrane vesicle-anchored protein, the ADP-ribosylation factor, to the surface of the phagosome (Nagai et al., 2002). LepA and LepB have homology to components of the mammalian SNARE system, which promotes membrane fusion of transport vesicles with cellular compartments (Chen et al., 2004). LepA and LepB are required for the release of intracellular legionellae from protozoan host cells and are thus essential for efficient dissemination of the pathogen (Chen et al., 2004). A yet unknown effector molecule secreted by the Dot-Icm system triggers apoptosis in the host cells through caspase-3 activation (Gao and Abu, 1999a, 1999b; Gao and Kwaik, 2000; Müller et al., 1996; Zink et al., 2002). Activated caspase-3 plays an essential role in the early infection phase. It prevents endocytic lysis of the pathogen-containing vacuole, allowing the bacteria to multiply intracellularly (Molmeret et al., 2004).

L. pneumophila encodes a second T4SS (the Lvh system), which has the ability to cause the conjugative transfer of plasmids (Segal et al., 1999b) and of chromosomal DNA (Miyamoto et al., 2003). The Lvh system functionally interferes with the Dot-Icm system and affects the efficiency of macrophage infection in a temperature-dependent manner (Ridenour et al., 2003).

C. burnetii is the cause of Q fever, a widely distributed zoonosis (Maurin and Raoult, 1999). In humans, it causes acute disease (self-limited febrile illness, pneumonia, and hepatitis) or chronic disease (endocarditis). *C. burnetii* replicates exclusively in large endocytic vacuoles inside eukaryotic cells. It belongs to the order *Legionellales* and encodes a Dot-Icm T4SS that functions similarly to the one encoded by *L. pneumophila* (Zamboni et al., 2003; Zusman et al., 2003).

Dot-Icm secretion systems are only distantly related to other T4SS but have high similarity to bacterial conjugation systems of the IncI plasmids CoIIb and R64. These secretion systems are therefore categorized as a subclass named type IVb secretion systems (T4bSS) (Christie and Vogel, 2000; Sexton and Vogel, 2002).

T4SS of *Bartonella* spp.

Bartonella spp. are pathogenic bacteria adapted to cause intraerythrocytic infection in their mammalian hosts. The hemotropic lifestyle of *Bartonella* spp. is related to their transmission mode via blood-sucking arthropods such as fleas, lice, or sand flies (Breitschwerdt and Kordick, 2000). In addition to invading erythrocytes, *Bartonella* spp. are able to colonize and invade vascular endothelial cells (Dehio, 2001). The bacteria induce the proliferation of human endothelial cells in vitro (Maeno et al., 1999) and, in infected humans, can lead to vasoproliferative tumor growth (pathological angiogenesis [Dehio, 2003; Manders, 1996]). The vascular endothelium is thought to be the primary niche for bartonellae. After invasion and intracellular replication in this niche, the bacteria are seeded to the bloodstream where they infect erythrocytes (Dehio, 2004).

The three major human pathogens are *Bartonella bacilliformis,* which causes Carrion's disease; *B. quintana,* which causes trench fever, bacillary angiomatosis, and endocarditis; and *B. henselae*, which causes intraerythrocytic bacteremia in the cat reservoir host and a wide range of clinical manifestations (including cat scratch disease) in incidentally infected humans (Karem et al., 2000). *B. henselae* can specifically colonize human vascular endothelial cells in vitro (Dehio, 2003) and is able to invade these cells by two separate mechanisms: either by classical endocytic uptake or by formation of invasomes. Invasomes are actin structures of the host cell that engulf and internalize large aggregates of bacteria (Dehio et al., 1997). Two distinct T4SS (Trw and VirB-VirD4, named according to their homology to the T4SS of plasmids R388 and Ti, respectively) are major pathogenicity factors of *B. henselae* (Padmalayam et al., 2000; Schulein and Dehio, 2002; Seubert et al., 2003). The VirB-VirD4 T4SS mediates most of the physiological changes associated with *B. henselae* infection of vascular endothelial cells, i.e., invasome-mediated uptake, proinflammatory activation, and antiapoptotic protection of endothelial cells (Schmid et al., 2004). Several protein substrates of the VirB-VirD4 T4SS of *B. henselae* were identified recently (Schulein et al., 2005). These substrates, termed Bep (for *"Bartonella*-translocated effector proteins"), contain a C-terminal domain mediating VirB-VirD4-dependent translocation.

In contrast to the VirB-VirD4 system, which is essential for infection of the vascular endothelium (the postulated primary niche), the Trw system seems to be essential only for infection of erythrocytes (Dehio, 2004). The Trw system of *B. henselae* has extensive homology to the Trw system of conjugative plasmid R388. The degree of homology is high enough for individual components to be functionally interchangeable between the two systems, suggesting a recent acquisition of the *B. henselae* Trw secretion system by horizontal gene transfer (Seubert et al., 2003). However, the *B. henselae* Trw system additionally encodes several variants of surface-exposed pilus components (homologues of VirB2, VirB5, and VirB7) and lacks a component otherwise essential for substrate transfer (the VirD4 homologue). It therefore is a working hypothesis that the Trw system, instead of transporting effector substrates, mediates bacterial attachment to erythrocytes. The use of different variants of surface-exposed pilus components may represent a general mechanism of immune evasion and/or may allow the interaction with different host cellular surface structures (Seubert et al., 2003).

The VirB System of *Brucella* spp.

Brucella is a zoonotic pathogen causing infectious abortion in animals and a febrile disease, known as Malta fever, in humans (Corbel, 1997; Samartino and Enright, 1993). The infectious disease, commonly referred to as brucellosis, is caused by intracellular parasitism by *Brucella* in mammalian macrophages (see Boschiroli et al. [2001] for a review). After invasion through wounds or mucosa, brucellae initially infect professional phagocytes. Subsequent infection occurs preliminarily in nonprofessional phagocytic cells in the tissue of reproductive organs, the placenta, or the fetus. Therefore, typical manifestations of brucellosis are abortion in pregnant females (which is the origin of the name of *Brucella abortus*) and sterility in males.

The pathogenic mechanisms of *Brucella* include invasion and intracellular replication (Gorvel and Moreno, 2002). A two-component regulatory system (BvrS-BvrR), consisting of a histidine kinase and a regulator protein, is essential for invasion and virulence (Sola-Landa et al., 1998). Intracellular survival and replication of *Brucella* are mediated by the VirB T4SS (O'Callaghan et al., 1999), which initiates a process involving the maturation of the pathogen-containing vacuole (Comerci et al., 2001; Delrue et al., 2001). Similar to the Dot-Icm system of *L. pneumophila,* the T4SS of *Brucella* spp. alters the endocytic pathway of the infected macrophage and blocks the fusion of the phagosome with the lysosome. The bacteria proliferate in this phagosome, which forms a compartment bound by the ER (Celli et al., 2003).

The Ptl System of *B. pertussis*

Bordetella pertussis is the causative agent of whooping cough. It colonizes the human respiratory tract by adhering to ciliated epithelial cells (Tuomanen and Weiss, 1985). One of the pathogenicity determinants of this bacterium is a T4SS called the pertussis toxin liberation (Ptl) system (Craig-Mylius and Weiss, 1999; Weiss et al., 1993), which transports its substrate, pertussis toxin (PT), to the extracellular milieu. PT is a multisubunit toxin of the A/B family and is composed of five subunits forming a hexameric structure (Stein et al., 1994; Tamura et al., 1982). The individual subunits are secreted into the periplasm by means of the Sec system (Locht and Keith, 1986; Nicosia et al., 1986) and assemble into the holotoxin, which is then secreted into the exocytoplasm by the Ptl system (Weiss et al., 1993). After secretion, the B domain of PT interacts with glycoprotein receptors of the mammalian host and targets the A domain of the toxin into the cytoplasm (Katada et al., 1983; Tamura et al., 1982). The A domain then ADP-ribosylates protein components of the host cell (G proteins), thus interfering with receptor-mediated activation and associated signaling pathways (Gilman, 1987; Katada and Ui, 1982).

The T4SS of *B. pertussis* is exceptional in the way that it delivers its substrate from the bacterial periplasm into the exocytoplasm instead of directly transporting it from the cytoplasm into the host cell. This unique mechanistic exception comes along with the lack of an otherwise essential and conserved T4SS component, the T4CP, giving a strong indication of the role of this component in the other T4SS ["VirD4 (T4CP): a Ring-Shaped Inner Membrane Pore Recruiting the Substrates to the Secretion Channel" below].

Other T4SS with a Possible Role in Pathogenicity

A number of genome-sequencing projects, several of which are still under way, have unearthed a multitude of putative T4SS based on sequence similarity. A complete list of T4SS, sorted by affiliation to (i) conjugation systems, (ii) DNA uptake and release, (iii) effector translocators, (iv) unknown function, and (v) type IVb secretion systems is available online at P. J Christie's homepage (http://mmg.uth.tmc.edu/webpages/faculty/supplements/pchristie/T4SS-updated.pdf). The criteria by which we have included several of these systems in the present list of putative pathogenicity-related T4SS are not completely arbitrary. First, we excluded the systems that seem to be part of bacterial conjugation systems or conjugative transposons. These systems are likely to be responsible for gene transfer but are probably not directly involved in the secretion of toxins or effector proteins. Next, we selected organisms having a high impact in the context of medical or economic significance. Thus, the following list is neither complete nor immune to changes, since it remains to be shown whether the mentioned T4SS are actually involved in pathogenicity-related mechanisms.

The order *Rickettsiales* comprises several pathogenic members, namely, *Rickettsia* spp., *Wolbachia* spp., and *Ehrlichia* spp., that encode predicted type IV secretion components (Andersson et al., 1998; Masui et al., 2000; Ogata et al., 2001; Ohashi et al., 2002). It is a common feature of these species that they obligatorily reside and replicate inside eukaryotic cells. Examples of human-pathogenic *Rickettsia* species encoding a putative T4SS are *R. prowazekii,* the agent of epidemic typhus; *R. conorii,* the agent of Mediterranean spotted fever; *R. typhi,* the agent of murine typhus; *R. rickettsii,* the agent of Rocky Mountain spotted fever; and *R. sibirica,* the agent of North Asian tick typhus. These bacteria reside inside arthropod vectors such as ticks, fleas, and lice, from which they are transmitted to mammalian hosts (Azad and Beard, 1998). *Wolbachia* species cause various sexual alterations in arthropods: cytoplasmic incompatibility, thelytokous parthenogenesis, feminization, and male killing (Stouthamer et al., 1999). *Ehrlichia* spp. are the cause of human ehrlichiosis, a tick-borne febrile illness (Anderson et al., 1991; Chen et al., 1994).

Actinobacillus actinomycetemcomitans is a pathogen associated with infective endocarditis (Chen et al., 1991), brain abscesses (Martin et al., 1967), prosthetic heart valve infections (Hamori and Slama, 1989), and certain forms of periodontal disease (Zambon et al., 1983). *A. actinomycetemcomitans* possesses a VirB-VirD4-like secretion system that is located either on a plasmid (pVT745 in strain VT745) or in the chromosome (strain VT747) (Novak et al., 2001). Whereas the plasmid-encoded T4SS mediates conjugative DNA transfer, the chromosome-encoded system seems to have a different function that may consist in secreting effector proteins.

A series of phytopathogens other than *Agrobacterium* and *Rhizobium* also encode T4SS. These include *Ralstonia solanacearum* (Salanoubat et al., 2002), several species of the genus *Xanthomonas* (da Silva et al., 2002), and *Pseudomonas syringae* pv. tomato (plasmids pDC3000A and pDC3000B) (Buell et al., 2003). However, involvement of these T4SS in pathogenicity is rather unlikely, since the pathogenicity-related functions have been attributed to a conserved type III secretion system (*hrp*) (Büttner and Bonas, 2002; Cornelis and Van Gijsegem, 2000). *Erwinia carotovora* subsp. *atroseptica* is the causative agent of soft rot and blackleg potato diseases. It contains a T4SS (VirB) located on a horizontally acquired genomic island (Bell et al., 2004). Mutational analysis has indicated that this VirB system is important for the virulence of *E. carotovora* subsp. *atroseptica* (Bell et al., 2004). *Xylella fastidiosa* causes a range of economically important plant diseases. It encodes a putative T4SS similar to that of nonvirulent conjugative plasmids pIP02 and pSB102, which are found in bacteria of the rhizosphere (Schneiker et al., 2001; Tauch et al., 2002). The *X. fastidiosa* T4SS is partially encoded on a plasmid (pXF51) and also on the chromosome (Marques et al., 2001; Simpson et al., 2000). It seems likely that this T4SS consists of a nonvirulent conjugation system that has been partially integrated into the chromosome of *X. fastidiosa*.

Novosphingobium aromaticivorans belongs to the sphingomonads, many members of which can degrade a wide variety of polycyclic aromatic hydrocarbons. Recently, *N. aromaticivorans* was proposed as a putative agent of primary biliary cirrhosis in humans, a disease that was previously thought to be an autoimmune disease. Ongoing genome sequencing of *N. aromaticivorans* has revealed the presence of a putative T4SS (VirB [Selmi et al., 2003]).

MOLECULAR COMPOSITION OF T4SS

The type IV secretion machinery is a multiprotein complex that spans the inner and outer membranes of the gram-negative bacterial donor cell (Grahn et al., 2000). As many as 12 or 13 proteins are engaged in a network of interactions that determines the overall architecture of the secretion apparatus. Genetic analysis has initiated the quest for unraveling the molecular composition of T4SS. The genetics of type IV secretion thus provide the fundamentals for the biochemistry and structural biology which have followed.

Genetics of T4SS: the Blueprint of VirB-Like Systems

Genetic studies of the conjugative plasmids F, pKM101, R388, RP4, and R27 have enabled us to define the genetic requirements for functional DNA secretion systems (Bolland et al., 1990; Lessl et al., 1992b; Sherburne et al., 2000; Willetts, 1981; Winans and Walker, 1985). Such systems are defined by their ability to mobilize non-self-transmissible plasmids like RSF1010 (Willetts and Crowther, 1981), which encode only DNA-processing related functions plus the origin of transfer (*oriT*). The structural integrity of the secretion apparatus (Mpf complex) is characterized by its ability to produce specific surface-exposed components such as sex pili and donor-specific phage receptors.

In analogy, other T4SS that are specialized in protein transfer are characterized by their functionality in protein transfer and/or by the production of T4SS-specific surface structures. The first such protein secretion system to be genetically characterized was the VirB-VirD4 system of the *A. tumefaciens* Ti plasmid (Stachel et al., 1986; Stachel and Nester,

1986; Ward et al., 1988). The growing number of available sequences of T4SS components has made it possible to predict the genetic framework of a given system by sequence comparison. This has facilitated efforts to characterize T4SS whose substrates are yet unknown.

A set of 11 or 12 genes are required for biosynthesis of the type IV secretion Mpf complex as defined by phage receptor and pilus production. These genes include *virB1* to *virB11* of the reference system of *A. tumefaciens.* For functionality of the secretion system, the *virD4* gene, encoding T4CP, is additionally required (Cabezón et al., 1997). Homologues of *virB1* to *virB11* and *virD4* are found in almost every T4SS, although a series of exceptions do exist. The set of VirB-like secretion systems comprises the conjugation systems encoded by plasmids of incompatibility groups IncN, IncW, IncX, and IncP and the pXF51/pSB102/pIPO2-like plasmids. Also included are the protein-secreting T4SS of *H. pylori, Bartonella* spp., *Brucella* spp., and *B. pertussis.* Notably, the *Brucella* spp. and *B. pertussis* systems exceptionally lack a *virD4* homologue, a fact that is discussed below in the section on VirD4. A subset of *virB* genes fulfilling a separate function has been found in the *A. tumefaciens* system: *virB1* to *virB4* and *virB7* to *virB10.* These genes are not sufficient for DNA secretion, but they strongly enhance DNA transfer efficiency when collectively expressed in recipient cells (Bohne et al., 1998; Liu and Binns, 2003)

Type IVb secretion systems, like the Dot-Icm system of *L. pneumophila,* as well as some of the bacterial conjugation systems (IncF, IncH, and IncI), are more divergent from VirB-like systems. These systems miss some of the well-conserved *virB* genes such as *virB11* and, conversely, include Mpf-encoding genes that do not have a counterpart in VirB-like systems.

Numerous genetic studies with individual components of T4SS have supported the efforts to biochemically and structurally characterize them. The relevant conclusions drawn from these studies are mentioned in the context of the biochemical and structural properties of T4SS components, as described in the following section. Special genetic tools, such as the yeast two-hybrid system, have furthermore been used to predict in vivo interactions between individual protein components of the type IV secretion complex. Together with the in vitro interactions detected by molecular biological methods, a complex network of interactions has begun to be untangled.

Structure and Function of T4SS Components: the Construction Plan of the Secretion Machine

VirB1 to VirB11 and VirD4 of the *A. tumefaciens* T4SS constitute a functional protein and DNA secretion machinery. These components are well conserved among the majority of T4SS, designated VirB-like T4SS (as defined in the previous section). VirB2, named "pilin," is the subunit of the T-pilus, which furthermore contains VirB5 and VirB7 as minor components (Eisenbrandt et al., 1999; Lai and Kado, 1998; Sagulenko et al., 2001a; Schmidt-Eisenlohr et al., 1999a). VirB7 to VirB10 are designated the core complex components, which probably form the transmembrane channel structure (Beaupré et al., 1997; Das and Xie, 2000; Fernandez et al., 1996a; Kumar and Das, 2001; Liu and Binns, 2003). VirB4, VirB11, and VirD4 are the putative motors of the secretion machine, since they might energize the translocation process by hydrolysis of nucleotides (Dang et al., 1999; Gomis-Rüth et al., 2001; Krause et al., 2000a; Sagulenko et al., 2001b).

Biochemical studies have aimed to define the enzymatic and molecular functions of VirB-like components. Thanks to the technical advances and the increasing efforts invested in the field of structural biology, information on the structure of VirB-like components has recently become available (reviewed by Yeo and Waksman [2004]). This structural and functional body of information, sorted by protein families VirB1 to VirB11 and VirD4, is summarized below. The role of an individual component with respect to its function in the secretion system as a whole is of particular interest, but this role is often difficult to assess. In vivo complementation analysis and protein interaction analysis (both in vitro and in vivo) have occasionally allowed the definition of subsets of VirB-like components that collectively mediate a specific function. In vitro reconstitution of a complete functional type IV secretion complex remains a future goal that would mark a milestone in this field of research.

VirB1: Perforation of the Peptidoglycan Cell Wall

VirB1 belongs to a widespread superfamily of lysozyme-like glycosylases (peptidoglycanases and muramidases) found in type II, III, and IV secretion systems, as well as in bacteriophages (Koraimann, 2003; Mushegian et al., 1996). Members of this family of enzymes are capable of locally disrupting the gram-negative peptidoglycan cell wall, as has been shown for the VirB1-like protein P19 of conjugative IncFII plasmid R1 (Bayer et al., 2001). P19 is not absolutely essential for conjugation, but deletion of the *p19* gene results in a 10-fold reduction of the conjugation rate of R1. Similarly, deletion of *virB1* of the *A. tumefaciens* VirB-VirD4 secretion system produces a 100-fold reduction of virulence (Berger and Christie, 1994). Such *virB1* mutants are furthermore impaired in pilus formation (Fullner et al., 1996) and express low levels of VirB4 and VirB10 (Berger and Christie, 1994).

VirB1-like proteins contain an N-terminal signal sequence mediating export into the periplasm by means of the Sec system (Bayer et al., 2000). VirB1 of *A. tumefaciens* undergoes further processing after export. Once in the periplasm, VirB1 is processed into two halves by an unknown protease (Baron et al., 1997a). The larger N-terminal half, harboring the glycosylase domain, probably induces local lysis of the peptidoglycan cell wall. Lysis of the cell wall may facilitate the assembly of the Mpf complex. The C-terminal half (VirB1*) is secreted into the exocytoplasm, where it remains loosely attached to the cell surface and to the Mpf complex (VirB9 has been identified as an interacting partner in cross-linking experiments) (Baron et al., 1997a). Processing of VirB1 and secretion of VirB1* are achieved in the absence of any other VirB component (Llosa et al., 2000). Both the N- and C-terminal halves of VirB1 independently support the functionality of the VirB-VirD4 secretion apparatus (Llosa et al., 2000). The function of VirB1* remains unknown, but it seems likely that it interacts with receptors of the target cell. In the Ptl secretion system of *B. pertussis,* a homologue of VirB1 is apparently missing. However, the VirB8-like protein of this secretion system (PtlE) contains an N-terminal domain mediating a VirB1-like peptidoglycanase activity (Rambow-Larsen and Weiss, 2002).

VirB2: the Structural Subunit of Pili

VirB2-like proteins (pilins) are the major components of conjugative sex pili and of the virulence pilus (T pilus) of *A. tumefaciens.* Prior to secretion and assembly into the pilus structure, pilins are processed in several steps starting from pilin precursor proteins. The precursors are hydrophobic peptides containing an N-terminal signal peptide sequence and two predicted transmembrane helices. In the F plasmid pilin (TraA), removal of the signal peptide is followed by N acetylation of the N terminus (Moore et al., 1993). The pilins of *A. tumefaciens* (VirB2) and RP4 (TrbC) undergo a different kind of modification after removal of the signal sequence: the peptides are cyclized via a specific cyclization reaction where the N and C termini are joined by a head-to-tail peptide bond (Eisenbrandt et al., 1999; Kalkum et al., 2002). TrbC furthermore contains a C-terminal sequence that is removed in a preceding step. The pilins of *B. pertussis* (PtlA), *Brucella* spp. (VirB1), and IncP plasmids (TrbC) probably follow a similar maturation process to that for TrbC, since they each display a high sequence similarity to the proposed processing motif of TrbC (X/AEIA/X, where X stands for any amino acid and slashes mark the two cleavage sites for C-terminal proteolysis and cyclization).

Mature pilin assembles in the periplasm and at the cell surface to form pili. Pili are long, tubular, filamentous appendages of the cell. Such structures have been detected in most bacterial conjugation systems (Frost, 1993) and in the *A. tumefaciens* VirB-VirD4 system (Lai and Kado, 2000). Much of the work to detect conjugative pili was done by D. E. Bradley and J. N. Coetzee and coworkers. Pili were characterized through their sensitivity to pilus-specific phages and were visualized by electron microscopy methods (Fig. 1). The pilus of the F plasmid (F pilus) has been characterized extensively (Frost, 1993). It is a flexible filament 1 to 2 μm in length and has a diameter of 8.5 to 9 nm and a lumen width of 2 to 2.5 nm. The F pilus mediates initial contact between donor and recipient cells and brings the cell surfaces into close proximity. Mechanistically, this is thought to function through retraction (depolymerization) of the pilus after binding to receptors of the target

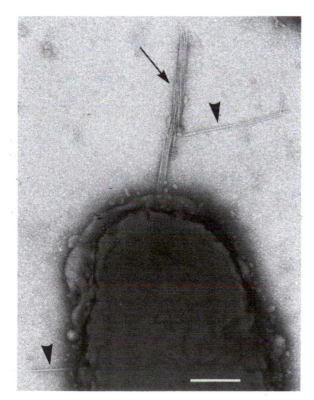

Figure 1. Electron microscopic image of P-pili. The image represents a section of an *E. coli* cell producing RP4-encoded P-pili. P-pili (arrowheads) are characteristically long and rigid and thus break off easily from the pilus-producing bacterium. When broken off, they assemble into large, filamentous bundles of pili (indicated by an arrow). Bar, 0.2 μm. This image was taken by Jana Haase and was provided by Gerhild Lüder and Rudi Lurz. Preparation and staining were as described previously (Haase et al., 1995), using *E. coli* strain JE2571 harboring plasmids pML123 and pWP471.

cell (Achtman et al., 1978). It is unknown whether the lumen of the pilus serves as a channel for the transport of substrates or signal molecules. Such a function seems improbable at least for sex pili of plasmid RP4, which are rigid and are only loosely attached to the cells. Also, the lumen of pili, 2 to 2.5 nm, is probably too narrow to allow the transport of macromolecules such as single-stranded DNA (ssDNA), proteins, or ssDNA-protein conjugates.

In the *A. tumefaciens* T pilus, VirB2 forms disulfide-cross-linked homodimers. The formation of disulfide bonds is not essential for T-pilus formation, but substitution of the responsible cysteine residue attenuates virulence and destabilizes the T pili (Sagulenko et al., 2001a). In the *B. pertussis* Ptl secretion system, the VirB2 homologue (PtlA) is likely to have a different function from building up a pilus. In fact, the production of pili is probably dispensable in this secretion system functioning without cell-cell contact and lacking an orthologue of VirB5, the postulated adhesin (see the discussion of VirB5 below). Further indications of a different role of VirB2 are given by the identification of VirB2 mutants (mutants of TrbC of plasmid RP4) that are proficient in substrate transfer although they are deficient in pilus production (Eisenbrandt et al., 2000). It is conceivable that these VirB2 mutants are unable to assemble into a pilus structure but do still form the base or "stump" of the pilus that is anchored in the outer membrane part of the type IV secretion channel. This stump structure is possibly sufficient for substrate transfer. VirB2 assembly and pilus elongation probably also depend on other factors, namely, VirB6 and VirB11 (see below).

VirB3: an Outer Membrane Component of T4SS

Little is known about VirB3, a conserved and essential component of the type IV secretion complex. VirB3 localizes to both the inner and outer membranes, with the majority of the protein present in the outer membrane (Shirasu and Kado, 1993). The stable expression and localization of VirB3 to the outer membrane is dependent on the presence of VirB4 (Jones et al., 1994), and VirB6 is also required for stable expression of VirB3 (Hapfelmeier et al., 2000). Possibly, both VirB4 and VirB6 interact with VirB3. In this context, it is worth noting that sequence analysis of the predicted T4SS component TriC of plasmid Cos100 revealed that TriC harbors both a VirB3-like domain and a VirB4-like domain, fused in a single protein (Strauch et al., 2003).

VirB4: a Motor of Secretion Embedded in the Heart of the Secretion Machinery?

VirB4-like proteins are one of the three so-called "traffic ATPases" that are conserved in T4SS: VirB4, VirB11, and VirD4, each of which contains a nucleotide-binding peptide motif (Walker A box) typically found in nucleotide hydrolases or synthetases such as the proton pump F_1F_0-ATPase (Walker et al., 1982). It is anticipated that one or several of these three putative traffic ATPases may function as the motor for type IV secretion by converting the chemical energy freed by nucleotide hydrolysis into the kinetic energy necessary for the process of secretion.

VirB4-like proteins are the largest of the VirB proteins (generally about 800 to 850 amino acids), and they localize to the inner membrane. Mutational analysis of several *virB4*-like genes has shown that the Walker A motif is essential for functionality of the secretion system (Berger and Christie, 1993; Cook et al., 1999; Fullner et al., 1994; Rabel et al., 2003). This implies that VirB4-like proteins bind to and possibly also hydrolyze nucleotides. However, experiments with purified VirB4-like proteins failed to detect such an activity in vitro (Rabel et al., 2003) and the previously reported ATPase activity of Ti plasmid VirB4 (Shirasu et al., 1994) remains a stand-alone report that has not been confirmed.

Genetic studies with *virB4*-like genes also show that strains producing mutant VirB4 in a wild-type background display a negative-dominant phenotype when the mutation is located in the Walker A motif (Berger and Christie, 1993; Fullner et al., 1994; Rabel et al., 2003). Apart from the Walker A motif, VirB4-like proteins contain three additional conserved motifs (motifs B, C, and D) that are critical for activity. Mutations in motifs B, C, and D equally display a dominant negative phenotype in the presence of a wild-type allele (Dang et al., 1999; Rabel et al., 2003). Negative dominance implies that the mutant proteins are strongly competitive with the wild-type protein and that they must participate in the formation of a protein complex whose function is abolished if a small number of nonfunctional mutant proteins are incorporated. Studies of the VirB4-like protein TrhC (plasmid R27) provide an additional line of evidence for this assumption. A TrhC-green fluorescent protein fusion was visualized at discrete sites of the cellular membrane (Gilmour et al., 2001). This localization was strictly dependent on the presence of other Mpf components but independent of a functional nucleotide-binding site (Gilmour and Taylor, 2004). Thus, VirB4-like proteins are likely to interact with other components of the type IV secretion apparatus and to assemble as multimers in the secretory membrane complex. Interactions of VirB4 with other T4SS components have so far not directly been shown. However, VirB4-VirB8 and VirB4-VirB10 interactions were observed in two independent two-hybrid screens of the *A. tumefaciens* and *Rickettsia sibirica* T4SS (Malek et al., 2004; Ward et al., 2002). The additionally detected VirB4-VirB11 (Ward et al., 2002) and VirB4-VirD4 (Malek et al., 2004) interactions remain unconfirmed since they were found only once.

VirB4 of the Ti plasmid is a polytopic inner membrane protein containing three cytoplasmic and two periplasmic domains, where the N and C termini are in the cytoplasm (Dang and Christie, 1997). The protein forms homomultimers, and a dimerization domain was localized to the amino terminus (residues 1 to 312) (Dang et al., 1999). The dimerization interface is possibly formed by one or both of the two transmembrane helices that enclose the first periplasmic loop. Alternatively, the periplasmic domain may mediate the intermolecular VirB4-VirB4 interactions (Dang et al., 1999). In contrast to VirB4 of *A. tumefaciens,* the VirB4-like proteins of plasmid R388 (TrwK) and RP4 (TrbE) are monomeric. The soluble domain of TrbE, which lacks a putative N-terminal transmembrane segment, is fully functional in DNA transfer, pilus production, and donor-specific phage propagation (Rabel et al., 2003). In the *Rhizobium* plasmid pNGR324a, the VirB4-like protein seems to be split in half: the plasmid encodes two proteins (TrbEa and TrbEb [Freiberg et al., 1997]) with homology to the N-terminal (residues 1 to 137) and C-terminal (residues 161 to 819) domains of TrbE of the Ti plasmid, respectively. The genetic organization of the *virB* and *virD4* genes on pNGR324a is otherwise colinear with the *virB-virD4* locus of the Ti plasmid. It seems likely that the two domains of VirB4 fulfill separate functions.

The presence of a subset of VirB proteins, including VirB4, in agrobacterial cells enhances the frequency of plasmid uptake when these cells serve as recipients in conjugation experiments with IncQ plasmid-mobilizing donors (Bohne et al., 1998). This enhancement of DNA uptake requires the presence of the oligomeric form of VirB4 in recipients, since VirB4 mutants that fail to multimerize also fail to increase the efficiency of plasmid uptake. In contrast, VirB4 mutants with an inactivated nucleotide-binding site, which are nonfunctional for secretion, promote DNA uptake to the same extent as wild-type VirB4 does. Hence, VirB4 homomultimers are probably important structural components necessary for the assembly of a constitutive membrane channel structure that is potentially capable of transferring DNA bidirectionally. The nucleotide-binding or hydrolyzing activity possibly configures this channel as a dedicated export machine.

VirB5: a Component of the Pilus Mediating Cell Adhesion?

VirB5-like proteins associate with T pili and conjugative sex pili (Schmidt-Eisenlohr et al., 1999a, 1999b). The proteins contain an N-terminal signal sequence mediating export to the periplasmic space. In *A. tumefaciens,* VirB5 cofractionates with the other pilus components VirB2 and VirB7 (Krall et al., 2002; Sagulenko et al., 2001a; Schmidt-Eisenlohr et al., 1999a). For stable expression of VirB5, the presence of VirB6 is required (Hapfelmeier et al., 2000). Apart from its association with the pilus, VirB5 is also found in cytoplasmic and inner membrane fractions (Thorstenson et al., 1993). The observation that VirB5 mutations in the pKM101 system can be complemented extracellularly (Winans and Walker, 1985), however, indicates that VirB5 functions are required only outside the cell, presumably at the pilus.

The determination of the crystal structure of the VirB5-like protein TraC (IncN plasmid pKM101) has revealed an elongated, α-helical structure (Yeo et al., 2003) (Color Plate 42 [see color insert]). Three long α-helices (α1, α2, and α3) form the bulk of the structure, which is accompanied by a satellite α-helical appendage formed by four short helices (αa, αb, αc, and αd). Helical bundles with three helices in the core domain are found in a number of proteins which exhibit diverse functions. Thus, the fold of TraC alone does not provide immediate clues regarding its possible function in bacterial type IV secretion. However, structure-based mutagenesis of TraC has provided important insights into the function of VirB5-like proteins. Since a large region of the surface of TraC forms the interface for crystal packing, it was initially thought that these extensive molecular interfaces lead to the formation of TraC dimers as the biologically relevant form. This is, however, not the case, since it was shown by ultracentrifugation experiments that TraC is a monomer in solution. Nevertheless, a patch (defined by residues Thr69, Asp142 and Val144) that coincides with the crystallographic molecular interface of two TraC molecules seems to be important for interactions of TraC with partner proteins in vivo (Yeo et al., 2003). The in vivo functionality of individual TraC point mutants was assessed by the capacity of the TraC mutant to complement a TraC-deficient strain for (i) localization of TraC to the pilus, (ii) propagation of pilus-specific phages, and (iii) conjugative DNA transfer. Generally, colocalization of TraC to the pilus was an essential requirement for transfer activity and susceptibility to phage attachment. However, remarkable exceptions were observed in two mutants (D142E and V144W), which were both functional with respect to pilus cofractionation but were altered in transfer activity and susceptibility for phage attachment. While the D142E mutation decreased transfer activity and prevented attachment of PRD1 and IKe phages, the V144W mutation was transfer proficient and discriminated between PRD1 (susceptible) and IKe (resistant) phages. Thus, the transfer-related function of the VirB5-like pilus component can be uncoupled from the ability to present receptors for phage attachment. Since IKe phages attach to the tip of pili and since PRD1 phages probably bind to the base of the pilus (Bradley et al., 1983; Frost, 1993; Haase et al., 1995; Kotilainen et al., 1993), TraC must be localized at both ends of the pilus.

In light of the ability of TraC to present receptors for bacteriophage attachment, the nature of the transfer-related function of VirB5-like proteins is thought to consist of bacterial attachment, similar to the function of adhesins (Yeo et al., 2003). Bacterial attachment by adhesins is a key process of most infectious diseases and occurs at early stages of the infection (Soto and Hultgren, 1999). The proposed function of VirB5 as an adhesin is supported by an additional line of argument relying on the fact that a VirB5 homologue exceptionally lacks in the *B. pertussis* Ptl system. The lack of VirB5 goes along with the exceptional ability of this T4SS to function without direct cell-cell contact. Instead, the translocated substrate, the pertussis toxin, is secreted into the extracellular milieu and becomes assimilated by host cell receptors (see "The Ptl System of *B. pertussis*" above). It seems logical that the presence of an adhesin is dispensable in such a secretion system.

VirB6: a Modulator of the Secretion Channel?

VirB6 is an inner membrane protein containing five transmembrane segments (Jakubowski et al., 2004). *A. tumefaciens* VirB6 is required for stable expression of VirB3 and VirB5 and promotes the formation of VirB7 dimers (Hapfelmeier et al., 2000). Detergent-solubilized VirB6 forms high-molecular-weight complexes that cofractionate with VirB7 (Krall et al., 2002). Additional evidence for a VirB6-VirB7 interaction is provided by the findings that VirB6 is required for formation of VirB7-VirB9 heterodimers and coimmunoprecipitates with both of these components (Jakubowski et al., 2003).

The finding that defined VirB6 insertion mutants are deficient in pilus production but proficient in DNA and VirE2 transfer (Jakubowski et al., 2003) is of particular interest. The uncoupling of pilus production and substrate transfer is an exclusive property of VirB6 and VirB11 mutants (Jakubowski et al., 2003; Sagulenko et al., 2001b) but is never observed in mutants of the other VirB components (except for mutations of the pilin itself [see the discussion on VirB2 above]). One possible explanation is that VirB6 and VirB11 mutations may not abolish the biosynthesis of the pilus altogether but may only prevent its elongation. Such mutants would still produce the "stump" of the pilus, which might be sufficient for the secretion process but insufficient for detection by electron microscopy or phage attachment. VirB5, the putative adhesin that is thought to mediate both phage attachment and bacterial attachment (see above), may play a role in the observed VirB6 or VirB11 mutant phenotypes. It is conceivable that VirB6 and VirB11 control the elongation and retraction of the pilus and can modulate the function of the secretion channel for either (i) secretion of pilus components such as VirB2 and VirB5, resulting in pilus elongation, or (ii) secretion of substrates such as VirE2 that become translocated into recipient cells.

VirB7: a Lipoprotein Connecting the Pilus to the Core Complex

VirB7 is a small (4.5-kDa) outer membrane lipoprotein that partly localizes at the periplasmic surface of the outer membrane (Fernandez et al., 1996b). A second fraction of VirB7 localizes exocellularly to the VirB2-composed T pilus (Sagulenko et al., 2001a). VirB7 is both monomeric and dimeric in solution, and the dimers are formed by intermolecular disulfide bridges (Sagulenko et al., 2001a; Spudich et al., 1996). Cellular VirB7 additionally forms stable heterodimers with VirB9 (Fernandez et al., 1996a), which can be isolated as disulfide-bridged protein complexes (Anderson et al., 1996; Baron et al., 1997b; Spudich et al., 1996). VirB7 stabilizes VirB4, VirB9, VirB10, and VirB11 during the assembly of the secretion apparatus (Fernandez et al., 1996a). The stabilizing effect on VirB10 is mediated by ViB7-VirB9 heterodimer formation (Beaupré et al., 1997). Exocellular VirB7 is proposed to contribute to the integrity of the T pilus by stabilizing the intermolecular disulfide bonds between pilin subunits (Sagulenko et al., 2001a).

VirB7-VirB9 dimers are also found with the homologues of *B. pertussis* Ptl (PtlI-PtlF) (Farizo et al., 1996) and the *E. coli* F-plasmid system (TraV-TraB) (Harris et al., 2001). Consistently, the VirB7 and VirB9 homologues of the *H. pylori* Cag secretion system, HP0532 and HP0528, both localize to an extracellular filament structure that is probably formed by other components of the Cag secretion system (Rohde et al., 2003; Tanaka et al., 2003).

VirB8: a Bridge over the Periplasm

VirB8 of *A. tumefaciens* localizes to the inner membrane (Thorstenson and Zambryski, 1994). The topology of the protein consists of a short cytoplasmic N-terminal tail followed by a membrane-spanning sequence and a large periplasmic C-terminal domain (residues 60 to 237) (Buhrdorf et al., 2003; Das and Xie, 1998).

VirB8 forms part of the three-component complex VirB8-VirB9-VirB10, as identified by yeast two-hybrid screens (Das and Xie, 2000; Kumar and Das, 2001) and by detergent extraction of membrane preparations (Krall et al., 2002). VirB8 is furthermore required for assembly of VirB9 and VirB10 to the T complex (Kumar et al., 2000). In a yeast two-hybrid screen with peptide fragments (peptide linkage mapping) of the *A. tumefaciens* VirB components, interactions between VirB8 and the peptidoglycanase VirB1 were detected (Ward et al., 2002). The hypothesized VirB8-VirB1 interaction is further supported by the fact that VirB8 and VirB1 form a single, bifunctional protein in the case of PtlE of the T4SS of *B. pertussis* (Rambow-Larsen and Weiss, 2002). It has been proposed that one of the functions of VirB8 may be that of recruiting VirB1 for localized peptidoglycan lysis at the site of pore assembly (Ward et al., 2002).

VirB9: an Outer Membrane Anchor of the Core Complex

VirB9 of *A. tumefaciens* colocalizes with VirB10 at discrete sites in the bacterial membrane probably representing the sites of T-complex assembly. The localization of VirB9 and VirB10 is dependent on the presence of VirB8, which probably recruits both proteins to these sites (Kumar et al., 2000). VirB9 is found in both the inner and outer membrane fractions when analyzed by sucrose density gradient centrifugation (Fernandez et al., 1996b; Finberg et al., 1995; Liu and Binns, 2003; Shirasu and Kado, 1993; Thorstenson et al., 1993). Immunoelectron microscopic detection of VirB9, however, shows that the protein localizes mainly to the outer membrane (Kumar et al., 2000). Furthermore, VirB9 forms disulfide-bridged complexes with outer membrane-associated VirB7 (see above). These VirB9-VirB7 complexes have a stabilizing effect on VirB10 (Beaupré et al., 1997). Homologues of VirB7, VirB9, and VirB10 in the Cag secretion system of *H. pylori* (HP0532, HP0528, and HP0527, respectively) localize to Cag-dependent surface structures of *H. pylori* (Rohde et al., 2003; Tanaka et al., 2003). While HP0532 (VirB7) localizes at the base of the filament structure, HP0527 (VirB10) localizes both on the filament and at discrete sites of the bacterial membrane (Rohde et al., 2003).

VirB10: a Nodal Point Connecting Inner and Outer Membrane Proteins of the Secretion Apparatus

As mentioned above, VirB10 forms part of a three-component membrane complex, VirB8-VirB9-VirB10, that is also linked to VirB7 through VirB7-VirB9 disulfide bridges. This complex probably spans the inner and outer membranes, since VirB8 localizes to the inner membrane, VirB7 localizes to the outer membrane, and VirB9 and VirB10 each localize to both membrane fractions (Fernandez et al., 1996b; Sagulenko et al., 2001a; Thorstenson and Zambryski, 1994). In addition to VirB7-VirB9, at least one of the three components VirB1, VirB2, and VirB3 is required for localization of VirB10 to the outer membrane (Liu and Binns, 2003).

VirB10-like proteins interact with T4CP (VirD4 of *A. tumefaciens*), which recruits the T4SS substrates to the cytoplasmic face of the secretion channel (see below). The T4CP interaction domain of the VirB10-like protein TrhB (IncHI1 plasmid R27) has been delineated to the N-terminal half. This domain equally mediates TrhB self-interactions (Gilmour et al., 2003). Likewise, the N-terminal half of the VirB10 homologue TrwE (IncW plasmid R388) interacts with the T4CP (TrwB) and with itself (Llosa et al., 2003). Remarkably, the VirB10-T4CP interaction is non-system-specific in the way that VirB10-like proteins can interact, although with lower affinity, with T4CPs of foreign T4SS. This has been demonstrated for VirB10-like proteins TrwE (IncW plasmid R388), TraF (IncN plasmid pKM101), and PilX10 (IncX plasmid R6K), which are each able to interact with their cognate T4CP as well as with the T4CPs of the other two systems, respectively (Llosa et al., 2003). The versatility of the VirB10-T4CP interaction largely explains the functionality of hybrid bacterial conjugation systems composed of a Mpf system (VirB1-VirB11) and a heterologous relaxosome system (the latter consisting of a T4CP, a relaxase, an *oriT* sequence, and, where required, other accessory Dtr proteins) (Bolland et al., 1990; Cabezón et al., 1997; Hamilton et al., 2000).

VirB11: a Ring-Shaped Cytoplasmic NTPase Fueling the Secretion Machinery

VirB11 belongs to the widespread family of PulE-like NTPases found in both type II and IV secretion systems (Motallebi-Veshareh et al., 1992; Planet et al., 2001; Whitchurch et al., 1991). In T4SS, it is one of the three potential NTPases (along with VirB4 and VirD4) that are thought to energize the secretion process and/or the assembly of the secretion machinery. The nucleotide-hydrolyzing activity (predominantly as an ATPase) has been confirmed biochemically for several members of the VirB11 protein family (Christie et al., 1989; Krause et al., 2000b; Rivas et al., 1997; Sexton et al., 2004). The in vitro NTPase activity that was determined for these proteins is rather weak (5 to 30 nmol of hydrolyzed nucleotides per min per mg of protein), lying in the range of that of chaperones like DnaK (15 to 20 nmol/min/mg) (Zylicz et al., 1983) and ClpA (5 to 6 nmol/min/mg) (Hwang et al., 1988). The substrate spectra of VirB11-like proteins can vary considerably: substrates for TrbB (plasmid RP4) are dATP, GTP, and ATP, whereas HP0525 (*H. pylori*) has a much wider spectrum with a preference for ATP and dATP but without favoritism for GTP (Krause et al., 2000b). Notably, the NTPase activity of VirB11-like proteins is enhanced by lipid binding (Krause et al., 2000b; Rivas et al., 1997), consistent with their localization to and/or partial association with the bacterial inner membrane. Moreover, the proteins undergo conformational changes in the presence of lipids (Krause et al., 2000b) and mutations in the nucleotide-binding site have effects on membrane localization (Sexton et al., 2004). Electron microscopic analyses of VirB11-like proteins have revealed that these proteins form hexameric rings, reminiscent of nucleotide-dependent molecular motors such as helicases and F_1-ATPases (Krause et al., 2000a; Patel and Picha, 2000; Yeo et al., 2000) and membrane fusion proteins such as *N*-ethylmaleimide-sensitive fusion protein (NSF) and p97 (Lenzen et al., 1998; Zhang et al., 2000). A further hint of the role of VirB11-like proteins is given by the finding that the VirB11-like protein TrwD (plasmid

R388) induces membrane destabilization and hemifusion of lipid vesicles (Machón et al., 2002). An earlier report stating that VirB11 of *A. tumefaciens* possesses an autophosphorylation activity (Christie et al., 1989) remains an isolated case. Such an activity was not detected for any other member of the family of VirB11-like proteins. The recent determination of the crystal structures of HP0525 (*H. pylori*), either in the unliganded form or in complex with ADP or the nonhydrolyzable substrate analogue ATPγS, has shed light on the structural determinants underlying the function of VirB11-like proteins.

HP0525 in complex with ADP and ATPγS: similar structures but different occupancy. The first molecular snapshot of a VirB11-like protein came from the crystal structure of HP0525 from *H. pylori* in complex with ADP (Yeo et al., 2000). It revealed the biologically relevant form of VirB11-like proteins, in which six subunits, containing an N-terminal domain (NTD) and a C-terminal domain (CTD), form a dome-like hexameric toroid closed at one end and open at the other (Color Plate 43 [see color insert]). The NTDs (residues 1 to 134) and the CTDs (residues 142 to 330) from each monomer sandwich the bound nucleotide and stack as separate rings around the hexameric assembly, with the NTDs defining the open side (internal diameter, 50 Å) and the CTDs contributing to a "six-claw grapple" at the closed end (internal diameter, 10 Å) (Color Plate 43). In general terms, the NTDs and the CTDs appear quite similar, with each featuring an extended central β-sheet flanked by accessory helices. Topologically, however, the two domains are very different in that the CTD adopts the well-known RecA fold (Story and Steitz, 1992), which is typical for ATPases, whereas the fold of the NTD is novel. Based on the molecular features of the HP0525-ADP complex, it was proposed that VirB11-like proteins carry out their function by cycling through closed and open forms that are regulated by ATP binding and hydrolysis and ADP release, respectively (Yeo et al., 2000).

The structure of HP0525 in complex with the nonhydrolyzable ATP analogue, ATPγS, was investigated to obtain a representative picture of the nucleotide-binding environment that most closely resembles that of the physiological substrate (Savvides et al., 2003). Compared to the nucleotide-binding site in ADP-HP0525, the active site in ATPγS-HP0525 undergoes further local rearrangements to accommodate the terminal γ-sulfate of ATPγS. In contrast to the expectation based on the canonical nucleotide-binding site of RecA, suggesting that Glu209 would be the water activator for the hydrolysis reaction, Glu248 emerged as the most plausible candidate to serve this function. Overall, however, the structure of ATPγS-HP0525 determined at 2.8-Å resolution is very similar to that of the ADP-HP0525 complex, implying that nucleotide binding and not hydrolysis is responsible for ATP-induced conformational changes. This appears to be a recurring theme across diverse families of oligomeric molecular machineries exhibiting NTPase activity, since similar observations have been made for GroEL-GroES (Ranson et al., 2001), F_1-ATPase (Abrahams et al., 1994; Yasuda et al., 2001), myosin (Houdusse et al., 2000), kinesin (Rice et al., 1999), HslUV protease (Wang et al., 2001), and p97 (Zhang et al., 2000). Surprisingly, however, ATPγS is present at full occupancy in only one of the two molecules of HP0525 in the asymmetric unit of the crystal (Color Plate 43D), in contrast to two fully occupied nucleotide-binding sites in the ADP-HP0525 complex. This feature of ATPγS binding is indeed quite intriguing considering that the two nucleotide-binding sites in ATPγS-HP0525 are almost equally accessible to solvent, suggesting that this observation may have mechanistic significance in the context of a sequential nucleotide-binding and hydrolysis mechanism.

Structure of unliganded HP0525: a glance at the structural dynamism of HP0525. The structure of unliganded HP0525 determined at 3-Å resolution shows a dramatically different picture from that of the nucleotide-bound protein (Savvides et al., 2003). While the CTD ring retains its previously observed six-claw grapple, which forms the apex of the dome-like internal chamber, the NTDs exhibit a medley of rigid-body rotations about the short linker region between the NTDs and the CTDs and away from the center of the chamber (Color Plate 43E). The observed asymmetry and structural variability among the NTDs in unliganded HP0525 support previous predictions that in the absence of a nucleotide, there is little to hold the CTDs and NTDs together, leading to destabilization of either the CTD or NTD ring in the hexamer (Yeo et al., 2000). The motion does not affect all subunits equally. The NTD of subunit A rotates by 2°, while the NTD of subunit F undergoes a 15° rotation. The extent of the rotation of the NTDs in all other subunits is between these two extremes (Color Plate 43F). Although the observed structural rearrangements in the NTDs vary in amplitude, it has been observed that subunits that are directly across from each other display similar behaviors, which may have mechanistic relevance.

The structure of unliganded HP0525 provides evidence that VirB11-like proteins are indeed dynamic molecular assemblies and that this property may facilitate their respective functions. The role of each of the two domain rings can be defined more precisely. Since the CTD ring remains unchanged in the unliganded or nucleotide-bound form of HP0525, a likely role for the CTDs is to drive oligomerization. In contrast, the NTD ring undergoes a large deformation and can rearrange according to the required function. To demonstrate that HP0525 undergoes equivalent conformational changes in solution, sedimentation velocity, a method more sensitive to macromolecular shape changes, has been employed (Savvides et al., 2003). Clear differences between unbound and nucleotide-bound HP0525 were observed, since binding of either ATPγS or 5′-adenylimidotriphosphate (AMP-PNP) caused a 5 to 7% increase in the sedimentation coefficient. This indicates that the protein adopts a more compact structure in the presence of nucleotides; the conformational changes needed to elicit such a transformation are, however, not dramatic enough to drastically change the hydrodynamic radius of the protein. Indeed, the structural studies show that the nucleotide-bound and unliganded forms of HP0525 have similar overall dimensions. Structural variability exists only among individual subunits of the hexamer, resulting in the breakdown of the molecular symmetry in unliganded HP0525.

Structural similarity to AAA proteins suggests analogous functions for VirB11-like proteins. Structural comparison of HP0525 with diverse hexameric ATPases has revealed a surprising connection to the p97 AAA ATPase despite the absence of any appreciable sequence homology outside their respective nucleotide-binding domains (Savvides et al., 2003). Like the related NSF, p97 plays a central role in organelle assembly and membrane fusion processes in the ER and the Golgi apparatus (Patel and Latterich, 1998; Ye et al., 2001). It has a three-domain structure consisting of a flexible N-terminal domain that undergoes nucleotide-dependent conformational changes and two C-terminal domains, D1 and D2, with folds very similar to that of the C-terminal domain of HP0525 but with only D1 supporting ATPase activity (Rouiller et al., 2000; Zhang et al., 2000). HP0525 cannot be readily classified as an AAA protein due to the absence of adequate sequence similarity in the fingerprint sequences for AAA proteins. However, striking structural similarities between HP0525 and p97 are found: (i) their overall shape is dome-like, (ii) their nucleotide-binding site is formed at the interface of the NTDs and the CTDs, (iii) their NTDs are flexible, (iv) their CTDs alone build a hexameric ring, and (v) their subtle active site architecture is remarkably similar. It has therefore been proposed that VirB11-like proteins, by analogy to p97 and NSF, could serve as mechanical transducers providing the necessary mechanical force for the recruitment, assembly, and disassembly of type IV secretion protein components, making them available to insert into the nascent secretion apparatus and/or to facilitate substrate translocation across the inner membrane (Savvides et al., 2003).

A yeast-two-hybrid analysis designed to elucidate candidate interactions between proteins in *H. pylori* has proposed a number of VirB11 partners, most of which are yet to be annotated to biological pathways (Rain et al., 2001). Furthermore, a high-resolution yeast two-hybrid study of *A. tumefaciens* identified VirB11 interaction partners encoded by the *vir* operon: VirB1, VirB4, and VirB9 (Ward et al., 2002). Of particular interest is the identification of the NTDs of VirB11, not the CTDs, as the mediators of these interactions, thus highlighting the crystallographically observed dynamic nature of the NTDs in facilitating insertion into target complexes and/or interaction with partner molecules. Subsequent nucleotide binding could lock the protein into a rigid conformation, which, following hydrolysis and ADP release, could generate the necessary mechanical force to pry open a macromolecular interaction. A four-step process combining sequential binding of nucleotides until all available sites are filled, followed by nucleotide release to return to unliganded VirB11, appears to be plausible (Color Plate 43G). Two lines of evidence support such a mechanism: (i) the structure of the ATPγS-HP0525 complex exhibits differential occupancy of nucleotide despite equal accessibility to solvent, and (ii) apo-HP0525 displays modular behavior between opposing subunits in the hexameric ring. Such nucleotide-binding patterns are reminiscent of those observed in T7 gene 4 helicase (Singleton et al., 2000) and in the HslU AAA protein (Wang et al., 2001). As the quest to discover a possible effector molecule for VirB11 ATPases continues, the mode of action of VirB11-like proteins can be described as follows (Color Plate 43G). In step 1, the nucleotide free form exists as an asymmetric hexamer exhibiting mobility of the NTD, while the CTDs are responsible for maintaining a pseudohexameric scaffold (VirB11 ATPases in this molecular state probably use their flexibility to bind effectively to target macromolecular complexes). In step 2, the binding of three ATP molecules locks three subunits into a rigid conformation. In step 3, hydrolysis of the three ATPs to ADP with concomitant binding of ATP to the remaining three nucleotide-free subunits results in a perfectly hexameric rigid form. In step 4, the structure retains its symmetry and rigidity until all ATP molecules are hydrolyzed, at which point HP0525 can return to its nucleotide-free form.

VirD4 (T4CP): a Ring-Shaped Inner Membrane Pore Recruiting the Substrates to the Secretion Channel

VirD4-like proteins, also referred to as T4CPs or TraG-like proteins, are putative NTPases which, along with VirB4 and VirB11, probably energize the secretion machinery. Whereas T4CPs are found in each conjugation system and in most other T4SS, in which they are essential for secretion, they are lacking in the T4SS of *Brucella* sp. and *B. pertussis*. T4CPs have two common motifs resembling the Walker A and B motifs of NTPases and ABC transporters (Lessl et al., 1992a; Llosa et al., 1994; Schneider and Hunke, 1998). These motifs are essential for transfer activity (Balzer et al., 1994; Kumar and Das, 2002; Moncalián et al., 1999), indicating that the postulated NTPase activity of T4CPs plays an important role in type IV secretion. Biochemical assays have so far failed to demonstrate NTPase activity of T4CPs in vitro. However, T4CPs were shown to bind nucleoside triphosphates (NTPs), as well as the products of nucleotide hydrolysis (NDPs) (Gomis-Rüth et al., 2001; Moncalián et al., 1999; Schröder et al., 2002; Schröder and Lanka, 2003). An unexplained feature of the nucleotide-binding activity of T4CPs is the effect of Mg^{2+}. Whereas the presence of Mg^{2+} is required for nucleotide-binding of most NTPases, it has a destabilizing effect on nucleotide binding of T4CPs (Schröder and Lanka, 2003). Possibly, binding and release of nucleotides, triggered by Mg^{2+}, has a mechanistic function related to substrate translocation.

The designation as "coupling proteins" originates from genetic studies carried out with T4CPs of bacterial conjugation systems. Although dispensable for pilus formation, coupling proteins were found to be the only component needed in addition to the pilus-forming Mpf system for transfer of non-self-transmissible plasmids (Beijersbergen et al., 1992). The coupling protein was identified as the factor that determines the specificity of a given Mpf system for transport of different mobilizable plasmids. T4CPs were thus proposed to mediate interactions between the Mpf complex and the DNA-processing Dtr functions (Cabezón et al., 1994; Cabezón et al., 1997; Hamilton et al., 2000; Lessl et al., 1993). This role of T4CPs has now been established. Physical interactions between Dtr proteins and T4CPs, as well as between Mpf proteins and T4CPs have been detected (Disqué-Kochem and Dreiseikelmann, 1997; Gilmour et al., 2003; Llosa et al., 2003; Pansegrau and Lanka, 1996a; Schröder et al., 2002; Szpirer et al., 2000).

On the side of the Dtr system, the main interacting component consists of the relaxase, which is the catalytic key component of the DNA-processing relaxosome (Pansegrau et al., 1990b). The interaction between the T4CP and the relaxase of the conjugative plasmid RP4 (IncP) has been characterized in vitro (Schröder et al., 2002; Schröder and Lanka, 2003). By using surface plasmon resonance technology, the kinetics of the binding reaction was determined and provided evidence of a strong but specific association of relaxase and T4CP. In addition to relaxase, DNA (preferentially ssDNA) is bound by the T4CP, giving a further indication of an interaction of the coupling protein with the DNA-processing relaxosome (Moncalián et al., 1999; Panicker and Minkley, 1992; Schröder et al., 2002; Schröder and Lanka, 2003). By analogy to its function in bacterial conjugation systems, the T4CP of A. tumefaciens (VirD4) interacts with the secreted substrate VirE2 (Atmakuri et al., 2003). The domain of VirE2 that mediates this interaction has been localized to the C terminus. On the side of the Mpf system, the T4CP interacts with the VirB10-like component of the membrane-spanning Mpf complex (Gilmour et al., 2003; Llosa et al., 2003). In contrast to the interaction of the T4CP with relaxases, the interaction with VirB10-like proteins seems to be less specific (Llosa et al., 2003).

T4CPs localize to the inner membrane, at the poles of the bacterial cell (Atmakuri et al., 2003; Kumar and Das, 2002). Membrane topology analysis of coupling proteins TraD (F plasmid), VirD4 (A. tumefaciens) and TraG (RP4) has shown that these proteins contain a short cytoplasmic N terminus followed by a periplasmic domain of 30 to 60 residues and a large cytoplasmic domain of 530 to 590 residues (Das and Xie, 1998; Lee et al., 1999; Schröder et al., 2002). Removal of the N-terminal membrane anchor increases the solubility of the protein and disrupts its ability to oligomerize in vitro (Hormaeche et al., 2002; Moncalián et al., 1999; Schröder and Lanka, 2003). The monomeric, cytoplasmic domain of the T4CP is sufficient for binding nucleotides as well as DNA (Moncalián et al., 1999;

Schröder and Lanka, 2003). In contrast, the binding to relaxase requires the presence of the N-terminal membrane anchor, probably because oligomerization of the T4CP is essential for relaxase binding (Schröder and Lanka, 2003). The nucleotide- and DNA-binding activities are partly competitive, and both activities show the same sensitivity to the presence of Mg^{2+}. These findings suggest that nucleotide and DNA binding functionally overlap and that Mg^{2+} induces conformational changes affecting both the nucleotide- and DNA-binding domains of the T4CP (Schröder and Lanka, 2003).

The soluble, cytoplasmic domain of the T4CP TrwB (TrwBΔN70) of plasmid R388 has been crystallized (Gomis-Rüth et al., 2001, 2002; Gomis-Rüth and Coll, 2001). A series of crystal structures of TrwBΔN70 (unliganded, in complex with the substrate analogues GDPNP and ADPNP, and in complex with ADP) have provided a comprehensive view of the structural properties underlying the function of T4CPs.

Structure of TrwBΔN70: a hexameric ring resembling molecular motor proteins. The structure of TrwBΔN70 reveals the biologically relevant form of the molecule as a rather large (110 by 90 Å), globular hexameric assembly composed of intimately associated subunits (Color Plate 44 [see color insert]). The TrwBΔN70 hexamer harbors a central channel that maintains a diameter of ~22 Å starting from the point where the structure is expected to emanate from the bacterial inner membrane almost all the way to the apex of the spherical hexamer facing the cytosolic milieu, at which point it constricts to ~7 Å. Hexamers of TrwB are also visible by electron microscopy of full-length TrwB, which carries an additional appendix, approximately 25 Å wide, corresponding to the transmembrane region of TrwB (Hormaeche et al., 2002).

The TrwBΔN70 subunits have the shape of an orange segment and are composed of two linked domains: a highly twisted RecA-like (Story and Steitz, 1992) α/β nucleotide-binding domain (NBD), and a smaller all-α helical domain (AAD) (Color Plate 44A). The nucleotide-binding sites are defined by superficial cavities at the interfaces between vicinal protomers and are readily accessible to solvent, unlike what has been observed in other ATPases that contain a RecA-like domain. The AAD has been proposed to be a putative DNA-binding domain and, in particular, to directly recruit the relaxosome. Although no experimental evidence is available to provide direct support for this hypothesis, some indirect correlations could be drawn from structural comparisons. While a number of hexameric-ring structures such as the helicases and the AAA proteins feature AADs, the AAD of TrwB bears significant similarities only to the nucleotide-binding portion of the site-specific recombinase XerD (belonging to the λ-integrase family) and to the DNA-binding domain of TraM of plasmid R1 (Gomis-Rüth et al., 2002). TraM is known to bind to the TrwB orthologue, TraD, of plasmid R1 and has been proposed to connect the relaxosome to the Mpf complex (Disqué-Kochem and Dreiseikelmann, 1997). Plasmid R388 does not have a TraM orthologue, implying that TrwB may contain a TraM-like module that interacts with the exported DNA. Like other T4CPs, TrwB interacts with the relaxase, the component of the relaxosome that is covalently bound to the DNA substrate (Llosa et al., 2003). Thus, T4CPs seem to interact with the relaxosome in two ways: through interaction with DNA and through interaction with the relaxase. Both interactions are probably mediated by the AAD domain, although this remains to be experimentally shown.

Structural comparison of TrwBΔN70 with diverse hexameric assemblies revealed that it is remarkably similar in shape and overall dimensions to the F_1-ATPase $\alpha_3\beta_3$ heterohexamer (Abrahams et al., 1994), suggesting that T4CPs may also act as molecular motors. In contrast to the dramatic rigid-body domain shifts between nucleotide and unliganded forms of HP0525 (VirB11), nucleotide-bound and nucleotide-free TrwBΔN70 exhibits local and more subtle conformational changes (Gomis-Rüth et al., 2001, 2002). These structural rearrangements originate at the nucleotide-binding sites close to the periphery of the hexameric assembly and propagate all the way to the internal channel of the structure. It has been proposed that this transmission of conformational changes can be a form of a molecular switch activated by nucleotide binding and hydrolysis, which could facilitate the binding and threading of ssDNA through the TrwB central channel during conjugation (Gomis-Rüth et al., 2001; Gomis-Rüth et al., 2002). The narrowing of the internal channel at the apex of the spherical hexameric assembly, however, suggests that additional conformational adjustments must be necessary to open up this end of the structure to a more accommodating aperture, so that passage of ssDNA (and probably also proteins) can take place unobstructed.

In summary, it is thought that the function of T4CPs consists of recruiting and translocating the dedicated substrates of T4SS through the inner membrane and into the lumen of the membrane-spanning type IV secretion pore. How this transport is achieved and how nucleotide binding and hydrolysis are involved in this process are questions that remain to be answered. The exceptional lack of a T4CP in the *B. pertussis* T4SS (Ptl) is associated with other unusual features of this secretion system. Whereas all other known T4SS substrates are secreted directly into attached or invaded host cells, the substrate of the Ptl system, the heterooligomeric PT, is secreted into the extracellular milieu, where it is taken up by host cell receptors. Moreover, secretion of PT is dependent on the Sec system, which delivers the PT subunits to the periplasm, where they assemble into the holotoxin. It seems likely that these differences between the *B. pertussis* Ptl system and the other T4SS exactly reflect the function of the T4CP.

T4SS Substrates: Diverse Molecules Transported along the Same Route

Since the first discovery of a T4SS substrate, PT of *B. pertussis* (see "The Ptl System of *B. pertussis*" above), a series of substrates that are transported by T4SS have been identified (Table 1). These substrates subvert functions of the host cell and were thus called "effector molecules." The structure of PT has been solved (Stein et al., 1994; Tamura et al., 1982). However, it is difficult to draw general conclusions concerning T4SS because of the uniqueness of the pertussis toxin secretion system in (i) taking advantage of the Sec system for export of the toxin subunits into the periplasm, (ii) assembling the holotoxin in the periplasm, and (iii) secreting the toxin into the extracellular milieu without the need for cellular attachment.

Recently, the crystal structures of a family of more widely distributed T4SS substrates, the conjugative relaxases, have been solved. Additionally, the pathway of a T4SS substrate that is always cotransferred with relaxases has been determined. We therefore confine ourselves to describing the features of conjugative relaxases and the pathway of a relaxase-associated partner molecule, the T-DNA of *A. tumefaciens*.

Model Substrates of T4SS: Conjugative Relaxases

Relaxases are conserved and essential components of bacterial conjugation systems that are required for processing of the DNA prior to its secretion. Relaxases are not part of the secretion apparatus itself, but a specific interaction between the relaxase and the T4SS is required for secretion of the DNA. Two key events mark the steps involved in bacterial conjugation: (i) generation of a copy of ssDNA as a result of rolling-circle-type replication (RCR) of the conjugative plasmid, and (ii) unidirectional transfer of the ssDNA by the plasmid-encoded T4SS. Conjugative relaxases initiate the first of these two key events by catalyzing a strand- and site-specific cleavage reaction at the *nic* cleavage site located within the origin of transfer (*oriT*) of the conjugative plasmid. The reaction relies on a nucleophilic attack by the hydroxyl group of a tyrosine residue on the 5′ side of the DNA phosphate. This transesterification reaction results in a stable phosphotyrosyl linkage plus a free 3′ hydroxyl group on the upper strand (transfer strand) of the plasmid (reviewed by Pansegrau and Lanka [1996b]). After RCR and transfer of a complete copy of the transfer strand into the recipient cell, the relaxase recognizes the reconstituted *nic* site and undergoes a second transesterification reaction, resulting in recircularization of the transfer strand. Mechanistic models imply that the relaxase may be actively secreted by the T4SS as a pilot protein into the recipient cell, trailing the covalently attached transfer strand. Subsequently, the relaxase would remain associated to the secretion pore on the recipients' side, scanning the incoming DNA for the reconstituted *nic* site in order to perform the recircularization reaction that terminates DNA transfer (Llosa et al., 2002; Pansegrau and Lanka, 1996b).

Recently, the structures of the relaxase domains of two conjugative enzymes have been elucidated: the TrwC relaxase of plasmid R388 and the TraI relaxase of F-factor plasmid (Datta et al., 2003; Guasch et al., 2003). An interesting aspect of the relaxase activity of these two proteins is that it has to be physically coupled to a helicase activity for efficient conjugative transfer of DNA (Llosa et al., 1996; Matson et al., 2001). The relaxase domains of TrwC and TraI are actually a small fraction of the full-length proteins: only about 300 amino acid residues out of 966 and 1,756, respectively (Llosa et al., 1996; Street et al., 2003). The structures reveal a conserved compact molecular scaffold possessing features that explain the high affinity and specificity for their respective DNA substrates. Particularly interesting is the structure of TrwC in complex with a 25-mer oligonucleotide, which provides valuable details of the interactions between conjugative relaxases and their substrate DNA.

Overall structures of the TraI and TrwC relaxase domains. TraI and TrwC are overall ellipsoidal structures built on a five-strand antiparallel β-sheet core flanked by α-helices (two-layer α/β or open sandwich folds) (Color Plate 45A [see color insert]). The structures show partial topological similarity to folds such as the "palm" domain of DNA polymerase, tomato yellow leaf curl virus Rep (Campos-Olivas et al., 2002), and adeno-associated virus Rep (Hickman et al., 2002). By analogy to the structural nomenclature of DNA polymerases, the central core of the structure can be referred to as a "palm" while the helical C-terminal domain corresponds to the "fingers" (Color Plate 45A).

DNA-binding site and relaxase-DNA interactions. TrwC was crystallized in complex with a 25-mer oligonucleotide corresponding to the sequence of the T-strand just upstream of the *nic* cleavage site (Color Plate 45B). The TrwC-DNA complex reveals that the enzyme is able to accommodate DNA in both double-stranded and single-stranded conformations over a surface of about 1,400 Å². The DNA sequence forms a double-stranded hairpin within the segment G1 to C16 followed by a single-stranded portion from G17 to T25. The TrwC interaction platform under the entire DNA oligomer is strongly electropositive and gradually tightens to a narrow cleft in the region of the ssDNA-binding site finally ending at a closed pocket that encompasses the active site (Color Plate 45B). TrwC thus appears to be able to recognize the cruciform of *oriT,* to melt the DNA, and to finally engulf the DNA strand that will be cleaved. The formation of a dead end at the site of *nic* implies that the relaxase has to undergo a conformational change in this region to allow the exchange of cleaved and uncleaved ssDNA. Indeed, a comparison of two crystal forms of TrwC bound to DNA shows that an extended loop between α10 and α11 in the "fingers" subdomain can adopt two conformations: a closed one, observed in one crystal form, and an open one, observed in the second crystal form (Color Plate 45A). This loop is disordered in the absence of DNA in the structure of TraI, lending further support to the involvement of the finger subdomain in controlling the entry and exit of ssDNA.

TraI has not been crystallized in a DNA-bound form, but correlation of available biochemical data describing the contribution of individual amino acid residues to ssDNA recognition (Harley and Schildbach, 2003) allowed a partial delineation of the ssDNA-TraI interaction surface (Color Plate 45C) (Datta et al., 2003). This indicated that the lining of the TraI ssDNA-binding cleft is made predominantly of nonaromatic hydrophobic residues and uncharged polar residues, in contrast to most ssDNA-binding proteins, which employ charged surfaces to bind to ssDNA. These findings suggest that TraI relies on surface complementarity rather than long-range electrostatic interactions to interact with its DNA substrate.

A special feature of the active sites of the TraI and TrwC relaxase domains is the key role of a catalytic tyrosine residue and a divalent metal ion. In the case of TraI relaxase, the metal has been identified as a Mg^{2+} ion, while in TrwC relaxase the metal is a Zn^{2+} ion. The available crystal structures, however, suggest that these metals do not act as direct activators of the tyrosine hydroxyl groups because they are localized too far away from the catalytic tyrosines. Possible roles for the metal cations may consist of polarizing the scissile phosphate to facilitate the nucleophilic attack by the catalytic tyrosine hydroxyl group and of stabilizing the ensuing transition state. On the other hand, Asp85 in TrwC has been proposed as a candidate general base that could activate Tyr18 by proton abstraction (Guasch et al., 2003).

TraI and TrwC each contain a second catalytic tyrosine residue (Tyr23 and Tyr26, respectively), which is thought to be involved in the second DNA strand transfer reaction that recircularizes the plasmid after one round of transfer (Pansegrau and Lanka, 1996a; Grandoso et al., 2000). This assumption is based on analogy to mechanisms observed for replication of ssDNA bacteriophages such as φX174 (van Mansfeld et al., 1986). Here, the replication protein GpA (gene A protein) contains a tandem arrangement of tyrosines that alternate in cleaving and recircularizing the ssDNA phage genome generated by RCR. This "flip-flop" mechanism allows the generation of multiple copies of the circular genome from a single replisome, without the need for reinitiation of the replication machinery (van Mansfeld et al., 1986; Hanai and Wang, 1993). An analogous flip-flop mechanism for conjugative relaxases would enable efficient recircularization of the ssDNA plasmid after transfer of a whole copy of the plasmid into the recipient cell. Moreover, it would allow the transfer of multiple plasmid copies through the same secretion pore, without the need for reinitiating DNA transfer from a new plasmid template. In the TrwC structure, the second catalytic tyrosine (Tyr26) is, unfortunately, part of a disordered loop, making it impossible to draw conclusions regarding the postulated flip-flop mechanism. However, the TraI structure can compensate for this misfortune. Here, the two catalytic tyrosine residues (Tyr16 and Tyr23), although not far separated, localize on opposite faces of the protein. This may be a necessary feature for enabling DNA cleavage by the second tyrosine residue, while the first tyrosine residue is still covalently attached to the DNA strand. The structure of TraI does not provide additional support for this hypothesis, since the molecular surroundings of Tyr23 do not display an obvious DNA-binding surface (Datta et al., 2004). Nevertheless, it is conceivable that structural rearrangements take place only under the conditions where a second *nic* cleavage is desired, i.e., after covalent association of the relaxase to the 5′ end of the transfer strand. Structure determination of such covalent protein-DNA adducts remains to be done.

Secretion Pathway of T-DNA: a Journey from VirD4 to VirB2

Investigations of individual components or subsets of the type IV secretion machinery (see "Structure and Function of T4SS Components: the Construction Plan of the Secretion Machine" above) have given indications of the possible pathway taken by T4SS substrates through the transport channel. Recently, a novel technique has made it possible to directly track down the road traveled by a T4SS substrate during secretion. This was accomplished by cross-linking the *A. tumefaciens* T-DNA substrate to components of the secretion channel in different mutant backgrounds (Cascales and Christie, 2004). T-DNA immunoprecipitation (TrIP) and PCR analysis allowed the identification of contacts between T-DNA and 6 out of the 12 VirB-VirD4 components of the *A. tumefaciens* T4SS. By performing TrIP analysis in different *virB* and *virD4* mutants, it was possible to trace the order in which the T-DNA meets these VirB and VirD4 components. The order is as follows: (i) VirD4, (ii) VirB11, (iii) VirB6 and VirB8, (iv) VirB2 and VirB9. Accordingly, VirD4, the coupling protein, is the point of departure for the T4SS substrate on its journey through the secretion pore. Subsequently, the substrate passes to VirB11, where it then enters the membrane-spanning channel, meeting the channel components VirB6 and VirB8. The final "way stations" before transfer into a recipient cell are the outer membrane and surface components VirB9 and VirB2. These data are strongly supportive of earlier models of the architecture of the secretion machinery and provide valuable new insights into the role of individual components (Lybarger and Sandkvist, 2004). A model view of a complete type IV secretion channel is proposed below.

CONCLUSIONS

The research on the molecular and structural aspects of type IV secretion, carried out over the past 3 to 4 decades by a large number of research groups, has provided a broad yet incomplete understanding of the structure and function of the type IV secretion machinery. Based on the current standard of knowledge, we are now able to propose a model view of the type IV secretion apparatus (Color Plate 46 [see color insert]). Three NTPases/NTP-binding proteins (VirB4, VirB11, and VirD4/T4CP), two of which form ring-like structures, seem to energize the translocation of T4SS substrates on the cytoplasmic/inner membrane face of the transport channel. The recruitment of these T4SS substrates to the channel is carried out by the coupling protein (VirD4/T4CP), which appears to play the role of the inner membrane gate of the secretion apparatus. A multiprotein complex spanning the inner and outer membranes of the gram-negative bacterium forms the transport channel. VirB3, VirB4, and VirB6 to VirB10 are the main components of this channel, which is topped by the pilus at the exocellular surface. The pilus is built up by pilin (VirB2) subunits, which, after transport through the VirB secretion channel, assemble to form the pilus structure. An additional pilus component is VirB5, which is thought to mediate cellular adhesion.

Although the picture of the architecture of the type IV secretion machinery becomes clearer, important questions regarding the mechanism of secretion and the contribution of individual components remain unanswered:

1. How do the T4SS substrates migrate through the secretion pore?
2. How does NTP hydrolysis actually drive the translocation of the substrates?
3. What molecular switch turns the secretion machinery on or off?
4. How is the membrane barrier of the targeted host cell penetrated?
5. How can we specifically and efficiently inhibit type IV secretion?

The list of questions could be continued, but we confine ourselves to the questions whose solutions are the most imminent. As is a general rule in research, the answer to any question involves a new list of questions. However, we have come close to answering at least the first and maybe also the second of the above questions.

Indeed, the type IV secretion pathway of a DNA substrate has recently been determined (see the previous section). Although this was the only report of such a pathway until now, it has given strong backing to earlier findings and has clarified the function of individual VirB components. In particular, the role of VirB11 as a central, cytoplasmic component of the secretion channel has now become established. Also, the engagement of VirB6 as a structural component of the secretion channel is a novel finding. With the TrIP technology, a powerful method to define the roles of individual proteins or protein domains has started to yield fruits.

Concerning the second of the above questions: the solving of the crystal structures of the NTPase VirB11 (HP0525) and the NTP-binding coupling protein (TrwB) has provided valuable insights, but important aspects of the functions of these energizers of T4SS remain uncertain. It remains a challenging task to capture molecular snapshots of these proteins in a state when they are in contact with a secretion substrate. Research in this area is under way.

Acknowledgments. We thank Gerhild Lüder and Rudi Lurz for providing the electron microscopic image of Fig. 1. G.S. and E.L. thank Hans Lehrach for generous support.

Financial support was given by the Swiss National Science Foundation (SNF) to G.S., by the European Molecular Biology Organization and Ghent University to S.N.S., and by Wellcome Trust grant 065932 to G.W.

Addendum in Proof. During the writing of this chapter, several important new findings were reported. VirB5 has been identified as an interaction partner of VirB3 (A. Shamaei-Tousi, R. Cahill, and G. Frankel, *J. Bacteriol.* **186:**4796–4801, 2004). The three NTP-binding/hydrolyzing T4SS components VirB4, VirB11, and VirD4 were found to interact with each other (K. Atmakuri, E. Cascales, and P. J. Christie, *Mol. Microbiol.* **54:**1199–1211, 2004). The crystal structure of the periplasmic domain of VirB8 was solved (L. Terradot, R. Bayliss, C. Oomen, G. Leonard, C. Baron, and G. Waksman, *Proc. Natl. Acad. Sci. USA,* in press). Finally, VirB10 was identified as an energy sensor responding to ATP utilization by the VirD4 and VirB11 ATP-binding subunits (Cascales and Christie, 2004).

REFERENCES

Abrahams, J. P., A. G. Leslie, R. Lutter, and J. E. Walker. 1994. Structure at 2.8 Å resolution of F_1-ATPase from bovine heart mitochondria. *Nature* **370:**621–628.

Achtman, M., G. Morelli, and S. Schwuchow. 1978. Cell-cell interactions in conjugating *Escherichia coli:* role of F pili and fate of mating aggregates. *J. Bacteriol.* **135:**1053–1061.

Anderson, B. E., J. E. Dawson, D. C. Jones, and K. H. Wilson. 1991. *Ehrlichia chaffeensis,* a new species associated with human ehrlichiosis. *J. Clin. Microbiol.* **29:**2838–2842.

Anderson, L. B., A. V. Hertzel, and A. Das. 1996. *Agrobacterium tumefaciens* VirB7 and VirB9 form a disulfide-linked protein complex. *Proc. Natl. Acad. Sci. USA* **93:**8889–8894.

Andersson, S. G., A. Zomorodipour, J. O. Andersson, T. Sicheritz-Ponten, U. C. Alsmark, R. M. Podowski, A. K. Naslund, A. S. Eriksson, H. H. Winkler, and C. G. Kurland. 1998. The genome sequence of *Rickettsia prowazekii* and the origin of mitochondria. *Nature* **396:**133–140.

Andrews, H. L., J. P. Vogel, and R. R. Isberg. 1998. Identification of linked *Legionella pneumophila* genes essential for intracellular growth and evasion of the endocytic pathway. *Infect. Immun.* **66:**950–958.

Asahi, M., T. Azuma, S. Ito, Y. Ito, H. Suto, Y. Nagai, M. Tsubokawa, Y. Tohyama, S. Maeda, M. Omata, T. Suzuki, and C. Sasakawa. 2000. *Helicobacter pylori* CagA protein can be tyrosine phosphorylated in gastric epithelial cells. *J. Exp. Med.* **191:**593–602.

Atmakuri, K., Z. Ding, and P. J. Christie. 2003. VirE2, a type IV secretion substrate, interacts with the VirD4 transfer protein at cell poles of *Agrobacterium tumefaciens. Mol. Microbiol.* **49:**1699–1713.

Azad, A. F., and C. B. Beard. 1998. Rickettsial pathogens and their arthropod vectors. *Emerg. Infect. Dis.* **4:**179–186.

Backert, S., Y. Churin, and T. F. Meyer. 2002. *Helicobacter pylori* type IV secretion, host cell signalling and vaccine development. *Keio J. Med.* **51**(Suppl. 2):6–14.

Backert, S., S. Moese, M. Selbach, V. Brinkmann, and T. F. Meyer. 2001. Phosphorylation of tyrosine 972 of the *Helicobacter pylori* CagA protein is essential for induction of a scattering phenotype in gastric epithelial cells. *Mol. Microbiol.* **42:**631–644.

Backert, S., E. Ziska, V. Brinkmann, U. Zimny-Arndt, A. Fauconnier, P. R. Jungblut, M. Naumann, and T. F. Meyer. 2000. Translocation of the *Helicobacter pylori* CagA protein in gastric epithelial cells by a type IV secretion apparatus. *Cell. Microbiol.* **2:**155–164.

Bacon, D. J., R. A. Alm, D. H. Burr, L. Hu, D. J. Kopecko, C. P. Ewing, T. J. Trust, and P. Guerry. 2000. Involvement of a plasmid in virulence of *Campylobacter jejuni* 81-176. *Infect. Immun.* **68:**4384–4390.

Bacon, D. J., R. A. Alm, L. Hu, T. E. Hickey, C. P. Ewing, R. A. Batchelor, T. J. Trust, and P. Guerry. 2002. DNA sequence and mutational analyses of the pVir plasmid of *Campylobacter jejuni. Infect. Immun.* **70:**6242–6250.

Balzer, D., W. Pansegrau, and E. Lanka. 1994. Essential motifs of relaxase (TraI) and TraG proteins involved in conjugative transfer of plasmid RP4. *J. Bacteriol.* **176:**4285–4295.

Baron, C., M. Llosa, S. Zhou, and P. C. Zambryski. 1997a. VirB1, a component of the T-complex transfer machinery of *Agrobacterium tumefaciens,* is processed to a C-terminal secreted product, VirB1*. *J. Bacteriol.* **179:**1203–1210.

Baron, C., Y. R. Thorstenson, and P. C. Zambryski. 1997b. The lipoprotein VirB7 interacts with VirB9 in the membranes of *Agrobacterium tumefaciens. J. Bacteriol.* **179:**1211–1218.

Bayer, M., K. Bischof, R. Noiges, and G. Koraimann. 2000. Subcellular localization and processing of the lytic transglycosylase of the conjugative plasmid R1. *FEBS Lett.* **466:**389–393.

Bayer, M., R. Iberer, K. Bischof, E. Rassi, E. Stabentheiner, G. Zellnig, and G. Koraimann. 2001. Functional and mutational analysis of p19, a DNA transfer protein with muramidase activity. *J. Bacteriol.* **183:**3176–3183.

Beaupré, C. E., J. Bohne, E. M. Dale, and A. N. Binns. 1997. Interactions between VirB9 and VirB10 membrane proteins involved in movement of DNA from *Agrobacterium tumefaciens* into plant cells. *J. Bacteriol.* **179:**78–89.

Beijersbergen, A., A. Den Dulk-Ras, R. A. Schilperoort, and P. J. J. Hooykaas. 1992. Conjugative transfer by the virulence system of *Agrobacterium tumefaciens. Science* **256:**1324–1327.

Bell, K. S., M. Sebaihia, L. Pritchard, M. T. Holden, L. J. Hyman, M. C. Holeva, N. R. Thomson, S. D. Bentley, L. J. Churcher, K. Mungall, R. Atkin, N. Bason, K. Brooks, T. Chillingworth, K. Clark, J. Doggett, A. Fraser, Z. Hance, H. Hauser, K. Jagels, S. Moule, H. Norbertczak, D. Ormond, C. Price, M. A. Quail, M. Sanders, D. Walker, S. Whitehead, G. P. Salmond, P. R. Birch, J. Parkhill, and I. K. Toth. 2004.

Genome sequence of the enterobacterial phytopathogen *Erwinia carotovora* subsp. *atroseptica* and character-ization of virulence factors. *Proc. Natl. Acad. Sci. USA* **101:**11105–11110.

Berger, B. R., and P. J. Christie. 1993. The *Agrobacterium tumefaciens virB4* gene product is an essential viru-lence protein requiring an intact nucleoside triphosphate-binding domain. *J. Bacteriol.* **175:**1723–1734.

Berger, B. R., and P. J. Christie. 1994. Genetic complementation analysis of the *Agrobacterium tumefaciens virB* operon: *virB2* through *virB11* are essential virulence genes. *J. Bacteriol.* **176:**3646–3660.

Berger, K. H., J. J. Merriam, and R. R. Isberg. 1994. Altered intracellular targeting properties associated with mutations in the *Legionella pneumophila dotA* gene. *Mol. Microbiol.* **14:**809–822.

Binns, A. N., and P. Costantino. 1998. The *Agrobacterium* oncogenes, p. 251–266. *In* H. P. Spaink, A. Kon-dorosi, and P. J. J. Hooykaas (ed.), *The* Rhizobiaceae: *Molecular Biology of Model Plant-Associated Bacte-ria.* Kluwer Academic Publishers, Dordrecht, The Netherlands.

Bohne, J., A. Yim, and A. N. Binns. 1998. The Ti plasmid increases the efficiency of *Agrobacterium tumefaciens* as a recipient in *virB*-mediated conjugal transfer of an IncQ plasmid. *Proc. Natl. Acad. Sci. USA* **95:**7057–7062.

Bolland, S., M. Llosa, P. Avila, and F. de la Cruz. 1990. General organization of the conjugal transfer genes of the IncW plasmid R388 and interactions between R388 and IncN and IncP plasmids. *J. Bacteriol.* **172:**5795–5802.

Boschiroli, M. L., V. Foulongne, and D. O'Callaghan. 2001. Brucellosis: a worldwide zoonosis. *Curr. Opin. Microbiol.* **4:**58–64.

Bradley, D. E., J. N. Coetzee, and R. W. Hedges. 1983. IncI2 plasmids specify sensitivity to filamentous bacte-riophage IKe. *J. Bacteriol.* **154:**505–507.

Brand, B. C., A. B. Sadosky, and H. A. Shuman. 1994. The *Legionella pneumophila icm* locus: a set of genes required for intracellular multiplication in human macrophages. *Mol. Microbiol.* **14:**797–808.

Breitschwerdt, E. B., and D. L. Kordick. 2000. *Bartonella* infection in animals: carriership, reservoir potential, pathogenicity, and zoonotic potential for human infection. *Clin. Microbiol. Rev.* **13:**428–438.

Buell, C. R., V. Joardar, M. Lindeberg, J. Selengut, I. T. Paulsen, M. L. Gwinn, R. J. Dodson, R. T. Deboy, A. S. Durkin, J. F. Kolonay, R. Madupu, S. Daugherty, L. Brinkac, M. J. Beanan, D. H. Haft, W. C. Nel-son, T. Davidsen, N. Zafar, L. Zhou, J. Liu, Q. Yuan, H. Khouri, N. Fedorova, B. Tran, D. Russell, K. Berry, T. Utterback, S. E. Van Aken, T. V. Feldblyum, M. D'Ascenzo, W. L. Deng, A. R. Ramos, J. R. Alfano, S. Cartinhour, A. K. Chatterjee, T. P. Delaney, S. G. Lazarowitz, G. B. Martin, D. J. Schneider, X. Tang, C. L. Bender, O. White, C. M. Fraser, and A. Collmer. 2003. The complete genome sequence of the *Arabidopsis* and tomato pathogen *Pseudomonas syringae* pv. *tomato* DC3000. *Proc. Natl. Acad. Sci. USA* **100:**10181–10186.

Buhrdorf, R., C. Förster, R. Haas, and W. Fischer. 2003. Topological analysis of a putative *virB8* homologue essential for the *cag* type IV secretion system in *Helicobacter pylori. Int. J. Med. Microbiol.* **293:**213–217.

Bundock, P., A. den Dulk-Ras, A. Beijersbergen, and P. J. J. Hooykaas. 1995. Trans-kingdom T-DNA transfer from *Agrobacterium tumefaciens* to *Saccharomyces cerevisiae. EMBO J.* **14:**3206–3214.

Büttner, D., and U. Bonas. 2002. Getting across—bacterial type III effector proteins on their way to the plant cell. *EMBO J.* **21:**5313–5322.

Cabezón, E., E. Lanka, and F. de la Cruz. 1994. Requirements for mobilization of plasmids RSF1010 and ColE1 by the IncW plasmid R388: *trwB* and RP4 *traG* are interchangeable. *J. Bacteriol.* **176:**4455–4458.

Cabezón, E., J. I. Sastre, and F. de la Cruz. 1997. Genetic evidence of a coupling role for the TraG protein fam-ily in bacterial conjugation. *Mol. Gen. Genet.* **254:**400–406.

Campos-Olivas, R., J. M. Louis, B. Clérot, B. Gronenborn, and A. M. Gronenborn. 2002. The structure of a replication initiator unites diverse aspects of nucleic acid metabolism. *Proc. Natl. Acad. Sci. USA* **99:**10310–10315.

Cascales, E., and P. J. Christie. 2003. The versatile bacterial type IV secretion systems. *Nat. Rev. Microbiol.* **1:**137–149.

Cascales, E., and P. J. Christie. 2004. Definition of a bacterial type IV secretion pathway for a DNA substrate. *Science* **304:**1170–1173.

Celli, J., C. de Chastellier, D. M. Franchini, J. Pizarro-Cerda, E. Moreno, and J. P. Gorvel. 2003. *Brucella* evades macrophage killing via VirB-dependent sustained interactions with the endoplasmic reticulum. *J. Exp. Med.* **198:**545–556.

Censini, S., C. Lange, Z. Xiang, J. E. Crabtree, P. Ghiara, M. Borodovsky, R. Rappuoli, and A. Covacci. 1996. *cag,* a pathogenicity island of *Helicobacter pylori,* encodes type I-specific and disease-associated viru-lence factors. *Proc. Natl. Acad. Sci. USA* **93:**14648–14653.

Charpentier, E., G. Gerbaud, and P. Courvalin. 1999. Conjugative mobilization of the rolling-circle plasmid pIP823 from *Listeria monocytogenes* BM4293 among gram-positive and gram-negative bacteria. *J. Bacteriol.* **181:**3368–3374.

Chen, J., K. S. de Felipe, M. Clarke, H. Lu, O. R. Anderson, G. Segal, and H. A. Shuman. 2004. *Legionella* effectors that promote nonlytic release from protozoa. *Science* **303:**1358–1361.

Chen, L., Y. Chen, D. W. Wood, and E. W. Nester. 2002. A new type IV secretion system promotes conjugal transfer in *Agrobacterium tumefaciens*. *J. Bacteriol.* **184:**4838–4845.

Chen, S. M., J. S. Dumler, J. S. Bakken, and D. H. Walker. 1994. Identification of a granulocytotropic *Ehrlichia* species as the etiologic agent of human disease. *J. Clin. Microbiol.* **32:**589–595.

Chen, Y. C., S. C. Chang, K. T. Luh, and W. C. Hsieh. 1991. *Actinobacillus actinomycetemcomitans* endocarditis: a report of four cases and review of the literature. *Q. J. Med.* **81:**871–878.

Christie, P. J., and J. P. Vogel. 2000. Bacterial type IV secretion: conjugation systems adapted to deliver effector molecules to host cells. *Trends Microbiol.* **8:**354–360.

Christie, P. J., J. E. Ward, Jr., M. P. Gordon, and E. W. Nester. 1989. A gene required for transfer of T-DNA to plants encodes an ATPase with autophosphorylating activity. *Proc. Natl. Acad. Sci. USA* **86:**9677–9681.

Churin, Y., E. Kardalinou, T. F. Meyer, and M. Naumann. 2001. Pathogenicity island-dependent activation of Rho GTPases Rac1 and Cdc42 in *Helicobacter pylori* infection. *Mol. Microbiol.* **40:**815–823.

Citovsky, V., M. L. Wong, and P. Zambryski. 1989. Cooperative interaction of *Agrobacterium* VirE2 protein with single-stranded DNA: implications for the T-DNA transfer process. *Proc. Natl. Acad. Sci. USA* **86:**1193–1197.

Citovsky, V., J. Zupan, D. Warnick, and P. Zambryski. 1992. Nuclear localization of *Agrobacterium* VirE2 protein in plant cells. *Science* **256:**1802–1805.

Comerci, D. J., M. J. Martínez-Lorenzo, R. Sieira, J. P. Gorvel, and R. A. Ugalde. 2001. Essential role of the VirB machinery in the maturation of the *Brucella abortus*-containing vacuole. *Cell. Microbiol.* **3:**159–168.

Conover, G. M., I. Derre, J. P. Vogel, and R. R. Isberg. 2003. The *Legionella pneumophila* LidA protein: a translocated substrate of the Dot/Icm system associated with maintenance of bacterial integrity. *Mol. Microbiol.* **48:**305–321.

Cook, D. M., K. M. Farizo, and D. L. Burns. 1999. Identification and characterization of PtlC, an essential component of the pertussis toxin secretion system. *Infect. Immun.* **67:**754–759.

Corbel, M. J. 1997. Brucellosis: an overview. *Emerg. Infect. Dis.* **3:**213–221.

Cornelis, G. R., and F. Van Gijsegem. 2000. Assembly and function of type III secretory systems. *Annu. Rev. Microbiol.* **54:**735–774.

Crabtree, J. E., Z. Xiang, I. J. Lindley, D. S. Tompkins, R. Rappuoli, and A. Covacci. 1995. Induction of interleukin-8 secretion from gastric epithelial cells by a cagA negative isogenic mutant of *Helicobacter pylori*. *J. Clin. Pathol.* **48:**967–969.

Craig-Mylius, K. A., and A. A. Weiss. 1999. Mutants in the *ptlA-H* genes of *Bordetella pertussis* are deficient for pertussis toxin secretion. *FEMS Microbiol. Lett.* **179:**479–484.

Dang, T. A., and P. J. Christie. 1997. The VirB4 ATPase of *Agrobacterium tumefaciens* is a cytoplasmic membrane protein exposed at the periplasmic surface. *J. Bacteriol.* **179:**453–462.

Dang, T. A., X. R. Zhou, B. Graf, and P. J. Christie. 1999. Dimerization of the *Agrobacterium tumefaciens* VirB4 ATPase and the effect of ATP-binding cassette mutations on the assembly and function of the T-DNA transporter. *Mol. Microbiol.* **32:**1239–1253.

Das, A., and Y. H. Xie. 1998. Construction of transposon Tn*3phoA*: its application in defining the membrane topology of the *Agrobacterium tumefaciens* DNA transfer proteins. *Mol. Microbiol.* **27:**405–414.

Das, A., and Y. H. Xie. 2000. The *Agrobacterium* T-DNA transport pore proteins VirB8, VirB9, and VirB10 interact with one another. *J. Bacteriol.* **182:**758–763.

**da Silva, A. C., J. A. Ferro, F. C. Reinach, C. S. Farah, L. R. Furlan, R. B. Quaggio, C. B. Monteiro-Vitorello, M. A. Van Sluys, N. F. Almeida, L. M. Alves, A. M. do Amaral, M. C. Bertolini, L. E. Camargo, G. Camarotte, F. Cannavan, J. Cardozo, F. Chambergo, L. P. Ciapina, R. M. Cicarelli, L. L. Coutinho, J. R. Cursino-Santos, H. El Dorry, J. B. Faria, A. J. Ferreira, R. C. Ferreira, M. I. Ferro, E. F. Formighieri, M. C. Franco, C. C. Greggio, A. Gruber, A. M. Katsuyama, L. T. Kishi, R. P. Leite, E. G. Lemos, M. V. Lemos, E. C. Locali, M. A. Machado, A. M. Madeira, N. M. Martinez-Rossi, E. C. Martins, J. Meidanis, C. F. Menck, C. Y. Miyaki, D. H. Moon, L. M. Moreira, M. T. Novo, V. K. Okura, M. C. Oliveira, V. R. Oliveira, H. A. Pereira, A. Rossi, J. A. Sena, C. Silva, R. F. de Souza, L. A. Spinola, M. A. Takita, R. E. Tamura, E. C. Teixeira, R. I. Tezza, M. Trindade dos Santos, D. Truffi, S. M. Tsai,

F. F. White, J. C. Setubal, and J. P. Kitajima. 2002. Comparison of the genomes of two *Xanthomonas* pathogens with differing host specificities. *Nature* **417:**459–463.

Datta, S., C. Larkin, and J. F. Schildbach. 2003. Structural insights into single-stranded DNA binding and cleavage by F factor TraI. *Structure* **11:**1369–1379.

de Groot, M. J., P. Bundock, P. J. J. Hooykaas, and A. G. Beijersbergen. 1998. *Agrobacterium tumefaciens*-mediated transformation of filamentous fungi. *Nat. Biotechnol.* **16:**839–842.

Dehio, C. 2001. *Bartonella* interactions with endothelial cells and erythrocytes. *Trends Microbiol.* **9:**279–285.

Dehio, C. 2003. Recent progress in understanding *Bartonella*-induced vascular proliferation. *Curr. Opin. Microbiol.* **6:**61–65.

Dehio, C. 2004. Molecular and cellular basis of *Bartonella* pathogenesis. *Annu. Rev. Microbiol.* **58:**365–390.

Dehio, C., M. Meyer, J. Berger, H. Schwarz, and C. Lanz. 1997. Interaction of *Bartonella henselae* with endothelial cells results in bacterial aggregation on the cell surface and the subsequent engulfment and internalisation of the bacterial aggregate by a unique structure, the invasome. *J. Cell Sci.* **110:**2141–2154.

de la Cruz, F., and E. Lanka. 1998. Function of the Ti-plasmid Vir proteins: T-complex formation and transfer to the plant cell, p. 281–301. *In* H. P. Spaink, A. Kondorosi, and P. J. J. Hooykaas (ed.), *The* Rhizobiaceae: *Molecular Biology of Model Plant-Associated Bacteria.* Kluwer Academic Publishers, Dordrecht, The Netherlands.

Delrue, R. M., M. Martínez-Lorenzo, P. Lestrate, I. Danese, V. Bielarz, P. Mertens, B. De, X, A. Tibor, J. P. Gorvel, and J. J. Letesson. 2001. Identification of *Brucella* spp. genes involved in intracellular trafficking. *Cell. Microbiol.* **3:**487–497.

Deng, W., and E. W. Nester. 1998. Determinants of host specificity of *Agrobacterium* and their function, p. 321–338. *In* H. P. Spaink, A. Kondorosi, and P. J. J. Hooykaas (eds.), *The* Rhizobiaceae: *Molecular Biology of Model Plant-Associated Bacteria.* Kluwer Academic Publishers, Dordrecht, The Netherlands.

Dillard, J. P., and H. S. Seifert. 2001. A variable genetic island specific for *Neisseria gonorrhoeae* is involved in providing DNA for natural transformation and is found more often in disseminated infection isolates. *Mol. Microbiol.* **41:**263–277.

Ding, Z., K. Atmakuri, and P. J. Christie. 2003. The outs and ins of bacterial type IV secretion substrates. *Trends Microbiol.* **11:**527–535.

Disqué-Kochem, C., and B. Dreiseikelmann. 1997. The cytoplasmic DNA-binding protein TraM binds to the inner membrane protein TraD in vitro. *J. Bacteriol.* **179:**6133–6137.

Duckely, M., and B. Hohn. 2003. The VirE2 protein of *Agrobacterium tumefaciens:* the Yin and Yang of T-DNA transfer. *FEMS Microbiol. Lett.* **223:**1–6.

Eisenbrandt, R., M. Kalkum, E. M. Lai, R. Lurz, C. I. Kado, and E. Lanka. 1999. Conjugative pili of IncP plasmids, and the Ti plasmid T pilus are composed of cyclic subunits. *J. Biol. Chem.* **274:**22548–22555.

Eisenbrandt, R., M. Kalkum, R. Lurz, and E. Lanka. 2000. Maturation of IncP pilin precursors resembles the catalytic dyad-like mechanism of leader peptidases. *J. Bacteriol.* **182:**6751–6761.

Farizo, K. M., T. G. Cafarella, and D. L. Burns. 1996. Evidence for a ninth gene, *ptlI,* in the locus encoding the pertussis toxin secretion system of *Bordetella pertussis* and formation of a PtlI-PtlF complex. *J. Biol. Chem.* **271:**31643–31649.

Fernandez, D., T. A. Dang, G. M. Spudich, X. R. Zhou, B. R. Berger, and P. J. Christie. 1996b. The *Agrobacterium tumefaciens virB7* gene product, a proposed component of the T-complex transport apparatus, is a membrane-associated lipoprotein exposed at the periplasmic surface. *J. Bacteriol.* **178:**3156–3167.

Fernandez, D., G. M. Spudich, X. R. Zhou, and P. J. Christie. 1996a. The *Agrobacterium tumefaciens* VirB7 lipoprotein is required for stabilization of VirB proteins during assembly of the T-complex transport apparatus. *J. Bacteriol.* **178:**3168–3176.

Fields, B. S., R. F. Benson, and R. E. Besser. 2002. *Legionella* and Legionnaires' disease: 25 years of investigation. *Clin. Microbiol. Rev.* **15:**506–526.

Finberg, K. E., T. R. Muth, S. P. Young, J. B. Maken, S. M. Heitritter, A. N. Binns, and L. M. Banta. 1995. Interactions of VirB9, -10, and -11 with the membrane fraction of *Agrobacterium tumefaciens:* solubility studies provide evidence for tight associations. *J. Bacteriol.* **177:**4881–4889.

Fischer, W., J. Püls, R. Buhrdorf, B. Gebert, S. Odenbreit, and R. Haas. 2001. Systematic mutagenesis of the *Helicobacter pylori cag* pathogenicity island: essential genes for CagA translocation in host cells and induction of interleukin-8. *Mol. Microbiol.* **42:**1337–1348.

Freiberg, C., R. Fellay, A. Bairoch, W. J. Broughton, A. Rosenthal, and X. Perret. 1997. Molecular basis of symbiosis between *Rhizobium* and legumes. *Nature* **387:**394–401.

Frost, L. S. 1993. Conjugative pili and pilus-specific phages, p. 189–221. *In* D. B. Clewell (ed.), *Bacterial Conjugation.* Plenum Press, New York, N.Y.

Fullner, K. J., J. C. Lara, and E. W. Nester. 1996. Pilus assembly by *Agrobacterium* T-DNA transfer genes. *Science* **273:**1107–1109.

Fullner, K. J., K. M. Stephens, and E. W. Nester. 1994. An essential virulence protein of *Agrobacterium tumefaciens,* VirB4, requires an intact mononucleotide binding domain to function in transfer of T-DNA. *Mol. Gen. Genet.* **245:**704–715.

Gao, L. Y., and K. Y. Abu. 1999a. Activation of caspase 3 during *Legionella pneumophila*-induced apoptosis. *Infect. Immun.* **67:**4886–4894.

Gao, L. Y., and K. Y. Abu. 1999b. Apoptosis in macrophages and alveolar epithelial cells during early stages of infection by *Legionella pneumophila* and its role in cytopathogenicity. *Infect. Immun.* **67:**862–870.

Gao, L. Y., and Y. A. Kwaik. 2000. The modulation of host cell apoptosis by intracellular bacterial pathogens. *Trends Microbiol.* **8:**306–313.

Gelvin, S. B. 2003. *Agrobacterium*-mediated plant transformation: the biology behind the "gene-jockeying" tool. *Microbiol. Mol. Biol. Rev.* **67:**16–37.

Gilman, A. G. 1987. G proteins: transducers of receptor-generated signals. *Annu. Rev. Biochem.* **56:**615–649.

Gilmour, M. W., J. E. Gunton, T. D. Lawley, and D. E. Taylor. 2003. Interaction between the IncHI1 plasmid R27 coupling protein and type IV secretion system: TraG associates with the coiled-coil mating pair formation protein TrhB. *Mol. Microbiol.* **49:**105–116.

Gilmour, M. W., T. D. Lawley, M. M. Rooker, P. J. Newnham, and D. E. Taylor. 2001. Cellular location and temperature-dependent assembly of IncHI1 plasmid R27-encoded TrhC-associated conjugative transfer protein complexes. *Mol. Microbiol.* **42:**705–715.

Gilmour, M. W., and D. E. Taylor. 2004. A subassembly of R27-encoded transfer proteins is dependent on TrhC nucleoside triphosphate-binding motifs for function but not formation. *J. Bacteriol.* **186:**1606–1613.

Gomis-Rüth, F. X., and M. Coll. 2001. Structure of TrwB, a gatekeeper in bacterial conjugation. *Int. J. Biochem. Cell Biol.* **33:**839–843.

Gomis-Rüth, F. X., G. Moncalián, F. de la Cruz, and M. Coll. 2002. Conjugative plasmid protein TrwB, an integral membrane type IV secretion system coupling protein. Detailed structural features and mapping of the active site cleft. *J. Biol. Chem.* **277:**7556–7566.

Gomis-Rüth, F. X., G. Moncalián, R. Pérez-Luque, A. González, E. Cabezón, F. de la Cruz, and M. Coll. 2001. The bacterial conjugation protein TrwB resembles ring helicases and F1-ATPase. *Nature* **409:**637–641.

Goodner, B., G. Hinkle, S. Gattung, N. Miller, M. Blanchard, B. Qurollo, B. S. Goldman, Y. Cao, M. Askenazi, C. Halling, L. Mullin, K. Houmiel, J. Gordon, M. Vaudin, O. Iartchouk, A. Epp, F. Liu, C. Wollam, M. Allinger, D. Doughty, C. Scott, C. Lappas, B. Markelz, C. Flanagan, C. Crowell, J. Gurson, C. Lomo, C. Sear, G. Strub, C. Cielo, and S. Slater. 2001. Genome sequence of the plant pathogen and biotechnology agent *Agrobacterium tumefaciens* C58. *Science* **294:**2323–2328.

Gormley, E. P., and J. Davis. 1991. Transfer of plasmid RSF1010 by conjugation from *Escherichia coli* to *Streptomyces lividans* and *Mycobacterium smegmatis. J. Bacteriol.* **173:**6705–6708.

Gorvel, J. P., and E. Moreno. 2002. *Brucella* intracellular life: from invasion to intracellular replication. *Vet. Microbiol.* **90:**281–297.

Grahn, A. M., J. Haase, D. H. Bamford, and E. Lanka. 2000. Components of the RP4 conjugative transfer apparatus form an envelope structure bridging inner and outer membranes of donor cells: implications for related macromolecule transport systems. *J. Bacteriol.* **182:**1564–1574.

Grandoso, G., P. Avila, A. Cayón, M. A. Hernando, M. Llosa, and F. de la Cruz. 2000. Two active-site tyrosyl residues of protein TrwC act sequentially at the origin of transfer during plasmid R388 conjugation. *J. Mol. Biol.* **295:**1163–1172.

Grohmann, E., G. Muth, and M. Espinosa. 2003. Conjugative plasmid transfer in gram-positive bacteria. *Microbiol. Mol. Biol. Rev.* **67:**277–301.

Guasch, A., M. Lucas, G. Moncalián, M. Cabezas, R. Pérez-Luque, F. X. Gomis-Rüth, F. de la Cruz, and M. Coll. 2003. Recognition and processing of the origin of transfer DNA by conjugative relaxase TrwC. *Nat. Struct. Biol.* **10:**1002–1010.

Haase, J., R. Lurz, A. M. Grahn, D. H. Bamford, and E. Lanka. 1995. Bacterial conjugation mediated by plasmid RP4: donor-specific phage propagation, RSF1010 mobilization and pili production require the same Tra2 core components of a proposed DNA transport structure. *J. Bacteriol.* **177:**4779–4791.

Hamilton, C. M., H. Lee, P. L. Li, D. M. Cook, K. R. Piper, S. B. von Bodman, E. Lanka, W. Ream, and S. K. Farrand. 2000. TraG from RP4 and TraG and VirD4 from Ti plasmids confer relaxosome specificity to the conjugal transfer system of pTiC58. *J. Bacteriol.* **182:**1541–1548.

Hamori, P. J., and T. G. Slama. 1989. *Actinobacillus* prosthetic valve endocarditis. *Am. Heart J.* **118:**853–854.

Hanai, R., and J. C. Wang. 1993. The mechanism of sequence-specific DNA cleavage and strand transfer by φX174 gene A* protein. *J. Biol. Chem.* **268:**23830–23836.

Hapfelmeier, S., N. Domke, P. C. Zambryski, and C. Baron. 2000. VirB6 is required for stabilization of VirB5 and VirB3 and formation of VirB7 homodimers in *Agrobacterium tumefaciens. J. Bacteriol.* **182:**4505–4511.

Harley, M. J., and J. F. Schildbach. 2003. Swapping single-stranded DNA sequence specificities of relaxases from conjugative plasmids F and R100. *Proc. Natl. Acad. Sci. USA* **100:**11243–11248.

Harris, R. L., V. Hombs, and P. M. Silverman. 2001. Evidence that F-plasmid proteins TraV, TraK, and TraB assemble into an envelope-spanning structure in *Escherichia coli. Mol. Microbiol.* **42:**757–766.

Heinemann, J. A., and G. F. Sprague, Jr. 1989. Bacterial conjugative plasmids mobilize DNA transfer between bacteria and yeast. *Nature* **340:**205–209.

Hickey, T. E., A. L. McVeigh, D. A. Scott, R. E. Michielutti, A. Bixby, S. A. Carroll, A. L. Bourgeois, and P. Guerry. 2000. *Campylobacter jejuni* cytolethal distending toxin mediates release of interleukin-8 from intestinal epithelial cells. *Infect. Immun.* **68:**6535–6541.

Hickman, A. B., D. R. Ronning, R. M. Kotin, and F. Dyda. 2002. Structural unity among viral origin binding proteins: crystal structure of the nuclease domain of adeno-associated virus Rep. *Mol. Cell* **10:**327–337.

Higashi, H., R. Tsutsumi, S. Muto, T. Sugiyama, T. Azuma, M. Asaka, and M. Hatakeyama. 2002a. SHP-2 tyrosine phosphatase as an intracellular target of *Helicobacter pylori* CagA protein. *Science* **295:**683–686.

Higashi, H., R. Tsutsumi, A. Fujita, S. Yamazaki, M. Asaka, T. Azuma, and M. Hatakeyama. 2002b. Biological activity of the *Helicobacter pylori* virulence factor CagA is determined by variation in the tyrosine phosphorylation sites. *Proc. Natl. Acad. Sci. USA* **99:**14428–14433.

Hofreuter, D., S. Odenbreit, and R. Haas. 2001. Natural transformation competence in *Helicobacter pylori* is mediated by the basic components of a type IV secretion system. *Mol. Microbiol.* **41:**379–391.

Hormaeche, I., I. Alkorta, F. Moro, J. M. Valpuesta, F. M. Goñi, and F. de la Cruz. 2002. Purification and properties of TrwB, a hexameric, ATP-binding integral membrane protein essential for R388 plasmid conjugation. *J. Biol. Chem.* **277:**46456–46462.

Horwitz, M. A. 1983. The Legionnaires' disease bacterium (*Legionella pneumophila*) inhibits phagosome-lysosome fusion in human monocytes. *J. Exp. Med.* **158:**2108–2126.

Houdusse, A., A. G. Szent-Györgyi, and C. Cohen. 2000. Three conformational states of scallop myosin S1. *Proc. Natl. Acad. Sci. USA* **97:**11238–11243.

Howard, E. A., B. A. Winsor, G. De Vos, and P. Zambryski. 1989. Activation of the T-DNA transfer process in *Agrobacterium* results in the generation of a T-strand-protein complex: tight association of VirD2 with the 5′ ends of T-strands. *Proc. Natl. Acad. Sci. USA* **86:**4017–4021.

Howard, E. A., J. R. Zupan, V. Citovsky, and P. C. Zambryski. 1992. The VirD2 protein of *A. tumefaciens* contains a C-terminal bipartite nuclear localization signal: implications for nuclear uptake of DNA in plant cells. *Cell* **68:**109–118.

Hwang, B. J., K. M. Woo, A. L. Goldberg, and C. H. Chung. 1988. Protease Ti, a new ATP-dependent protease in *Escherichia coli,* contains protein-activated ATPase and proteolytic functions in distinct subunits. *J. Biol. Chem.* **263:**8727–8734.

Jakubowski, S. J., V. Krishnamoorthy, E. Cascales, and P. J. Christie. 2004. *Agrobacterium tumefaciens* VirB6 domains direct the ordered export of a DNA substrate through a type IV secretion system. *J. Mol. Biol.* **341:**961–977.

Jakubowski, S. J., V. Krishnamoorthy, and P. J. Christie. 2003. *Agrobacterium tumefaciens* VirB6 protein participates in formation of VirB7 and VirB9 complexes required for type IV secretion. *J. Bacteriol.* **185:**2867–2878.

Johnson, T. M., and A. Das. 1998. Organization and regulation of expression of the *Agrobacterium* virulence genes, p. 267–279. *In* H. P. Spaink, A. Kondorosi, and P. J. J. Hooykaas (ed.), *The Rhizobiaceae: Molecular Biology of Model Plant-Associated Bacteria.* Kluwer Academic Publishers, Dordrecht, The Netherlands.

Jones, A. L., K. Shirasu, and C. I. Kado. 1994. The product of the *virB4* gene of *Agrobacterium tumefaciens* promotes accumulation of VirB3 protein. *J. Bacteriol.* **176:**5255–5261.

Kalkum, M., R. Eisenbrandt, R. Lurz, and E. Lanka. 2002. Tying rings for sex. *Trends Microbiol.* **10:**382–387.

Karem, K. L., C. D. Paddock, and R. L. Regnery. 2000. *Bartonella henselae, B. quintana,* and *B. bacilliformis:* historical pathogens of emerging significance. *Microbes. Infect.* **2:**1193–1205.

Katada, T., M. Tamura, and M. Ui. 1983. The A protomer of islet-activating protein, pertussis toxin, as an active peptide catalyzing ADP-ribosylation of a membrane protein. *Arch. Biochem. Biophys.* **224:**290–298.

Katada, T., and M. Ui. 1982. Direct modification of the membrane adenylate cyclase system by islet-activating protein due to ADP-ribosylation of a membrane protein. *Proc. Natl. Acad. Sci. USA* **79:**3129–3133.

Kerr, A. 1992. The genus *Agrobacterium,* p. 2214–2235. *In* K. Balows (ed.), *The Prokaryotes,* vol. III, Springer-Verlag KG, Berlin, Germany.

Konkel, M. E., B. J. Kim, V. Rivera-Amill, and S. G. Garvis. 1999. Bacterial secreted proteins are required for the internalization of *Campylobacter jejuni* into cultured mammalian cells. *Mol. Microbiol.* **32:**691–701.

Koraimann, G. 2003. Lytic transglycosylases in macromolecular transport systems of Gram-negative bacteria. *Cell Mol. Life Sci.* **60:**2371–2388.

Kotilainen, M. M., A. M. Grahn, J. K. Bamford, and D. H. Bamford. 1993. Binding of an *Escherichia coli* double-stranded DNA virus PRD1 to a receptor coded by an IncP-type plasmid. *J. Bacteriol.* **175:**3089–3095.

Krall, L., U. Wiedemann, G. Unsin, S. Weiss, N. Domke, and C. Baron. 2002. Detergent extraction identifies different VirB protein subassemblies of the type IV secretion machinery in the membranes of *Agrobacterium tumefaciens. Proc. Natl. Acad. Sci. USA* **99:**11405–11410.

Krause, S., M. Bárcena, W. Pansegrau, R. Lurz, J. M. Carazo, and E. Lanka. 2000a. Sequence-related protein export NTPases encoded by the conjugative transfer region of RP4 and by the *cag* pathogenicity island of *Helicobacter pylori* share similar hexameric ring structures. *Proc. Natl. Acad. Sci. USA* **97:**3067–3072.

Krause, S., W. Pansegrau, R. Lurz, F. de la Cruz, and E. Lanka. 2000b. Enzymology of type IV macromolecule secretion systems: the conjugative transfer regions of plasmids RP4 and R388 and the *cag* pathogenicity island of *Helicobacter pylori* encode structurally and functionally related nucleoside triphosphate hydrolases. *J. Bacteriol.* **182:**2761–2770.

Kumar, R. B., and A. Das. 2001. Functional analysis of the *Agrobacterium tumefaciens* T-DNA transport pore protein VirB8. *J. Bacteriol.* **183:**3636–3641.

Kumar, R. B., and A. Das. 2002. Polar location and functional domains of the *Agrobacterium tumefaciens* DNA transfer protein VirD4. *Mol. Microbiol.* **43:**1523–1532.

Kumar, R. B., Y. H. Xie, and A. Das. 2000. Subcellular localization of the *Agrobacterium tumefaciens* T-DNA transport pore proteins: VirB8 is essential for the assembly of the transport pore. *Mol. Microbiol.* **36:**608–617.

Kunik, T., T. Tzfira, Y. Kapulnik, Y. Gafni, C. Dingwall, and V. Citovsky. 2001. Genetic transformation of HeLa cells by *Agrobacterium. Proc. Natl. Acad. Sci. USA* **98:**1871–1876.

Kurenbach, B., C. Bohn, J. Prabhu, M. Abudukerim, U. Szewzyk, and E. Grohmann. 2003. Intergeneric transfer of the *Enterococcus faecalis* plasmid pIP501 to *Escherichia coli* and *Streptomyces lividans* and sequence analysis of its *tra* region. *Plasmid* **50:**86–93.

Lai, E. M., and C. I. Kado. 1998. Processed VirB2 is the major subunit of the promiscuous pilus of *Agrobacterium tumefaciens. J. Bacteriol.* **180:**2711–2717.

Lai, E. M., and C. I. Kado. 2000. The T-pilus of *Agrobacterium tumefaciens. Trends Microbiol.* **8:**361–369.

Lederberg, J., and E. Tatum. 1946. Gene recombination in *E. coli. Nature* **158:**558.

Lee, M. H., N. Kosuk, J. Bailey, B. Traxler, and C. Manoil. 1999. Analysis of F factor TraD membrane topology by use of gene fusions and trypsin-sensitive insertions. *J. Bacteriol.* **181:**6108–6113.

Lenzen, C. U., D. Steinmann, S. W. Whiteheart, and W. I. Weis. 1998. Crystal structure of the hexamerization domain of *N*-ethylmaleimide-sensitive fusion protein. *Cell* **94:**525–536.

Lessl, M., W. Pansegrau, and E. Lanka. 1992a. Relationship of DNA-transfer-systems: essential transfer factors of plasmids RP4, Ti and F share common sequences. *Nucleic Acids Res.* **20:**6099–6100.

Lessl, M., D. Balzer, R. Lurz, V. L. Waters, D. G. Guiney, and E. Lanka. 1992b. Dissection of IncP conjugative plasmid transfer: definition of the transfer region Tra2 by mobilization of the Tra1 region in *trans. J. Bacteriol.* **174:**2493–2500.

Lessl, M., D. Balzer, K. Weyrauch, and E. Lanka. 1993. The mating pair formation system of plasmid RP4 defined by RSF1010 mobilization and donor-specific phage propagation. *J. Bacteriol.* **175:**6415–6425.

Lessl, M., and E. Lanka. 1994. Common mechanisms in bacterial conjugation and Ti-mediated T-DNA transfer to plant cells. *Cell* **77:**321–324.

Li, P. L., D. M. Everhart, and S. K. Farrand. 1998. Genetic and sequence analysis of the pTiC58 *trb* locus, encoding a mating-pair formation system related to members of the type IV secretion family. *J. Bacteriol.* **180:**6164–6172.

Lilley, A., P. Young, and M. Bailey. 2000. Bacterial population genetics: do plasmids maintain bacterial diversity and adaptation? p. 287–300. *In* C. M. Thomas (ed.), *The Horizontal Gene Pool.* Harwood Academic Publishers, Amsterdam, The Netherlands.

Liu, Z., and A. N. Binns. 2003. Functional subsets of the *virB* type IV transport complex proteins involved in the capacity of *Agrobacterium tumefaciens* to serve as a recipient in *virB*-mediated conjugal transfer of plasmid RSF1010. *J. Bacteriol.* **185:**3259–3269.

Llosa, M., S. Bolland, and F. de la Cruz. 1994. Genetic organization of the conjugal DNA processing region of the IncW plasmid R388. *J. Mol. Biol.* **235:**448–464.

Llosa, M., and F. de la Cruz. 2005. Bacterial conjugation: a potential tool for genomic engineering. *Res. Microbiol.* **156:**1–6.

Llosa, M., F. X. Gomis-Rüth, M. Coll, and F. de la Cruz. 2002. Bacterial conjugation: a two-step mechanism for DNA transport. *Mol. Microbiol.* **45:**1–8.

Llosa, M., G. Grandoso, M. A. Hernando, and F. de la Cruz. 1996. Functional domains in protein TrwC of plasmid R388: dissected DNA strand transferase and DNA helicase activities reconstitute protein function. *J. Mol. Biol.* **264:**56–67.

Llosa, M., S. Zunzunegui, and F. de la Cruz. 2003. Conjugative coupling proteins interact with cognate and heterologous VirB10-like proteins while exhibiting specificity for cognate relaxosomes. *Proc. Natl. Acad. Sci. USA* **100:**10465–10470.

Llosa, M., J. Zupan, C. Baron, and P. Zambryski. 2000. The N- and C-terminal portions of the *Agrobacterium* VirB1 protein independently enhance tumorigenesis. *J. Bacteriol.* **182:**3437–3445.

Locht, C., and J. M. Keith. 1986. Pertussis toxin gene: nucleotide sequence and genetic organization. *Science* **232:**1258–1264.

Luo, Z. Q., and R. R. Isberg. 2004. Multiple substrates of the *Legionella pneumophila* Dot/Icm system identified by interbacterial protein transfer. *Proc. Natl. Acad. Sci. USA* **101:**841–846.

Lybarger, R. L., and M. Sandkvist. 2004. A hitchhiker's guide to type IV secretion. *Science* **304:**1122–1123.

Machón, C., S. Rivas, A. Albert, F. M. Goñi, and F. de la Cruz. 2002. TrwD, the hexameric traffic ATPase encoded by plasmid R388, induces membrane destabilization and hemifusion of lipid vesicles. *J. Bacteriol.* **184:**1661–1668.

Maeno, N., H. Oda, K. Yoshiie, M. R. Wahid, T. Fujimura, and S. Matayoshi. 1999. Live *Bartonella henselae* enhances endothelial cell proliferation without direct contact. *Microb. Pathog.* **27:**419–427.

Malek, J. A., J. M. Wierzbowski, W. Tao, S. A. Bosak, D. J. Saranga, L. Doucette-Stamm, D. R. Smith, P. J. McEwan, and K. J. McKernan. 2004. Protein interaction mapping on a functional shotgun sequence of *Rickettsia sibirica. Nucleic Acids Res.* **32:**1059–1064.

Manders, S. M. 1996. Bacillary angiomatosis. *Clin. Dermatol.* **14:**295–299.

Marques, M. V., A. M. da Silva, and S. L. Gomes. 2001. Genetic organization of plasmid pXF51 from the plant pathogen *Xylella fastidiosa. Plasmid* **45:**184–199.

Martin, B. F., B. M. Derby, G. N. Budzilovich, and J. Ransohoff. 1967. Brain abscess due to *Actinobacillus actinomycetemcomitans. Neurology* **17:**833–837.

Masui, S., T. Sasaki, and H. Ishikawa. 2000. Genes for the type IV secretion system in an intracellular symbiont, *Wolbachia,* a causative agent of various sexual alterations in arthropods. *J. Bacteriol.* **182:**6529–6531.

Matson, S. W., J. K. Sampson, and D. R. Byrd. 2001. F plasmid conjugative DNA transfer: the TraI helicase activity is essential for DNA strand transfer. *J. Biol. Chem.* **276:**2372–2379.

Maurin, M., and D. Raoult. 1999. Q fever. *Clin. Microbiol. Rev.* **12:**518–553.

Mimuro, H., T. Suzuki, J. Tanaka, M. Asahi, R. Haas, and C. Sasakawa. 2002. Grb2 is a key mediator of *Helicobacter pylori* CagA protein activities. *Mol. Cell* **10:**745–755.

Miyamoto, H., S. Yoshida, H. Taniguchi, and H. A. Shuman. 2003. Virulence conversion of *Legionella pneumophila* by conjugal transfer of chromosomal DNA. *J. Bacteriol.* **185:**6712–6718.

Molmeret, M., S. D. Zink, L. Han, A. Abu-Zant, R. Asari, D. M. Bitar, and K. Y. Abu. 2004. Activation of caspase-3 by the Dot/Icm virulence system is essential for arrested biogenesis of the *Legionella*-containing phagosome. *Cell. Microbiol.* **6:**33–48.

Moncalián, G., E. Cabezón, I. Alkorta, M. Valle, F. Moro, J. M. Valpuesta, F. M. Goñi, and F. de la Cruz. 1999. Characterization of ATP and DNA binding activities of TrwB, the coupling protein essential in plasmid R388 conjugation. *J. Biol. Chem.* **274:**36117–36124.

Moore, D., C. M. Hamilton, K. Maneewannakul, Y. Mintz, L. S. Frost, and K. Ippen-Ihler. 1993. The *Escherichia coli* K-12 F plasmid gene *traX* is required for acetylation of F pilin. *J. Bacteriol.* **175:**1375–1383.

Motallebi-Veshareh, M., D. Balzer, E. Lanka, G. Jagura-Burdzy, and C. M. Thomas. 1992. Conjugative transfer functions of broad-host-range plasmid RK2 are coregulated with vegetative replication. *Mol. Microbiol.* **6:**907–920.

Mushegian, A. R., K. J. Fullner, E. V. Koonin, and E. W. Nester. 1996. A family of lysozyme-like virulence factors in bacterial pathogens of plants and animals. *Proc. Natl. Acad. Sci. USA* **93:**7321–7326.

Müller, A., J. Hacker, and B. C. Brand. 1996. Evidence for apoptosis of human macrophage-like HL-60 cells by *Legionella pneumophila* infection. *Infect. Immun.* **64:**4900–4906.

Nagai, H., J. C. Kagan, X. Zhu, R. A. Kahn, and C. R. Roy. 2002. A bacterial guanine nucleotide exchange factor activates ARF on *Legionella* phagosomes. *Science* **295:**679–682.

Nagai, H., and C. R. Roy. 2001. The DotA protein from *Legionella pneumophila* is secreted by a novel process that requires the Dot/Icm transporter. *EMBO J.* **20:**5962–5970.

Nicosia, A., M. Perugini, C. Franzini, M. C. Casagli, M. G. Borri, G. Antoni, M. Almoni, P. Neri, G. Ratti, and R. Rappuoli. 1986. Cloning and sequencing of the pertussis toxin genes: operon structure and gene duplication. *Proc. Natl. Acad. Sci. USA* **83:**4631–4635.

Novak, K. F., B. Dougherty, and M. Peláez. 2001. *Actinobacillus actinomycetemcomitans* harbours type IV secretion system genes on a plasmid and in the chromosome. *Microbiology* **147:**3027–3035.

Oberhelman, R. A., and D. N. Taylor. 2000. *Campylobacter* infections in developing countries, p. 139–153. *In* I. Nachamkin and M. J. Blaser (ed.), Campylobacter, 2nd ed. ASM Press, Washington, D.C.

O'Callaghan, D., C. Cazevieille, A. Allardet-Servent, M. L. Boschiroli, G. Bourg, V. Foulongne, P. Frutos, Y. Kulakov, and M. Ramuz. 1999. A homologue of the *Agrobacterium tumefaciens* VirB and *Bordetella pertussis* Ptl type IV secretion systems is essential for intracellular survival of *Brucella suis. Mol. Microbiol.* **33:**1210–1220.

Odenbreit, S., J. Püls, B. Sedlmaier, E. Gerland, W. Fischer, and R. Haas. 2000. Translocation of *Helicobacter pylori* CagA into gastric epithelial cells by type IV secretion. *Science* **287:**1497–1500.

Ogata, H., S. Audic, P. Renesto-Audiffren, P. E. Fournier, V. Barbe, D. Samson, V. Roux, P. Cossart, J. Weissenbach, J. M. Claverie, and D. Raoult. 2001. Mechanisms of evolution in *Rickettsia conorii* and *R. prowazekii. Science* **293:**2093–2098.

Ohashi, N., N. Zhi, Q. Lin, and Y. Rikihisa. 2002. Characterization and transcriptional analysis of gene clusters for a type IV secretion machinery in human granulocytic and monocytic ehrlichiosis agents. *Infect. Immun.* **70:**2128–2138.

Padmalayam, I., K. Karem, B. Baumstark, and R. Massung. 2000. The gene encoding the 17-kDa antigen of *Bartonella henselae* is located within a cluster of genes homologous to the *virB* virulence operon. *DNA Cell Biol.* **19:**377–382.

Panicker, M. M., and E. G. Minkley, Jr. 1992. Purification and properties of the F sex factor TraD protein, an inner membrane conjugal transfer protein. *J. Biol. Chem.* **267:**12761–12766.

Pansegrau, W., G. Ziegelin, and E. Lanka. 1990a. Covalent association of the *traI* gene product of plasmid RP4 with the 5′-terminal nucleotide at the relaxation nick site. *J. Biol. Chem.* **265:**10637–10644.

Pansegrau, W., D. Balzer, V. Kruft, R. Lurz, and E. Lanka. 1990b. *In vitro* assembly of relaxosomes at the transfer origin of plasmid RP4. *Proc. Natl. Acad. Sci. USA* **87:**6555–6559.

Pansegrau, W., and E. Lanka. 1996a. Mechanisms of initiation and termination reactions in conjugative DNA processing. Independence of tight substrate binding and catalytic activity of relaxase (TraI) of IncPα plasmid RP4. *J. Biol. Chem.* **271:**13068–13076.

Pansegrau, W., and E. Lanka. 1996b. Enzymology of DNA transfer by conjugative mechanisms. *Prog. Nucleic Acid Res. Mol. Biol.* **54:**197–251.

Pansegrau, W., F. Schoumacher, B. Hohn, and E. Lanka. 1993. Site-specific cleavage and joining of single-stranded-DNA by VirD2 protein of *Agrobacterium tumefaciens* Ti plasmids: analogy to bacterial conjugation. *Proc. Natl. Acad. Sci. USA* **90:**11538–11542.

Parkhill, J., B. W. Wren, K. Mungall, J. M. Ketley, C. Churcher, D. Basham, T. Chillingworth, R. M. Davies, T. Feltwell, S. Holroyd, K. Jagels, A. V. Karlyshev, S. Moule, M. J. Pallen, C. W. Penn, M. A.

Quail, M. A. Rajandream, K. M. Rutherford, A. H. van Vliet, S. Whitehead, and B. G. Barrell. 2000. The genome sequence of the food-borne pathogen *Campylobacter jejuni* reveals hypervariable sequences. *Nature* **403:**665–668.

Patel, S., and M. Latterich. 1998. The AAA team: related ATPases with diverse functions. *Trends Cell Biol.* **8:**65–71.

Patel, S. S., and K. M. Picha. 2000. Structure and function of hexameric helicases. *Annu. Rev. Biochem.* **69:**651–697.

Piers, K. L., J. D. Heath, X. Liang, K. M. Stephens, and E. W. Nester. 1996. *Agrobacterium tumefaciens*-mediated transformation of yeast. *Proc. Natl. Acad. Sci. USA* **93:**1613–1618.

Planet, P. J., S. C. Kachlany, R. DeSalle, D. H. Figurski. 2001. Phylogeny of genes for secretion NTPases: identification of the widespread *tadA* subfamily and development of a diagnostic key for gene classification. *Proc. Natl. Acad. Sci. USA* **98:**2503–2508.

Purcell, M., and H. A. Shuman. 1998. The *Legionella pneumophila icmGCDJBF* genes are required for killing of human macrophages. *Infect. Immun.* **66:**2245–2255.

Püls, J., W. Fischer, and R. Haas. 2002. Activation of *Helicobacter pylori* CagA by tyrosine phosphorylation is essential for dephosphorylation of host cell proteins in gastric epithelial cells. *Mol. Microbiol.* **43:**961–969.

Rabel, C., A. M. Grahn, R. Lurz, and E. Lanka. 2003. The VirB4 family of proposed traffic nucleoside triphosphatases: common motifs in plasmid RP4 TrbE are essential for conjugation and phage adsorption. *J. Bacteriol.* **185:**1045–1058.

Rain, J. C., L. Selig, H. De Reuse, V. Battaglia, C. Reverdy, S. Simon, G. Lenzen, F. Petel, J. Wojcik, V. Schächter, Y. Chemama, A. Labigne, and P. Legrain. 2001. The protein-protein interaction map of *Helicobacter pylori. Nature* **409:**211–215.

Rambow-Larsen, A. A., and A. A. Weiss. 2002. The PtlE protein of *Bordetella pertussis* has peptidoglycanase activity required for Ptl-mediated pertussis toxin secretion. *J. Bacteriol.* **184:**2863–2869.

Ranson, N. A., G. W. Farr, A. M. Roseman, B. Gowen, W. A. Fenton, A. L. Horwich, and H. R. Saibil. 2001. ATP-bound states of GroEL captured by cryo-electron microscopy. *Cell* **107:**869–879.

Rice, S., A. W. Lin, D. Safer, C. L. Hart, N. Naber, B. O. Carragher, S. M. Cain, E. Pechatnikova, E. M. Wilson-Kubalek, M. Whittaker, E. Pate, R. Cooke, E. W. Taylor, R. A. Milligan, and R. D. Vale. 1999. A structural change in the kinesin motor protein that drives motility. *Nature* **402:**778–784.

Ridenour, D. A., S. L. Cirillo, S. Feng, M. M. Samrakandi, and J. D. Cirillo. 2003. Identification of a gene that affects the efficiency of host cell infection by *Legionella pneumophila* in a temperature-dependent fashion. *Infect. Immun.* **71:**6256–6263.

Rivas, S., S. Bolland, E. Cabezón, F. M. Goñi, and F. de la Cruz. 1997. TrwD, a protein encoded by the IncW plasmid R388, displays an ATP hydrolase activity essential for bacterial conjugation. *J. Biol. Chem.* **272:**25583–25590.

Rohde, M., J. Püls, R. Buhrdorf, W. Fischer, and R. Haas. 2003. A novel sheathed surface organelle of the *Helicobacter pylori* cag type IV secretion system. *Mol. Microbiol.* **49:**219–234.

Rossi, L., B. Tinland, and B. Hohn. 1998. Role of virulence proteins of *Agrobacterium* in the plant, p. 303–320. *In* H. P. Spaink, A. Kondorosi, and P. J. J. Hooykaas (ed.), *The* Rhizobiaceae: *Molecular Biology of Model Plant-Associated Bacteria.* Kluwer Academic Publishers, Dordrecht, The Netherlands.

Rouiller, I., V. M. Butel, M. Latterich, R. A. Milligan, and E. M. Wilson-Kubalek. 2000. A major conformational change in p97 AAA ATPase upon ATP binding. *Mol. Cell* **6:**1485–1490.

Roy, C. R., and L. G. Tilney. 2002. The road less traveled: transport of *Legionella* to the endoplasmic reticulum. *J. Cell Biol.* **158:**415–419.

Sagulenko, V., E. Sagulenko, S. Jakubowski, E. Spudich, and P. J. Christie. 2001a. VirB7 lipoprotein is exocellular and associates with the *Agrobacterium tumefaciens* T pilus. *J. Bacteriol.* **183:**3642–3651.

Sagulenko, E., V. Sagulenko, J. Chen, and P. J. Christie. 2001b. Role of *Agrobacterium* VirB11 ATPase in T-pilus assembly and substrate selection. *J. Bacteriol.* **183:**5813–5825.

Salanoubat, M., S. Genin, F. Artiguenave, J. Gouzy, S. Mangenot, M. Arlat, A. Billault, P. Brottier, J. C. Camus, L. Cattolico, M. Chandler, N. Choisne, C. Claudel-Renard, S. Cunnac, N. Demange, C. Gaspin, M. Lavie, A. Moisan, C. Robert, W. Saurin, T. Schiex, P. Siguier, P. Thébault, M. Whalen, P. Wincker, M. Levy, J. Weissenbach, and C. A. Boucher. 2002. Genome sequence of the plant pathogen *Ralstonia solanacearum. Nature* **415:**497–502.

Salmond, G. P. C. 1994. Secretion of extracellular virulence factors by plant pathogenic bacteria. *Annu. Rev. Phytopathol.* **32:**181–200.

Samartino, L. E., and F. M. Enright. 1993. Pathogenesis of abortion of bovine brucellosis. *Comp. Immunol. Microbiol. Infect. Dis.* **16**:95–101.

Savvides, S. N., H. J. Yeo, M. R. Beck, F. Blaesing, R. Lurz, E. Lanka, R. Buhrdorf, W. Fischer, R. Haas, and G. Waksman. 2003. VirB11 ATPases are dynamic hexameric assemblies: new insights into bacterial type IV secretion. *EMBO J.* **22**:1969–1980.

Schmid, M. C., R. Schulein, M. Dehio, G. Denecker, I. Carena, and C. Dehio. 2004. The VirB type IV secretion system of *Bartonella henselae* mediates invasion, proinflammatory activation and antiapoptotic protection of endothelial cells. *Mol. Microbiol.* **52**:81–92.

Schmidt-Eisenlohr, H., N. Domke, C. Angerer, G. Wanner, P. C. Zambryski, and C. Baron. 1999a. Vir proteins stabilize VirB5 and mediate its association with the T pilus of *Agrobacterium tumefaciens*. *J. Bacteriol.* **181**:7485–7492.

Schmidt-Eisenlohr, H., N. Domke, and C. Baron. 1999b. TraC of IncN plasmid pKM101 associates with membranes and extracellular high-molecular-weight structures in *Escherichia coli. J. Bacteriol.* **181**:5563–5571.

Schneider, E., and S. Hunke. 1998. ATP-binding-cassette (ABC) transport systems: functional and structural aspects of the ATP-hydrolyzing subunits/domains. *FEMS Microbiol. Rev.* **22**:1–20.

Schneiker, S., M. Keller, M. Dröge, E. Lanka, A. Pühler, and W. Selbitschka. 2001. The genetic organization and evolution of the broad host range mercury resistance plasmid pSB102 isolated from a microbial population residing in the rhizosphere of alfalfa. *Nucleic Acids Res.* **29**:5169–5181.

Schrammeijer, B., A. Den Dulk-Ras, A. C. Vergunst, E. Jurado Jacome, and P. J. J. Hooykaas. 2003. Analysis of Vir protein translocation from *Agrobacterium tumefaciens* using *Saccharomyces cerevisiae* as a model: evidence for transport of a novel effector protein VirE3. *Nucleic Acids Res.* **31**:860–868.

Schröder, G., S. Krause, E. L. Zechner, B. Traxler, H. J. Yeo, R. Lurz, G. Waksman, and E. Lanka. 2002. TraG-like proteins of DNA transfer systems and of the *Helicobacter pylori* type IV secretion system: inner membrane gate for exported substrates? *J. Bacteriol.* **184**:2767–2779.

Schröder, G., and E. Lanka. 2003. TraG-like proteins of type IV secretion systems: functional dissection of the multiple activities of TraG (RP4) and TrwB (R388). *J. Bacteriol.* **185**:4371–4381.

Schulein, R., and C. Dehio. 2002. The VirB/VirD4 type IV secretion system of *Bartonella* is essential for establishing intraerythrocytic infection. *Mol. Microbiol.* **46**:1053–1067.

Schulein, R., P. Guye, T. A. Rhomberg, M. C. Schmid, G. Schröder, A. C. Vergunst, I. Carena, and C. Dehio. 2005. A bipartite signal mediates the transfer of type IV secretion substrates of *Bartonella henselae* into human cells. *Proc. Natl. Acad. Sci. USA* **102**:856–861.

Segal, E. D., J. Cha, J. Lo, S. Falkow, and L. S. Tompkins. 1999a. Altered states: involvement of phosphorylated CagA in the induction of host cellular growth changes by *Helicobacter pylori. Proc. Natl. Acad. Sci. USA* **96**:14559–14564.

Segal, G., M. Purcell, and H. A. Shuman. 1998. Host cell killing and bacterial conjugation require overlapping sets of genes within a 22-kb region of the *Legionella pneumophila* genome. *Proc. Natl. Acad. Sci. USA* **95**:1669–1674.

Segal, G., J. J. Russo, and H. A. Shuman. 1999b. Relationships between a new type IV secretion system and the *icm/dot* virulence system of *Legionella pneumophila. Mol. Microbiol.* **34**:799–809.

Segal, G., and H. A. Shuman. 1997. Characterization of a new region required for macrophage killing by *Legionella pneumophila. Infect. Immun.* **65**:5057–5066.

Selbach, M., S. Moese, T. F. Meyer, and S. Backert. 2002. Functional analysis of the *Helicobacter pylori cag* pathogenicity island reveals both VirD4-CagA-dependent and VirD4-CagA-independent mechanisms. *Infect. Immun.* **70**:665–671.

Selmi, C., D. L. Balkwill, P. Invernizzi, A. A. Ansari, R. L. Coppel, M. Podda, P. S. Leung, T. P. Kenny, J. Van De Water, M. H. Nantz, M. J. Kurth, and M. E. Gershwin. 2003. Patients with primary biliary cirrhosis react against a ubiquitous xenobiotic-metabolizing bacterium. *Hepatology* **38**:1250–1257.

Seubert, A., R. Hiestand, F. de la Cruz, and C. Dehio. 2003. A bacterial conjugation machinery recruited for pathogenesis. *Mol. Microbiol.* **49**:1253–1266.

Sexton, J. A., J. S. Pinkner, R. Roth, J. E. Heuser, S. J. Hultgren, and J. P. Vogel. 2004. The *Legionella pneumophila* PilT homologue DotB exhibits ATPase activity that is critical for intracellular growth. *J. Bacteriol.* **186**:1658–1666.

Sexton, J. A., and J. P. Vogel. 2002. Type IVB secretion by intracellular pathogens. *Traffic* **3**:178–185.

She, Q., H. Phan, R. A. Garrett, S. V. Albers, K. M. Stedman, and W. Zillig. 1998. Genetic profile of pNOB8 from *Sulfolobus:* the first conjugative plasmid from an archaeon. *Extremophiles* **2**:417–425.

Sherburne, C. K., T. D. Lawley, M. W. Gilmour, F. R. Blattner, V. Burland, E. Grotbeck, D. J. Rose, and D. E. Taylor. 2000. The complete DNA sequence and analysis of R27, a large IncHI plasmid from *Salmonella typhi* that is temperature sensitive for transfer. *Nucleic Acids Res.* **28:**2177–2186.

Shirasu, K., and C. I. Kado. 1993. Membrane location of the Ti plasmid VirB proteins involved in the biosynthesis of a pilin-like conjugative structure on *Agrobacterium tumefaciens*. *FEMS Microbiol. Lett.* **111:**287–294.

Shirasu, K., Z. Koukolíková-Nicola, B. Hohn, and C. I. Kado. 1994. An inner-membrane-associated virulence protein essential for T- DNA transfer from *Agrobacterium tumefaciens* to plants exhibits ATPase activity and similarities to conjugative transfer genes. *Mol. Microbiol.* **11:**581–588.

Simpson, A. J., F. C. Reinach, P. Arruda, F. A. Abreu, M. Acencio, R. Alvarenga, L. M. Alves, J. E. Araya, G. S. Baia, C. S. Baptista, M. H. Barros, E. D. Bonaccorsi, S. Bordin, J. M. Bové, M. R. Briones, M. R. Bueno, A. A. Camargo, L. E. Camargo, D. M. Carraro, H. Carrer, N. B. Colauto, C. Colombo, F. F. Costa, M. C. Costa, C. M. Costa-Neto, L. L. Coutinho, M. Cristofani, E. Dias-Neto, C. Docena, H. El Dorry, A. P. Facincani, A. J. Ferreira, V. C. Ferreira, J. A. Ferro, J. S. Fraga, S. C. França, M. C. Franco, M. Frohme, L. R. Furlan, M. Garnier, G. H. Goldman, M. H. Goldman, S. L. Gomes, A. Gruber, P. L. Ho, J. D. Hoheisel, M. L. Junqueira, E. L. Kemper, J. P. Kitajima, J. E. Krieger, E. E. Kuramae, F. Laigret, M. R. Lambais, L. C. Leite, E. G. Lemos, M. V. Lemos, S. A. Lopes, C. R. Lopes, J. A. Machado, M. A. Machado, A. M. Madeira, H. M. Madeira, C. L. Marino, M. V. Marques, E. A. Martins, E. M. Martins, A. Y. Matsukuma, C. F. Menck, E. C. Miracca, C. Y. Miyaki, C. B. Monteriro-Vitorello, D. H. Moon, M. A. Nagai, A. L. Nascimento, L. E. Netto, A. Nhani, Jr., F. G. Nobrega, L. R. Nunes, M. A. Oliveira, M. C. de Oliveira, R. C. de Oliveira, D. A. Palmieri, A. Paris, B. R. Peixoto, G. A. Pereira, H. A. Pereira, Jr., J. B. Pesquero, R. B. Quaggio, P. G. Roberto, V. Rodrigues, A. J. De M. Rosa, V. E. De Rosa, Jr., R. G. de Sá, R. V. Santelli, H. E. Sawasaki, A. C. da Silva, A. M. da Silva, F. R. da Silva, W. A. Da Silva, Jr., J. F. da Silveira, M. L. Silvestri, W. J. Siqueira, A. A. de Souza, A. P. de Souza, M. F. Terenzi, D. Truffi, S. M. Tsai, M. H. Tsuhako, H. Vallada, M. A. Van Sluys, S. Verjovski-Almeida, A. L. Vettore, M. A. Zago, M. Zatz, J. Meidanis, and J. C. Setubal. 2000. The genome sequence of the plant pathogen *Xylella fastidiosa*. *Nature* **406:**151–157.

Singleton, M. R., M. R. Sawaya, T. Ellenberger, and D. B. Wigley. 2000. Crystal structure of T7 gene 4 ring helicase indicates a mechanism for sequential hydrolysis of nucleotides. *Cell* **101:**589–600.

Smith, E. F. 1907. A plant-tumor of bacterial origin. *Science* **25:**671–673.

Sola-Landa, A., J. Pizarro-Cerdá, M. J. Grilló, E. Moreno, I. Moriyón, J. M. Blasco, J. P. Gorvel, and I. López-Goñi. 1998. A two-component regulatory system playing a critical role in plant pathogens and endosymbionts is present in *Brucella abortus* and controls cell invasion and virulence. *Mol. Microbiol.* **29:**125–138.

Soto, G. E., and S. J. Hultgren. 1999. Bacterial adhesins: common themes and variations in architecture and assembly. *J. Bacteriol.* **181:**1059–1071.

Spudich, G. M., D. Fernandez, X. R. Zhou, and P. J. Christie. 1996. Intermolecular disulfide bonds stabilize VirB7 homodimers and VirB7/VirB9 heterodimers during biogenesis of the *Agrobacterium tumefaciens* T-complex transport apparatus. *Proc. Natl. Acad. Sci. USA* **93:**7512–7517.

Stachel, S. E., and E. W. Nester. 1986. The genetic and transcriptional organization of the *vir* region of the A6 Ti plasmid of *Agrobacterium tumefaciens*. *EMBO J.* **5:**1445–1454.

Stachel, S. E., B. Timmerman, and P. Zambryski. 1986. Generation of single-stranded T-DNA molecules during the initial-stages of T-DNA transfer from *Agrobacterium tumefaciens* to plant cells. *Nature* **322:**706–712.

Stedman, K. M., Q. She, H. Phan, I. Holz, H. Singh, D. Prangishvili, R. Garrett, and W. Zillig. 2000. pING family of conjugative plasmids from the extremely thermophilic archaeon *Sulfolobus islandicus:* insights into recombination and conjugation in *Crenarchaeota. J. Bacteriol.* **182:**7014–7020.

Stein, M., F. Bagnoli, R. Halenbeck, R. Rappuoli, W. J. Fantl, and A. Covacci. 2002. c-Src/Lyn kinases activate *Helicobacter pylori* CagA through tyrosine phosphorylation of the EPIYA motifs. *Mol. Microbiol.* **43:**971–980.

Stein, M., R. Rappuoli, and A. Covacci. 2000. Tyrosine phosphorylation of the *Helicobacter pylori* CagA antigen after *cag*-driven host cell translocation. *Proc. Natl. Acad. Sci. USA* **97:**1263–1268.

Stein, P. E., A. Boodhoo, G. D. Armstrong, S. A. Cockle, M. H. Klein, and R. J. Read. 1994. The crystal structure of pertussis toxin. *Structure* **2:**45–57.

Story, R. M., and T. A. Steitz. 1992. Structure of the RecA protein-ADP complex. *Nature* **355:**374–376.

Stouthamer, R., J. A. Breeuwer, and G. D. Hurst. 1999. *Wolbachia pipientis:* microbial manipulator of arthropod reproduction. *Annu. Rev. Microbiol.* **53:**71–102.

Strauch, E., G. Goelz, D. Knabner, A. Konietzny, E. Lanka, and B. Appel. 2003. A cryptic plasmid of *Yersinia enterocolitica* encodes a conjugative transfer system related to the regions of CloDF13 Mob and IncX Pil. *Microbiology* **149:**2829–2845.

Street, L. M., M. J. Harley, J. C. Stern, C. Larkin, S. L. Williams, D. L. Miller, J. A. Dohm, M. E. Rodgers, and J. F. Schildbach. 2003. Subdomain organization and catalytic residues of the F factor TraI relaxase domain. *Biochim. Biophys. Acta* **1646:**86–99.

Szpirer, C. Y., M. Faelen, and M. Couturier. 2000. Interaction between the RP4 coupling protein TraG and the pBHR1 mobilization protein Mob. *Mol. Microbiol.* **37:**1283–1292.

Tamura, M., K. Nogimori, S. Murai, M. Yajima, K. Ito, T. Katada, M. Ui, and S. Ishii. 1982. Subunit structure of islet-activating protein, pertussis toxin, in conformity with the A-B model. *Biochemistry* **21:**5516–5522.

Tanaka, J., T. Suzuki, H. Mimuro, and C. Sasakawa. 2003. Structural definition on the surface of *Helicobacter pylori* type IV secretion apparatus. *Cell. Microbiol.* **5:**395–404.

Tauch, A., S. Schneiker, W. Selbitschka, A. Pühler, L. S. van Overbeek, K. Smalla, C. M. Thomas, M. J. Bailey, L. J. Forney, A. Weightman, P. Ceglowski, T. Pembroke, E. Tietze, G. Schröder, E. Lanka, and J. D. van Elsas. 2002. The complete nucleotide sequence and environmental distribution of the cryptic, conjugative, broad-host-range plasmid pIPO2 isolated from bacteria of the wheat rhizosphere. *Microbiology* **148:**1637–1653.

Thorstenson, Y. R., G. A. Kuldau, and P. C. Zambryski. 1993. Subcellular localization of seven VirB proteins of *Agrobacterium tumefaciens:* implications for the formation of a T-DNA transport structure. *J. Bacteriol.* **175:**5233–5241.

Thorstenson, Y. R., and P. C. Zambryski. 1994. The essential virulence protein VirB8 localizes to the inner membrane of *Agrobacterium tumefaciens. J. Bacteriol.* **176:**1711–1717.

Tinland, B., B. Hohn, and H. Puchta. 1994. *Agrobacterium tumefaciens* transfers single-stranded transferred DNA (T-DNA) into the plant cell nucleus. *Proc. Natl. Acad. Sci. USA* **91:**8000–8004.

Top, E. M., Y. Moënne-Loccoz, T. Pembroke, and C. M. Thomas. 2000. Phenotypic traits conferred by plasmids, p. 249–285. *In* C. M. Thomas (ed.), *The Horizontal Gene Pool.* Harwood Academic Publishers, Amsterdam, The Netherlands.

Trieu-Cuot, P., C. Carlier, P. Martin, and P. Courvalin. 1987. Plasmid transfer by conjugation from *Escherichia coli* to Gram-positive bacteria. *FEMS Microbiol. Lett.* **48:**289–294.

Tuomanen, E., and A. Weiss. 1985. Characterization of two adhesins of *Bordetella pertussis* for human ciliated respiratory-epithelial cells. *J. Infect. Dis.* **152:**118–125.

Tzfira, T., and V. Citovsky. 2000. From host recognition to T-DNA integration: the function of bacterial and plant genes in the *Agrobacterium*-plant cell interaction. *Mol. Plant Pathol.* **1:**201–212.

van Mansfeld, A. D., H. A. van Teeffelen, P. D. Baas, and H. S. Jansz. 1986. Two juxtaposed tyrosyl-OH groups participate in φX174 gene A protein catalysed cleavage and ligation of DNA. *Nucleic Acids Res.* **14:**4229–4238.

Vergunst, A. C., B. Schrammeijer, A. Dulk-Ras, C. M. de Vlaam, T. J. Regensburg-Tuink, and P. J. J. Hooykaas. 2000. VirB/D4-dependent protein translocation from *Agrobacterium* into plant cells. *Science* **290:**979–982.

Vergunst, A. C., M. C. van Lier, A. den Dulk-Ras, T. A. Grosse Stuve, A. Ouwehand, and P. J. J. Hooykaas. 2005. Positive charge is an important feature of the C-terminal transport signal of the VirB/D4-translocated proteins of *Agrobacterium. Proc. Natl. Acad. Sci. USA* **102:**832–837.

von Bodman, S. B., J. E. McCutchan, and S. K. Farrand. 1989. Characterization of conjugal transfer functions of *Agrobacterium tumefaciens* Ti plasmid pTiC58. *J. Bacteriol.* **171:**5281–5289.

Walker, J. E., M. Saraste, M. J. Runswick, and N. J. Gay. 1982. Distantly related sequences in the α- and β-subunits of ATP synthase, myosin, kinases, and other ATP-requiring enzymes and a common nucleotide binding fold. *EMBO J.* **1:**945–951.

Wang, J., J. J. Song, I. S. Seong, M. C. Franklin, S. Kamtekar, S. H. Eom, and C. H. Chung. 2001. Nucleotide-dependent conformational changes in a protease-associated ATPase HsIU. *Structure* **9:**1107–1116.

Ward, D. V., O. Draper, J. R. Zupan, and P. C. Zambryski. 2002. Peptide linkage mapping of the *Agrobacterium tumefaciens vir*-encoded type IV secretion system reveals protein subassemblies. *Proc. Natl. Acad. Sci. USA* **99:**11493–11500.

Ward, J. E., D. E. Akiyoshi, D. Regier, A. Datta, M. P. Gordon, and E. W. Nester. 1988. Characterization of the *virB* operon from an *Agrobacterium tumefaciens* Ti plasmid. *J. Biol. Chem.* **263:**5804–5814.

Waters, V. L. 1999. Conjugative transfer in the dissemination of beta-lactam and aminoglycoside resistance. *Front. Biosci.* **4:**D433–D456.

Waters, V. L. 2001. Conjugation between bacterial and mammalian cells. *Nat. Genet.* **29:**375–376.

Weiss, A. A., F. D. Johnson, and D. L. Burns. 1993. Molecular characterization of an operon required for pertussis toxin secretion. *Proc. Natl. Acad. Sci. USA* **90:**2970–2974.

Whitchurch, C. B., M. Hobbs, S. P. Livingston, V. Krishnapillai, and J. S. Mattick. 1991. Characterisation of a *Pseudomonas aeruginosa* twitching motility gene and evidence for a specialised protein export system widespread in eubacteria. *Gene* **101:**33–44.

White, F. F., and E. W. Nester. 1980. Relationship of plasmids responsible for hairy root and crown gall tumorigenicity. *J. Bacteriol.* **144:**710–720.

Whitehouse, C. A., P. B. Balbo, E. C. Pesci, D. L. Cottle, P. M. Mirabito, and C. L. Pickett. 1998. *Campylobacter jejuni* cytolethal distending toxin causes a G_2-phase cell cycle block. *Infect. Immun.* **66:**1934–1940.

Willetts, N. 1981. Sites and systems for conjugal DNA transfer in bacteria, p. 207–215. *In* S. B. Levy, R. C. Clowes, and E. L. Koenig (ed.), *Molecular Biology, Pathogenicity and Ecology of Bacterial Plasmids.* Plenum Press, New York, N.Y.

Willetts, N., and C. Crowther. 1981. Mobilization of the non-conjugative IncQ plasmid RSF1010. *Genet. Res.* **37:**311–316.

Winans, S. C., and G. C. Walker. 1985. Conjugal transfer system of the IncN plasmid pKM101. *J. Bacteriol.* **161:**402–410.

Yasuda, R., H. Noji, M. Yoshida, K. Kinosita, Jr., and H. Itoh. 2001. Resolution of distinct rotational substeps by submillisecond kinetic analysis of F_1-ATPase. *Nature* **410:**898–904.

Ye, Y., H. H. Meyer, and T. A. Rapoport. 2001. The AAA ATPase Cdc48/p97 and its partners transport proteins from the ER into the cytosol. *Nature* **414:**652–656.

Yeo, H. J., S. N. Savvides, A. B. Herr, E. Lanka, and G. Waksman. 2000. Crystal structure of the hexameric traffic ATPase of the *Helicobacter pylori* type IV secretion system. *Mol. Cell* **6:**1461–1472.

Yeo, H. J., Q. Yuan, M. R. Beck, C. Baron, and G. Waksman. 2003. Structural and functional characterization of the VirB5 protein from the type IV secretion system encoded by the conjugative plasmid pKM101. *Proc. Natl. Acad. Sci. USA* **100:**15947–15952.

Yeo, H. J., and G. Waksman. 2004. Unveiling molecular scaffolds of the type IV secretion system. *J. Bacteriol.* **186:**1919–1926.

Yusibov, V. M., T. R. Steck, V. Gupta, and S. B. Gelvin. 1994. Association of single-stranded transferred DNA from *Agrobacterium tumefaciens* with tobacco cells. *Proc. Natl. Acad. Sci. USA* **91:**2994–2998.

Zambon, J. J., L. A. Christersson, and J. Slots. 1983. *Actinobacillus actinomycetemcomitans* in human periodontal disease. Prevalence in patient groups and distribution of biotypes and serotypes within families. *J. Periodontol.* **54:**707–711.

Zamboni, D. S., S. McGrath, M. Rabinovitch, and C. R. Roy. 2003. *Coxiella burnetii* express type IV secretion system proteins that function similarly to components of the *Legionella pneumophila* Dot/Icm system. *Mol. Microbiol.* **49:**965–976.

Zechner, E. L., F. de la Cruz, R. Eisenbrandt, A. M. Grahn, G. Koraimann, E. Lanka, G. Muth, W. Pansegrau, C. M. Thomas, B. M. Wilkins, and M. Zatyka. 2000. Conjugative DNA transfer processes, p. 87–173. *In* C. M. Thomas (ed.), *The Horizontal Gene Pool.* Harwood Academic Publishers, Amsterdam, The Netherlands.

Zhang, X., A. Shaw, P. A. Bates, R. H. Newman, B. Gowen, E. Orlova, M. A. Gorman, H. Kondo, P. Dokurno, J. Lally, G. Leonard, H. Meyer, M. van Heel, and P. S. Freemont. 2000. Structure of the AAA ATPase p97. *Mol. Cell* **6:**1473–1484.

Zink, S. D., L. Pedersen, N. P. Cianciotto, and Y. Abu-Kwaik. 2002. The Dot/Icm type IV secretion system of *Legionella pneumophila* is essential for the induction of apoptosis in human macrophages. *Infect. Immun.* **70:**1657–1663.

Zusman, T., G. Yerushalmi, and G. Segal. 2003. Functional similarities between the *icm/dot* pathogenesis systems of *Coxiella burnetii* and *Legionella pneumophila. Infect. Immun.* **71:**3714–3723.

Zylicz, M., J. H. LeBowitz, R. McMacken, and C. Georgopoulos. 1983. The DnaK protein of *Escherichia coli* possesses an ATPase and autophosphorylating activity and is essential in an in vitro DNA replication system. *Proc. Natl. Acad. Sci. USA* **80:**6431–6435.

Structural Biology of Bacterial Pathogenesis
Edited by G. Waksman et al.
© 2005 ASM Press, Washington, D.C.

Chapter 11

Injectosomes in Gram-Positive Bacteria

Rodney K. Tweten and Michael Caparon

Gram-positive bacteria are enclosed by a single membrane and therefore have apparently not evolved or acquired secretion systems similar to the type I, III, and IV systems found in the gram-negative bacteria as described elsewhere in this volume. The type I secretion system is used to secrete certain classes of proteins into the extracellular milieu, whereas the type III and IV secretion systems are specialized systems that translocate proteins directly from the bacterial cytosol into the eukaryotic cytosol. Many of the proteins translocated by the type III and IV systems are involved in the pathogenesis of various bacterial species. Although type III and IV systems are not found in gram-positive bacterial pathogens, these pathogens still translocate effectors into eukaryotic cells. The mechanism by which gram-positive bacterial pathogens accomplish effector translocation is varied and to a large extent involves self-translocating toxins that are typified by the A-B toxins that include diphtheria toxin, anthrax toxin, and the clostridial neurotoxins. However, more recently, Madden et al. (2001) have shown that *Streptococcus pyogenes* injects at least one effector, NAD glycohydrolase (SPN), into the cytoplasm of keratinocytes by a novel mechanism. This system is functionally analogous but structurally unrelated to the type III and IV secretions systems of the gram-negative bacteria. Translocation of SPN requires the combination of the Sec-dependent secretion system to secrete SPN across the single membrane of the gram-positive bacterial cell and the utilization of the pore-forming toxin streptolysin O (SLO) to translocate the SPN across the membrane of the keratinocyte. SLO is a member of the cholesterol-dependent cytolysin (CDC) family of pore-forming toxins (reviewed by Alouf [1999], Heuck et al. [2001], and Tweten et al. [2001]). Hence, the combination of the Sec-dependent secretion pathway and the pore-forming activity of SLO form an "injectosome," a novel system for the injection of proteins into a eukaryotic cell that is analogous in function to the type III and IV translocation systems of gram-negative bacteria, as well as the A-B toxins. In this chapter we examine the structure-function aspects of the CDCs and how the CDC mechanism has been adapted for use by *S. pyogenes* to form its injectosome.

Rodney K. Tweten • Microbiology and Immunology, BMSB-1053, 940 Stanton L. Young Blvd., The University of Oklahoma Health Sciences Center, Oklahoma City, OK 73104. *Michael Caparon* • Department of Molecular Microbiology, Washington University School of Medicine, St. Louis, MO 63110-1093.

CDCs

Expression and Secretion

The gene for the CDCs is probably the most widely dispersed toxin gene currently known; nearly 25 different gram-positive bacterial species have been identified thus far to carry a CDC gene (Alouf, 1999). To date, genes for CDCs have been found only in pathogenic strains, suggesting that these molecules play an important role in virulence. Originally the CDCs were termed the thiol-activated cytolysins because of their sensitivity to oxidation. This description is no longer accurate since these toxins have been determined not to be thiol activated (Pinkney et al., 1989; Saunders et al., 1989; Shepard et al., 1998). The CDCs are typically secreted by the bacterial cell and in most cases are probably secreted across the membrane via a Sec-dependent system since, except for pneumolysin from *Streptococcus pneumoniae,* they contain a Sec-dependent amino-terminal signal peptide (Walker et al., 1987). Pneumolysin lacks a signal peptide, and therefore it was thought that it was released only by cell lysis. However, more recent work has suggested that release of pneumolysin from *S. pneumoniae* is not dependent on cell autolysins (Balachandran et al., 2001), suggesting that another mechanism may be responsible for its release. The CDCs are highly soluble proteins once released from the secretion system of the bacterial cell; however, as described below, they undergo a remarkable transition that results in their conversion from soluble molecules into a supramolecular pore-forming membrane complex.

The Primary Structure

The primary structures of the CDCs exhibit 40 to 70% identity, suggesting that the basic molecular mechanisms of these toxins are similar due to the high degree of similarity in the primary structures. Several CDCs exhibit unique features which are encoded in their primary structures and that appear to serve unique functions for those CDCs. Hence, it is clear that evolutionary changes have taken place which have adapted the basic architecture of the CDC to the pathogenic program of individual pathogens. To understand the mechanism of the CDCs and their role in pathogenesis, it is necessary to explore both their structural similarities and their unique features.

Perhaps the most recognizable feature of the CDC primary structure is the conserved undecapeptide that is located near the carboxy terminus of the CDCs. The consensus sequence of this 11-mer peptide is ECTGLAWEWWR. This sequence is very highly conserved; only three CDCs with undecapeptide structures that vary from this sequence have been identified. The undecapeptide forms a loop at the tip of domain 4 (D4) (Color Plate 47 [see color insert]) and is known to penetrate the surface of the membrane, but it does not penetrate deeply into the bilayer (Heuck et al., 2000). Three other loops at the tip of D4 that include residues A401, V403, A437, and L491 also penetrate the surface of the bilayer (Ramachandran et al., 2002). Therefore, along with the undecapeptide loop, these loops anchor D4 to the membrane. Interestingly, the rest of D4 is exposed to the aqueous milieu in the oligomer and is therefore not in contact with either the membrane or the neighboring monomers of the oligomeric complex (Ramachandran et al., 2002).

The undecapeptide contains the only cysteine in the primary structure of most CDCs; only two members of the CDC family contain an alanine at this position. It is clear that this cysteine is located at a sensitive structural location since it cannot be chemically modified via the sulfhydryl group without the loss of cytolytic activity (Iwamoto et al., 1987). Therefore, the original designation of these toxins as "thiol activated" arose because in crude preparations this cysteine was probably oxidized by its reaction with small thiols to form a disulfide. Since it is at a structurally sensitive site, this modification caused the loss of activity that could be regained by the reduction of the disulfide. However, the sulfhydryl group of the cysteine itself is not necessary for the activity of the CDCs since it can be replaced by alanine without a significant loss in cytolytic activity (Pinkney et al., 1989; Saunders et al., 1989; Shepard et al., 1998) and the CDCs pyolysin (Billington et al., 1997) and intermedilysin (ILY) (Nagamune et al., 1996) have alanine naturally substituted at this site. Why the majority of the CDCs retain a cysteine at this location, a residue that makes the CDCs susceptible to inactivation by its oxidation, remains unclear. It is possible that the cysteine sulfhydryl acts as an environmental sensor and perhaps only under conditions where it is not oxidized does the CDC function properly.

The undecapeptide is also conspicuous for the presence of three closely spaced tryptophan residues that have also been shown to be sensitive to conservative substitutions in perfringolysin O (PFO) (Sekino-Suzuki et al., 1996). As mentioned above, the undecapeptide was found to enter the membrane, a discovery that was based on the observation that the fluorescence of one or more of these tryptophans was collisionally quenched with membrane-restricted nitroxides (Sekino-Suzuki et al., 1996; Heuck et al., 2000). The highly conserved nature of the undecapeptide structure suggests a specialized function for this region. Modification of the cysteine or mutation of the tryptophans can alter the affinity of PFO for the membrane (Iwamoto et al., 1987; Sekino-Suzuki et al., 1996).

A unique peptide sequence is present at the amino terminus of the CDCs produced by various *Listeria* species (Decatur et al., 2000; Lety et al., 2001; Lety et al., 2002; Frehel et al., 2003), of which the most extensively studied is listeriolysin O (LLO) produced by *Listeria monocytogenes*. LLO plays an essential role in the escape of this facultative intracellular parasite from the phagosome of the macrophage (Portnoy et al., 1988). LLO contains a PEST (proline-glutamate-serine-threonine) consensus sequence at its amino terminus (Decatur et al., 2000). PEST sequences play a role in targeting eukaryotic proteins for proteolytic degradation (Rechsteiner and Rogers, 1996). The listerial PEST sequence appears to be important in regulating the activity of LLO, but the precise mechanism by which it does so remains ambiguous.

The only other conspicuous feature in the primary structure of a CDC is the presence of a peptide of approximately 75 amino acids at the amino terminus of SLO, a peptide sequence that is not present in the primary structure of any other CDC. This peptide does not exhibit similarity to any other known or hypothetical protein, suggesting that its function may be specific to *S. pyogenes*. It is notable that this peptide is particularly rich in acidic and, to a lesser extent, basic amino acids. The cytolytic mechanism of SLO is unchanged whether or not this peptide is present on the molecule (Bhakdi et al., 1984), and so the function of this additional peptide sequence remains unknown.

Perfringolysin O Crystal Structure and Cytolytic Mechanism

The soluble monomer structure of PFO (Color Plate 47) was determined by Rossjohn et al. (1997); this was a major step in the progress toward understanding the cytolytic mechanism of the CDCs. The availability of the PFO structure has greatly facilitated the study of the cytolytic mechanism of the CDCs and has helped to reveal details about the mechanism that would have not been otherwise possible. As can be seen from the crystal structure, PFO is an elongated, four-domain molecule that is composed primarily of β-strands with some α-helices present in domains 1 and 2 (D1 and D2) (Color Plate 47). D4 forms a β-sandwich, whereas D1 and D3 contain both β-sheet and α-helices. D2 is primarily a single β-sheet that connects D4 with D1 and D3.

An unusual structural feature of the solution monomer is the D3-D2 interface. Unlike most domain interfaces, this interface is not complementary, suggesting that it is unstable. Furthermore, there is a pronounced bend in the core β-sheet that spans D1 and D3 (Rossjohn et al., 1997). These observations suggest that this unstable interface may be broken and the bend may be straightened during the transition of the molecule into the membrane. As described below, this scenario seems likely since one of the transmembrane regions, which was identified after the crystal structure had been solved, is buried in the D3-D2 interface and therefore it is necessary that it extricate itself from this position so that it can insert into the membrane.

Membrane Recognition

The CDCs appear to be uniquely targeted to the eukaryotic membrane since their cytolytic mechanism is absolutely dependent on the presence of membrane cholesterol. The importance of cholesterol to the CDC mechanism was determined when it was discovered that preincubation of the CDC soluble monomers with small amounts of cholesterol inhibited their cytolytic activity on erythrocytes (Prigent and Alouf, 1976). Although the role of the cholesterol molecule in the mechanism of the CDCs remains incompletely understood, there is a significant body of evidence that one of the CDCs, PFO, can directly bind to and use cholesterol as a receptor (Ohno-Iwashita et al., 1988, 1990, 1991; Iwamoto et al., 1990, 1993, 1997; Nakamura et al., 1995; Waheed et al., 2001; Mobius et al., 2002; Shimada et al., 2002). Binding of the cholesterol molecule appears to occur via D4 (Color Plate 47) (Shimada et al., 2002) of PFO and probably requires an intact 3-β-hydroxy group of the cholesterol (Prigent and Alouf, 1976).

The fusion of green fluorescence protein and the D4 of PFO results in a hybrid protein that binds to membranes in a cholesterol-dependent fashion (Shimada et al., 2002). The interaction of PFO with cholesterol must occur near the tip of D4, since Ramachandran et al. (2002) have shown that in addition to the undecapeptide, only the short hydrophobic loops at the tip of D4 interact with and penetrate the membrane (Nakamura et al., 1995; Heuck et al., 2000) (Color Plate 47). However, as observed with the undecapeptide, these loops do not penetrate deeply into the membrane. Nitroxide quenchers located near the center of the bilayer did not collisionally quench fluorescent probes placed at the tips of these loops, whereas if the nitroxides were moved nearer to the bilayer surface the fluorescent probes were efficiently quenched. Ramachandran et al. (2002) also showed that all sides of the D4 β-sandwich appeared to be surrounded by the aqueous phase and therefore that it was not in contact with the membrane surface or peptide regions of neighboring molecules of the oligomer. This latter observation was surprising since the results of earlier work (Iwamoto et al., 1990; Tweten et al., 1991) had suggested that a tryptic peptide from PFO encompassing residues between K304 and the carboxy terminus could inhibit the oligomerization of PFO on the membrane. This fragment prevented oligomerization by forming inactive dimers of the fragment and intact toxin that prevented further oligomerization. Since this fragment contained all of D4 and parts of D2 and D3, it was thought that D4 mediated the interaction with the native toxin molecules. This now appears unlikely; presumably, the interaction was actually mediated via the residues of this tryptic peptide that were outside of D4. Therefore, as shown in the model of the oligomer in Color Plate 48 (see color insert), D4 functions to anchor PFO to the membrane and acts as a platform for the subsequent conformational changes that occur in D1 to D3 during pore formation.

However, other studies have suggested that the role of cholesterol is more complex than suggested by the studies with PFO. Rottem et al. (1982) determined that efficient binding of tetanolysin to liposomes required at least 30 mol% cholesterol. More recently Heuck et al. (2000) showed that quantitative binding of PFO to cholesterol-phosphotidylcholine liposomes required high concentrations of cholesterol (45 to 50 mol%). They also showed that the transition from no detectable binding to full binding of PFO occurred within a very narrow range of cholesterol concentrations (45 to 50 mol%), suggesting that at lower concentrations cholesterol was not available for binding, even though it was present in vast molar excess over the toxin. Also, Jacobs et al. (1998) showed that pore formation by the CDC LLO was inhibited by preincubation with small amounts of cholesterol but that binding of the toxin to natural membranes was not inhibited. They suggested that the interaction of LLO with cholesterol was a post-binding event required for downstream events that were necessary for toxin activity.

Recently, Giddings et al. (2003) showed that the basis for the extreme sensitivity of the CDCs to membrane cholesterol was not based primarily on the loss of membrane binding as generally believed. They examined the effect of cholesterol depletion from the membranes of erythrocytes on the mechanism of PFO, SLO, and ILY, a CDC produced by *Streptococcus intermedius* that exhibits a unique specificity for human cells. The depletion of 90% of the membrane cholesterol from human erythrocytes did not significantly affect the binding of SLO and ILY and decreased the binding of PFO by only 10-fold. However, in each case the hemolytic activity of each toxin was reduced by >99.9%, and therefore changes (or lack thereof) in their binding to the cholesterol-depleted membranes could not account for the dramatic loss in activity. These authors also found that cholesterol depletion of the membranes trapped all three toxins in the prepore complex (Color Plate 48), the stage at which the toxin forms the membrane oligomer but has not yet inserted its transmembrane β-barrel. Therefore, the hallmark sensitivity of the CDCs to membrane cholesterol is due primarily to the inability of the prepore to convert to the pore complex. How cholesterol facilitates the prepore-to-pore transition remains unclear; it may require the direct interaction of cholesterol with the CDC, or cholesterol may alter the structure of the membrane (i.e., the generation of an inverted hexagonal bilayer phase) such that it facilitates the insertion of the CDC β-barrel.

Pore Formation

After membrane binding, the CDC monomers interact with each other on the membrane to generate an oligomeric prepore complex that is then converted to the pore complex by the insertion of the transmembrane β-barrel (Shepard et al., 2000; Hotze et al., 2001) (Color Plate 48). It has been estimated that between 30 and 50 CDC monomers comprise the membrane oligomer (Olofsson et al., 1993; Bonev et al., 2000; Bonev et al., 2001). The fact that the oligomer size seems to vary suggests that certain factors can influence the size of the ring and therefore the angle at which monomers interact with one another. It is possible that the monomers of the various CDCs may interact at slightly different angles in the oligomer, thus increasing or decreasing the size of the circular oligomer. Another possibility is that the membrane composition exerts certain stresses on the oligomer as it grows and may force closure of the ring at an earlier or later stage of growth.

The pathway to the formation of the oligomeric pore is not completely understood, although some of the major features have been elucidated in recent years. One of the most intensely studied aspects of the CDC mechanism has been the nature of the structural changes that occur as these toxins make the transition from a membrane-bound monomer to a membrane-inserted pore-forming oligomeric complex. Initial studies of this transition involved the identification of the membrane-spanning domains for PFO. Shepard et al. (1998) revealed the location and structure of the first transmembrane structure of PFO. They developed and applied multiple independent fluorescence techniques to analyze derivatives of PFO that contained strategically placed unique cysteine residues that were modified with an environmentally sensitive fluorescence probe via the cysteine sulfhydryl. They determined that PFO, like *Staphylococcus aureus* alpha-toxin (Gouaux et al., 1994), utilized a transmembrane β-hairpin structure to cross the membrane and contribute to the formation of the transmembrane β-barrel. However, on inspection of the crystal structure of the soluble monomer of PFO (Color Plate 47), it was clear that the residues that formed the transmembrane β-hairpin of PFO were packed in the soluble monomer as three short α-helices in D3. Therefore, to form the transmembrane β-hairpin (TMH1), these residues not only undergo an α-helix to β-strand transition, a comparatively rare structural transition in proteins, but also must extricate themselves from their buried position with the D3-D2 interface. Consistent with this model is that the D3-D2 interface is not complementary and therefore appears unstable. Thus, it is likely that this interface is more easily disrupted to allow the conversion of the α-helices to form the extended transmembrane β-hairpin.

As is apparent from the crystal structure, a second set of three short α-helices are present in D3 (Color Plate 47). Shatursky et al. (1999) subsequently determined that these helices form a second transmembrane hairpin (TMH2). In contrast to TMH1, TMH2 is largely solvent exposed but retains the same amphipathic structure as TMH1. The presence of two transmembrane β-hairpins per monomer is novel and has not been shown to exist in any other group of pore-forming toxins. Due to the size of the CDC oligomer, which is composed of up to 50 CDC monomers (Olofsson et al., 1993), up to 100 β-hairpins form the transmembrane β-barrel of the CDCs. The coordination of the insertion of so many hairpins represents a conceptual problem in comparison with the insertion of the much smaller β-barrel of the smaller heptameric pore formers such as alpha-hemolysin or anthrax protective antigen. The coordinated insertion of the hairpins may be facilitated by the assembly of a prepore complex. An important intermediate in the formation of the pore for smaller pore-forming toxins is the formation of a prepore intermediate (Walker et al., 1995; Fang et al., 1997; Sellman et al., 1997; Miller et al., 1999). The prepore complex is the oligomerized complex that has not yet inserted its transmembrane β-barrel.

Initial studies into the mechanism of pore formation were guided by two models (Rossjohn et al., 1997; Palmer et al., 1998). The first model was based on the prepore model for alpha-hemolysin (Walker et al., 1995; Fang et al., 1997) for which Rossjohn et al. (1997) proposed that a CDC prepore structure was generated prior to the insertion of the transmembrane β-barrel. The second model, proposed by Palmer et al. (1998), suggested that a small CDC oligomer, such as a dimer, inserted into the membrane first and formed a small channel that grew into a larger oligomer as additional monomers added to the growing pore complex. Understanding the structure of the transmembrane domain of PFO has played a critical role in distinguishing between these models, and the majority of evidence collected so far has supported the prepore model. Thus, the CDCs probably follow the same prepore-to-pore pathway as do the small pore formers. Shepard et al. (2000) initially showed that PFO formed a prepore complex based on the observation that oligomerization could be separated from the insertion of the transmembrane β-barrel by lowering the temperature. Therefore, oligomerization on the membrane was not dependent on the insertion of the β-hairpins of the individual monomers. They also showed that PFO formed only large channels on planar bilayers, indicating that a rapid insertion of the transmembrane β-barrel of the fully assembled prepore oligomer was responsible for the formation of the large pore (Shepard et al., 2000). Knowledge of the PFO transmembrane structure was subsequently used to show that PFO can be trapped in the prepore oligomer both by the introduction of a disulfide that restricts the movement of transmembrane β-hairpins (Hotze et al., 2001) and by point mutations (Hotze et al., 2002). It has been recently shown that depleting a membrane of cholesterol traps PFO in the prepore complex (Giddings et al., 2003) and that the size of the pore generated by PFO and SLO is independent of the toxin concentration (Heuck et al., 2003). At limiting concentrations of toxin, both toxins formed only large channels (Heuck et al., 2003). Taken together, these studies have provided compelling evidence for the prepore model by demonstrating that the prepore complex forms and that only prepore complexes of a sufficiently large size are competent for insertion of the transmembrane β-barrel.

Not only has it been shown that the prepore complex exists, but also the investigation into the prepore mechanism has yielded information on the functional necessity of the prepore. A specific point mutation in which Y181 of PFO was converted to alanine was found to trap PFO in a sodium dodecyl sulfate-sensitive prepore complex (Hotze et al., 2002). However, it was determined that native and mutant toxin monomers could form mixed oligomers and that if sufficient native toxin was present in the oligomer, it could force the Y181 mutant monomers to insert their transmembrane β-hairpins. Therefore, cooperation of the monomers in the prepore complex was necessary to drive the insertion of the transmembrane β-barrel. It was also shown that the conversion of the prepore to the pore was an all-or-none phenomenon; i.e., prepore-to-pore conversion occurred only if there was sufficient native monomer in the mixed oligomer to overcome the energetic barrier posed by the Y181 mutation in the mutant monomers. If insufficient native monomer was present in the mixed prepore oligomer to overcome this barrier, then the oligomer remained in the prepore state. These data were consistent with the recent studies by Heuck et al. (2003) that are described above and suggest that only when the prepore is of sufficient size is it capable of inserting its transmembrane β-barrel.

Therefore, as shown in Color Plate 48, there is evidence for the following mechanism of membrane insertion of the CDCs. The first step is membrane recognition and binding (Color Plate 48A). This stage of the mechanism remains incompletely understood, although PFO can apparently utilize membrane cholesterol as a receptor; however, as described above, ILY from *S. intermedius* is specific for human cells, which suggests that it uses another molecule for membrane targeting (although its mechanism still remains dependent on membrane cholesterol). Once the monomer is bound to the membrane, it diffuses laterally to interact with other bound monomers to initiate oligomer formation to form the prepore complex (Color Plate 48B). The tip of D4 is inserted superficially into the membrane surface, and all sides of D4 are exposed to the aqueous milieu (Ramachandran et al., 2002). Once the oligomer reaches an insertion-competent size, the transmembrane β-hairpins of each monomer cooperate to insert into the membrane (Hotze et al., 2002). Whether the β-hairpins organize into a β-sheet prior to insertion, as shown in Color Plate 48D, remains unknown, but this would be energetically favorable since it would satisfy much of the hydrogen bond potential of the β-hairpins and therefore lessen the energy requirement for insertion of the β-sheet. The two regions of D3 that form the β-hairpins must exist in a disordered state while making the transition from α-helical structure in the monomer to the extended β-sheet in the oligomer. Presumably the inserted pore exhibits a fully organized transmembrane β-sheet (Color Plate 48E).

THE STREPTOCOCCAL INJECTOSOME

The size of the CDC pore has always been a conspicuous feature of these toxins. It is not clear why a pore that has a diameter of 25 to 30 nm is generated if eukaryotic cell lysis is the primary function of these toxins during bacterial disease. A much smaller pore would function adequately for cell lysis in vivo, and therefore the cell would expend a considerably smaller amount of energy for the production of a smaller pore. However, the widespread distribution of the CDC toxin gene and the conservation of the large pore diameter suggest that the CDC pore may be used in other, unexpected ways that require the large pore diameter. The discovery by Madden et al. (2001) that SLO of *S. pyogenes* is involved in the translocation of another protein has made it necessary to reassess the function and use of the large CDC pore during bacterial infections. No other known pore-forming toxin has a pore of sufficient size to facilitate the unrestricted diffusion of another protein through its pore. The CDCs have been used for years by scientists to permeabilize eukaryotic cells in order to introduce proteins into the cell interior (reviewed by Bhakdi et al. [1993]); therefore, it is no surprise that a bacterial cell utilizes a CDC for the same purpose. The ability of *S. pyogenes* to combine the Sec-dependent secretion system and SLO to form an injectosome to directly translocate a protein into the cytosol of a eukaryotic cell is, to date, unique.

The *S. pyogenes* injectosome is composed of two known components, SLO and the Sec-dependent secretion pathway; in combination, they are necessary for the translocation of the NAD^+ glycohydrolase (SPN) from *S. pyogenes* into the cytoplasm of human keratinocytes. It is also possible that there are other components of this system that have not been discovered. The discovery of the injectosome pathway by Madden et al. (2001) resulted in part from the coincidental presence of anti-SPN antibodies in anti-SLO antisera. It is likely that the anti-SPN antibody arose from contamination of the SLO antigen preparation with SPN that was used for the generation of the anti-SLO antisera. In retrospect, the contamination of the SLO antigen with SPN is not surprising since SLO and SPN were originally thought to be the same protein (Fehrenbach, 1972), suggesting that complete separation of these two proteins is not always achieved. Whether these early observations suggest some weak association of the two proteins in solution is unknown.

Keratinocytes and Streptococcal Disease

Before the discovery of the injectosome by Madden et al., Wang et al. (1997) and Ruiz et al. (1998) had studied the effects of adherent and nonadherent and SLO^+ and SLO^- derivatives of *S. pyogenes* on the proinflammatory response of keratinocytes. *S. pyogenes* causes a large number of suppurative and inflammatory infections of cutaneous tissue, some of which can become systemic. Cutaneous infections can range from mild (impetigo) to life-threatening (necrotizing fasciitis) (Stevens, 1996). The keratinocytes are therefore important in the detection of cutaneous infections and the generation of a proinflammatory response. The ability of keratinocytes to detect cutaneous infections and generate a proinflammatory response has become apparent only in recent years. Human keratinocytes are now known to express functional CD14 and Toll-like receptors (Song et al., 2002). Wang et al. (1997) found that adherent and nonadherent *S. pyogenes* strains resulted in conspicuously altered patterns of expression of proinflammatory mediators by keratinocytes. Increased expression of interleukin-1α (IL-α), IL-1β, and IL-8 and a more rapid release of prostaglandin E_2 were induced in keratinocytes with adherent *S. pyogenes* but not nonadherent *S. pyogenes* strains. The expression of these mediators was dependent on protein expression of the bacterial cell, suggesting that preexisting molecules on the cell surface of the bacteria were not sufficient to promote this response. Subsequently, Ruiz et al. (1998) observed that SLO and the adherence of *S. pyogenes* to keratinocytes were synergistic in the modulation of the keratinocyte proinflammatory response. SLO knockouts of *S. pyogenes* resulted in a significant decrease in the expression of IL-1β, IL-6, and IL-8.

Madden et al. (2001) continued the study of the involvement of SLO in the stimulation of the proinflammatory response in keratinocytes, and it was during these studies that the salient observation was made that led to the discovery of the injectosome. An immunoblot of the cell fractions from keratinocytes infected with *S. pyogenes,* probed with anti-SLO antiserum, revealed a 52-kDa protein that was present in the Triton X-100-soluble fraction. However, the SLO band (70 kDa) was present only in the insoluble fraction, presumably because it associates with the lipid raft fraction of the cell membrane similar to the related PFO (Shimada et al., 2002). The 52-kDa protein was determined to be the *S. pyogenes* NAD glycohydrolase (SPN), whose gene is located immediately upstream of the SLO gene and may be cotranscribed with SLO (Madden et al., 2001). *S. pyogenes* has been known for over 40 years to express NAD glycohydrolase (Bernheimer et al., 1957), yet its contribution to disease had not been well understood since there did not appear to be a means by which it gained entry to the eukaryotic cytosol. SPN has been studied largely as a disease-related marker, and its presence has recently been associated with severe invasive group A streptococcal infections (Stevens et al., 2000).

Streptococcal NAD Glycohydrolase

SPN is a 52-kDa protein that contains a typical signal peptide for a Sec-dependent protein and therefore is found in the supernatant of cultured organisms. No SPN-related enzymes have been identified in other gram-positive bacteria, and SPN is not structurally related to any known protein. The NAD glycohydrolase activity of SPN splits the common intracellular molecule NAD^+ into nicotinamide and ADP-ribose. A-B toxins, such as diphtheria toxin, also exhibit NAD glycohydrolase activity; however, the ADP-ribose moiety is typically transferred to a protein substrate, if present, which results in its covalent modification and the alteration of its activity. It is unknown whether SPN has intracellular substrates that can serve as acceptors for the ADP-ribose moiety. The search for an intracellular substrate for SPN has not been extensively investigated, since SPN did not apparently exert a cytotoxic effect on eukaryotic cells by itself. However, SPN can transfer ADP-ribose to the synthetic substrate poly-L-arginine (Stevens et al., 2000), suggesting that it may be capable of ADP-ribosylating one or more intracellular substrates. SPN also generates cADP-ribose in addition to APD-ribose, a known signaling molecule in eukaryotic cells (reviewed by Galione [1994], Lee [1994], and Galione et al. [2002]). The primary effect of cADP-ribose appears to be in triggering Ca^{2+} release from the endoplasmic reticulum.

The Injectosome Pathway

The studies by Madden et al. (2001) revealed key characteristics of the injectosome that showed that this system functioned as an analog to the type III secretory systems found in some gram-negative bacteria. The critical observation was made when it was observed that the majority of SPN was secreted into the media in a SLO⁻ strain of S. pyogenes, even though it was in tight association with the keratinocyte. However, when SLO and SPN were coexpressed, approximately 75% of the SPN was found in the cytoplasm of the keratinocyte. This single observation revealed a critical feature of the injectosome: the translocation of SPN was highly biased to the cytoplasm of the keratinocyte in the presence of SLO, and therefore SLO was a critical component of the injectosome. Furthermore, isogenic knockouts of S. pyogenes that expressed either SPN or SLO could not complement one another. Hence, a second important feature of the injectosome was revealed: both SPN and SLO had to be expressed by the same cell for a functioning injectosome to be present.

These two observations impose certain characteristics on the injectosome. The fact that 75% of the SPN is present in the keratinocyte cytoplasm demonstrates that SPN does not equilibrate between the aqueous milieu and the cytoplasm of the eukaryotic cell via the SLO pore. If the SLO channel was exposed on one side to the keratinocyte cytoplasm and on the other to the aqueous milieu, then at equilibrium 50% of the SPN would be present in the cytoplasm, assuming the same volume inside and outside of the cell. In reality, the volume of the extracellular milieu was probably much larger than the intracellular volume of the keratinocytes, and so a considerably smaller fraction of the total SPN would have been present in the cytoplasm if SPN simply equilibrated across the SLO pore. Furthermore, the large size of the SLO pore would have allowed the outward flow of the concentrated cytoplasmic contents of the keratinocyte, and therefore SPN levels would not have equilibrated until most of the cytoplasmic contents of the keratinocyte were lost. Hence, the SLO pore cannot simply be open on both sides of the keratinocyte membrane; instead, it is likely to be capped on the extracellular side to prevent outward flow of the cytoplasmic contents.

Madden et al. (2001) also found that deletion of the S. pyogenes adhesin for keratinocytes prevented translocation of SPN into the eukaryotic cell. Therefore, a tight association between the prokaryotic cell and the eukaryotic cell was necessary for injectosome function, possibly because the prokaryotic cell forms a tight seal with the extracellular side of the SLO pore. Curiously, in the adhesin-deficient strain of S. pyogenes, not only was SPN not translocated to the cytoplasm of the keratinocyte but also SLO did not appear to insert into the membrane. Presumably, tight association of the bacterial cell with the keratinocyte was also necessary for the efficient insertion and pore formation by SLO. This aspect of the interaction of SLO with keratinocytes warrants further investigation. Walev et al. (1995) showed that, compared to human erythrocytes, keratinocytes are relatively resistant to SLO. Approximately 100-fold more SLO was required for 60% lysis of the keratinocytes than was required for lysis of the erythrocytes. This observation suggests that the concentration of SLO at the interface of the bacterial and keratinocyte surfaces may be required for pore formation by SLO in the injectosome.

Cytochalasin D, an inhibitor of microfilament formation and therefore of endosome formation, did not affect the translocation of SPN into the keratinocyte. This observation eliminated the possibility that the translocation of SPN into the eukaryotic cytosol did not involve endosomal uptake of the bacterial cell, perforation of the endosomal membrane by SLO, and diffusion of SPN from within the endosomal vesicle into the cytoplasm. Hence, the mechanism of biased secretion occurs on the surface of the eukaryotic cell.

A General Model for the Injectosome

Based on the studies by Madden et al. (2001) described above, a model of the *S. pyogenes* injectosome assembly and function can be proposed (Color Plate 49 [see color insert]). First, *S. pyogenes* binds to the keratinocyte via the bacterial adhesin, thus bringing the two cells into close contact. SLO is then secreted by the Sec-dependent secretion system, possibly at the site of contact between the two cells, and assembles into its large pore-forming homo-oligomeric complex. At this point, the SLO pore complex is presumably in tight contact with the bacterial cell surface and therefore forms a continuous pore from the surface of *S. pyogenes* to the cytoplasm of the keratinocyte. Concomitantly, SPN is also secreted across the bacterial cell membrane, probably at the location of the SLO pore, by the Sec-dependent secretion system. Both the secretion of SLO and the secretion of SPN are necessarily biased to the region of cell-to-cell contact. SPN is then translocated via the large SLO pore into the eukaryotic cell cytoplasm. The docking of the streptococcal cell onto the pore may serve as a cap to prevent the outward flow of the cytoplasmic contents of the eukaryotic cell.

Although this model is consistent with the findings of Madden et al. (2001), it leaves unanswered several highly intriguing questions. One of the most interesting questions is how the bacterial cell biases the localization of SLO to the membrane site at which the bacterial cell is attached and then targets the secretion of SPN to the SLO pore. The most likely solution to this problem, that of endocytosis, has already been ruled out by the studies by Madden et al. (2001). Therefore, the bacterial cell appears to polarize the secretion of SPN to the location of the SLO pore in another way. Evidence exists that the type II secretion pathway can be biased to a polar location in the gram-negative pathogen *Vibrio cholerae* (Scott et al., 2001), although the mechanism by which this is accomplished is not understood. A similar mechanism may function in *S. pyogenes* to bias the secretion of SPN to SLO pore.

Physiological Consequences of the Injectosome

The general effect of translocating SPN into the keratinocyte appears to be the acceleration of keratinocyte death (Madden et al., 2001) compared with bacterial cells expressing only SLO or SPN or neither protein. Bricker et al. (2002) have now determined that the combination of SLO and SPN inhibits the internalization of the bacteria cell by the keratinocyte, thus preventing the bacterium from entering a host compartment where it probably cannot multiply (Greco et al., 1995). The absence of either SLO or SPN results in an increased rate of internalization of the bacteria by the keratinocyte; therefore, both proteins are required to decrease the rate of internalization. Interestingly, the cytolytic effects of SLO alone were not sufficient to reduce bacterial uptake, indicating that while the CDCs are highly cytolytic in vitro, S. pyogenes does not apparently secrete large amounts of SLO into the culture supernatant when attached to the keratinocyte. If it did so it would seem likely that SLO could insert and permeabilize the keratinocyte membrane, thus lysing and killing the cell. Therefore, S. pyogenes appears to carefully regulate its use of the SLO pore for the translocation of SPN and does not seem to utilize it for general cell lysis.

Bricker et al. (2002) also found that the combination of SLO and SPN induced apoptosis in the keratinocyte whereas each alone had no significant effect. This observation was consistent with that of the accelerated cell death reported by Madden et al. (2001) and suggests that the accelerated death of the keratinocytes is at least partially due to apoptosis. The molecular basis by which SPN prevents bacterial uptake and induces apoptosis remains unclear. However, since SPN is capable of generating cADP-ribose, it is possible that interference of one or more host cell signaling systems may occur.

Are other S. pyogenes proteins secreted into the keratinocyte cytoplasm via the injectosome? At this time, no proteins other than SPN have been identified that are translocated by the S. pyogenes injectosome. However, it is possible that any Sec-dependent protein that is normally secreted by S. pyogenes is a potential candidate for translocation. This is primarily due to the large size of the SLO pore and the fact that it does not appear to exhibit any selectivity in vitro (Heuck et al., 2003). Most proteins have molecular diameters that easily pass through the 25-nm CDC pore. It has recently been shown that molecules as large as antibodies diffuse through SLO and PFO pores as rapidly as small molecules such as glutathione (Heuck et al., 2003).

Unique Structural Features of SLO and the Injectosome

It is important to note that, as mentioned above, the SLO structure contains an additional amino-terminal extension of about 75 amino acids that is not present on other known CDCs. This sequence is conspicuous both for its presence and because it does not exhibit any similarity to any known protein. This latter aspect suggests that its function may be specific to SLO and S. pyogenes. Assuming that the three-dimensional structure of SLO does not differ significantly from that of PFO, this additional peptide would be positioned near the top of the molecule and therefore the oligomer (Color Plate 47). This would place this peptide in an optimal location to participate in an interaction with the surface of the bacterial cell, if this is its purpose. Furthermore, it is possible that the interaction of this region and/or other regions of the SLO oligomer with the bacterial cell may signal the cell to localize secretion of the SPN to the site of the SLO pore.

ARE THERE OTHER INJECTOSOME SYSTEMS?

Although there is no evidence that a CDC is used as part of an injectosome in any other bacterial systems, it is possible that the CDC produced by L. monocytogenes, LLO, functions in an analogous manner to the S. pyogenes injectosome. L. monocytogenes is a facultative intracellular parasite that can invade macrophages and then replicate within their cytosol (reviewed by Cossart [2002] and Portnoy et al. [2002]). LLO is an essential virulence factor that is necessary for the escape of *L. monocytogenes* from the macrophage phagosome (Cossart et al., 1989; Bielecki et al., 1990; Portnoy et al., 1992; Jones and Portnoy, 1994; Glomski et al., 2002). *L. monocytogenes* escapes the phagosome and replicates in the protected environment of the eukaryotic cytosol. The cytolytic activity of LLO is pH sensitive such that it is active at the acidic pH of the phagosome but is inactive at the neutral pH of the eukaryotic cytosol. Therefore, LLO can form pores in the phagosomal membrane but does not permeabilize the plasma membrane of the macrophage, in part due to its loss of activity at the neutral pH of the cytoplasm. *L. monocytogenes* mutants of LLO that lack pH sensitivity or have a pH-insensitive CDC instead of LLO experience a loss of intracellular replication and virulence (Jones and Portnoy, 1994; Jones et al., 1996; Glomski et al., 2002).

It is accepted that LLO permeabilizes the endosome membrane and therefore allows the bacterial cell to escape into the host cell cytoplasm. However, the way LLO facilitates the escape of *L. monocytogenes* is not entirely clear. The bacterial cell is too large to fit through the pore of LLO, and the phagosome does not lyse by a colloid-osmotic effect. The equilibration of osmotic pressure between the cytoplasm of the host cell and the phagosome would result in the shrinkage of the phagosome, not its uncontrolled expansion as is seen when a CDC pore is placed into the plasma membrane of a eukaryotic cell. Therefore, the problem of how LLO actually facilitates the escape of *L. monocytogenes* from the phagocytic vacuole remains an enigma. One possible explanation is that so much LLO inserts into the endosomal membrane that eventually it becomes unstable and fragments, allowing the escape of the bacteria cell. However, another intriguing possibility is that the LLO pore facilitates the transfer of proteins in and/or out of the phagosome that then aid in the escape of the bacterial cell.

L. monocytogenes secretes at least two phospholipase C enzymes (phosphatidylinositol PC and phosphatidylcholine PC) that are also involved in the escape of *L. monocytogenes* from both primary and secondary vacuoles (Smith et al., 1995). There is evidence that the LLO pore in the phagosome might allow these phospholipase enzymes to have access to both sides of the phagocytic vesicle and other cellular membranes prior to the escape of the bacteria cell from the phagosome (Wadsworth and Goldfine, 2002). Therefore, LLO may function in a manner analogous to the injectosome of *S. pyogenes* in that one function may be to allow passage of certain proteins out of the phagocytic vacuole that are essential for the escape of the bacterial cell. Also, the large size of the CDC pore appears to preclude a unidirectional transfer of proteins, and so the LLO pore may allow certain host factors *into* the phagosome that also facilitate the escape of the bacterium.

CONCLUDING REMARKS

The discovery of the injectosome by Madden et al. (2001) has changed our view of how a CDC pore can be used during a bacterial infection. The injectosome of *S. pyogenes* is a paradigm that will impact several areas of biology because of the many interesting questions posed by this system. These include questions about the structure of the SLO pore and the function of the amino-terminal peptide, how *S. pyogenes* targets SLO and SPN to sites at the bacterial interface with the eukaryotic cell, and the molecular basis of the effect of SPN on the eukaryotic cell. Also, is the injectosome a common feature of other bacterial pathogens that produce a CDC, or is the *S. pyogenes* injectosome unique? If the latter is true, then how are other CDCs utilized? The answers to these intriguing questions will probably come from a combination of studies that continue to investigate the pathogenic mechanisms of bacterial species that produce CDCs and that explore the structural similarities and unique features of the CDCs produced by these pathogens.

Acknowledgments. We thank Michael Parker and Jamie Rossjohn for the preparation of Color Plate 48.

REFERENCES

Alouf, J. E. 1999. Introduction to the family of the structurally related cholesterol-binding cytolysins ('sulfhydryl-activated toxins'), p. 443–456. *In* J. Alouf and J. Freer (ed.), *Bacterial Toxins: a Comprehensive Sourcebook.* Academic Press, Ltd., London, United Kingdom.

Balachandran, P., S. K. Hollingshead, J. C. Paton, and D. E. Briles. 2001. The autolytic enzyme LytA of *Streptococcus pneumoniae* is not responsible for releasing pneumolysin. *J. Bacteriol.* **183:**3108–3116.

Bernheimer, A. W., P. D. Lazarides, and A. T. Wilson. 1957. Diphosphopyridine nucleotidase as an extracellular product of streptococcal growth and its possible relationship to leukotoxicity. *J. Exp. Med.* **106:**27–37.

Bhakdi, S., M. Roth, A. Sziegoleit, and J. J. Tranum. 1984. Isolation and identification of two hemolytic forms of streptolysin-O. *Infect. Immun.* **46:**394–400.

Bhakdi, S., U. Weller, I. Walev, E. Martin, D. Jonas, and M. Palmer. 1993. A guide to the use of pore-forming toxins for controlled permeabilization of cell membranes. *Med. Microbiol. Immunol.* **182:**167–175.

Bielecki, J., P. Youngman, P. Connelly, and D. A. Portnoy. 1990. *Bacillus subtilis* expressing a haemolysin gene from *Listeria monocytogenes* can grow in mammalian cells. *Nature* **345:**175–176.

Billington, S. J., B. H. Jost, W. A. Cuevas, K. R. Bright, and J. G. Songer. 1997. The *Arcanobacterium (Actinomyces) pyogenes* hemolysin, pyolysin, is a novel member of the thiol-activated cytolysin family. *J. Bacteriol.* **179:**6100–6106.

Bonev, B., R. Gilbert, and A. Watts. 2000. Structural investigations of pneumolysin/lipid complexes. *Mol. Membr. Biol.* **17:**229–235.

Bonev, B. B., R. J. Gilbert, P. W. Andrew, O. Byron, and A. Watts. 2001. Structural analysis of the protein/lipid complexes associated with pore formation by the bacterial toxin pneumolysin. *J. Biol. Chem.* **276:**5714–5719.

Bricker, A. L., C. Cywes, C. D. Ashbaugh, and M. R. Wessels. 2002. NAD^+-glycohydrolase acts as an intracellular toxin to enhance the extracellular survival of group A streptococci. *Mol. Microbiol.* **44:**257–269.

Cossart, P. 2002. Molecular and cellular basis of the infection by *Listeria monocytogenes:* an overview. *Int. J. Med. Microbiol.* **291:**401–409.

Cossart, P., M. F. Vincente, J. Mengaud, F. Baquero, J. C. Perez-Diaz, and P. Berche. 1989. Listeriolysin O is essential for the virulence of *Listeria monocytogenes:* direct evidence obtained by gene complementation. *Infect. Immun.* **57:**3629–3639.

Decatur, A. L., and D. A. Portnoy. 2000. A PEST-like sequence in listeriolysin O essential for *Listeria monocytogenes* pathogenicity. *Science* **290:**992–995.

DeLano, W. L. 2002. *The PyMOL Molecular Graphics System.* DeLano Scientific, San Carlos, Calif.

Fang, Y., S. Cheley, H. Bayley, and J. Yang. 1997. The heptameric prepore of a staphylococcal alpha-hemolysin mutant in lipid bilayers imaged by atomic force microscopy. *Biochemistry* **36:**9518–9522.

Fehrenbach, F. J. 1972. NAD-glycohydrolase (streptolysin-O), ec 3225 and its role in cytolysis. *Biochem. Biophys. Res. Commun.* **48:**828–832.

Frehel, C., M. A. Lety, N. Autret, J. L. Beretti, P. Berche, and A. Charbit. 2003. Capacity of ivanolysin O to replace listeriolysin O in phagosomal escape and in vivo survival of *Listeria monocytogenes. Microbiology* **149:**611–620.

Galione, A. 1994. Cyclic ADP-ribose, the ADP-ribosyl cyclase pathway and calcium signalling. *Mol. Cell. Endocrinol.* **98:**125–131.

Galione, A., and G. C. Churchill. 2002. Interactions between calcium release pathways: multiple messengers and multiple stores. *Cell Calcium* **32:**343–354.

Giddings, K. S., A. E. Johnson, and R. K. Tweten. 2003. Redefining cholesterol's role in the mechanism of the cholesterol-dependent cytolysins. *Proc. Natl. Acad. Sci. USA* **100:**11315–11320.

Glomski, I. J., M. M. Gedde, A. W. Tsang, J. A. Swanson, and D. A. Portnoy. 2002. The *Listeria monocytogenes* hemolysin has an acidic pH optimum to compartmentalize activity and prevent damage to infected host cells. *J. Cell Biol.* **156:**1029–1038.

Gouaux, J. E., O. Braha, M. R. Hobaugh, L. Song, S. Cheley, C. Shustak, and H. Bayley. 1994. Subunit stoichiometry of staphylococcal alpha-hemolysin in crystals and on membranes: a heptameric transmembrane pore. *Proc. Natl. Acad. Sci. USA* **91:**12828–12831.

Greco, R., L. De Martino, G. Donnarumma, M. P. Conte, L. Seganti, and P. Valenti. 1995. Invasion of cultured human cells by *Streptococcus pyogenes*. *Res. Microbiol.* **146:**551–560.

Heuck, A. P., E. Hotze, R. K. Tweten, and A. E. Johnson. 2000. Mechanism of membrane insertion of a multimeric β-barrel protein: perfringolysin O creates a pore using ordered and coupled conformational changes. *Mol. Cell* **6:**1233–1242.

Heuck, A. P., R. K. Tweten, and A. E. Johnson. 2001. Beta-barrel pore-forming toxins: intriguing dimorphic proteins. *Biochemistry* **40:**9065–9073.

Heuck, A. P., R. K. Tweten, and A. E. Johnson. 2003. Assembly and topography of the prepore complex in cholesterol-dependent cytolysins. *J. Biol. Chem.* **278:**31218–31225.

Hotze, E. M., A. P. Heuck, D. M. Czajkowsky, Z. Shao, A. E. Johnson, and R. K. Tweten. 2002. Monomer-monomer interactions drive the prepore to pore conversion of a beta-barrel-forming cholesterol-dependent cytolysin. *J. Biol. Chem.* **277:**11597–11605.

Hotze, E. M., E. M. Wilson-Kubalek, J. Rossjohn, M. W. Parker, A. E. Johnson, and R. K. Tweten. 2001. Arresting pore formation of a cholesterol-dependent cytolysin by disulfide trapping synchronizes the insertion of the transmembrane beta-sheet from a prepore intermediate. *J. Biol. Chem.* **276:**8261–8268.

Iwamoto, M., I. Morita, M. Fukuda, S. Murota, S. Ando, and Y. Ohno-Iwashita. 1997. A biotinylated perfringolysin O derivative: a new probe for detection of cell surface cholesterol. *Biochim. Biophys. Acta* **1327:** 222–230.

Iwamoto, M., M. Nakamura, K. Mitsui, S. Ando, and Y. Ohno-Iwashita. 1993. Membrane disorganization induced by perfringolysin O (theta-toxin) of *Clostridium perfringens*—effect of toxin binding and self-assembly on liposomes. *Biochim. Biophys. Acta* **1153:**89–96.

Iwamoto, M., Y. Ohno-Iwashita, and S. Ando. 1987. Role of the essential thiol group in the thiol-activated cytolysin from *Clostridium perfringens*. *Eur. J. Biochem.* **167:**425–430.

Iwamoto, M., Y. Ohno-Iwashita, and S. Ando. 1990. Effect of isolated C-terminal fragment of theta-toxin (perfringolysin-O) on toxin assembly and membrane lysis. *Eur. J. Biochem.* **194:**25–31.

Jacobs, T., A. Darji, N. Frahm, M. Rohde, J. Wehland, T. Chakraborty, and S. Weiss. 1998. Listeriolysin O: cholesterol inhibits cytolysis but not binding to cellular membranes. *Mol. Microbiol.* **28:**1081–1089.

Jones, S., and D. A. Portnoy. 1994. Characterization of *Listeria monocytogenes* pathogenesis in a strain expressing perfringolysin O in place of listeriolysin O. *Infect. Immun.* **62:**5608–5613.

Jones, S., K. Preiter, and D. A. Portnoy. 1996. Conversion of an extracellular cytolysin into a phagosome-specific lysin which supports the growth of an intracellular pathogen. *Mol. Microbiol.* **21:**1219–1225.

Lee, H. C. 1994. Cyclic ADP-ribose: a new member of a super family of signalling cyclic nucleotides. *Cell Signal* **6:**591–600.

Lety, M. A., C. Frehel, P. Berche, and A. Charbit. 2002. Critical role of the N-terminal residues of listeriolysin O in phagosomal escape and virulence of *Listeria monocytogenes*. *Mol. Microbiol.* **46:**367–379.

Lety, M. A., C. Frehel, I. Dubail, J. L. Beretti, S. Kayal, P. Berche, and A. Charbit. 2001. Identification of a PEST-like motif in listeriolysin O required for phagosomal escape and for virulence in *Listeria monocytogenes*. *Mol. Microbiol.* **39:**1124–1139.

Madden, J. C., N. Ruiz, and M. Caparon. 2001. Cytolysin-mediated translocation (CMT): a functional equivalent of type III secretion in gram-positive bacteria. *Cell* **104:**143–152.

Miller, C. J., J. L. Elliot, and R. L. Collier. 1999. Anthrax protective antigen: prepore-to-pore conversion. *Biochemistry* **38:**10432–10441.

Mobius, W., Y. Ohno-Iwashita, E. G. van Donselaar, V. M. Oorschot, Y. Shimada, T. Fujimoto, H. F. Heijnen, H. J. Geuze, and J. W. Slot. 2002. Immunoelectron microscopic localization of cholesterol using biotinylated and non-cytolytic perfringolysin O. *J. Histochem. Cytochem.* **50:**43–55.

Nagamune, H., C. Ohnishi, A. Katsuura, K. Fushitani, R. A. Whiley, A. Tsuji, and Y. Matsuda. 1996. Intermedilysin, a novel cytotoxin specific for human cells secreted by *Streptococcus intermedius* UNS46 isolated from a human liver abscess. *Infect. Immun.* **64:**3093–3100.

Nakamura, M., N. Sekino, M. Iwamoto, and Y. Ohno-Iwashita. 1995. Interaction of theta-toxin (per-fringolysin O), a cholesterol-binding cytolysin, with liposomal membranes: change in the aromatic side chains upon binding and insertion. *Biochemistry* **34:**6513–6520.

Ohno-Iwashita, Y., M. Iwamoto, S. Ando, K. Mitsui, and S. Iwashita. 1990. A modified θ-toxin produced by limited proteolysis and methylation: a probe for the functional study of membrane cholesterol. *Biochim. Biophys. Acta* **1023:**441–448.

Ohno-Iwashita, Y., M. Iwamoto, K. Mitsui, S. Ando, and S. Iwashita. 1991. A cytolysin, theta-toxin, prefer-entially binds to membrane cholesterol surrounded by phospholipids with 18-carbon hydrocarbon chains in cholesterol-rich region. *J. Biochem.* **110:**369–375.

Ohno-Iwashita, Y., M. Iwamoto, K. Mitsui, S. Ando, and Y. Nagai. 1988. Protease nicked q-toxin of *Clostridium perfringens,* a new membrane probe with no cytolytic effect, reveals two classes of cholesterol as toxin-binding sites on sheep erythrocytes. *Eur. J. Biochem.* **176:**95–101.

Olofsson, A., H. Hebert, and M. Thelestam. 1993. The projection structure of perfringolysin-O (*Clostridium perfringens* theta-toxin). *FEBS Lett.* **319:**125–127.

Palmer, M., R. Harris, C. Freytag, M. Kehoe, J. Tranum-Jensen, and S. Bhakdi. 1998. Assembly mechanism of the oligomeric streptolysin O pore: the early membrane lesion is lined by a free edge of the lipid membrane and is extended gradually during oligomerization. *EMBO J.* **17:**1598–1605.

Pinkney, M., E. Beachey, and M. Kehoe. 1989. The thiol-activated toxin streptolysin O does not require a thiol group for activity. *Infect. Immun.* **57:**2553–2558.

Portnoy, D., P. S. Jacks, and D. Hinrichs. 1988. The role of hemolysin for intracellular growth of *Listeria monocytogenes. J. Exp. Med.* **167:**1459–1471.

Portnoy, D. A., V. Auerbuch, and I. J. Glomski. 2002. The cell biology of *Listeria monocytogenes* infection: the intersection of bacterial pathogenesis and cell-mediated immunity. *J. Cell Biol.* **158:**409–414.

Portnoy, D. A., R. K. Tweten, M. Kehoe, and J. Bielecki. 1992. The capacity of of listeriolysin O, streptolysin O, and perfringolysin O to mediate growth of *Bacillus subtilis* within mammalian cells. *Infect. Immun.* **60:**2710–2717.

Prigent, D., and J. E. Alouf. 1976. Interaction of streptolysin O with sterols. *Biochem. Biophys. Acta* **433:**422–428.

Ramachandran, R., A. P. Heuck, R. K. Tweten, and A. E. Johnson. 2002. Structural insights into the mem-brane-anchoring mechanism of a cholesterol-dependent cytolysin. *Nat. Struct. Biol.* **9:**823–827.

Rechsteiner, M., and S. W. Rogers. 1996. PEST sequences and regulation by proteolysis. *Trends Biochem. Sci.* **21:**267–271.

Rossjohn, J., S. C. Feil, W. J. McKinstry, R. K. Tweten, and M. W. Parker. 1997. Structure of a cholesterol-binding thiol-activated cytolysin and a model of its membrane form. *Cell* **89:**685–692.

Rottem, S., R. M. Cole, W. H. Habig, M. F. Barile, and M. C. Hardegree. 1982. Structural characteristics of tetanolysin and its binding to lipid vesicles. *J. Bacteriol.* **152:**888–892.

Ruiz, N., B. Wang, A. Pentland, and M. Caparon. 1998. Streptolysin O and adherence synergistically modulate proinflammatory responses of keratinocytes to group A streptococci. *Mol. Microbiol.* **27:**337–346.

Saunders, K. F., T. J. Mitchell, J. A. Walker, P. W. Andrew, and G. J. Boulnois. 1989. Pneumolysin, the thiol-activated toxin of *Streptococcus pneumoniae,* does not require a thiol group for in vitro activity. *Infect. Immun.* **57:**2547–2552.

Scott, M. E., Z. Y. Dossani, and M. Sandkvist. 2001. Directed polar secretion of protease from single cells of Vibrio cholerae via the type II secretion pathway. *Proc. Natl. Acad. Sci. USA* **98:**13978–13983.

Sekino-Suzuki, N., M. Nakamura, K. I. Mitsui, and Y. Ohno-Iwashita. 1996. Contribution of individual tryp-tophan residues to the structure and activity of theta-toxin (perfringolysin O), a cholesterol-binding cytolysin. *Eur. J. Biochem.* **241:**941–947.

Sellman, B. R., B. L. Kagan, and R. K. Tweten. 1997. Generation of a membrane-bound, oligomerized pre-pore complex is necessary for pore formation by *Clostridium septicum* alpha toxin. *Mol. Microbiol.* **23:**551–558.

Shatursky, O., A. P. Heuck, L. A. Shepard, J. Rossjohn, M. W. Parker, A. E. Johnson, and R. K. Tweten. 1999. The mechanism of membrane insertion for a cholesterol dependent cytolysin: a novel paradigm for pore-forming toxins. *Cell* **99:**293–299.

Shepard, L. A., A. P. Heuck, B. D. Hamman, J. Rossjohn, M. W. Parker, K. R. Ryan, A. E. Johnson, and R. K. Tweten. 1998. Identification of a membrane-spanning domain of the thiol-activated pore-forming toxin *Clostridium perfringens* perfringolysin O: an α-helical to β-sheet transition identified by fluorescence spec-troscopy. *Biochemistry* **37:**14563–14574.

Shepard, L. A., O. Shatursky, A. E. Johnson, and R. K. Tweten. 2000. The mechanism of assembly and insertion of the membrane complex of the cholesterol-dependent cytolysin perfringolysin O: formation of a large prepore complex. *Biochemistry* **39:**10284–10293.

Shimada, Y., M. Maruya, S. Iwashita, and Y. Ohno-Iwashita. 2002. The C-terminal domain of perfringolysin O is an essential cholesterol-binding unit targeting to cholesterol-rich microdomains. *Eur. J. Biochem.* **269:**6195–6203.

Smith, G. A., H. Marquis, S. Jones, N. C. Johnston, D. A. Portnoy, and H. Goldfine. 1995. The two distinct phospholipases C of *Listeria monocytogenes* have overlapping roles in escape from a vacuole and cell-to-cell spread. *Infect. Immun.* **63:**4231–4237.

Song, P. I., Y. M. Park, T. Abraham, B. Harten, A. Zivony, N. Neparidze, C. A. Armstrong, and J. C. Ansel. 2002. Human keratinocytes express functional CD14 and toll-like receptor 4. *J. Investig. Dermatol.* **119:**424–432.

Stevens, D. L. 1996. Invasive group A streptococcal disease. *Infect. Agents Dis.* **5:**157–166.

Stevens, D. L., D. B. Salmi, E. R. McIndoo, and A. E. Bryant. 2000. Molecular epidemiology of nga and NAD glycohydrolase/ADP-ribosyltransferase activity among *Streptococcus pyogenes* causing streptococcal toxic shock syndrome. *J. Infect. Dis.* **182:**1117–1128.

Tweten, R. K., R. W. Harris, and P. J. Sims. 1991. Isolation of a tryptic fragment from *Clostridium perfringens* θ-toxin that contains sites for membrane binding and self-aggregation. *J. Biol. Chem.* **266:**12449–12454.

Tweten, R. K., M. W. Parker, and A. E. Johnson. 2001. The cholesterol-dependent cytolysins. *Curr. Top. Microbiol. Immunol.* **257:**15–33.

Wadsworth, S. J., and H. Goldfine. 2002. Mobilization of protein kinase C in macrophages induced by *Listeria monocytogenes* affects its internalization and escape from the phagosome. *Infect. Immun.* **70:**4650–4660.

Waheed, A. A., Y. Shimada, H. F. Heijnen, M. Nakamura, M. Inomata, M. Hayashi, S. Iwashita, J. W. Slot, and Y. Ohno-Iwashita. 2001. Selective binding of perfringolysin O derivative to cholesterol-rich membrane microdomains (rafts). *Proc. Natl. Acad. Sci. USA* **98:**4926–4931.

Walev, I., M. Palmer, A. Valeva, U. Weller, and S. Bhakdi. 1995. Binding, oligomerization, and pore formation by streptolysin O in erythrocytes and fibroblast membranes: detection of nonlytic polymers. *Infect. Immun.* **63:**1188–1194.

Walker, B., O. Braha, S. Cheley, and H. Bayley. 1995. An intermediate in the assembly of a pore-forming protein trapped with a genetically-engineered switch. *Chem. Biol.* **2:**99–105.

Walker, J. A., R. L. Allen, P. Falmagne, M. K. Johnson, and G. J. Boulnois. 1987. Molecular cloning, characterization, and complete nucleotide sequence of the gene for pneumolysin, the sulfhydryl-activated toxin of *Streptococcus pneumoniae. Infect. Immun.* **55:**1184–1189.

Wang, B., N. Ruiz, A. Pentland, and M. Caparon. 1997. Keratinocyte proinflammatory responses to adherent and nonadherent group A streptococci. *Infect. Immun.* **65:**2119–2126.

Structural Biology of Bacterial Pathogenesis
Edited by G. Waksman et al.
© 2005 ASM Press, Washington, D.C.

Chapter 12

Toll/Interleukin-1 Receptors and Innate Immunity

Liang Tong

Innate immunity represents the sole mechanism of host defense against microbial infections in invertebrates. In mammals and other vertebrates, innate immunity provides the first line of host defense against these infections. In addition, the innate immune response is crucial for the stimulation and specificity of the adaptive immune response, which is present only in the vertebrates. Recent studies have shown that there are remarkable similarities between the innate immune systems of vertebrates and invertebrates, suggesting that innate immunity may have a common ancestry and may pre-date the emergence of adaptive immunity.

Innate immune responses are dependent on germ line-encoded receptors that recognize the conserved molecular structures from a large collection of pathogenic microorganisms. These conserved structures are known as pathogen-associated molecular patterns (PAMPs), and the receptors that recognize them are known as pattern-recognition receptors (PRRs) (Janeway, 1989; Medzhitov and Janeway, 1997). In insects and animals, there are a limited number of PRRs encoded by the host genome. This necessitates that the PRRs recognize patterns that are conserved among a large number of pathogens, so that each PRR can be used to sense an entire class of microbial organisms. At the same time, the conserved nature of these features indicates that they are indispensable for the survival of the pathogens, which minimizes the degree of mutational variability of these patterns. For example, lipopolysaccharide (LPS), also known as endotoxin, is the unique cell wall component of most gram-negative bacteria and is recognized as a PAMP by the innate immune system. It is also likely that each pathogen possesses several different PAMPs and potentially can activate several different PRRs simultaneously. This may be a mechanism for achieving specificity in the immune response to a given pathogen. In plants, on the other hand, the germ line encodes a much larger set of PRRs (plant disease resistance proteins), thus enabling the direct recognition of many specific products from pathogens (gene-for-gene recognition).

Studies over the past few years have shown that Toll-like receptors (TLRs) are crucial molecules in the recognition of various PAMPs, and therefore represent a class of PRRs (Fig. 1). In contrast to earlier presumptions that innate immunity is mostly a nonspecific response to pathogens, the recognition of PAMPs by the TLRs has exquisite specificity,

---

Liang Tong • Department of Biological Sciences, Columbia University, New York, NY 10027.

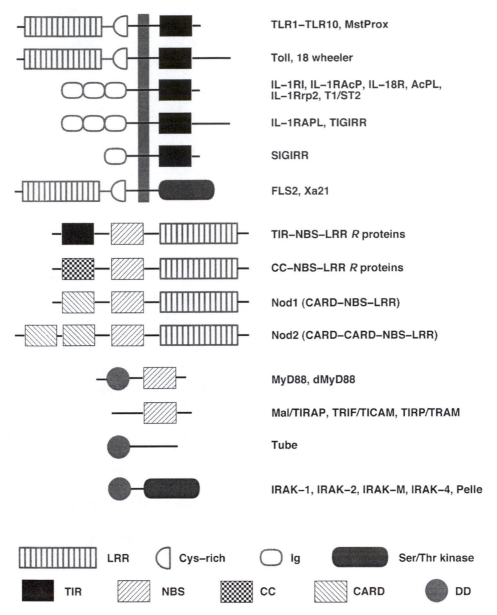

Figure 1. Domain organization of TLRs, IL-1Rs, and associated molecules. The individual domains are indicated by the shapes of their symbols. For all the proteins, the N terminus is on the left.

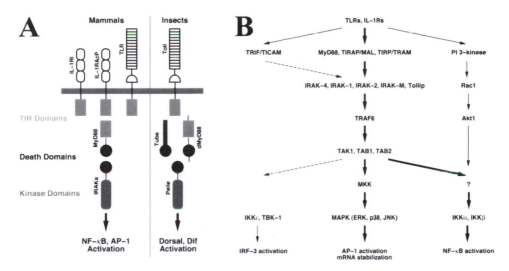

Figure 2. Signaling pathways for TLRs and IL-1Rs. (A) Membrane-proximal events in the signal transduction. The TIR domains are shown as solid rectangles, and the death domains are shown as solid circles. (B) Overall signaling events in the TLR and IL-1R pathways. Abbreviations: TRAF, TNF-α receptor-associated factor; TAK, transforming growth factor β-activated kinase; MKK, MAP kinase kinase; IKK, I-κB kinase; TBK, TANK-binding kinase.

with each TLR responding only to a unique set of microbial products. This specific recognition of pathogens by the innate immune PRRs is also crucial for the stimulation of the proper adaptive immune response in the vertebrates, thereby ensuring the correct specificity for the overall host immune defense. Ten TLRs have so far been identified from the human genome, indicating the limited repertoire of this set of PRRs.

The signal transduction pathways of the TLRs are shared with those of the interleukin-1 receptor superfamily (IL-1Rs), since these receptors share a conserved intracellular domain, known as the Toll/IL-1 receptor (TIR) domain (Fig. 2). Like the TLRs, IL-1Rs also play key roles in host responses to infection and inflammation. This chapter describes the current state of knowledge of these two superfamilies of receptors and their ligands, as well as the molecules that mediate the membrane-proximal events of their signaling (Table 1). Space limitations unfortunately prevent the citation of most of the primary literature prior

Table 1. IL-1 receptors and TLRs

Receptor	Ligands	Adaptors
IL-1RI/IL-1RAcP	IL-1α, IL-1β, IL-1ra	MD88, TIRP/TRAM
IL-18R/AcPL	IL-18	MyD88, TIRP/TRAM
TLR2, TLR2-TLR1, TLR2-TLR6	Peptidoglycan, lipoproteins, zymosan, GPI anchor (*T. cruzi*), lipoarabinomannan, MALP-2, HSPs	MyD88, TIRAP/MAL
TLR3	dsRNA, poly(I-C)	TRIF/TICAM, MyD88
TLR4	LPS, taxol, HSPs	MyD88, TIRAP/MAL, TRIF/TICAM
TLR5	Flagellin	MyD88
TLR7, TLR8	Antiviral compounds	MyD88
TLR9	Unmethylated CpG-DNA	MyD88

to 2002. These publications can be found in the many excellent reviews on innate immunity, TLRs, IL-1Rs, and plant disease resistance (Aderem and Ulevitch, 2000; Anderson, 2000; Beutler, 2000; Dunne and O'Neill, 2003; Hoffmann et al., 1999; Janeway, 1989; Medzhitov and Janeway, 1997; O'Neill, 2003; Sims, 2002; Takeda and Akira, 2003; Underhill and Ozinsky, 2002), as well as the collection of monographs in *Current Topics in Microbiology and Immunology* (Beutler and Wagner, 2002).

DROSOPHILA Toll AND ITS HOMOLOGS

Drosophila Toll was discovered in 1984 as a maternal-effect gene that is crucial for the establishment of dorsoventral polarity in developing embryos. Recessive mutations of Toll produce dorsalized embryos, whereas dominant mutations produce ventralized embryos. The name Toll comes from German slang, in which it means crazy, fantastic, or "far-out", which was used to describe the ventralized embryos (Anderson, 2000). The Toll gene was cloned in 1988 by using P-element transposon tagging. It encodes a type I transmembrane receptor of 1,097 amino acid residues, with a large extracellular region of 803 residues, a single membrane-spanning segment (with 25 residues), and an intracellular segment of 269 residues.

The extracellular region of Toll contains the leucine-rich repeat (LRR) domain and a cysteine-rich domain. LRRs are amino acid motifs of about 24 residues each that contain abundant numbers of leucine residues. Crystal structures of LRRs from other proteins show that each repeat forms a β-α or α-α structural unit, and the copies of the repeat are arranged side by side, forming a curved, horseshoe-like structure (Color Plate 50A [see color insert]). The LRR domain is involved mostly in protein-protein interactions and could also bind other ligands. A Toll mutant lacking the LRR domain produces a pronounced ventral cell fate, suggesting that the LRR domain may be a negative regulator of Toll signaling. The cysteine-rich domain is frequently found at the end of the LRR domain, just prior to the transmembrane segment. Disulfide bonding in this domain is important for the proper function of the receptor. Mutations that replace cysteine residues in this domain with tyrosine produce constitutively active, ventralizing receptors.

The intracellular region of Toll contains two domains (Fig. 1). The first, membrane-proximal domain has about 150 residues and bears sequence homology to the intracellular domain of the IL-1 receptor (IL-1R). This domain is known as the TIR domain and is discussed in more detail below. The second domain covers the 100 residues at the C terminus of the receptor, and it may play an inhibitory role in signal transduction by this receptor.

Genetic studies identify Toll as one of 12 genes that are involved in the establishment of the dorsoventral axis in *Drosophila* embryos. These genes can be arranged in a linear pathway, suggesting a signal transduction process through the Toll receptor. Seven of these genes are upstream of Toll, whereas four are downstream. Immediately upstream of Toll, the cystine knot protein Spätzle is the ligand of Toll (Weber et al., 2003). Spätzle must be processed by a proteolytic cascade (Gastrulation defective, Snake, and Easter) before it can activate Toll. Downstream of Toll, the proteins Tube and Pelle function as signaling adaptors. The activation of Toll for dorsoventral patterning ultimately leads to the degradation of Cactus, the *Drosophila* homolog of the mammalian I-κB subunit. This allows the Dorsal protein, a *Drosophila* homolog of the mammalian NF-κB transcription factor, to migrate into the nucleus. The dorsoventral polarity is established by a concentration gradient of Dorsal in the nucleus. These downstream molecules in the Toll signal transduction pathway (Tube, Pelle, and Dorsal) and their mammalian homologs are discussed in more detail below.

In adult fruit flies, Toll and its signaling pathway have important functions in the innate immune responses to infections by fungi and gram-positive bacteria. These responses include the production of drosomycin and other antimicrobial peptides in the fat body, the equivalent of the liver in mammals. Toll does not directly contact the microbial pathogens. Instead, it is still activated by the Spätzle ligand. However, there are significant differences between the Toll signaling pathways for development and immunity. Although the Spätzle, Tube, Pelle, and Cactus proteins are required for the immune response, a different protease, Persephone, is required to cleave the Spätzle ligand. Moreover, the ultimate transcription factor that is activated to produce antimicrobial peptides is not Dorsal but the related protein Dif. A different pathway, the immune deficiency (*imd*) pathway, is used by *Drosophila* to recognize infections by gram-negative bacteria.

The *Drosophila* genome contains a total of eight Toll homologs, including 18 wheeler, Tehao, and MstProx. Tehao is involved in the innate immune response. The 18 wheeler homolog may have an important function in embryonic development. It is not involved in the innate immune response to gram-negative bacterial infections in *Drosophila,* despite earlier reports. The functions of the other homologs are not known, and neither are the identities of the ligands of these Toll homologs.

Toll-LIKE RECEPTORS IN MAMMALS

A human homolog of the *Drosophila* Toll receptor was cloned from similarity searches in the expressed-sequence tag (EST) database and characterized in 1997. This receptor has the same domain architecture as *Drosophila* Toll, with a large extracellular domain, a single transmembrane segment, and an intracellular segment, and is now known as Toll-like receptor 4 (TLR4). In contrast to Toll, TLR4 and most other TLRs do not have the second domain in the intracellular segment (Fig. 1). Overexpression of a constitutively active mutant of this receptor, obtained by fusing the extracellular domains of CD4 to the transmembrane and intracellular domains of TLR4, can activate the NF-κB transcription factor, suggesting that the signal transduction pathways may be similar between the mammalian and insect receptors. TLR4 can also induce the production of the costimulatory molecule B7.1, suggesting that TLR4 may be involved in the activation of adaptive immunity.

The functional role of TLR4 in innate immunity was revealed from genetic characterizations of mice that are hyposensitive to LPS, a cell wall component of gram-negative bacteria. This proved that TLR4 is the receptor for LPS and led to the realization that the TLRs are PRRs for PAMPs. This produced an explosion of research on these receptors and their signal transduction pathways. More than 1,000 papers have been published on the TLRs alone over the past 5 years. A total of 10 TLRs have been identified from humans, and the ligands for most of these TLRs have now been characterized. In contrast to *Drosophila,* where Toll and its homologs recognize the pathogens indirectly, it is thought that the mammalian TLRs recognize their microbial ligands directly.

TLR4 and the Recognition of LPS

LPS is the unique cell wall component of most gram-negative bacteria and therefore represents a PAMP for these pathogens. At the core of LPS is the lipid A moiety, which is a phosphorylated and multiply acylated glucosamine disaccharide. LPS is a potent activator of host immunity and promotes the secretion of tumor necrosis factor (TNF) and other cytokines for the antimicrobial response. Early recognition of this PAMP is crucial for the host to mount an effective immune response, which leads to the clearance of infections by gram-negative bacteria. Mice that cannot respond to LPS are resistant to the toxic effects of this compound but succumb easily to infections by gram-negative bacteria. On the other hand, an unregulated response to LPS can ultimately lead to sepsis and other serious diseases.

Two strains of mice, C3H/HeJ and C57BL/10ScCr, are known to have a defective immune response to LPS. Genetic studies show that the defects are due to mutations in the same gene. Positional cloning identified this gene as murine TLR4 and proves that TLR4 is the receptor for LPS. The Lps^d (d for "defective") allele of C3H/HeJ mice carries the Pro712His single-point mutation, located in the intracellular TIR domain of TLR4 (to be described in more detail below). The C57BL/10ScCr mice are naturally occurring TLR4 knock out mutants. While the null mutation in C57BL/10ScCr mice is recessive, the Lps^d Pro712His mutation in C3H/HeJ mice has dominant negative effects on TLR4 signaling. However, the intermediate LPS response in $Lps^n:Lps^d$ heterozygotes is due to monoallelic expression of the TLR4 gene rather than to a reduced activity of the wild-type/mutant heterocomplex (Pereira et al., 2003).

The recognition of LPS by TLR4 requires the presence of several accessory proteins, including LBP (LPS-binding protein), CD14, and MD-2. LBP is a plasma protein that is produced by the liver and helps to transfer LPS to CD14, which is linked to the membrane through a glycosylphosphoinositol (GPI) anchor. The CD14-LBP-LPS complex cannot transduce a signal since it does not have an intracellular component. TLR4 directly contacts LPS and also provides the intracellular component for this signaling complex. MD-2 is a small, extracellular protein that is tightly associated with the extracellular region of TLR4. Mice that are deficient in MD-2 do not respond to LPS. Interestingly, the trafficking of TLR4 to the cell membrane is also disrupted in the absence of MD-2, the receptor being mostly retained by the Golgi apparatus in MD-$2^{-/-}$ mice. MD-2 is a homolog of MD-1, which is tightly associated with the RP105 receptor. RP105 has LRR in its extracellular region, similar to the TLRs, but has only a small intracellular region.

The human TLR4/MD-2 complex can recognize hexa-acylated LPS, whereas the murine TLR4/MD-2 complex cannot (Hajjar et al., 2002). This specificity is determined by a region in TLR4 that is hypervariable between humans and mice. This observation is consistent with the direct recognition of LPS by TLR4.

In mice, the TLR4 and MD-2 molecules also mediate the immuno stimulatory activities of Taxol, which can mimic LPS in the activation of host immunity. However, the human TLR4 and MD-2 proteins are not sensitive to Taxol, providing evidence that Taxol, like LPS, contacts these receptor molecules directly.

TLR2-(TLR1 or TLR6) and the Recognition of Bacterial Peptidoglycan, Lipoproteins, and Yeast Zymosan

For gram-positive bacteria, mycobacteria, and spirochetes, several different cell wall components are recognized by the host immune system. These include peptidoglycan, lipoteichoic acid, glycoproteins, lipoproteins, and outer surface proteins. TLR2 is the PRR for these and many other PAMPs, including lipopeptides, GPI anchors from *Trypanosoma cruzi*, lipoarabinomannan from *Myobacterium tuberculosis*, special variants of LPS, and zymosan from yeast. The *Yersinia* pathogen secretes a protein, LcrV, that engages TLR2, leading to immunosuppression by IL-10 induction and evasion of host immunity (Sing et al.,

2002). Therefore, in contrast to TLR4, TLR2 can recognize a wide array of different PAMPs. In addition, the recognition of bacterial lipoproteins may mediate the initiation of apoptosis.

The apparent ability of TLR2 to recognize this large collection of microbial products may be related to the fact that it functions as a heterooligomer with TLR1 or TLR6. The TLR2-TLR6 heterocomplex is required for the recognition of the macrophage-activating lipopeptide (MALP-2), a diacylated lipoprotein, but TLR6 is not required for the recognition of a triacylated lipoprotein. In comparison, TLR4 and TLR5 are thought to signal as homodimers or homo-oligomers.

TLR3 and the Recognition of Double-Stranded Viral RNA

Double-stranded RNA (dsRNA) is produced by most viruses during their life cycle in the host cell and therefore is a PAMP for the detection of viral infections. Polyinosine-polycytidylic acid [poly(I-C)], a synthetic analog of dsRNA, can elicit similar immune responses in mammalian cells, including the production of alpha and beta interferons (IFN-α and IFN-β), and other cytokines such as IL-6 and IL-12, activation of NF-κB and mitogen-activated protein (MAP) kinases, and maturation of dendritic cells.

TLR3 is the PRR for poly(I-C) and dsRNA. TLR3$^{-/-}$ mice have a normal appearance but have a reduced immune response to poly(I-C) and are resistant to the lethal effect of this compound. Human 293 cells do not respond to poly(I-C) unless they have been transfected with TLR3, whereas transfection with other TLRs do not sensitize these cells to poly(I-C). Murine TLR3 is most abundantly expressed in the lungs, brain, and kidneys, and the expression is strongly induced by the presence of LPS. Interestingly, LPS stimulation of TLR4 can also lead to the activation of IRF-3.

TLR5 and the Recognition of Bacterial Flagellin

Flagellin is the principal component of the bacterial flagellar filament, which confers motility to many gram-positive and gram-negative bacteria. It is highly conserved among bacteria but does not have homologs in higher eukaryotes. Therefore, flagellin is a PAMP for the detection of bacterial infections.

Flagellin is a potent activator of immune responses in the vertebrates, insects, and plants. In a screen for human TLR5 ligands, flagellin was identified by tandem mass spectrometry as the PAMP for this PRR. TLR5 can recognize monomeric, denatured flagellin, and the engagement of this receptor leads to the activation of NF-κB and the production of TNF-α.

In *Drosophila,* flagellin activates the production of the antimicrobial peptide cecropin, but it is not known whether the Toll pathway, or the *imd* pathway, is involved in this recognition. In plants, flagellin is recognized by a receptor, FLS2 in *Arabidopsis,* that contains extracellular LRRs. However, the intracellular region of this receptor contains a protein Ser/Thr kinase domain (Fig. 1), in contrast to the TIR domain of the TLRs.

TLR7, TLR8, and the Recognition of Small Antiviral Compounds

Imidazoquinolines are a class of low-molecular-weight synthetic compounds that are potent activators of immune cells. These compounds can be used for antiviral and anticancer therapy. For example, one such compound, imiquimod (Color Plate 50B), is in clinical use for the treatment of external genital warts caused by human papillomavirus infection. A related compound, R-848 (also known as resiquimod [Color Plate 50B]), is more potent than imiquimod. Although these compounds are not derived from microbial organisms, their effects on the host immune system are similar to those of PAMPs.

TLR7 is a receptor for these small antiviral compounds. TLR7$^{-/-}$ mouse macrophages do not respond to these compounds but respond normally to LPS, PGN, CpG-DNA, and other microbial PAMPs. Human 293 cells become responsive to these compounds after transfection with TLR7 but not with TLR2 or TLR4. Interestingly, human 293 cells transfected with human TLR8 can also respond to R-848, suggesting that TLR8 may be a receptor for these compounds as well. However, murine TLR8 does not respond to R-848. These compounds are unlikely to be the natural ligands of these receptors.

TLR9 and the Recognition of Unmethylated CpG-DNA

Unmethylated CpG motifs occur only in bacteria and are potent stimulators of the mammalian immune response. Bacterial CpG-DNA can be used as adjuvants in vaccines against various human diseases. In comparison, CpG sequences in mammalian DNA are less frequent and are mostly methylated. This makes unmethylated CpG motifs a PAMP for the detection of bacterial infections in general.

TLR9 is the PRR for CpG motifs. TLR9$^{-/-}$ mice do not respond to CpG and are highly resistant to the toxic effects of CpG-DNA. The CpG-DNA undergoes endocytosis and is recognized by TLR9 in endosomes. In addition to the LRRs, the extracellular region of TLR9 contains two DNA-binding motifs, which may be the mechanism for the recognition of CpG motifs by this receptor. There is exquisite species specificity in this recognition. While the mouse TLR9 recognizes the core sequence GACGTT, human TLR9 recognizes GTCGTT, a difference of only a single base, even though the human and mouse TLR9s have 76% amino acid sequence identity.

TLRs and the Recognition of Host Proteins

TLRs are normally receptors that recognize exogenous molecules, the PAMPs. At the same time, they can also recognize endogenous proteins from the host. For example, TLR2 and TLR4 mediate immune responses to endogenous heat shock proteins (HSPs), and possibly other molecules, that are released to the extracellular environment by necrotic cells. In this regard, TLRs can be considered more generally to be receptors for danger signals, which are derived either from invading pathogens or from host cells that have undergone necrotic death. Interestingly, HSPs from many pathogens are also recognized by the host immune system as a PAMP.

PLANT DISEASE RESISTANCE (*R*) PROTEINS

While animals recognize conserved patterns (PAMPs) from infecting microbes, plants directly recognize distinct effectors from pathogenic organisms. These effectors are known as pathogen avirulence factors and are crucial for the infection and colonization of plant tissues by the pathogens. Plant disease resistance proteins (*R* proteins) recognize these avirulence factors in a "gene-for-gene" manner. The innate immune responses include rapid oxidative bursts that can lead to apoptosis-like cell death at the site of infection, a phenomenon known as the hypersensitive response. Additional responses include increased expression of defense genes and the production of salicylate and ion fluxes.

The direct recognition of pathogen avirulence factors necessitates the presence of a large repertoire of host *R* proteins. These *R* proteins can be transmembrane receptors with extracellular LRRs and intracellular protein Ser/Thr kinase domains, such as FLS2, which recognizes flagellin (Fig. 1). There are also transmembrane *R* proteins with extracellular LRRs but only a short intracellular segment, reminiscent of the mammalian RP105 protein. Cytoplasmic protein Ser/Thr kinases, such as the Pto protein, can also be used for disease resistance.

The largest family of *R* proteins includes those that contain nucleotide-binding site (NBS)-LRR motifs. The NBS domain is involved mostly in protein-protein interactions. It may be associated with ATP or GTP binding, although it remains to be demonstrated whether the NBS domains in *R* proteins can actually bind nucleotide. At the N termini of these *R* proteins, there is either a TIR domain or a coiled-coil (CC) domain, giving the TIR-NBS-LRR and the CC-NBS-LRR *R* proteins (Fig. 1). The presence of both LRR and TIR domains may suggest some evolutionary relationships between the plant *R* proteins and mammalian TLRs. However, there are significant differences between them. The NBS-LRR proteins are mostly cytoplasmic, and the locations of the TIR and LRR domains are at opposite ends of these proteins from the TLRs (Fig. 1). In addition, the TIR domain in *R* proteins can directly recognize avirulence factors.

The completion of the sequencing of the *Arabidopsis* genome allows a complete annotation of the NBS-LRR-containing *R* proteins. There are 92 TIR-NBS-LRR and 51 CC-NBS-LRR proteins in the genome, representing about 0.5% of all the open reading frames (Meyers et al., 2003). In addition to these receptors, there are a large number of putative adaptor molecules that contain only the TIR-NBS domains (23 total), only the CC-NBS domains (5), or only the TIR domain (30). In the rice genome, there are more than 500 CC-NBS-LRR proteins but, interestingly, no TIR-NBS-LRR proteins.

The downstream events of signal transduction through the *R* proteins are still poorly understood. A MAP kinase cascade has been identified that functions downstream of the FLS2 receptor for innate immune response to flagellin (Asai et al., 2002).

Nod PROTEINS

In contrast to the TLRs, which are PRRs that recognize extracellular PAMPs, the Nod proteins, Nod1 and Nod2, mediate the recognition of pathogen patterns within the cell. Nod1 recognizes diaminopimelate-containing muropeptides in the peptidoglycan of gram-negative bacteria, but it does not recognize LPS, despite earlier reports to the contrary (Chamaillard et al., 2003; Girardin et al., 2003). Mutations in Nod2 are associated with Crohn's disease (inflammatory bowel disease) and Blau syndrome. Overexpression of Nod1 or Nod2 can activate NF-κB.

Both Nod proteins are cytoplasmic and contain caspase recruitment domains (CARDs), NBS, and LRRs (Fig. 1). The domain organization of these proteins has some similarity to that of the NBS-LRR plant disease resistance *R* proteins, except that CARD domains in Nods replace the TIR or CC domains in *R* proteins. One difference between Nod1 and Nod2 is that the latter has two CARDs (Fig. 1). These proteins belong to the CATER-PILLER superfamily of proteins, and detailed sequence analysis shows that the NBS could be broken into two subdomains, NACHT and NACHT-associated domain (Tschopp et al., 2003).

IL-1 RECEPTORS

IL-1 is a proinflammatory cytokine and has key functions in a variety of cellular processes including the acute-phase response, which is a type of innate immune response. IL-1 activates the transcriptional factors NF-κB and AP-1, thereby controlling the expression of many genes involved in host responses to infection, injury, and inflammation. The effects of IL-1 are mediated by two weakly homologous proteins, IL-1α and IL-1β, which have 24% amino acid sequence identity. The IL-1 superfamily also includes IL-1ra, IL-18, and several other members such as IL-1δ and IL-1ε. IL-1α and IL-1β have the same receptor, IL-1RI, while IL-1ra is a receptor antagonist for the functions of IL-1α and IL-1β.

Ten IL-1R homologs are currently known from humans: IL-1RI, IL-1RAcP, IL-1RII, IL-18R, AcPL, IL-1Rrp2, T1/ST2, IL-1RAPL, TIGIRR, and SIGIRR (Fig. 1). The genes for six of these receptors are clustered on chromosome 2q12. Of the remaining four, the IL-1RAcP gene is located on chromosome 3q28, the SIGIRR gene is located on 11p15, and the IL-1RAPL and TIGIRR genes are located on the X chromosome.

Functional studies have classified these receptors into three categories: receptor-like molecules whose intracellular domain can substitute for that of IL-1RI in the activation of NF-κB (IL-1RI, IL-18R, T1/ST2, and IL-1Rrp2), accessory protein-like molecules which can replace IL-1RAcP (IL-1RAcP and AcPL), and receptors that cannot substitute for either of the IL-1 receptor chains (IL-1RAPL, TIGIRR, and SIGIRR).

IL-1RI and IL-1RAcP

IL-1 type I receptor (IL-1RI) was identified by expression cloning in 1988 (interestingly, the same year when Toll was cloned). The receptor is expressed mostly on T cells and fibroblasts. It is a type I transmembrane protein, with three extracellular immunoglobulin-like (Ig) domains (D1, D2, and D3), a single membrane-spanning segment, and an intracellular domain with homology to that of Toll (the TIR domain [Fig. 1]). IL-1RI$^{-/-}$ mice have impaired IL-1 responses, confirming the importance of the receptor for IL-1 function and showing that the IL-1 cytokines are important pyrogens in response to viral and other infections.

IL-1α and IL-1β, as well as IL-1ra, bind IL-1RI with high affinity. The affinity of IL-1RI for IL-1α is about 10-fold higher than that for IL-1β. The crystal structure of IL-1RI in complex with IL-1β shows that the two N-terminal, membrane-distal Ig domains (D1 and D2) of the receptor are closely associated and coupled with an interdomain disulfide bridge whereas the C-terminal, membrane-proximal Ig domain (D3) is connected to the two N-terminal domains through a flexible linker (Color Plate 50C). The three Ig domains of the receptor wrap around the IL-1β ligand. IL-1β makes two equally important contacts with the receptor. The first contact is at the interface between domains D1 and D2, burying 780 Å^2 of the surface area of the receptor in this region. A small peptide that can block the association of IL-1 with the receptor is also bound at this interface. The second contact is with the membrane-proximal Ig domain (D3), burying 810 Å^2 of the surface area of the receptor in this region.

There is a large difference in the position of the D3 domain of the receptor when it is in complex with IL-1ra. As a result, IL-1ra has a major contact with the two membrane-distal domains of the receptor (burying 1,030 Å^2 of the surface area) but the contact with the C-terminal Ig domain is much weaker (burying 450 Å^2 of the surface area). The structural observations are confirmed by binding studies. The D1-D2 domains of the receptor bind IL-1ra with high affinity (K_d of 28 nM), whereas their affinity for IL-1α is much lower (K_d of 7 μM).

Signal transduction by IL-1 requires a second receptor molecule, IL-1R accessory protein (IL-1RAcP), which has the same domain organization as IL-1RI (Fig. 1). This molecule does not have high affinity for IL-1 itself, but it binds to the IL-1RI/IL-1 complex. The presence of the accessory protein also increases the affinity of IL-1RI for the ligand. A monoclonal antibody against IL-1RAcP can block signaling by the receptor. IL-1 may promote the hetero-oligomerization of the two receptor chains, and the formation of this oligomer may be the signal for activated IL-1 receptors. On the other hand, IL-1ra cannot recruit IL-1RAcP into the signaling complex, which may explain its antagonistic effect for IL-1 signaling. The structural difference of the D3 domain of the receptor in complexes with IL-1β and IL-1ra may be the molecular basis for the failed recruitment of IL-1RAcP by the IL-1ra complex (Color Plate 50D). The receptor-ligand complexes leave open a surface patch of the ligand, which contains the Asp145 residue in IL-1β (Lys145 in IL-1ra). Mutation of this residue can affect the agonist-antagonist activity of the ligand, suggesting that this open surface patch of the ligand might be used for binding the accessory protein. Therefore, the ligand may be a direct determinant in the recruitment of the accessory protein.

Besides IL-1ra, IL-1 signaling is also negatively regulated by a decoy receptor, IL-1RII, expressed mostly on B cells and neutrophils, and a soluble form of IL-1RAcP (sIL-1RAcP). IL-1RII has three extracellular Ig domains and has high affinity for IL-1β. However, it has only 29 amino acid residues in the intracellular domain and therefore cannot mediate the signal transduction. In addition, proteolytic cleavage of IL-1RII produces a soluble form of this receptor, which can sequester IL-1β from activating IL-1RI as well. IL-1RII can also associate with IL-1RAcP, another mechanism for this decoy receptor to modulate IL-1 signaling. Interestingly, several viruses encode homologs of this decoy receptor, for example, the B15R protein from vaccinia virus, which may be important for inhibiting the host immune response to viral infections. IL-1RII has low affinity for IL-1ra, in agreement with the notion that both molecules function as inhibitors of IL-1 signaling. sIL-1RAcP is derived from alternative splicing. It can interact with IL-1RI and functions as an inhibitor of IL-1R signaling.

IL-18R and AcPL

IL-18 was originally characterized as IFN-γ-inducing factor, because it can potently induce the production of IFN-γ from T cells. It also has a variety of other functions in innate and adaptive immune responses, including promotion of Th1-cell differentiation and induction of natural killer cell activity. The IL-18 receptor complex contains IL-18R and the accessory protein AcPL, similar to the complex for IL-1. Both IL-18R and AcPL have similar domain organizations to IL-1RI, with three extracellular Ig domains and an intracellular TIR domain (Fig. 1).

IL-18R was purified by affinity chromatography using a monoclonal antibody and was shown to be identical to an orphan receptor known as IL-1Rrp. AcPL was identified by a sequence homology search in the EST database. IL-18R has moderate affinity for IL-18, with a K_d of about 50 nM. In the presence of AcPL, the affinity becomes much higher, with a K_d of about 0.3 nM. Overexpression of both proteins is required for the activation of NF-κB and JNK in response to IL-18 in Cos7 cells.

Signaling by IL-18 is inhibited by the IL-18-binding protein (IL-18bp), which may function as a soluble decoy receptor. Members of the poxvirus family carry homologs of this protein. In contrast to IL-1RII and other IL-1R receptors, IL-18bp contains only one Ig domain. Besides IL-18, another ligand for IL-18R has been identified. IL-1H has the highest sequence similarity to IL-1ra and could be a receptor antagonist for IL-18R.

T1/ST2

T1/ST2 was originally identified as a late response gene that was induced by serum or overexpression of oncogenes in NIH 3T3 cells. T1/ST2 is abundantly expressed on Th2 cells but not on Th1 cells. It is also expressed on mast cells, which are crucial for Th2-mediated effector responses. These observations suggest that T1/ST2 is important for the Th2 immune response, but this is only partially supported by observations from T1/ST2-deficient mice.

The receptor form of the T1/ST2 protein has the same domain organization as IL-1RI. By alternative splicing, the T1/ST2 gene can also produce a secreted, soluble form of the protein, which lacks the transmembrane and the intracellular segment. The extracellular domain has no affinity for IL-1, and the natural ligand of this receptor is currently not known. Two putative ligands have been reported, but neither has been proven convincingly.

Homo-oligomerization of this receptor, induced by overexpression or antibody cross-linking, can activate MAP kinases (ERK, p38, and JNK), the AP-1 transcription factor, and IL-4 production, but it cannot activate NF-κB (Brint et al., 2002). Interestingly, a chimera containing the extracellular domain of IL-1RI and the intracellular domain of T1/ST2 is able to activate NF-κB in response to IL-1 stimulation. This is probably mediated by a hetero-oligomer of the TIR domains of T1/ST2 and IL-1RAcP.

IL-1RAPL and TIGIRR

IL-1RAPL (for "IL-1 receptor accessory protein-like") was identified by positional cloning as the gene responsible for X-linked, nonspecific mental retardation (MRX). Nonoverlapping deletions and a nonsense mutation in the middle of the TIR domain of IL-1RAPL are found in MRX patients. This protein is highly expressed in postnatal brain structures that are important for memory and learning. IL-1RAPL was independently cloned by homology searching in the EST database, which also identified TIGIRR (for "three Ig domain-containing IL-1 receptor-related") as a close homolog of IL-1RAPL.

IL-1RAPL and TIGIRR are both encoded on the X chromosome in humans. They have three Ig domains in the extracellular region and a TIR domain in the intracellular region. In contrast to IL-1R, these two proteins also contain a 100-residue segment at the C terminus (Fig. 1). The function of this unique extension is not known. Recent studies suggest that residues in this region may interact with neuronal calcium sensor 1 (NCS-1), and IL-1RAPL may play a role in regulating the exocytosis of secretory and neurotransmitter substances (Bahi et al., 2003). NCS-1 is upregulated in patients with schizophrenia and bipolar disease (Koh et al., 2003).

IL-1Rrp2 and SIGIRR

Of the remaining two IL-1R superfamily members, IL-1Rrp2 (for "IL-1 receptor-related protein 2") may be the receptor for the novel IL-1 superfamily member IL-1F9 and can mediate the activation of NF-κB by this ligand. SIGIRR (for "single Ig IL-1R-related") was identified by homology searching in the EST database. As the name implies, it contains only a single Ig domain in the extracellular region, in contrast to the other IL-1Rs (Fig. 1). However, this is probably not very unusual, since IL-18bp also has only one Ig domain. Another feature of this receptor is that it also contains a 100-residue extension at the C terminus. The amino acid sequence of this extension in SIGIRR is different from that in IL-1RAPL and TIGIRR. A chimera containing the extracellular domain of IL-1RI and the intracellular domain of SIGIRR (with or without the C-terminal extension) cannot activate the IL-8 promoter in response to IL-1. Recent studies with SIGIRR$^{-/-}$ mice show that it is a negative regulator of signaling by IL-1R as well as some TLRs (Wald et al., 2003).

THE Toll/INTERLEUKIN-1 RECEPTOR DOMAIN

Sequence homology between the intracellular domains of IL-1RI and *Drosophila* Toll was recognized in 1991. This domain is now known as the Toll/interleukin-1 receptor (TIR) domain, and it has been identified in the IL-1R superfamily, the TLR superfamily, disease resistance *R* protein superfamily, and many other cytoplasmic molecules in the vertebrates, insects, plants, viruses, and other organisms. Molecules containing this domain are generally involved in host immune response, development, and other processes. More recently, putative TIR domains have also been found in some prokaryotes (O'Neill, 2003), such as *Bacillus subtilis,* but the function of the domain in these organisms is unknown.

Sequence Conservation of TIR Domains

The TIR domains contain between 150 and 200 amino acid residues. They can be roughly divided into four subfamilies: (i) the TLR subfamily, containing the mammalian TLRs and the *Drosophila* Toll homologs; (ii) the IL-1R subfamily, containing all receptors of the IL-1R superfamily; (iii) the adaptor subfamily, containing the molecules that mediate signaling by the receptors, including mammalian MyD88, TIRAP/MAL, TRIF/TICAM, TIRP/TRAM, and *Drosophila* dMyD88; and (iv) the plant disease resistance *R* protein subfamily. The sequences of these domains are generally weakly conserved, with about 25% amino acid identity between most pairs of the domains (Color Plate 51 [see color insert]). Higher sequence conservation among TIR domains of the same subfamily is also observed; for example, the TIR domains of human TLR1 and TLR2 have 50% amino acid sequence identity. The strongest sequence conservation is between TLR1 and TLR6, which have 87% amino acid identity for their TIR domains. This sequence homology also extends to the extracellular domains of these two receptors; interestingly, both of them function as hetero-oligomers with TLR2.

The amino acid sequences of most TIR domains are marked by a (F/Y)DA motif near the N-terminal end (also known as the box 1 motif), and an FW motif near the C-terminal end (also known as the box 3 motif) (Color Plate 51). Within the domain itself, there is another highly conserved motif, RDxxPG (box 2 motif). Newer members of the TIR domain superfamily have more divergent sequences. For example, TRIF/TICAM and TIRP/TRAM have significant deviations from all three sequence motifs (Color Plate 51).

Functions of TIR Domains

The integrity and authenticity of the TIR domains are required for signal transduction through the TLRs and the IL-1Rs. A chimera consisting of extracellular IL-1RI and intracellular TLR1 cannot function in response to IL-1. Mutagenesis studies showed that residues that are conserved between the TLRs and the IL-1Rs are required for signal transduction through the IL-1Rs. In addition, deletion of the C-terminal portion of the TIR domains led to ablation of the signaling process.

Most importantly, the Lps^d mutation of TLR4, Pro712His, is located in the middle of the primary sequence of the TIR domain. This Pro residue is part of the box 2 motif (Color Plate 51) and is conserved among all mammalian TLRs except for TLR3. Mutation of the equivalent residue in TLR2, Pro681His, also blocks the signal transduction in response to stimulation by yeast and gram-positive bacteria. In fact, mutation of this residue to His in most TIR domains make them dominant negative inhibitors of receptor signaling, and this property can be used to map the signaling pathways of the various receptors and their associated adaptors.

TIR domains are thought to be protein-protein interaction modules for signal transduction by the IL-1Rs, TLRs, and R proteins. They do not possess any enzymatic activities, although some of them may become phosphorylated during the signaling process (see below). It is expected that ligand-induced homo- or hetero-oligomerization of the extracellular domains of the receptors will lead to oligomerization of their intracellular TIR domains. This receptor-TIR domain complex can then recruit downstream, cytoplasmic adaptor molecules through homotypic interactions with their TIR domains. Therefore, there may exist a TIR domain signaling complex that contains TIR domains from both the receptors and the adaptors. The adaptors can in turn transduce the signal to molecules further downstream, ultimately leading to the nuclear translocation and/or activations of transcription factors.

Although biological experiments suggest the importance of a TIR domain signaling complex, most of the TIR domains that have been purified so far behave as monomers by themselves. The affinity for homo- or hetero-oligomerization of these isolated TIR domains is very low, possibly in the millimolar range. It is possible that the other domains of these TIR-containing molecules also play an important role in the oligomerization process on receptor activation.

Structures of TIR Domains

Crystal structures of the TIR domains of human TLR1 and TLR2 have been determined. The TIR domain structures contain a central five-strand fully parallel β-sheet (βA through βE) that is surrounded by seven helices (αA through αE, with the insertion of two additional helices, αC′ and αC″, between αC and βD) on both sides (Color Plate 52A [see color insert]). Remarkably, the backbone fold of the domain is similar to that of the bacterial chemotaxis protein CheY, although TIR domains do not contain Mg^{2+}-binding sites (Color Plate 52B). It is worth noting that a domain that is used by eukaryotes to sense extracellular stimuli has the same backbone fold as that used by bacteria to sense extracellular environment.

Despite having 50% amino acid sequence identity, the structures of the two TIR domains contain regions with substantial differences (Color Plate 52C). Since the sequence conservation among most TIR domains is in the 25% range, the structural studies suggest that the TIR domains may have a significant degree of structural (and sequence) diversity. This diversity may be important for achieving specificity in the signaling processes from the various receptors.

The BB Loop

Of the three conserved sequence motifs among the TIR domains, the first motif is located just before the first β-strand of the structure whereas the third is located at the beginning of the last helix (αE) of the structure. It is likely that residues in both of these motifs contribute to the structural stability of the domain. The side chains of the FW residues in the third motif are buried in the hydrophobic core of the structure.

The second motif corresponds to a loop linking strand βB and helix αB, hence, this loop is known as the BB loop. Residues in this loop are projected away from the rest of the TIR domain (Color Plate 52A) and form a highly prominent feature on the surface of the TIR domain (Color Plate 52D). The third residue of this loop, the BB3 residue, is the Arg residue of the motif. It has ion pair interactions with a highly conserved Glu residue in helix αA. The Asp residue of the motif, the BB4 residue, also interacts with the BB3 Arg residue. Most importantly, the Lps^d mutation (Pro712His in TLR4) corresponds to the seventh residue in this loop (the BB7 residue), at the tip of the loop and farthest away from the rest of the molecule. This indicates that the BB loop is crucial for the function of TIR domains, which is confirmed by mutagenesis studies.

It is expected that this highly conserved surface patch on the TIR domains may be used to recruit the MyD88 adaptor molecule by the TLRs and IL-1Rs. MyD88 is the adaptor for signaling from most of these receptors; for all these receptors to recognize this single adaptor molecule, they should present a conserved surface feature. Biochemical experiments confirm that there are interactions between the TIR domains of TLR2 and MyD88 and that the Lps^d mutation in the TIR domain of TLR2 can abolish this interaction. This suggests that the surface patch formed by the BB loop is probably used to recruit the MyD88 adaptor molecule. Structural studies with the Pro681His mutant of the TIR domain of TLR2 (equivalent to the Lps^d mutation in TLR4) show that the mutation does not affect the overall structure or stability of the TIR domain, consistent with the notion that the BB loop may be important for protein-protein interactions.

Covalent Modifications of TIR Domains

Accumulating evidence suggests that TIR domains can become covalently modified during the signaling process by the TLRs and the IL-1Rs. The Tyr residue in the sequence motif Tyr-Glu-X-Met of the TIR domain of IL-1RI is rapidly phosphorylated on IL-1 stimulation (Marmiroli et al., 1998). This phosphorylated sequence is recognized by the SH2 domain of the p85 subunit of phosphatidylinositol 3-kinase (PI 3-kinase), which mediates the recruitment and activation of the PI 3-kinase by the IL-1Rs. Mutation of this Tyr residue to Phe abolished the recruitment of PI 3-kinase.

TLR2 also becomes transiently phosphorylated on Tyr residues in the TIR domain on stimulation (Arbibe et al., 2000). The phosphorylated receptor TIR domain mediates the recruitment of PI 3-kinase, as well as the small Rho GTPase Rac1. Mutations of two Tyr residues to Ala in the TIR domain of TLR2 abolished NF-κB activation and p85 association by the receptor, and the mutations also reduced the degree of phosphorylation of the TIR domain on receptor activation.

Biochemical studies have shown that the protein Ser/Thr kinase Pelle can phosphorylate the TIR domain of the Toll receptor in *Drosophila*. This phosphorylation may be important to disrupt the Toll-Tube-Pelle complex after receptor activation, freeing Pelle from the membrane to interact with or phosphorylate downstream cytoplasmic molecules (Shen and Manley, 1998).

Crystal structures of the TIR domains of TLR1 and TLR2 show that the Cys residue at the beginning of the αC' helix (Color Plate 51) is prone to oxidation and covalent modification. This residue is conserved among all the human TLRs (Color Plate 51). Mutation of this Cys residue to Ser in TLR2 abolished the function of the receptor. It remains to be demonstrated whether oxidative modifications of TIR domains also play a role in TLR signaling.

ADAPTOR MOLECULES

Many molecules have been identified in the signaling pathways of the IL-1Rs and TLRs. This section focuses on the membrane-proximal events of these pathways here. These adaptor molecules couple the activated receptors to downstream molecules. Most of the adaptor molecules contain TIR domains, supporting the notion that a receptor-adaptor homotypic TIR domain complex mediates the signal transduction of these receptors. At the same time, adaptor molecules that lack TIR domains have also been identified. They interact directly with the receptors or the TIR-containing adaptors and are important for the functions of the receptors as well.

MyD88

The MyD88 gene was first cloned in 1990 as a myeloid differentiation primary response gene, hence the name "myeloid differentiation factor 88." Later work identified this molecule as a crucial adaptor in signaling by many IL-1Rs and TLRs, including IL-1RI/IL-1RAcP, IL-18R/AcPL, TLR2, TLR3, TLR4, TLR5, TLR7, and TLR9. MyD88-deficient mice have no IL-1- and IL-18-mediated immune and inflammatory functions and cannot produce cytokines in response to LPS. These mice are refractory to the toxic effects of LPS.

Human MyD88 has 296 amino acid residues. MyD88 contains a death domain (DD) at the N terminus and a TIR domain at the C terminus, with a linker segment between the two domains that is also required for function (Fig. 1). On receptor activation, MyD88 is recruited to the IL-1Rs and TLRs via its TIR domain, and the DD in MyD88 can in turn recruit the downstream signaling molecule IRAK via DD-DD interactions. The TIR and DD domains of MyD88 also mediate homodimerization and possibly oligomerization of this protein. Overexpression of MyD88 produces potent activation of NF-κB, whereas the TIR domain of MyD88 is a dominant negative inhibitor of this activation.

An alternatively spliced form of MyD88 lacks the linker domain between the DD and the TIR domain. This short form of the protein, MyD88s, is a dominant negative regulator of signaling through the TLRs and IL-1Rs, since it cannot recruit the crucial kinase IRAK-4 (Burns et al., 2003). MyD88s is induced on prolonged stimulation of cells by LPS and TNF-α, and the production of this short form may be a mechanism for the host to downregulate its immune response.

The *Drosophila* genome encodes a MyD88 homolog, CG2078 (also known as dMyD88). dMyD88 contains 537 amino acid residues, and its larger size is due to the presence of a 150-residue C-terminal segment and a 70-residue N-terminal segment. This protein is required for signaling by Toll for dorsoventral polarity as well as the innate immune response to fungal and gram-positive bacterial infections. dMyD88 interacts with the Toll receptor, as well as the downstream Pelle kinase.

TIRAP/MAL

Although cytokine production induced by LPS and poly(I-C) is abolished in MyD88$^{-/-}$ mice, NF-κB and MAP kinases are still activated, but with delayed kinetics. Moreover, IRF-3-dependent genes (IP-10 and other IFN-sensitive genes) are still induced by LPS stimulation in MyD88-deficient mice. This suggests the presence of MyD88-independent pathway(s) of signaling from the TLRs. Homology searches in the EST databases led to the identification of TIRAP (for "TIR domain-containing adaptor protein"), also known as MAL (for "MyD88 adaptor-like"). TIRAP/MAL contains 235 amino acid residues, with a TIR domain at the C terminus. In contrast to MyD88, TIRAP/MAL does not have a death domain at the N terminus (Fig. 1).

Overexpression of TIRAP/MAL can activate NF-κB and the MAP kinases (p38, JNK, and ERK). This protein can form homodimers, as well as heterodimers with MyD88. Mice that are deficient in this gene have impaired signaling by TLR1, TLR2, TLR4, and TLR6. However, these mice, as well as those that are doubly deficient in TIRAP/MAL and MyD88, have normal induction of IRF-3. This shows that TIRAP/MAL is not the MyD88-independent adaptor but, rather, another adaptor that functions in association with MyD88 in TLR2 and TLR4 signaling. TIRAP/MAL is not in the signaling pathways of IL-1R, IL-18R, TLR3, TLR5, TLR7, or TLR9.

TRIF/TICAM and TIRP/TRAM

Recent studies have identified and characterized additional adaptors containing a TIR domain. These include TRIF (for "TIR domain-containing adaptor inducing IFN-β"), also known as TICAM (for "TIR-containing adaptor molecule"), and TIRP (for "TIR-containing protein") (Bin et al., 2003), also known as TRAM (for "TRIF-related adaptor molecule") (Fig. 1). The amino acid sequences of these adaptors are highly divergent from those of MyD88, TIRAP/MAL, and the TIR domains of IL-1Rs and TLRs (Color Plate 51).

TRIF/TICAM was found by homology searches in the EST database, as well as by yeast two-hybrid screening with the TIR domain of TLR3 as bait. TLR3 is selected for the screening since neither MyD88 nor TIRAP/MAL plays an important role in TLR3 signaling, indicating the presence of additional adaptors. Studies with mice deficient in this gene confirm that TRIF/TICAM is the adaptor for MyD88-independent signaling through TLR3 and TLR4 and the activation of IRF-3 (Hoebe et al., 2003; Yamamoto et al., 2003).

Human TRIF/TICAM contains 712 amino acid residues. The TIR domain is located in the C-terminal one-third of the molecule, and there is a large N-terminal domain (with about 390 residues) and a smaller C-terminal domain (with about 170 residues). Overexpression of this protein leads to potent activation of IRF-3 and IFN-β production and weaker activation of NF-κB. The N-terminal domain is required for IFN-β induction but dispensable for NF-κB activation. The TIR domain itself functions as a dominant negative inhibitor of TLR3-mediated activation of IRF-3, as well as signaling through TLR2, TLR4, and TLR7. However, in an independent study, the TIR domain was found to potentiate IRF-3 activation by TLR3.

TIRP/TRAM is a close homolog of TRIF/TICAM in the TIR domain, with the two proteins having 37% amino acid sequence identity for that domain. However, TIRP/TRAM has only 235 residues and so is much smaller than TRIF/TICAM. The TIR domain is located at the C terminus, and there is a small (70-amino-acid) unique domain prior to the TIR domain. In comparison to TRIF/TICAM, TIRP/TRAM appears to be involved in IL-1R signaling. It can interact with IL-1RI/IL-1RAcP, the adaptors MAL and TRIF (but not MyD88), IRAK, and TRAF6 (Bin et al., 2003). Overexpression of this protein can activate NF-κB. Conflicting results have been presented about whether TIRP/TRAM can also activate IRF-3.

Other TIR-Containing Adaptors

The human genome encodes at least one more adaptor that contains a TIR domain. This protein was named SARM since it contains two sterile α-motif (SAM) domains and an Armadillo repeat motif (ARM). SAM domains mediate protein-protein interactions by forming homotypic oligomers, while the Armadillo repeat mediates interactions with β-catenin and other proteins. The TIR domain in SARM has weak sequence homology to other TIR domains and is located at the extreme C terminus of this 690-residue protein. The functional relevance of this protein to IL-1R/TLR signaling, or other biological processes, remains to be characterized.

In some plants, there are many proteins that contain only a TIR motif, as well as proteins that contain only the TIR-NBS motif. These are probably the adaptor molecules for the signaling cascades that are initiated by the disease resistance *R* proteins.

Several viruses carry small proteins that contain TIR domains. For example, the vaccinia virus has a TIR domain homolog, A46R, and possibly a second one, A52R. These proteins can inhibit host immune responses that are mediated by TIR domains and may be part of an important strategy by the virus to neutralize host immunity.

Tube

Tube is required for the establishment of dorsoventral polarity in *Drosophila* embryos, and genetic studies place it downstream of Toll but upstream of the protein Ser/Thr kinase Pelle. Tube contains 462 amino acid residues but does not have a TIR domain. It has a DD at the N terminus (Fig. 1). The C-terminal domain contains five copies of an eight-residue motif, N(V/I/L)PX(L/I)(T/S)XL, but this domain is not required for the function of Tube. The first 180 residues of Tube is sufficient to restore dorsoventral polarity to embryos lacking maternal Tube activity, but a construct containing only the first 154 residues cannot rescue any of these mutants. A vertebrate homolog of Tube has yet to be identified.

IRAKs and Pelle

Genetic studies of *Drosophila* identified Pelle as a protein Ser/Thr kinase that is required for the establishment of dorsoventral polarity in embryos as well as for innate immune response in adults. The mammalian homolog of Pelle is known as IRAK ("for IL-1 receptor-associated kinase"). Four different IRAKs have so far been identified from humans: IRAK-1, IRAK-2, IRAK-M (or IRAK-3), and IRAK-4. These kinases all contain a

DD at the N terminus, followed by a protein Ser/Thr kinase domain (Fig. 1). IRAK-1 (712 amino acid residues), IRAK-2 (590 residues), and IRAK-M (596 residues) also contain a long C-terminal segment that may be important for the recruitment of downstream signaling molecules, such as TRAF6. In comparison, IRAK-4 (460 residues) lacks this tail and is therefore the closest homolog of *Drosophila* Pelle (501 residues).

IRAK-2 and IRAK-M do not have protein kinase activity, since they lack key catalytic residues in their kinase domains. Moreover, the kinase activity of IRAK-1 is not required for its signaling activity, suggesting that IRAK-1, IRAK-2, and IRAK-M function mostly as adaptor molecules in IL-1R and TLR signal transduction. However, hyperphosphorylation of IRAK-1 is a crucial step in this signaling process, and recent experiments implicate IRAK-4 as the protein kinase that is responsible for this phosphorylation after receptor activation. IRAK-M is a negative regulator of this signaling pathway.

Pellino

Drosophila Pellino was identified by yeast two-hybrid screening using Pelle as the bait (Grosshans et al., 1999). It contains 424 amino acid residues and interacts with the activated (phosphorylated) protein kinase domain of Pelle. Two homologs of this protein, Pellino1 and Pellino2, have been identified in mice and humans and are important for signal transduction by the IL-1Rs and TLRs through their interactions with IRAK-4, IRAK-1, and the downstream signaling molecule TRAF6 and TAK1 (Jensen and Whitehead, 2003; Jiang et al., 2003; Yu et al., 2002). There may also be a third homolog of Pellino in humans.

Tollip

Tollip (for "Toll-interacting protein") was identified by yeast two-hybrid screening using the intracellular domain of IL-1RAcP as bait. Murine Tollip contains 274 amino residues, with a C2 domain in the middle of the sequence (residues 54 to 186). This C2 domain is unlikely to be regulated by calcium, since the ligands for binding calcium are absent in the sequence. Tollip forms a complex with the downstream signaling molecule IRAK and mediates its recruitment to activated IL-1R, TLR2, and TLR4 complexes. Overexpression of Tollip can inhibit the IL-1-induced NF-κB activation by sequestering the IRAK adaptor and blocking its phosphorylation.

SIGNALING PATHWAYS OF TLRs AND IL-1Rs

The signal transduction through the TLRs and the IL-1Rs ultimately leads to the activation of the transcription factor NF-κB, the MAP kinases (ERK, p38, and JNK) and transcription factors of the AP-1 family. Some of these receptors (TLR3 and TLR4) can also activate the IRF-3 transcription factor (IFN-β production). Many molecules have been identified that participate in the signaling process leading from receptors to transcriptional activation. There may be several parallel signaling pathways for these receptors, recruiting different subsets of the adaptor molecules. There may also be a significant degree of overlap and cross talk among the different pathways.

Studies with the IL-1 receptors provide a glimpse into the molecular mechanism for receptor activation. On ligand binding, IL-1RI can recruit an accessory protein, IL-1RAcP, and this should lead to the heterodimerization or oligomerization of IL-1RI and IL-1RAcP molecules. Such an oligomeric complex may be the signal for the activation of the IL-1Rs, and a similar mechanism may also apply to the IL-18Rs. For the TLRs, TLR4 is thought to function as homo-oligomers whereas TLR2 may function as hetero-oligomers in association with TLR1 or TLR6. Overall, the general scenario is that the activation of these receptors is through their homo- and/or hetero-oligomerization, which can be induced by ligand binding or overexpression.

The oligomerization of the receptors on activation brings their intracellular TIR domains into close proximity, and this enables the activated receptors to recruit TIR-containing downstream adaptor molecules through homotypic TIR-TIR interactions among the receptor and the adaptor TIR domains (Fig. 2A). Three types of interaction interfaces can be envisioned in such a TIR domain signaling complex: the R (receptor) face, which mediates interactions among the receptor TIR domains; the A (adaptor) face, which mediates interactions among the adaptor TIR domains; and, finally, the S (signaling) face, which mediates interactions between the receptor and adaptor TIR domains (Color Plate 52E).

Structural and biochemical studies of the TIR domains suggest that the BB loop may be involved in the recruitment of the MyD88 adaptor molecule by the TLRs. Therefore, the BB loop may correspond to the S face in this putative TIR domain signaling complex and the Lps^d mutation is located in the center of this interface (Color Plate 52D). The molecular details of the interactions at the R and A faces are currently not known.

MyD88 is the most important TIR-containing adaptor molecule and is involved in the signaling by most of these receptors. Additional adaptors of this class include TIRAP/MAL, TRIF/TICAM, and TIRP/TRAM, which participate in signaling by selected receptors. MyD88 can recruit IRAK via the DD that is present in both molecules. Observations with the MyD88s splice variant suggest that the linker segment between the DD and TIR domain is also important for IRAK recruitment. Both IRAK-1 and IRAK-2 are required for signaling by MyD88. Additional modes of IRAK recruitment must exist as well, since IRAK-2 is required for the signaling by TIRAP/MAL, which lacks a DD. The dominant negative form of IRAK-2 inhibits signaling by MyD88 and TIRAP/MAL, whereas the dominant negative form of IRAK-1 can inhibit only MyD88, indicating the difference between the two adaptors. IRAK-4 is the primary protein kinase that is needed to phosphorylate IRAK-1, and it functions upstream of the other IRAKs.

Direct interactions between the IRAKs and the receptors are also observed. This may be an additional mechanism for ensuring specificity in the signal transduction pathway. For example, IL-1RI is thought to help recruit IRAK-2 in the IL-1R complex, whereas IL-1RAcP can help recruit IRAK-1. In addition, the recruitment of Tollip, together with IRAKs, occurs only for some of the IL-1Rs and TLRs.

In *Drosophila,* the Pelle kinase can directly bind to the intracellular domain of Toll. In addition, a heterotrimeric complex of the DD in Tube, Pelle, and dMyD88 is required for Toll signaling (Sun et al., 2002). One of the functions of dMyD88 is to mediate the interactions between Toll and Tube, since Tube may not directly contact the Toll receptor. The molecular details of the interactions between the DD of Tube and Pelle have been observed by crystallographic analysis. The two domains have the canonical DD fold, with a six-helix bundle (α1 through α6 [Color Plate 50E]). The Tube DD also has an insertion of two helices (αA and αB) between α2 and α3. The two DDs are arranged in a head-to-tail fashion in the complex, forming an extensive interface between them. The structural analysis shows there is some degree of flexibility in the Pelle-Tube interface. A crucial component of the interface is the docking of a C-terminal segment of Tube (residues 164 to 173) onto the surface of the Pelle DD (Color Plate 50E). This is consistent with biological studies showing that the first 154 residues of Tube cannot rescue mutants lacking Tube activity whereas the first 180 residues of Tube can do so. The amino acid sequence of this C-terminal segment is, however, unique to Tube, and it is probably unlikely that the DD complex between MyD88 and IRAK in the mammalian TLR/IL-1R signaling pathway can have a similar mechanism of association.

Many additional protein molecules that mediate the signal transduction by these receptors have been identified downstream of the IRAKs (Fig. 2B). Activated IRAK leaves the membrane and interacts with the TNF-α receptor-associated factor 6 (TRAF6). This in turn activates the kinase TAK1 (transforming growth factor β-activated kinase 1). Here, the signaling pathway diverges, leading to the activation of NF-κB and the MAP kinases, respectively. The exact relationship among the various molecules and the various pathways, as well as the molecular details of the interactions among these molecules, are still generally poorly understood. Moreover, it is likely that many protein molecules that are important for the functions of these receptors remain to be identified.

PERSPECTIVES

Studies of the TLRs and IL-1Rs over the past few years have greatly increased our understanding of the general mechanisms of the host innate immune response. The microbial pathogens are recognized through features that are conserved among a large number of these organisms (PAMPs), with the limited set of TLRs encoded by the host genome. The TLR-PAMP recognition is direct and specific, with each TLR responding to only a unique set of PAMPs. Tremendous progress has been made in the identification of the ligands of each of the TLRs and the membrane-proximal adaptors for their signaling. Once the receptor is engaged with PAMPs, the downstream signaling pathways are activated. A crucial component of this may be the formation of a TIR domain signaling complex between the receptors and the membrane-proximal adaptors.

Significant questions still remain regarding the detailed molecular mechanisms of host innate immunity. Many more biological and biochemical investigations are needed to carefully identify and define the roles of the many proteins in this immune system and, most importantly, the interactions among these proteins during the signal transduction process.

Structural biology is the most important experimental technique for addressing issues regarding the molecular basis of innate immunity. Structural studies of the TIR domains, DDs, and others are beginning to shed light on the molecular interactions among these proteins. An important question that remains to be answered by structural analyses is the molecular mechanism of specificity in the signaling process. To this end, obtaining structural information on the putative TIR domain signaling complex will be extremely helpful in elucidating these signaling pathways and their specificity. Such structural biology studies, complemented by other biophysical as well as mutagenesis and biological studies, should lead to a much deeper understanding of the functions of these receptors and the immune system in general.

Acknowledgments. I thank Xiao Tao and Javed Khan for critical reading of the manuscript. My research is supported by grant AI49475 from the National Institutes of Health.

REFERENCES

Aderem, A., and R. J. Ulevitch. 2000. Toll-like receptors in the induction of the innate immune response. *Nature* **406:**782–787.

Anderson, K. V. 2000. Toll signaling pathways in the innate immune response. *Curr. Opin. Immunol.* **12:**13–19.

Arbibe, L., J.-P. Mira, N. Teusch, L. Kline, M. Guha, N. Mackman, P. J. Godowski, R. J. Ulevitch, and U. G. Knaus. 2000. Toll-like receptor2-mediated NF-κB activation requires a Rac1-dependent pathway. *Nat. Immunol.* **1:**533–540.

Asai, T., G. Tena, J. Plotnikova, M. R. Willmann, W.-L. Chiu, L. Gomez-Gomez, T. Boller, F. M. Ausubel, and J. Sheen. 2002. MAP kinase signalling cascade in *Arabidopsis* innate immunity. *Nature* **415:**977–983.

Bahi, N., G. Friocourt, A. Carrie, M. E. Graham, J. L. Weiss, P. Chafey, F. Fauchereau, R. D. Burgoyne, and J. Chelly. 2003. IL1 receptor accessory protein like, a protein involved in X-linked mental retardation, interacts with neuronal calcium sensor-1 and regulates exocytosis. *Human Mol. Gen.* **12:**1415–1425.

Beutler, B. 2000. Tlr4: central component of the sole mammalian LPS sensor. *Curr. Opin. Immunol.* **12:**20–26.

Beutler, B., and H. Wagner (ed.) 2002. *Toll-Like Receptor Family Members and Their Ligands.* Springer-Verlag KG, Berlin, Germany.

Bin, L.-H., L.-G. Xu, and H.-B. Shu. 2003. TIRP: a novel Toll/interleukin-1 receptor (TIR) domain containing adaptor protein involved in TIR signaling. *J. Biol. Chem.* **278:**24526–24532.

Brint, E. K., K. A. Fitzgerald, P. Smith, A. J. Coyle, J.-C. Gutierrez-Ramos, P. G. Fallon, and L. A. J. O'Neill. 2002. Characterization of signaling pathways activated by the interleukin 1 (IL-1) receptor homologue T1/ST2. *J. Biol. Chem.* **277:**49205–49211.

Burns, K., S. Janssens, B. Brissoni, N. Olivos, R. Beyaert, and J. Tschopp. 2003. Inhibition of interleukin 1 receptor/Toll-like receptor signaling through the alternatively spliced, short form of MyD88 is due to its failure to recruit IRAK-4. *J. Exp. Med.* **197:**263–268.

Chamaillard, M., M. Hashimoto, Y. Horie, J. Masumoto, S. Qiu, L. Saab, Y. Ogura, A. Kawasaki, K. Fukase, S. Kusumoto, M. A. Valvano, S. J. Foster, T. W. Mak, G. Nunez, and N. Inohara. 2003. An essential role for NOD1 in host recognition of bacterial peptidoglycan containing diaminopomelic acid. *Nat. Immunol.* **4:**702–707.

Dunne, A., and L. A. J. O'Neill. 2003. The interleukin-1 recepor/Toll-like receptor superfamily: signal transduction during inflammation and host defense. *Sci. STKE* **2003:**re3.

Girardin, S. E., I. G. Boneca, L. A. M. Carneiro, A. Antignac, M. Jehanno, J. Viala, K. Tedin, M.-K. Taha, A. Labigne, U. Zahringer, A. J. Coyle, P. S. DiStefano, J. Bertin, P. J. Sansonetti, and D. J. Philpott. 2003. Nod1 detects a unique muropeptide from Gram-negative bacterial peptidoglycan. *Science* **300:**1584–1587.

Grosshans, J., F. Schnorrer, and C. Nusslein-Volhard. 1999. Oligomerization of Tube and Pelle leads to nuclear localisation of Dorsal. *Mech. Dev.* **81:**127–138.

Hajjar, A. M., R. K. Ernst, J. H. Tsai, C. B. Wilson, and S. I. Miller. 2002. Human Toll-like receptor 4 recognizes host-specific LPS modifications. *Nat. Immunol.* **3:**354–359.

Hoebe, K., K. Du, P. Georgel, E. Janssen, K. Tabeta, S. O. Kim, J. Goode, P. Lin, N. Mann, S. Mudd, K. Crozat, S. Sovath, J. Han, and B. Beutler. 2003. Identification of Lps2 as a key transducer of MyD88-independent TIR signalling. *Nature* **424:**743–748.

Hoffmann, J. A., F. C. Kafatos, C. A. Janeway Jr., and R. A. B. Ezekowitz. 1999. Phylogenetic perspectives in innate immunity. *Science* **284:**1313–1318.

Janeway, C. A., Jr. 1989. Approaching the asymptote? Evolution and revolution in immunology. *Cold Spring Harbor Symp. Quant. Biol.* **LIV:**1–13.

Jensen, L. E., and A. S. Whitehead. 2003. Pellino 2 activates the mitogen activated protein kinase pathway. *FEBS Lett.* **545:**199–202.

Jiang, Z., H. J. Johnson, H. Nie, J. Qin, T. A. Bird, and X. Li. 2003. Pellino 1 is required for interleukin-1 (IL-1)-mediated signaling through its interaction with the IL-1 receptor-associated kinase 4 (IRAK4)-IRAK-tumor necrosis factor receptor-associated factor 6 (TRAF6) complex. *J. Biol. Chem.* **278:**10952–10956.

Koh, P. O., A. S. Undie, N. Kabbani, R. Levenson, P. S. Goldman-Rakic, and M. S. Lidow. 2003. Up-regulation of neuronal calcium sensor-1 (NCS-1) in the prefrontal cortex of schizophrenic and bipolar patients. *Proc. Natl. Acad. Sci. USA* **100:**313–317.

Marmiroli, S., A. Bavelloni, I. Faenza, A. Sirri, A. Ognibene, V. Cenni, J. Tsukada, Y. Koyama, M. Ruzzene, A. Ferri, P. E. Auron, A. Toker, and N. M. Maraldi. 1998. Phosphotidylinositol 3-kinase is recruited to a specific site in the activated IL-1 receptor 1. *FEBS Lett.* **438:**49–54.

Medzhitov, R., and C. A. Janeway, Jr. 1997. Innate immunity: the virtues of a nonclonal system of recognition. *Cell* **91:**295–298.

Meyers, B. C., A. Kozik, A. Griego, H. Kuang, and R. W. Michelmore. 2003. Genome-wide analysis of NBS-LRR-encoding genes in *Arabidopsis. Plant Cell* **15:**809–834.

O'Neill, L. A. J. 2003. The role of MyD88-like adaptors in Toll-like receptor signal transduction. *Biochem. Soc. Trans.* **31:**643–647.

Pereira, J. P., R. Girard, R. Chaby, A. Cumano, and P. Vieira. 2003. Monoallelic expression of the murine gene encoding Toll-like receptor 4. *Nat. Immunol.* **4:**464–470.

Shen, B., and J. L. Manley. 1998. Phosphorylation modulates direct interactions between the Toll receptor, Pelle kinase and Tube. *Development* **125:**4719–4728.

Sims, J. E. 2002. IL-1 and IL-18 receptors, and their extended family. *Curr. Opin. Immunol.* **14:**117–122.

Sing, A., D. Rost, N. Tvardovskaia, A. Roggenkamp, A. Wiedemann, C. J. Kirschning, M. Aepfelbacher, and J. Heesemann. 2002. *Yersinia* V-antigen exploits Toll-like receptor 2 and CD14 for interleukin 10-mediated immunosupression. *J. Exp. Med.* **196:**1017–1024.

Sun, H., B. N. Bristow, G. Qu, and S. A. Wasserman. 2002. A heterotrimeric death domain complex in Toll signaling. *Proc. Natl. Acad. Sci. USA* **99:**12871–12876.

Takeda, K., and S. Akira. 2003. Toll receptors and pathogen resistance. *Cell. Microbiol.* **5:**143–153.

Tschopp, J., F. Martinon, and K. Burns. 2003. NALPs: a novel protein family involved in inflammation. *Nat. Rev. Mol. Cell Biol.* **4:**95–104.

Underhill, D. M., and A. Ozinsky. 2002. Toll-like receptors: key mediators of microbe detection. *Curr. Opin. Immunol.* **14:**103–110.

Wald, D., J. Qin, Z. Zhao, Y. Qian, M. Naramura, L. Tian, J. Towne, J. E. Sims, G. Stark, and X. Li. 2003. SIGIRR, a negative regulator of Toll-like receptor-interleukin 1 receptor signaling. *Nat. Immunol.* **4:**920–927.

Weber, A. N. R., S. Tauszig-Delamasure, J. A. Hoffmann, E. Lelievre, H. Gascan, K. P. Ray, M. A. Morse, J. L. Imler, and N. J. Gay. 2003. Binding of the *Drosophila* cytokine Spatzle to Toll is direct and establishes signaling. *Nat. Immunol.* **4:**794–800.

Yamamoto, M., S. Sato, H. Hemmi, K. Hoshino, T. Kaisho, H. Sanjo, O. Takeuchi, M. Sugiyama, M. Okabe, K. Takeda, and S. Akira. 2003. Role of adaptor TRIF in the MyD88-independent Toll-like receptor signaling pathway. *Science* **301:**640–643.

Yu, K.-Y., H.-J. Kwon, D. A. M. Norman, E. Vig, M. G. Goebl, and M. A. Harrington. 2002. Mouse Pellino-2 modulates IL-1 and lipopolysaccharide signaling. *J. Immunol.* **169:**4075–4078.

INDEX

AAA proteins, similarity to VirB11-like proteins, 199–200
ActA protein, 64
Actin cytoskeleton
  modulation by effectors of type III secretion system, 157–165, Color Plate 39
  reorganization, 62–64
Actin signaling, 62–64
  Tir-based, 64–65
α-Actinin, 60, 63
Actinobacillus actinomycetemcomitans, type IV secretion components, 187
Actin-rich pedestals, 60, 63–64
Adaptor molecules, TIR-containing, 253–256
  MyD88, 256–257, 260–261
  TIRAP/MAL, 257, 260
  TIRP/TRAM, 257–258, 260
  TRIF/TICAM, 257–258, 260
Adherence, *H. influenzae* to host epithelia, 129–148, Color Plates 28–34
Adhesins, 37
  comparisons among, 44–45, Color Plate 18
  H. influenzae, 129–148, Color Plates 28–34
  host receptors of bacterial origin, 49–68, Color Plate 19
  lectins, 40–43
  pilins, 69–79, 194–195
  vaccines based on, 45
ADP-ribosylation factor, 184
A/E pathogens, 50, 65
Agr system, 28, 31
Agrobacterium tumefaciens, T-DNA transfer system, 180–182, 189–206, Color Plate 46
AIP, *see* Autoinducing peptide
Akt/PKBα pathway, 165–166
Alginate production system, *P. aeruginosa*, 30
AlgR1-AlgR2 system, 30
Alpha-hemolysin, 228
Alpha-toxin, *S. aureus*, 228
Anti-σ factors, 1–16
  FlgM, 7–8, 10, Color Plate 4
  general themes of anti-σ regulation, 8–10
  RseA, 5–7, 9–10, Color Plate 3
  SpoIIAB, 3–5, Color Plate 2

Anti-anti-σ factors, SpoIIAA, 3–5, 10
Antimicrobial drugs
  adhesins as targets, 45
  sortases as targets, 103, 121–123
  two-component systems as targets, 17, 26–28
Antiviral compounds, recognition by TLR7 and TLR8, 247–248, Color Plate 50
AP-1 transcription factor, 250, 252, 259
Apoptosis, 234
Arc proteins
  ArcA, 22
  ArcB, 19
Arp2/3 complex, 60, 63–64
AsiA protein, 9
Asialoglycolipid receptors, 88
ATPase
  assembly and retraction, 95–96
  p97 AAA, 199–200
  traffic, 192–194
  type II secretion, 95
  type III secretion, 55, 161
  type IV secretion, 95
  VirB11, 199–200
ATPγS-HP0525 complex, 198, 200, Color Plate 43
Attaching and effacing lesion, 49–51
Autoinducing peptide (AIP), 28, 31
Autotransporter, 130–138
  trimeric, 138–140
AvhB system, 182
AvrA protein, 165, 167

B7.1 protein, 245
Bacillus, sporulation, 3–5, Color Plate 2
Bacterial attachment, 37–48
  adhesins, *see* Adhesins
Bacterial virulence
  convergent evolution, 167–169
  functional mimicry, 166–167, Color Plate 41
  horizontal acquisition, 167–169, Color Plate 41
  two-component systems and, 27–28
Bartonella, type IV secretion system, 180, 185–186
BB loop, 255, Color Plate 52
Bep proteins, 185

265